D1709779

Commercial Management of Projects
Defining the Discipline

Edited by

David Lowe

University of Manchester

With

Roine Leiringer

The University of Reading

Editorial offices:
Blackwell Publishing Ltd, 9600 Garsington Road, Oxford OX4 2DQ, UK
Tel: +44 (0)1865 776868
Blackwell Publishing Inc., 350 Main Street, Malden, MA 02148-5020, USA
Tel: +1 781 388 8250
Blackwell Publishing Asia Pty Ltd, 550 Swanston Street, Carlton, Victoria 3053, Australia
Tel: +61 (0)3 8359 1011

First published 2006 by Blackwell Publishing Ltd

ISBN-13: 978–1–4051–2450–8
ISBN-10: 1–4051–2450–4

Library of Congress Cataloging-in-Publication Data
Commercial management of projects : defining the discipline / edited by David Lowe with Roine Leiringer.
p. cm.
Includes bibliographical references and index.
ISBN-13: 978–1–4051–2450–8 (alk. paper)
ISBN-10: 1–4051–2450–4 (alk. paper)
1. Project management. 2. Industrial management. I. Lowe, David. II. Leiringer, Roine.

HD69.P75L685 2006
658.4'04—dc22

2005019143

A catalogue record for this title is available from the British Library

Set in 10/12.5 pt Palatino
by Integra Software Services Pvt. Ltd, Pondicherry, India
Printed and bound in India
by Replika Press Pvt, Ltd, Kundli

The publisher's policy is to use permanent paper from mills that operate a sustainable forestry policy, and which has been manufactured from pulp processed using acid-free and elementary chlorine-free practices. Furthermore, the publisher ensures that the text paper and cover board used have met acceptable environmental accreditation standards.

For further information on Blackwell Publishing, visit our website:
www.thatconstructionsite.com

Contents

Contributors

Professor Jan Bröchner — Department of Service Management, Chalmers University of Technology, SE-412 96 Gothenburg, Sweden

Michael Brown — White Young Green plc, Arndale Court, Headingley, Leeds LS6 2UJ, UK

Professor Andrew Cox — Centre for Business Strategy and Procurement, Birmingham Business School, The University of Birmingham, Edgbaston, Birmingham B15 2TT, UK

Dr Andrew Davies — The Innovation Studies Centre, Tanaka Business School, Imperial College, London, SW7 2AZ

Dr Richard Fellows — Faculty of Architecture, Department of Real Estate and Construction, The University of Hong Kong, Pokfulam Road, Hong Kong

Dr Peter Fenn — School of Mechanical, Aerospace and Civil Engineering, The University of Manchester, Manchester M60 1QD, UK

Professor Marc Goergen — Management School, The University of Sheffield, 9 Mappin Street, Sheffield S1 4DT, UK

Professor Michael Hobday — The CoPS Innovation Centre, The Freeman Centre, The University of Sussex, Brighton, East Sussex BN1 9QE, UK

Professor Malcolm Horner — Department of Civil Engineering, The University of Dundee, Nethergate, Dundee DD1 4HN, UK

Dr Will Hughes — School of Construction Management and Engineering, The University of Reading, Whiteknights, PO Box 219, Reading RG6 6AW, UK

Dr David Hugill — EC Harris LLP, The Royal Exchange, Manchester M2 7EH, UK

Dr Paul Ireland — Centre for Business Strategy and Procurement, Birmingham Business School, The University of Birmingham, Edgbaston, Birmingham B15 2TT, UK

Dr Kalle Kähkönen — VTT Building Technology, Technical Research Centre of Finland, PO Box 1800, FIN-02044 VTT, Finland

Professor John Kelly School of the Built and Natural Environment, Glasgow Caledonian University, City Campus, Cowcaddens Road, Glasgow G4 0BA, UK

Dr Malik Khalfan Salford Centre for Research and Innovation (SCRI) in the Built and Human Environment, The University of Salford, Bridgewater Building, Salford, Manchester M7 1NU, UK

Professor David Langford Department of Architecture, The University of Strathclyde, 131 Rottenrow, Glasgow G4 0NG, UK

Dr Roine Leiringer School of Construction Management and Engineering, The University of Reading, Whiteknights, PO Box 219, Reading RG6 6AW, UK

Dr David Lowe School of Mechanical, Aerospace and Civil Engineering, The University of Manchester, M60 1QD, UK

Dr Peter McDermott Salford Centre for Research and Innovation (SCRI) in the Built and Human Environment, The University of Salford, Bridgewater Building, Salford, Manchester M7 1NU, UK

Mr Krisen Moodley Construction Management Group, School of Civil Engineering, The University of Leeds, Leeds LS2 9JT, UK

Dr Mike Murray Department of Architecture, The University of Strathclyde, 131 Rottenrow, Glasgow G4 0NG, UK

Dr Christopher Preece Construction Management Group, School of Civil Engineering, The University of Leeds, Leeds LS2 9JT, UK

Dr Luc Renneboog CentER, Faculty of Economics and Business Administration, Tilburg University, P.O. Box 90153, 5000 LE Tilburg, The Netherlands

Dr Andrew Ross School of the Built Environment, Liverpool John-Moores University, Byrom Street, Liverpool L3 3AF, UK

Professor Martin Skitmore School of Construction Management and Property, Queensland University of Technology, 2 George Street, GPO Box 2434, Brisbane Q4001, Australia

Dr Hedley Smyth Construction and Project Management, Bartlett School of Graduate Studies, University College London, Gower Street, (Torrington Place Site), London WC1E 6BT, UK

Dr Will Swan Salford Centre for Research and Innovation (SCRI) in the Built and Human Environment, The University of Salford, Bridgewater Building, Salford, Manchester M7 1NU, UK

Professor Graham M. Winch Operations, Innovation and Technological Management, Manchester Business School, The University of Manchester, Booth Street West, Manchester M15 6PB, UK

Preface

The term commercial management has been used for some time, not least in construction, while the job title commercial manager can be found across a spectrum of industries, especially those that are predominantly project based, for example, Aerospace, Construction, IT, Pharmaceutical and Telecommunications. Despite this, to date, these terms have no acknowledged definition and, as a direct consequence, there is a paucity of literature categorised as commercial management; this book seeks to address this omission.

Commercial organisations increasingly use projects as tools, contributing to the execution of organisational strategies, to address corporate objectives, goals and missions. Based on the fundamental assumption that the key objective for any commercial organisation is to generate profit, commercial issues, and thus commercial management, play an integral part in this. In particular, the commercial management function is vital in linking operations at the project level and multiple projects (portfolios/programmes) at the organisational level with the organisational (corporate) core of the company. The function, therefore, can be found at the interfaces between organisations as well as between divisions within an organisation. Within the context of this book, commercial management is defined as:

The management of contractual and commercial issues relating to projects, from project inception to completion.

There are compelling arguments for considering commercial management, not solely as a task undertaken by commercial managers, but as a discipline in itself, which bridges traditional project management and organisational theories. While differences in approach and application both between and within industry sectors have been established, sufficient similarity and synergy in practice have been ascertained to identify a specific role of commercial management within project-based organisations. These areas of similarity encompass contract management and dispute resolution, whereas the areas of divergence include a greater involvement in financial and value management within construction and bid management within defence/aerospace.

This book aims to present the current state of knowledge concerning the commercial management of projects. It is not a prescriptive text; rather the book focuses on clarifying and reviewing the current understanding, based upon existing research, of a wide range of issues relevant to commercial management practice in project-based industries. Collectively, the chapters constitute a step in the direction of systematic knowledge building: a step towards the creation of a body of knowledge and a research agenda for the development of an underlining theory relevant to commercial management. Thus, the book seeks to contribute to the growing literature that sees the necessity of widening project management research to include also the management of project-based organisations, project-based industries and even project-based careers.

The book is divided into four parts:

- *The external milieu*: The interfaces between the organisation and the environment within which it is active
- *The corporate milieu*: The interfaces between the organisation (at a corporate level) and both the projects milieu (the environment within which individual projects collectively reside) in which it is active and external milieu
- *The projects milieu*: The environment within which individual projects collectively reside and the interfaces between the corporate level milieu of an organisation and the individual projects in which it is active
- *The project milieu*: This last category deals with the interfaces on the specific project.

Whilst all the chapters address topics that are common for project-based organisations regardless of industry, most of the chapters are oriented towards construction and use this industry to illustrate specific points. We make no excuses for this as construction could lay claim to be the oldest project-based industry. Indeed, project management literature in general is heavily influenced by studies undertaken in this industry.

David Lowe and Roine Leiringer

Acknowledgements

We would like to thank all those who have contributed chapters to this book: Jan Bröchner, Michael Brown, Andrew Davies, Andrew Cox, Richard Fellows, Peter Fenn, Marc Goergen, Michael Hobday, Malcolm Horner, Will Hughes, David Hughill, Paul Ireland, Kalle Kähkönen, John Kelly, Malik Khalfan, David Langford, Peter McDermott, Krisen Moodley, Mike Murray, Chris Preece, Luc Renneboog, Andrew Ross, Martin Skitmore, Hedley Smyth, Will Swan, Graham M. Winch. It has been a pleasure working with all of you in the production of this book.

David Lowe and Roine Leiringer

Additionally, Peter Fenn has been my sounding board throughout the process of compiling this book. Pam Hyde, sometime administrator of the Project Management Division of the Manchester Centre for Civil and Construction Engineering at UMIST, deserves recognition, having helped with the formatting of the text and for aiding and abetting me while I 'made' time to complete my part of the enterprise.

To my wife, Ruth, for her patience, love and encouragement: I am indebted.

David Lowe

Dedication

I would like to dedicate this book to Andrew (my stepson) who died suddenly and unexpectedly while this book was being written.

David Lowe

1 Commercial Management – Defining a Discipline?

David Lowe and Roine Leiringer

Introduction

The term commercial management has been used for some time, similarly the job title commercial manager. However, as of yet, little emphasis has been placed on either defining the role of commercial managers or assessing the importance of commercial management as a discipline. We maintain that there are compelling arguments for considering commercial management, not solely as a task undertaken by commercial managers, but as a discipline in itself.

In accordance with the subsequent chapters in the book, this chapter focuses on commercial management within the context of project oriented and predominantly business-to-business (b2b) centric organisations. Consequently, it is concerned with the commercial management of projects.

This chapter seeks to establish commercial management as a discrete discipline. It delineates the context within which commercial managers operate; defines the terms 'commercial management' and 'commercial manager'; presents a cross-industry sector analysis of the function of commercial management, as a first step in establishing a body of knowledge for commercial management; and provides a rationale for the scope and content of the following chapters.

The context in brief

Projects

The past two decades have seen a change in how commercial organisations go about their day-to-day business. Within most areas of commerce, projects are increasingly used to undertake a variety of tasks and working assignments of most types and sizes. It is common to find simple definitions of projects as 'one-off activities', and all too often projects are studied as single entities. This might have been relevant for the large-scale projects for which techniques such as Critical Path Method (CPM) and Program Evaluation and Review Technique (PERT) were developed, but for the modern business environment it is perhaps no longer quite so valid. We do not wish to impose any new definition of projects or project management; this is a long running debate that is dealt with at length elsewhere and it would serve little or no point to revisit the arguments here. Instead, the view taken is that projects are predominantly initiated and/or executed within a larger organisational setting, that is, a commercial environment. Acknowledging, therefore, the development of portfolio and programme management (see, for example, Gareis 2004; Artto & Dietrich 2004; Archer & Ghasemzadeh 2004; Thiry 2004).

While some industries, for example construction, are inherently project based, this trend is also emerging in other industry sectors which traditionally have been less inclined toward projects. Some commentators have even gone as far as stating that projects are the future for global business, due to increasing technical and product complexity, shortening time to market windows and the need for cross-functional integration, and responding to changes in purchaser (client) needs (Lundin & Söderholm 1998). Whether or not this will turn out to be the case is for time to decide. However, in contemporary organisational practice, projects represent a growing percentage of an organisation's ordinary operations, while temporary organisational structures are no longer used solely to handle undertakings of extraordinary nature, such as large-scale engineering projects. Moreover, in many industry sectors there is also a clear trend towards a project-based approach to organisational design (see, for example, Miles *et al.* 1997; Hughes 1998; Whittington *et al.* 1999) as more and more companies adopt project oriented work methods in their businesses. This development is acknowledged in the *Guide to the Project Management Body of Knowledge* (PMI 2000) which defines the 'projectized organization' as an organisation where individuals are grouped together based on project commitments, notwithstanding their functional background or expertise. Project management has, therefore, gradually become a core business process for most firms and projects are increasingly considered as key vehicles for the strategic management of the company (see, for example, Jamieson & Morris 2004; Cleland 2004). Thus, commercial organisations use projects as tools contributing to the execution of organisational strategies with the aim of fulfilling corporate objectives, goals and missions.

Project-based organisations

In this evolving business environment, organisations will, if they haven't already done so, have to adjust to being project-based. The project-based firm's work consists of undertaking certain activities on projects or, if they are mainly operationally oriented, undertaking projects as an important part of their overall activity. Thus, they are either part of external projects, execute internal projects, or active in varying combinations of the two. While some organisations engage in one project at a time, others could well be part of numerous projects, all in various stages of completion. This is not, however, the same as stating that all project-based firms have organisational forms that fall within Galbraith's (1977) definition of the project-based organisation. Indeed, some firms might choose to have a matrix organisational design, while others might opt for functional organisational designs. Nonetheless, whichever organisational design strategy is employed, the organisation is, at any point in time, part of one or several project networks, all with interfaces of varying degrees of complexity. Exactly how organisations choose to deal with these interfaces, be they external or internal, varies. Yet, for most, it is essential to be able to organise their activities towards functioning in these networks.

Commercial management interfaces

Even though a different rhetoric might be used on individual projects, there is no escaping the fact that the various organisations involved in projects have their own agendas and vested interests. It follows logically, therefore, that even if projects are considered as stand alone entities, most decisions taken on any particular project will in one way or the other have repercussions on the organisations involved. For the project-based firm this means that there is an obvious connection between issues that have to be resolved at both the project and the corporate level. Hence, it is fundamental to strike a balance between the short-term goals of the project and the long-term objectives of the company. This includes obvious activities such as balancing the allocation of resources (of various kinds) between projects, as well as finding ways of distributing risk and, potentially, profit between projects and business units. Based on the fundamental assumption that the key objective for any commercial organisation is to generate a profit (alternatively, a prime objective of a not-for-profit organisation is to achieve best value), commercial issues, and thus commercial management, play an integral part in how most of the interfaces referred to in the previous section are managed. A fundamental challenge to the successful management of projects is, therefore, interrelationship management; for example, between the purchaser and their principal supplier, and the various firms involved in the project supply chain.

We contend that within a commercial organisation it is the commercial management function that links operations at a project level, multiple projects (portfolios/programmes) at the organisational level and the organisational core of the company.

The commercial management function and the commercial manager

The term 'commercial management' and the title 'commercial manager' are by no means new; indeed they can be traced back at least to the middle of the twentieth century. During the two decades both before and after World War II, commercial managers could be found at the very top of companies, and were typically responsible for acquiring new business and developing company strategies. This has subsequently changed. During the past two decades there has been an increase in the number of individuals assuming this title in organisations. Thus, what once used to be a board level activity has become more of a middle-to-senior management role in companies. In the UK construction industry, for example, a clear trend can be identified: the role of the traditional contractor's quantity surveyor evolving into that of a commercial manager. Other industries, likewise, have gone through their own evolution, and the trend towards an increasing number of commercial managers (also called contract or commercial contract managers) is equally unmistakable. As a result, commercial managers can now be found across a spectrum of industries, especially those that are predominantly project based.

While the term 'commercial management' and the title 'commercial manager' are both widely used and accepted, there is presently no commonly recognised definition of either. Moreover, it is not apparent that they mean the same thing both between and within different organisations and industry sectors.

What's in a name?

It would be naïve to state that the specific tasks undertaken by individual commercial managers are identical across industry sectors, or even between divisions of large organisations. However, a growing recognition of commercial management as a discipline and an associated skill set can be detected as evidenced by the emergence of various professional bodies in the UK and USA eager to promote commercial management as part of their portfolios.

Towards a definition of commercial management

As previously mentioned, the term 'commercial management' has been used for quite some time, not least in construction. Cottrell (1978), for example, refers to the job title 'commercial manager' in builders' organisations. Yet, research has shown that there is no consensus over what commercial management is or, for that matter, what a commercial manager does (Lowe *et al.* 1997). In their literature review, Lowe *et al.* failed to unearth any definitive statements of what was meant by commercial management or descriptions of the tasks and skills associated with the role of a commercial manager. This paucity of literature categorised as commercial management was again confirmed by Lowe and Leiringer (2005). This is not to say that research has not been conducted in this domain. On the contrary, there is an abundance of literature addressing issues associated with commercial management (e.g. Scott 1974; Gilbreath 1983; Turner 1995).

It is clear that commercial management is, in its literal interpretation, ambiguous, while its wide applicability in practice makes it difficult to provide a real boundary for the phenomenon. Two main interpretations tend to stand out in the construction related literature: the first sees commercial management as a subset of project management on individual projects, and a second uses 'quantity surveying' and 'commercial management' interchangeably.

Thomas and Adams (1995) consider the commercial manager to be subsumed within the project team, to be subservient to the project manager but responsible for the planning engineer, contract manager and cost engineer/quantity surveyor. Moreover, the current Association for Project Management Body of Knowledge (BOK), developed by the Centre for Research in the Management of Projects at UMIST (Morris *et al.* 2000; Morris 2001) contains a section labelled 'Commercial', thus placing commercial management within the context of project management and at the project level. It is contended that commercial issues may drive the conduct of a project, while the 'Commercial' section, as defined by Morris and the APM team, involves the business case, marketing and sales, financial management, procurement and a legal awareness.

Following the second line of interpretation of commercial management, Cottrell (1978) and more recently Walker and Wilkie (2002) hold the title 'commercial manager' to be synonymous with that of the builder's/contractor's quantity surveyor: a role that is integrated into the management of a project and involves the valuation of work, change management, financial accounting and subcontractor administration. This line of reasoning is further supported by the fact that it is not uncommon for contemporary private quantity surveying practices to advertise their services as commercial management. Furthermore, other commentators have noted that the contractor's quantity surveyor is now cast in the role of a financial manager, providing financial control of projects (see, for example, Pearson 1996). Research has been conducted in the construction sector into the role of the quantity surveyor both in consultancy and construction companies (see, for example, Cottrell 1978; Bennett 1989; Male 1990; RICS 1984, 1991, 1992; Cornick & Osbon 1994; Fortune & Skitmore 1994).

Looking more closely at the two ambiguous terms that make up commercial management a literal definition can be derived from the following definitions found in *The Oxford Modern English Dictionary* (Swannell 1992):

Commercial: '...of, engaged in, or concerned with commerce; having profit as a primary aim rather than artistic, etc.'
Commerce: 'financial transactions'
Manage: 'organize; regulate; be in charge of (a business, etc.)'
Management: '...the process or an instance of managing or being managed'
Manager: 'a person controlling or administering a business or part of a business'.

Adapting these definitions commercial management can be defined as follows:

The process of controlling or administering the financial transactions of an organisation with the primary aim of making a profit.

Likewise a commercial manager could be defined as follows:

A person controlling or administering the financial transactions of an organisation with the primary aim of generating a profit.

Unfortunately, these definitions, in their simplicity, leave much for interpretation. Yet, they reflect the findings from a prior investigation into commercial management in construction. In their study, Lowe *et al.* (1997) and Lowe and Fenn (1997) concluded that the most common description of commercial management in the construction industry was 'looking after the profits of the company...by keeping costs to a minimum and maximising income'. Further, they established the importance of the following attributes to the function of commercial management:

- The ability to take a broad perspective
- The ability to think laterally
- The ability to analyse risks
- The ability to convert work required on a project into packages
- The ability to negotiate
- The ability to communicate.

However, these traits are far from being unique to commercial management. Eraut (1994), for example, considers deliberate processes such as planning, problem solving, analysing, evaluating and decision making to be the essence of all professional practice.

The lack of generally acknowledged definitions for both the job title 'commercial manager' and the term 'commercial management', despite their widespread usage, prompted an investigation into both the role and function of commercial managers within project-based organisations (Lowe & Leiringer 2005). This research is seen as a first step in developing a Body of Knowledge for Commercial Management. The following section presents the results obtained from this study.

Commercial management as a discipline – a cross sector analysis

Lowe and Leiringer (2005) investigated the role of commercial managers across three industry sectors, construction, telecommunications/ICT and defence/aerospace, in order to establish whether or not commercial management could be identified as a distinct discipline. The empirical research that underpinned their investigation was conducted in four separate stages. The first stage consisted of a literature review and analysis of over 100 job advertisements for commercial managers. Stage two comprised a pilot questionnaire sent to commercial managers in a large telecommunications company, which was further corroborated through in-depth interviews with commercial managers from construction companies. The third stage involved a large-scale questionnaire survey generating 106 responses distributed between construction (42), telecommunications/ICT (33) and defence/aerospace (31). The fourth and final stage involved the presentation and validation of the questionnaire results by means of a focus group comprising eight representative senior commercial managers.

Findings

Commercial management

The majority of the respondents worked on multiple projects or were not allocated to specific projects; of those allocated to only one project the majority were from the construction sector. The involvement of commercial managers in projects varied between the industries. A tendency was revealed for individuals in the telecoms/ICT and defence/aerospace sectors to be appointed earlier in the project life cycle than those in the construction sector. This was evident from the fact that commercial managers in the construction sector were mainly involved in contract administration activities with project execution dominating their time, whilst commercial managers in the other sectors were predominantly involved in pre-contract activities with project development dominating their time. This difference between the sectors is perhaps a result of the traditional division of the commercial function in the construction industry, the estimating department being responsible for pre-contract activities with the quantity surveying section taking responsibility for contract administration activities (Brown 1997).

Functions of the commercial manager

Twenty-eight functions were initially identified as potential components of the role of the commercial manager. The respondents were asked to (1) indicate whether these functions formed part of their job, they provided support to the function, they provided a counterbalance to the function, or had little or no contact with the function; and (2) indicate the importance, to their day-to-day work, of those functions deemed to form part of their job.

Table 1.1 presents the functions ranked in order of the mode/median response to (1) above, and indicates the ranked importance given to them, in accordance with

Table 1.1 Functions of a commercial manager – ranked in order of the mode/median response.

	Involvement with function			Importance
Function	**Mode response**	**Median response**	**Rank**	**Rank**
Contract formulation	Job	Job	1	3
Contract negotiation	Job	Job	2	1
Risk management	Job	Job	3	2
Dispute resolution	Job	Job	4	5=
Bidding	Job	Job	5	4
Price formulation	Job	Support	6	7
Claim formulation	Job	Support	7	23=
Payments	Job	Support	8	9=
Cash flow management	Job	Support	9	11
Creation of outline proposals	Job	Support	10	15
Cost management	Job	Support	11	5=
Cost value reconciliation	Job	Support	12	8
Value management	Job	Support	13	13=
Subcontracting administration	Job	Support	14	9=
Procurement strategy	Job	Support	15	16
Acquiring approvals and permits	Job	Support	16	13=
Marketing	Support	Support	17	27
Supply chain management	Support	Support	18	22
Business plan development	Support	Support	19	12
Sales	Support	Support	20	18=
Estimating	Support	Support	21	20=
Quality assessment	Support	Support	22	25=
Development appraisal	Support	Support	23	25=
Supplier evaluation	Support	Support	24	23=
Image/reputation management	Support	Support	25	20=
Performance measurement	Support	Support	26	17
Innovation management	Support	Support	27	18=
R&D	No contact	No contact	28	–

NB: **Job** = This function forms part of my job; **Support** = I provide support to the function; **No contact** = I have little or no contact with this function; **Importance** = Functions ranking based on their importance in the day-to-day work of a commercial manager

(2). Overall, the results confirm the contribution of: contract negotiation, contract formulation, risk management, dispute resolution and bidding to the function of commercial management. They also indicate that commercial management can entail: price formulation, claim formulation, payments, cash flow management, creation of outline proposals, cost management, cost value reconciliation, value management, subcontracting administration, procurement strategy, and acquiring approvals and permits. Further, the results reveal that the commercial management function provides support to: marketing, supply chain management, business plan development, sales, estimating, quality assessment, development appraisal, supplier evaluation, image/reputation management, performance measurement, and innovation management.

Hierarchical cluster analysis was applied to the 28 functions initially held to contribute to the role of the commercial manager, to identify relatively homogeneous groups of activities. The resulting clusters were then interpreted and labelled (see Fig. 1.1) this graphical representation was held to represent genuine clusters. Figure 1.1 also shows, on an industry sector basis, whether these functions formed part of their job, they provided support or counterbalance to the function, or had little or no contact with the function.

The three industries show both similarity and divergence. Contract management (comprising contract negotiation, contract formulation and risk management) and the functions dispute resolution and bidding were generally consistently seen as forming part of the role of the commercial manager, although bidding and contract formulation were not seen as being quite so important in the construction sector. The latter could be explained in part by the greater tendency to use standardised contracts in construction than in the other two sectors.

Another area that showed convergence between the three industries was research and development, which was stated by virtually all to be outside the remit of commercial management. Indeed, the similar areas of opportunity development (in particular marketing, image/reputation management and development appraisal) and innovation management were consistently viewed as areas where commercial managers provided a support role. Thus, whilst all these are important for the survival of commercial firms in competitive markets and, therefore, justifiably could be claimed to be an integral part of commercial management, they are not primarily dealt with by commercial managers.

Areas of particular interest in which the three industries differ include bid management and financial management. Bid management was seen as a part of the role of a commercial manager in the defence/aerospace sector, whereas there was a tendency in construction and telecoms/ICT to view bid management more as a support function. Financial management (comprising functions such as cost management and cost value reconciliation), value management and supplier management (particularly subcontracting administration) were seen to be integral parts of the role of the commercial manager in the construction sector, while in telecoms/ICT and defence/aerospace involvement in these areas was generally in a support capacity. Again, this could be explained by peculiarities within the industry sectors.

Finally, there was a general lack of agreement between the sectors on the role commercial managers take concerning the tasks of estimating and acquiring approvals and permits.

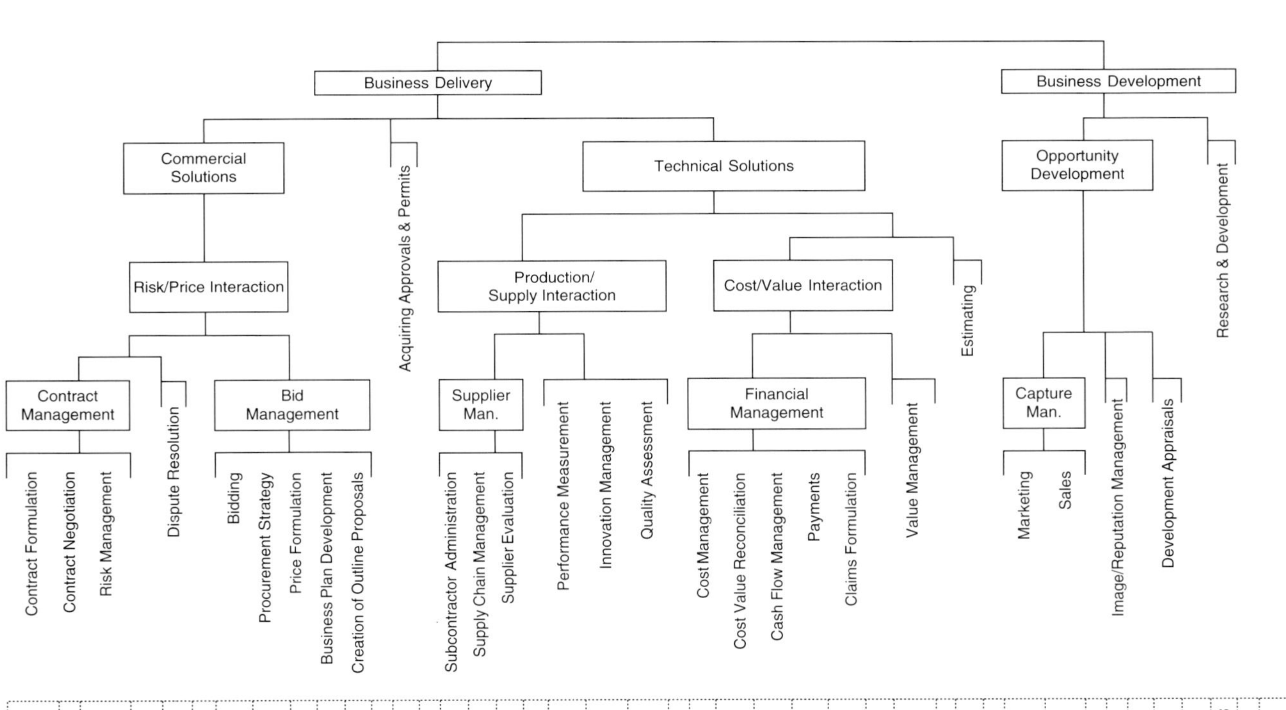

Function	D	C	T
Contract Formulation	J	J	J
Contract Negotiation	J	J	J
Risk Management	J/S	J	J
Dispute Resolution	J	J	S
Bidding	J	J/S	J/S
Procurement Strategy	J	J/S	S
Price Formulation	J	J/S	S
Business Plan Development	S	S	S
Creation of Outline Proposals	J	S	J/S
Acquiring Approvals & Permits	J	S/N	J/S
Subcontractor Administration	S	J	S/N
Supply Chain Management	S	S	S/N
Supplier Evaluation	S/N	S	N
Performance Measurement	S	J/S	S/N
Innovation Management	S	S	S
Quality Assessment	S/N	S/N	S/N
Cost Management	S	J	S
Cost Value Reconciliation	S	J	S
Cash Flow Management	J/S	J	S
Payments	J/S	J	S/N
Claims Formulation	J	J	S/N
Value Management	S	J	S
Estimating	J/S	S	S/N
Marketing	S	S/N	S/N
Sales	S	S/N	S
Image/Reputation Management	S	S	S
Development Appraisals	S	S	S
Research & Development	S/N	N	N

Figure 1.1 Functions of a commercial manager (J, function part of job; S, provides support/counter balance to this function; N, little or no contact with this function; T, telecoms/ICT; C, construction; D, defence/aerospace).

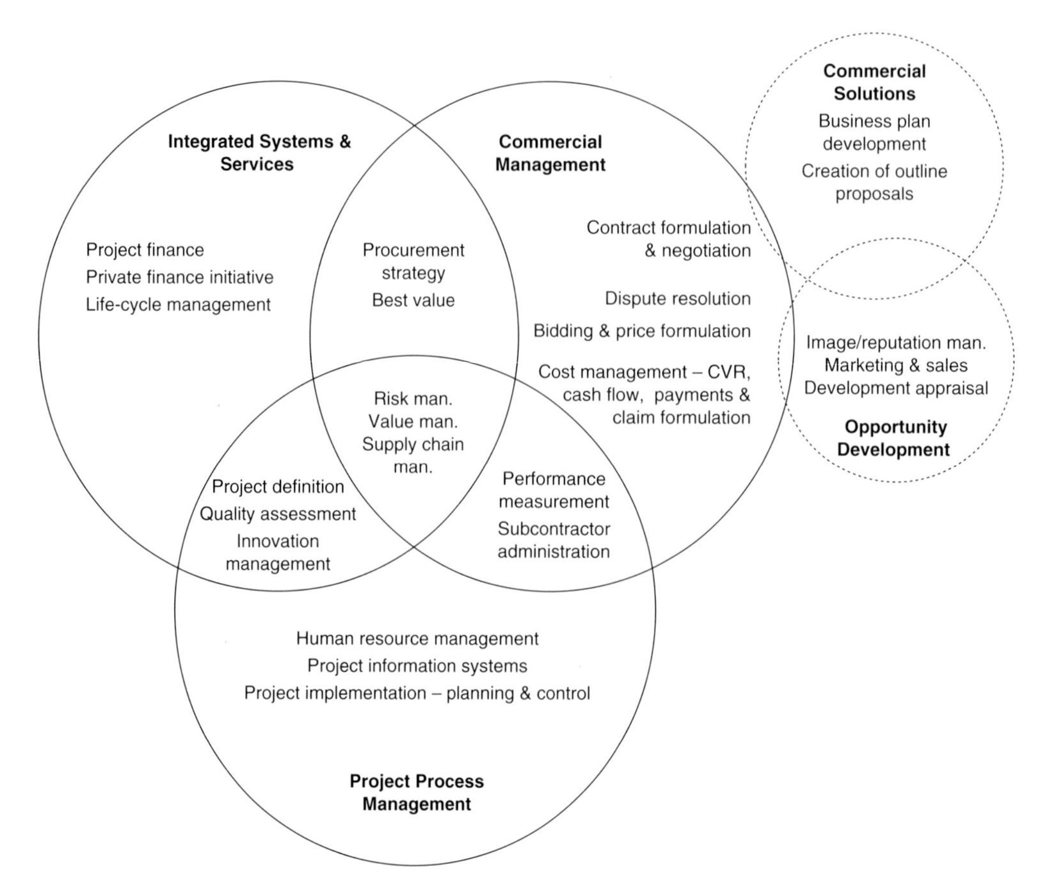

Figure 1.2 Commercial management with the context of managing complex projects (adapted from: Research Interests – Project Management Division, Manchester Centre for Civil and Construction Engineering 2003).

Figure 1.2 illustrates how the function of commercial management is linked to the overlapping areas of 'integrated systems and services' and 'project process management'. Likewise, it indicates its association with the areas of 'commercial solutions' and 'opportunity development'. Importantly, it shows the centrality of risk management, value management and supply chain management to the management of projects.

Interactions

Commercial management is concerned with managing interfaces. Overall, the most frequent interactions were found to be between the commercial manager and specialist divisions in their organisation, the client's representative and the client. However, differences in interaction did occur within the industry sectors; for example, in the telecoms/ICT sector interaction was significantly more frequent between commercial managers and specialist divisions in their organisation but less frequent with the client's representative, end users and the client.

Further, in the construction sector, interaction was significantly more frequent with subcontractors compared with both the telecoms/ICT and defence/aerospace sectors.

Working definition of commercial management

Keeping in mind that the above discussion deals exclusively with projects and the project-based organisation, a modified definition of commercial management can be deduced:

> *The management of contractual and commercial issues relating to projects, from project inception to completion.*

This definition entails specific activities and functions as well as whole processes within individual firms additional to the project organisations of which they might be a part. Also, implicit in this definition is an acceptance of the centrality of risk management to commercial management.

Professional standing of commercial managers

Predominantly, in the study described above, the commercial managers across all three sectors held at least a first degree, with a proportion holding higher degrees. In the construction sector, there is a tendency for commercial managers to be corporate members of the Royal Institution of Chartered Surveyors (RICS) and/or the Chartered Institute of Building (CIOB). Outside construction, however, there is a distinct lack of professional representation and recognition. While there are a number of aspiring institutions and associations active in this and in related disciplines, these apparently have not appealed to practising commercial managers within the targeted organisations of telecoms/ICT and defence/aerospace. This lack of status for commercial management led a blue-chip telecommunications organisation to approach a leading UK university in 2000 to develop an MSc programme in Commercial Management and more recently to establish an MBA for Commercial Executives. Both these programmes were the first of their kind.

Notwithstanding, it would appear that we are observing the embryonic stages of the professionalisation[i] of the commercial management function, which parallel the emergence of project management in the later part of the twentieth century. Whilst there is some debate over what constitutes a profession, the importance of a profession's knowledge base is widely acknowledged (Eraut 1994). Moreover, as Eraut (1994) states, the power and status of professional personnel is considerably influenced by the degree to which they can lay claim to unique forms of expertise and the value placed on that expertise.

Frequently, mature professions, to protect their position, seek to define themselves in terms of their core skills/competencies and/or knowledge base [see, for example, RICS (1992) which sought to establish the core skills and knowledge base of the Chartered Quantity Surveyor]. Likewise, project management associations, for example, the International Project Management Association (Caupin *et al.* 1998), the Project Management Institute (PMI 2000) and the Association for Project Management

(Dixon 2000), have defined bodies of knowledge (BOKs) to establish in what a project management practitioner should be knowledgeable (i.e. a set of professional competences) to underpin professional certification. However, the purpose of BOKs extends beyond providing a basis for establishing competence to practice. According to Morris (2001), they should reflect the purpose and provide a set of topics, relationships and definitions of a particular domain, reflecting, therefore, a distinct professional discipline.

An obvious development would entail the creation of a generic institute or association to represent commercial managers, disseminate good practice and certify competency to practice. However, these formal structures might perhaps represent nineteenth century ideals more than current aspirations. Possibly a more progressive response would be to utilise the Internet to establish loosely based 'communities of practice' (Lave & Wenger 1991) as a vehicle to disseminate good practice and cutting edge research pertinent to commercial managers. In terms of developing a research agenda to further underpin the development of commercial management as a distinct discipline, there is a need to co-ordinate the disparate strands of research currently being undertaken.

Aims of this book

Project management literature all too often concentrates on the single project with its established goals and time-line. This book is an attempt to address this imbalance by drawing attention to the vast array of commercial issues that companies have to contend with before, during and after being involved in projects. These issues, which are sometimes embedded, have to be understood and properly managed, regardless of whether or not the organisation 'owns' the project or is merely contributing to it as a minor subcontractor. Thus, the book aims to contribute to the growing literature that acknowledges the necessity of widening project management research to include also the management of project-based organisations, project-based industries and even project-based careers.

While commercial managers and the function commercial management evidently exist, so far, no attempt has been made to define the practice or establish an underlying theory (or theories) of commercial management. This book aims to present the current state of knowledge concerning the commercial management of projects. To overcome potential personality bias, attention will be drawn away from the individual commercial manager to focus instead on commercial management as a discipline. This is based on the belief that the commercial management function will be undertaken irrespective of the titles held by individuals in the process. However, it seems quite obvious that the efficiency and success with which the tasks are undertaken are dependent on the level of importance accredited to them. It may well be that the designation of an individual with the title of commercial manager could improve on an organisation's performance, but it would be naïve to expect one person to be able to master all functions and tasks.

Each chapter addresses a specific area pertinent to commercial management. The purpose is not to provide detailed descriptions of how to go about these specific tasks: this is not a prescriptive text that provides the reader with 'do's' and 'don'ts' or lengthy explanations of techniques and procedures. Instead the book focuses on clarifying and reviewing the current understanding based upon existing research in a wide range of commercial issues relevant to the undertaking of projects. Together the chapters constitute a step in the direction of systematic knowledge building, that is, a step towards the creation of a body of knowledge and a research agenda for the development of an underlining theory (or theories) relevant to commercial management practice.

Scope and content

Evidently it would not be possible to do justice in one book to all commercial issues that an organisation faces. Therefore, while it is acknowledged that commercial managers can be found in a diversity of industries, the book, in line with the research initiative reported in this chapter, is limited in scope to commercial management practice in project-based industries. Thus, it deals with firms whose work consists of undertaking certain activities on projects or, if the organisations are mainly operationally oriented, undertaking projects as an important part of their overall activity.

As stated in the introduction to this chapter, the position taken within the book is that of the supplier within business-to-business (b2b) relationships. Therefore, involving interfaces both up- and downstream in the supply chain. These interfaces can be divided into four general categories, depending at which level of abstraction they arise:

- The external milieu
- The corporate milieu
- The projects milieu
- The project milieu.

While it is acknowledged that various specialist research disciplines may use other terms, the structure of the book is based upon this rudimentary categorisation. The chapters presented under these headings represent core areas within which commercial managers are active. It can be argued that certain topics could equally be attributed to several, if not all, of the levels of interaction, for example, the areas of trust and dispute resolution. However, we have attempted to place the discussion of each area within the category most applicable to the exposition provided by the authors.

External milieu

All commercial organisations are subject to an external environment within which they have to abide, the most apparent being the legal and regulatory framework in which such organisations operate. This category deals with the interfaces between the organisation and the environment within which it is active. This will include, for example, purchasers, direct and indirect competitors, and suppliers (subcontractors). Three specific topics have been selected for further explanation:

- Competition
- Culture
- Procurement systems.

Corporate milieu

All firms exist for a reason – commonly to make profit. Thus, they will have a governance structure and a strategy (even if this strategy is to have no strategy at all) to achieve this goal. This category deals with the interfaces between the organisation (at a corporate level) and both the projects milieu (the environment within which individual projects collectively reside within the organisation) in which it is active and external milieu described above. Five specific topics have been selected for further explanation:

- Corporate governance
- Strategy
- Marketing
- Trust
- Outsourcing.

Projects milieu

This category deals with the environment within which individual projects collectively reside and the interfaces between the corporate level milieu of an organisation and the individual projects in which it is active. Four specific topics have been selected for further explanation:

- Management of uncertainty
- Conflict management and dispute resolution
- Performance measurement
- Value management.

Project milieu

This last category deals with the interfaces at the specific project level. The following five specific topics have been selected for further explanation:

- Project governance
- Contract management
- Bidding
- Purchasing, logistics and supply
- Cost–value reconciliation.

Whilst we argue that all the chapters seek to present topics and generic issues relevant to commercial management across all project-based organisations and industry sectors, most of the chapters are orientated towards construction and use this industry to illustrate specific points. We make no excuses for this as construction could lay claim to be the oldest project-based industry. Indeed, project management literature in general is heavily influenced by studies undertaken in this industry.

Conclusion

This introductory chapter has endeavoured to validate commercial management as a discrete discipline within project-based organisations. It has outlined the context within which commercial managers operate and defined the term 'commercial management' and the job title 'commercial manager'. Further, it has compared and contrasted the role of commercial managers from a range of organisations and across industry sectors as a first step in developing a Body of Knowledge for Commercial Management. While differences in approach and application both between and within industry sectors have been established, sufficient similarity and synergy in practice have been ascertained to identify a specific role of commercial management in project-based organisations. Finally, it has sought to present a rationale for the scope and content of the following chapters.

References

Archer, N. & Ghasemzadeh, F. (2004) Project portfolio selection and management. In: *The Wiley Guide to Managing Projects* (eds P.W.G. Morris & J.K. Pinto). John Wiley & Sons, Hoboken, NJ; pp. 237–255.

Artto, K.A. & Dietrich, P.H. (2004) Strategic business management through multiple projects. In: *The Wiley Guide to Managing Projects* (eds P.W.G. Morris & J.K. Pinto). John Wiley & Sons, Hoboken, NJ; pp. 144–176.

Bennett, J. (1989) *The Contractor's Quantity Surveyor and Reporting from Site*. Technical Information Service, No. 106. Chartered Institute of Building, Englemere.

Brown, G. (1997) A rose by any other name. *Construction Manager*, **3**(8), 20–21.

Caupin, G., Knofel, H., Morris, P.W.G., Motzel, E. & Pannenbacker, O. (1998) *ICB IPMA Competence Baseline*. International Project Management Association, Zurich.

Cleland, D. (2004) Strategic management: the project linkages. In: *The Wiley Guide to Managing Projects* (eds P.W.G. Morris & J.K. Pinto). John Wiley & Sons, Hoboken, NJ; pp. 206–222.

Cornick, T. & Osbon, K. (1994) A study of the contractor's quantity surveying practice during the construction process. *Construction Management and Economics*, **12**, 107–111.

Cottrell, G.P. (1978) *The Builder's Quantity Surveyor*. Surveying Information Service No. 1. Chartered Institute of Building, Englemere.

Dixon, M. (2000) Project Management Body of Knowledge (4th edn). Association of Project Management, High Wycombe.

Eraut, M. (1994) *Developing Professional Knowledge and Competence*. Falmer Press, London.

Fortune, C. & Skitmore, M. (1994) Quantification skills in the construction industry. *Construction Management and Economics*, **12**, 79–88.

Galbraith, J. (1977) *Organizational Design*. Addison-Wesley, Reading, MA.

Gareis, R. (2004) Management of the project-oriented company. In: *The Wiley Guide to Managing Projects* (eds P.W.G. Morris & J.K. Pinto). John Wiley & Sons, Hoboken, NJ; pp. 123–143.

Gilbreath, R.D. (1983) *Managing Contracts: Operational Controls for Commercial Risks*. Wiley, New York.

Hughes, T.P. (1998) *Rescuing Prometheus*. Pantheon Books, New York.

Jamieson, A. & Morris, P.W.G. (2004) Moving from corporate strategy to project strategy. In: *The Wiley Guide to Managing Projects* (eds P.W.G. Morris & J.K. Pinto). John Wiley & Sons, Hoboken, NJ; pp. 177–205.

Johnson, T.J. (1984) Professionalism: occupation or ideology. In: *Education for the Professions: Quis Custodiet?* (ed. S. Goodlad). SRHE and NFER – Nelson, Guildford; pp. 17–25.

Lave, J. & Wenger, E. (1991) *Situated Learning: Legitimate Peripheral Participation*. Oxford University Press, Oxford.

Lowe, D.J. & Fenn, P. (1997) Commercial management and quantity surveying: defining the name. *Construction Manager*, **3**(8), 20–21.

Lowe, D.J. & Leiringer, R. (2005) Commercial management in project-based organisations. *Journal of Financial Management of Property and Construction*, **10**(1), 4–18.

Lowe, D.J., Fenn, P. & Roberts, S. (1997) Commercial management: an investigation into the role of the commercial manager within the UK Construction Industry. *CIOB Construction Papers*, **81**, 1–8.

Lundin, R. & Söderholm, A. (1998). Conceptualizing a projectified society – discussion of an eco-institutional approach to a theory on temporary organizations. In: *Projects as Arenas for Renewal and Learning Processes* (eds R.A. Lundin & C. Midler). Kluwer Academic, Boston, MA.

Male, S. (1990) Professional authority, power and emerging forms of 'profession' in quantity surveying. *Construction Management and Economics*, **8**, 191–204.

Miles, R., Snow, C., Mathews, J., Miles, G. & Coleman, H. (1997) Organizing in the knowledge age: anticipating the cellular form. *Academy of Management Executive*, **11**, 7–24.

Mills, R. & Turner, R. (1995) Projects for shareholder value. In: *The Commercial Project Manager* (ed. J.R. Turner). McGraw-Hill, London; pp. 3–17.

Morris, P.W.G., Patel, M.B. & Wearne, S.H. (2000) Research into revising the APM Project Management Body of Knowledge. *International Journal of Project Management*, **18**(3), 155–164.

Morris, P.W.G. (2001) Updating the project management bodies of knowledge. *Project Management Journal*, **32**, 21–30.

Pearson, D. (1996) Contractor's QS or financial manager: what's in a name? *Construction Manager*, **2**(7), 13.

PMI (2000) *A Guide to the Project Management Body of Knowledge* (2000 edn). Project Management Institute, Newtown Square, PA.

Research Interests – Project Management Division, Manchester Centre for Civil and Construction Engineering (2003) www.umist.ac.uk/departments/civil/research/management/interests.htm (accessed July 2004).

RICS (1984) *A Study of Quantity Surveying Practice and Client Demand*. Prepared by Building Design Partnership on behalf of the RICS. Royal Institution of Chartered Surveyors, London.

RICS (1991) *QS 2000 – The Future of the Chartered Surveyor*. Prepared by Davis, Langdon and Everest on behalf of the RICS. Royal Institution of Chartered Surveyors, London.

RICS (1992) *The Core Skills and Knowledge Base of the Quantity Surveyor*. Royal Institution of Chartered Surveyors, London.

Scott, P. (1974) *The Commercial Management of Engineering Projects*. Gower Press, Epping.

Swannell, J. (ed.) (1992) *The Oxford Modern English Dictionary*. Clarendon Press, Oxford.
Thiry, M. (2004) Program management: strategic decision management process. In: *The Wiley Guide to Managing Projects* (eds P.W.G. Morris & J.K. Pinto). John Wiley & Sons, Hoboken, NJ; pp. 257–287.
Thomas, F. & Adams, T. (1995) Dispute resolution. In: *The Commercial Project Manager* (ed. J.R. Turner). McGraw-Hill, London; pp. 366–386.
Turner, J.R. (ed.) (1995) *The Commercial Project Manager*. McGraw-Hill, London.
Walker, I. & Wilkie, R. (2002) *Commercial Management in Construction*. Blackwell Science, Oxford.
Whittington, R., Pettigrew, A., Peck, S., Fenton, E. & Conyon, M. (1999) Change and complementarities in the new competitive landscape: a European Panel study, 1992–1996. *Organization Science*, **10**(5), 583–600.

Endnote

[i] 'Professionalisation': the route by which occupations attempt to obtain status and privilege in line with the ideology of professionalism (Johnson 1984).

Part 1
External Milieu

Introduction

All commercial organisations are exposed to and dependent on an external environment within which they operate, the most obvious being the influence of legal and regulatory frameworks. This part covers the interfaces between the organisation, at the corporate, projects and project level, and the environment within which the organisation is active. This external environment is all encompassing, including, for example, government (both central and local) and its agencies, purchasers (clients), competitors and suppliers. It also contains the general public, for example, in the form of pressure groups such as Greenpeace and Friends of the Earth. The influence of government is extensive, including competition law and regulatory mechanisms; in the business environment, purchasers instigate the procurement system and generally impose contract conditions. Further, as markets become global, organisations have to consider the impact of culture on these interfaces. From this array of potential topics, the areas of competition, culture and procurement have been selected for further explanation.

Competition

Few commercial organisations have the luxury of being monopolies; in one form or another, firms will have to compete with others that provide similar products and services, be it on certain specific aspects of operation or at a business level. Thus, the competitive environment within any particular market and industry will, to some degree, shape the way that an organisation conducts its business. This will influence its relationships with purchasers (clients), suppliers and all other relevant parties, as well as intra-organisational structures and relationships. In Chapter 2, Hedley Smyth illustrates how competition, in particular its inherent risk and uncertainty, shapes the behaviour of commercial entities. Using the construction industry as an example he shows how the competitive environment forces firms to constantly reappraise their strategies concerning market and organisational structure.

Culture

Culture is a widely applicable term used in a variety of contexts. It affects the way that people consider issues, analyse and solve problems, make decisions and take actions in response to opportunities and threats. Thus, culture impacts on how individuals and groups think and act. For example, the structural framework of the project, of which contracts are part, impacts on formal and informal procedures, i.e. the behavioural manifestations of culture, but does not necessarily affect the underpinning values and beliefs they may hold. Indeed, as Richard Fellows points out in Chapter 3, *culture is the constant practice of daily existence*. Drawing on a wide range of cross-industry studies on culture he discusses the basic knowledge blocks of culture, including ethics, and how these issues could manifest themselves on projects.

Procurement systems

Procurement – the process of acquiring new services or products – is fundamental to most project-based organisations. It covers a vast array of activities, ranging from the financial appraisal of the various options available to pricing, purchasing and administration of contracts. While the concept of procurement may well differ between various industry sectors, organisations still face similar strategic issues of how to adapt to different procurement systems. In this chapter, David Langford and Mike Murray, using the construction industry as an example, aptly illustrate how different procurement routes suit different types of projects and, thus, different types of actors. They show how social and political forces constantly influence procurement systems, which in turn shape how work is implemented on projects.

The contributors' profiles

Hedley Smyth

Dr Hedley Smyth is a Senior Lecturer in Construction and Project Management in the Bartlett School of Graduate Studies. He has worked in the built environment field for 30 years, working in volume house building and development, contracting, marketing architectural and design professional practices, and in academia. He has worked at the University of Bristol and Oxford Brookes University where he initiated the Centre for Construction Marketing, as well as University College London. Consultancy has been carried out for trade bodies, including the British Cement Association, contractors and both engineering and architectural practices. Current research is centred on the development and management of trust in relationships, especially at the client–contractor interface. This emanates out of a longstanding interest in relationship marketing. Other recent research activity includes case study work on information and knowledge management, supplied from external providers into contracting and project environments. Hedley is responsible for the enterprise management stream of teaching on two Masters programmes: MSc Construction, Economics and Management and MSc Project and Enterprise Management.

Richard Fellows

Dr Richard Fellows is Associate Professor in the Department of Real Estate and Construction, The University of Hong Kong, and Associate Dean of Architecture. Previously he was Reader in Construction Management at the University of Bath, UK. Richard has over 25 years' experience of teaching and research following several years of working for construction companies in UK. He has carried out many research contracts, has published 14 books and contributed chapters to six more, over 30 papers in leading international refereed journals and over 60 conference papers and reports. Richard has lectured in several universities around the world, including West Indies, USA, Australia and China, and has served on several committees and boards of professional institutions, notably RICS and CIOB. He is an editorial correspondent for *Construction Management and Economics* and referees papers for many leading journals and international conferences. Currently, Richard's main research interests lie in the field of culture in construction, for which he is joint co-ordinator of CIB Task Group TG-23. That research is progressing to include ethics, organisational citizenship behaviour, and corporate social responsibility.

David Langford

Professor David Langford has published widely, and books to which he has contributed include: *Construction Management in Practice, Direct Labour Organisations in Construction, Construction Management Vols I and II, Strategic Management in Construction, Human Resource Management in Construction* and *Managing Overseas Construction.* He has co-edited a history of government interventions in the UK construction industry since World War II. He has contributed to seminars on the field of construction management in all five continents. David Langford holds the Barr Chair of Construction in the Department of Architecture at the University of Strathclyde in Glasgow. He has published widely in the field of construction management, co-authoring eight books and editing three volumes on construction research. He is a regular visiting lecturer at universities around the world. His interests are travel, theatre and cricket, and he plays golf with more enthusiasm than skill.

Mike Murray

Dr Mike Murray is a lecturer in construction management within the Department of Architecture at the University of Strathclyde. He completed his PhD research in June 2003 and also holds a first class honours degree in Building Engineering and Management and an MSc in Construction Management. He has lectured at three Scottish universities (The Robert Gordon University, Heriot Watt and, currently, Strathclyde) and has developed a pragmatic approach to both research and lecturing. He has delivered research papers to academics and practitioners at UK and overseas symposiums and workshops. Mike began his career in the construction industry with an apprenticeship in the building services sector and was later to lecture in this topic at several further education colleges. He is a member of the Chartered Institute of Building and is co-editor of *Construction Industry Reports 1944–1998* (2003) and the *RIBA Handbook of Construction Project Management* (2004).

2 Competition

Hedley Smyth

Introduction

A chapter on competition that generalises about complex project markets would be ideal; in reality this might be fruitless as the experience of competition differs in each sector. Complex projects are delivered in a number of sectors and are also inherently diverse: the factors of significance change according to sector and project type. This chapter seeks to challenge thinking about competition in complex projects generally by showing how it operates in construction. The reader should use this as a basis to challenge thinking and to reinterpret the analysis in a way that fits experience in other project sectors.

The factors of competition evident in other sectors do not always apply for projects in construction as many factors have been managed out of the construction sector. The responsibility has been 'thrown over the wall'. The causes are characteristics found in many projects, especially complex projects: *uncertainty* and *risk*. In other words, the levels of project uncertainty and risk mean that contractors will try to insulate themselves as far as possible from the full impact of competitive forces in the market, hence market uncertainty and risk. The way in which this has been achieved is twofold:

(1) *Structural solutions*: Main contractors create 'walls' between themselves and the market to the extent that issues organised as processes in most industries are organised as structures in construction, the walls dividing procurement routes, which the client and their representative are left to select
(2) *Transaction cost management*: Main contractors largely compete between each other in terms of minimising their management costs, especially transactions costs, which constricts investment and constrains continuous improvement.

Competition is therefore considered at two levels in this chapter: the level of the sector and within the sector as contractors compete for work. Figure 2.1 shows the primary structure of the contracting enterprise positioned between clients and suppliers. It will be shown how different this position is compared to two of the competitive forces cited by Porter (1980). The 'walls' are shown in Fig. 2.2 by thick vertical lines.

Figure 2.3 shows the 'walls' main contractors tend to develop within their organisation. Some are impenetrable and are shown as thick verticals, acting as 'Chinese walls', while others are permeable to varying degrees and are shown as thick yet broken. These boundaries are typically divisions between procurement routes. Frequently a barrier exists between main offices and projects in terms of systems and management, which is indicated by the horizontal line.

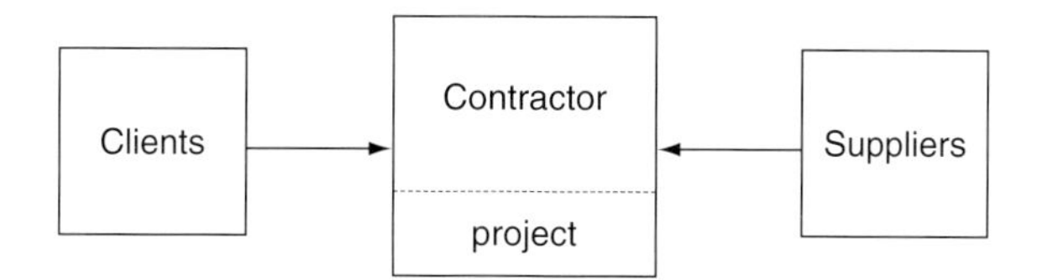

Figure 2.1 Market structure in construction: client–contractor–supplier structure.

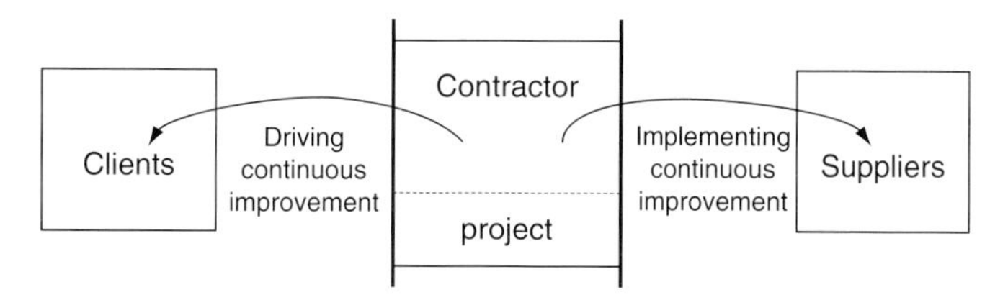

Figure 2.2 Market structure in construction: throwing continuous improvement over the walls.

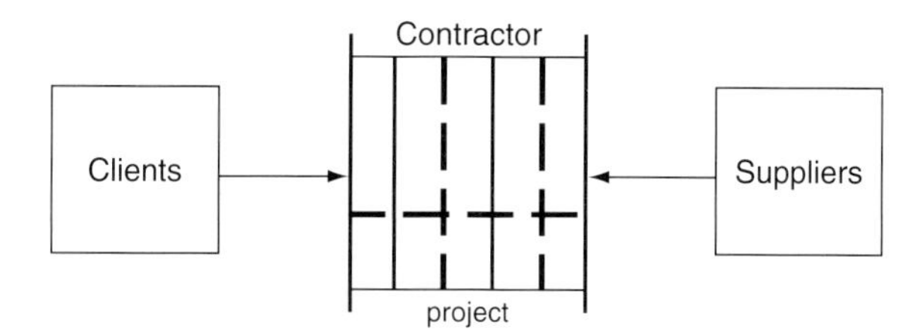

Figure 2.3 Divisional structure by procurement route.

The market is not static, nor is competition. Contractors continue to restructure at sector level. Yet client expectations and hence demands are driving forces for change – continuous improvement. It will be argued the effect of high uncertainty and risk is twofold:

(1) Contractors are concentrating resources into *restructuring*
(2) Contractors are minimising investment into *continuous improvement*.

The argument will be developed by defining dimensions of the industry and defining competition. An examination of management thinking concerning competition within market structures will be presented first, by exploring the forces at work in the marketplace through the *five forces of competition* (Porter, 1980). This will be evaluated in the construction context, developing an analysis of defensive adoption of structural solutions to the market.

Then the argument will be developed by examining the *core competency* model (Hamel & Prahalad 1994), which identifies resources controlled by an enterprise that are hard or costly to emulate. Core competencies will be evaluated for construction, particularly concerning continuous improvement. Competitive dynamics in the market are affecting and can affect the quality of the buildings and service experience. This in turn can affect market structure. Finally, a summary and a research agenda will be set out from the analysis for ten years hence.

Defining the dimensions of contracting and competition

What are the dimensions of project markets? Market competition can be conducted through the open market. Market competition can also be selective or closed, that is restricted to certain firms. The general pattern in construction is selective tendering. This selectivity is no different to other sectors, for buyers do not normally consider the full range of suppliers. Selectivity is overt and formalised in contracting. Most other goods and many services are produced ahead of sale or 'speculatively'. Construction projects are produced under contract; in other words, the sale is secured first.

Projects have become an established means to deliver complex requirements across a range of sectors (cf. Morris 1994; Morris & Pinto 2004). Projects are characterised by high degrees of *uncertainty*, which provides motivation to contract out work in order to pass attendant risk onto a specialist provider, in this case the construction contractor (Winch 2002).

Competition is generally present under two conditions. Competition occurs where there is more than one party wanting to provide something, in this case a project, for another party. Suppliers compete between each other to make the provision. Competition also occurs where there is more than one party wanting to secure something from another party. Customers compete between each other to secure the scarce provision. Intense contractor competition to secure a project is sometimes witnessed for projects, giving the client sufficient choice to prevent a scarcity of providers.

Complex projects usually require diverse resources and skills to deliver a specific asset. Hence, many contractors are primarily or exclusively service providers. Projects are subdivided into a series of mini-projects or packages of work, which are subcontracted in competition. The higher the level of subcontracting present in a contract, the less the contractor is producing. The less the contractor is producing, the greater the management focus upon the various inputs; hence, the more the main contractor becomes a manager of costs, especially *transaction costs* [see Chapter 14 and also Winch (2002), Gruneberg & Ive (2000)]. In essence, an exchange in the market does not come free. There is a cost to an exchange – searching out information about the product or service, about the providers, and the time agreeing a price, and so on throughout the project life cycle. These are the costs of the transaction. The client decision to produce or outsource is thus a question of comparative costs. This is tempered by the uncertainty and risk management tactics, in other words, the extent to which there are additional 'hidden costs' and which party is most able to manage these (Williamson 1985; Douma & Schreuder 1991). The contractor emphasis upon competition is price: managing prices received from subcontractors and managing internal costs in order to achieve price advantage in bidding for work.

In reality, market conditions change. For example, in World War II, projects such as airfields, Mulberry Harbours, the Fog Investigation Dispersal Operations (FIDO) and Pipelines Under The Ocean (PLUTO) were governed more by time than price, which frequently led to cost plus and target cost contracts (Smyth 1985). When time is the main focus, and thus clients make allowances in their expectations concerning price, it is possible to invest in new or different ways of managing projects and technologies, hence reducing the impact of transaction costs.

In recent years, clients in many sectors have turned their attention to quality and scope. Adding value has elevated just-in-time, lean and agile production and supply chain management to procurement strategies in many sectors in order to improve the value-for-money equation. This has occurred in construction, for example, in the UK through *Rethinking Construction* (Egan 1998). Construction clients have become generally used to continuous improvement through cost reductions and added value in procuring from other sectors. Procurement has driven improvement in other sectors. Clients wish to see improvements in construction. They have been dissatisfied with both poor service quality and poor product quality on projects. The question is whether client desire to improve quality means that contractor transaction cost management is subsumed under investment for continuous improvement. It has been argued in construction that the client is the only party with the motive, means and opportunity to drive quality improvements and innovation (Ive 1995; Gann 2000). Are contractors responding and, if so, to what extent are they able to respond?

Competition and management thinking

Competition in complex project delivery can be understood and analysed using concepts in mainstream business management literature. This extensive literature [see, for example, the review of Johnson & Scholes (2002)] provides a valuable overview. The conceptual understanding can broadly be divided into two streams: *structure* and *process*: the *structural* approach provides concepts to understand forces at work in the marketplace; the *process* stream provides concepts to understand how enterprises operate in the market in order to secure competitive advantage.

Therefore, understanding market structure concerns the external context for the enterprise or firm. Enterprise strategies help determine the structure of the market (Ive 1990). The strategies and tactics for implementation are the commercial management function, securing competitive advantage in the market. Competitive advantage switches focus from the market to the firm. Competitive advantage is being more efficient and effective than others. The focus for efficiency and effectiveness may vary across sectors and within firms. For some enterprises profitability and return on capital employed (ROCE) will dominate, while other firms may emphasise growing turnover and market share, others investment and service development, yet others client management and satisfaction. The choice or combination will depend upon strategy, essentially determining whether to compete on price or service differentiation (Porter 1985). ROCE and turnover are important for many contractors, reflecting that size is a key dimension of competitive advantage in construction. Contractors minimise working capital requirements, manage cashflow and recycle working capital; hence ROCE tends to be more important than profit margin for the project, and hence the firm.

Thus, it is disadvantageous to tie up resources in overheads and investment. Improvement demands from clients will only invoke a defensive response when absolutely necessary, rather than competition *per se*. Clients may not always exert demands in a way that stimulates investment. There has been a recent trend amongst major corporate clients to appoint 'project teams' rather than 'contractors'. While the contract is with the main contractor, the contract terms require that the same team is kept in place during the project life cycle. Client concern is lack of service continuity as team members are changed by contractors during a project and also between projects in a programme. Contractor motivation is to keep transaction costs to the lowest level *vis-a-vis* securing new contracts in the market on the one hand and the absence of systems at the contractor–project team interface to secure continuity of service on the other hand.

While management literature does provide an overview, the detailed application to construction contractors faces problems. The literature tends to assume that competition is for consumer products and services or the more standardised products and services purchased by other businesses. While contractor corporate strategies can apply the same decision criteria concerning finance, human resources and marketing across corporate functions, the construction client–contractor interface is structured in a different way to many buyer–seller interfaces. There are limits to the relevance of mainstream management thinking on competition for construction firms, being reached sooner than for most enterprises. The limits arise because project work can be erratic, discontinuous and 'lumpy', making the allocation and management of resources critical in the face of *market risk* as well as project uncertainties and risks. This situation is not so much the product of poor management theory, but rather that more conceptual effort has been put into manufacturing and consumer markets for products and services, competition in project markets requiring greater understanding.

Construction challenges the models, which will be considered through one aspect of *structure* and one aspect of *process* experienced among contractors working on complex projects. Therefore analysis is developed in subsequent sections that there is some evidence of improvement to the client service experience; hence, contractors are beginning to occupy different positions, managing the market operationally in differentiated ways. This is not particularly in response to continuous improvement drivers from clients and policies. Low contractor investment is the constraint, yet contractors need to progress trends further. The conclusion will assert that there is a tension between increasing investment, increasing both overheads and transaction costs, yet providing opportunity to create higher barriers to entry and hence increase capital concentration in the sector, especially at the upper end of the market. This will reduce intensity of competition, yet increase uncertainty. In the market, contractors will be less flexible in the wake of trade and economic cycles. Firms will be vulnerable to price competition until new norms are established for the minimum demand levels, i.e. where new investment to provide *added value* is 'displaced' into *value added*. This will be a dynamic process of competition and inherent risk if the demands for service differentiation and improvement become established in the long term to secure continuous improvement.

Competition and structure

The *structural* approach to competition in the management literature provides concepts and models for understanding marketplace forces. The *five forces of competition* (Porter 1980) is one prime structural model (Fig. 2.4). Key structural features of

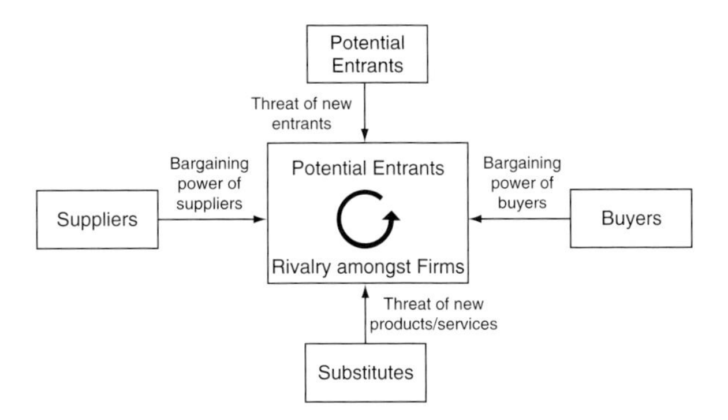

Figure 2.4 Five forces of competition (adapted from: Porter 1980).

competition are determined by the relationship of an enterprise to the five forces (Fig. 2.3). *Competition amongst firms* concerns those in immediate and current competition in the market. Porter (1985) states there are two generic strategies – price leadership and product differentiation. Customers are generally price conscious, so price is a key. However, price leadership is difficult to establish and sustain. Simply cutting price leads to instability – price cuts can be matched, resulting in lower revenues for all in the long term – intense *rivalry* without advantage. This option remains tempting for contractors in order to secure a contract. On securing the contract, contractors endeavour to make up low margins on variations, claims and outturn cost differentials, and by passing price cuts down subcontract supply chains. Intense competition, especially in times of recession, encourages price cutting. All are inhibitors to investment by main contractors.

Buying power arises when enterprises are in a commanding position over supply chains (Campbell 1985). The buyer is able to exert leverage or market power, usually by virtue of the size or volumes of purchase, especially in comparison to its sales. Switching costs, that is the transaction costs of switching suppliers, need also to be low. Main contractors can use buying power on subcontractors as they can employ direct labour if necessary, and typically low profit margins motivate hard bargaining. However, clients exert similar forces over contractors, hence nullifying some of the competitive benefits. The power position shifts once the contract is signed – greater scope for opportunism arises as switching costs for clients increase dramatically (Williamson 1985; Smyth 2000; Gruneberg & Ive 2000). One example can be found on infrastructure project, the Boston Big Dig (2004).

Figure 2.1 shows that the primary structure of the contracting enterprise is positioned between clients and suppliers. The analysis of Porter (1980) regarding buyers and suppliers concerns market power and leveraging the value for money. This may take the form of price reductions or added value, invoking the responses of either price leadership or product differentiation strategies (Porter 1985). In the case of construction, forces tend to act in the opposite direction – instead forces are exerted outwards from the enterprise as the procurement approach is passed down the chain [Fig. 2.2; see also Smyth (2005)]. Industry structure has focused upon price rather than adding value; thus continuous improvement is displaced outside the structure of the contracting firm towards the client on the one side and the suppliers on the other.

Whether processes developing *competencies* for service improvement are able to override or counteract this trend will be considered in the next section.

The *threat of substitutes* is the final category in the five forces model. The threat arises when the price is reduced because similar products or the same product with a new technology is introduced in competition. For projects, uniqueness largely renders this obsolete; however, procurement routes, which are equivalent to 'place' or delivery channel in marketing (Smyth 2000), can pose a threat. Procurement options determine variance in service value and hence threaten main contractors that do not offer a particular route. This is particularly pertinent for complex projects. Contractors tend to set up different structural divisions for each procurement route. Once personnel have internalised the requirements of a route, management input–output ratios concerning transaction costs are at their most efficient, although this does inhibit responsiveness to clients (Smyth 2000).

The markets for large construction projects tend to be dominated by a few clients with considerable market leverage until the contract is signed and power shifts. Pre-contract contractors occupy weak *supply* bargaining positions. The client has the initiative for innovation and performance development (Ive 1995; Gann 2000). Clients drive changes in design, in technology or in procurement route. Transaction cost management theoretically dominates contractor tactics; therefore competition concerning performance enhancement is weak.

In the areas of buying, supplying and substitution, main contractors tend to be passive recipients of market forces rather than proactively seeking competitive advantage and hence stimulating greater *rivalry* amongst themselves.

The final force, the *threat of new entrants*, carries considerable power in construction. A highly fragmented industry minimising capital investment creates low barriers and hence invokes high competition from this threat. Gruneberg and Ive (2000) cited small firm proliferation, partly due to extensive subcontracting; they also noted that the largest firms have grown larger, squeezing firms occupying the middle ground. Size seems to be of importance in construction, a threshold of competition needing to be crossed to enter the next level and hence obtain access to a different size of project. Large construction contractors do not necessarily undertake more projects than small ones. Economies of scale are achieved, complexity and size of contract being the key features. Therefore entry to the complex project market tends to be restricted by size of enterprise rather than by entry at the bottom of the market, which is easy. Size also tends to be a feature required by clients when cost is not the main determinant. World War II projects have already been cited. The same holds true when faced with an emergency, such as 9/11, when contractors of size and reputation were appointed to undertake Ground Zero recovery (MPA 2003). Although the competition is limited by geographical coverage, the larger the contractor, the greater the geographical spread of projects undertaken. Contractors wish to protect their markets of large, complex projects from entry by smaller contractors. This competition, coupled with high project risk, renders contractors vulnerable in the market. In construction, Laing experienced a number of unrelated project problems and failures whilst undertaking diversification into investment-led projects; company rating shrank to such an extent that it was bought by specialist subcontractor, O'Rourke.

The five forces model provides a useful overview. The analysis Porter (1985) provides shows how real situations depart from the idealised neoclassical market laws. The model's strength is to answer 'what' questions, but it fails to answer 'how' to address the forces. It is the 'how' questions that commercial managers wish to answer in developing strategy. Porter (1980, p. 32) does say:

Definition of an industry is not the same as definition of where the firm wants to compete.

However, the underlying assumption is the long-term tendency towards equilibrium and industry maturity. It has also been shown that the definitions relating to construction are very different compared to other sectors. It has been shown here that application of the five forces model to construction at a detailed level is highly constricted.

A deeper consideration of the structural features in the main contractor market is therefore necessary. Two strong structural trends in complex project markets are prevalent, leading to greater capital concentration and hence fewer larger firms (Gruneberg & Ive 2000). The first is for main contractors to adopt structural solutions to *market management*. The second concerns market *restructuring*.

Market management is considered first. Many contractors adopt structural solutions to marketing, hence managing their market. Separate divisions or business units are established for major procurement routes, operating like 'subsidiaries'. Each division is grown to undertake the same number of contracts as a separate company. By this means main contractors enter markets for larger contracts, giving client confidence in those markets as the critical size of the whole enterprise grows.

Second, this approach aids transaction cost management. For example, construction contractors developed new divisions as clients developed new procurement routes – design and build (D&B) in the 1980s, management contracting and partnering in the 1990s and private finance initiative (PFI) this century. As each procurement route emerges, the structural solution to market management is favoured; so the management requirements of each route, especially cost control, are instilled into personnel dedicated to that route. The management requirements are absorbed into the culture of the division. The divisions operate separately, 'Chinese walls' developing. Hence the client and their representative have to source the most appropriate procurement route rather than the contractor understanding the client requirements and mobilising resources to serve them (Fig. 2.3). As the market for each route matures, the financial return declines. Management reallocates resources; sometimes amalgamating procurement routes with similarities (see the broken vertical lines in Fig. 2.3), and usually concentrating business development activities in the zero sum game of following growth markets [see Chapter 7 and Smyth (2000)]. Clients learn to exercise their buying power with each route, and squeeze costs. Thus, procurement routes mature quickly. The process only reinvigorates itself as new routes and variants emerge, usually initiated by a client with considerable market leverage (Cox & Townsend 1998).

However, contractors have been driven by clients to improve performance. Clients have come to experience improved value for money in purchasing standardised products and services in other sectors. Improved value for money has come from lean production and other production-related tools, such as just-in-time production and delivery in order to eliminate waste, reduce costs and increase speed of response.

Improved value has also come from added value to products. Clients have come to expect more from construction projects too. Partnering and output specifications for design, build, finance and operate (DBFO) projects have been used as levers to facilitate tools such as supply chain management and lean production. Some improvements have arisen from these means. However, all require investment, which cuts across the thrust of minimising transaction costs and overheads.

A new trend is emerging. Many main contractors have invested more resources into market *restructuring* than continuous improvement. Instead of product differentiation, which requires investment to improve performance, a new structural solution is being negotiated through the market as an indirect means of market management. Broadly, contractors are restructuring along the following lines:

(1) Generating asset-related activities
(2) Creating income streams
(3) Committing to 'traditional' contracting, in essence a shift towards main contractors becoming large or mega subcontractors.

As procurement routes mature, too many players exist in these markets for average profit margins to be earned and an adequate ROCE. This echoes what is happening across many maturing industries – low rates of return or breakeven. Therefore contracting becomes a door to other sources of more profitable work for some firms, leaving a rump focusing upon 'traditional' contracting.

In the first category, asset generating activities, large contractors use construction as access to development profits through urban regeneration and commercial speculative development. DBFO contracts provide access to building up assets through consortium companies, which can be sold at a good return later. Laing sold its construction arm – admittedly in forced circumstances – with the remnant focusing on what it had come to see as its core business, investment-related construction such as property development, urban regeneration and PFI/DBFO schemes. Shareholders forced further specialisation, demanding the sale of its speculative housebuilding arm (reflecting the 20 year UK trend of the formation of a specialist house building sector, almost exclusively separate from mainstream contracting).

In the second category, contracting is a door to long-term income streams. Construction contractors are entering into long-term maintenance contracts or undertaking PFI to gain access to other income generating streams in facilities management (FM). New build is short term and not the primary source of profit. Mowlem has focused more upon maintenance and FM work, sometimes related to PFI. The former Tilbury Douglas, now called Interserve, has FM as a core business. Amey is another organisation that is now mainly a maintenance and FM enterprise. Other contractors have this area as one of their core business markets, such as Balfour Beatty.

In both these categories, these other sources for profit offer opportunity for main contractors to be rerated on the stock market. This is not new, for a number of leading contractors achieved this through property development in the late 1960s and early 1970s (Smyth 1985). However, property portfolios were compromised through the recession of the late 1970s, and the demand for working capital in subsequent expansion times was at odds with capital intensive development. Contractors did not depart from contracting as their core business and hence profit source. This time around it may be different (Hillebrandt *et al.* 1995).

Where does this leave 'traditional' contracting, the third category? Contracting is being transformed into a mega-subcontracting market, where main contractors simply become large scale subcontractors, thus behaving as such. The symbol of this transformation is embodied by Laing selling its construction division to O'Rourke, a subcontractor. The market power or leverage of the new mega-subcontractor will be less than the old-style traditional contractor (cf. Campbell 1985; Smyth 2000). Such mega-subcontractors will be required to act collaboratively, as demanded by clients that have pushed alliances, partnering and driven supply chain management, yet these contractors will have not made the investment in the corresponding marketing response. They simply drive the procurement solution down the chain (Smyth 2004, 2005).

Significantly this restructuring to some extent cuts across the structural solutions of Chinese walls within the firm. It begins to differentiate contractors by the type of work they do, not by procurement route or building type, but the characteristic of the *service*. This is the case for all three categories. Asset- and income-based firms can manage their markets in accordance with the demands of the related market. However, stock market rerating will permit further investment to improve construction services in order to secure work, i.e. to open the gate to their respective markets. For the mega-subcontractor, service differentiation will only occur where contractors make the marketing response and invest in service quality, which begins to address the agenda for continuing improvement. In other words, the current restructuring, whilst clearly being distinct from continuous improvement and indeed in conflict with it at present, may link to it in future.

However, the restructuring process is far from complete. Determining the core business is not always straightforward. For example, Carillion has placed more focus upon internal restructuring, creating new divisions for a mix of perceived growth segments and procurement routes. In 2001, Carillion had divisions for:

- Private finance
- Infrastructure management
- Services
- Capital projects
- Building.

Within a year this was refocused into:

- Business services
- Construction services infrastructure management
- Investments.

At this stage it appeared to have a foot in all three camps, although 'traditional' contracting was the mainstay in turnover. Within another year (2003) a further refocus yielded:

- Road services
- Building
- Capital projects.

Capital projects are the investment focus. Road services have the FM and maintenance focus, reflecting the demise of rail maintenance contracts on the open market. The Carillion strategy has refined markets by procurement route, while also making efforts to engage with the primary market divisions, although the process exhibits some confusion. This is a reflection of the sector as a whole, yet maturation continues, especially at the top end of project markets. This restructuring is producing ways to secure:

- *Profitability*: through acceptable margins or return on capital employed (ROCE)
- *Reduction in competition*: through focusing core business in primary market divisions, hence diversifying the scope of contracting towards greater specialisation
- *Scope to grow in size*: undertake the largest, complex projects.

Structural divisions, or 'walls' in the market, plus restructuring could prove to be the beginning of significant differentiation. This is the result of defensive action rather than outright competition, yet market forces are shaping these trends which could yield greater competition around quality issues of added value and hence continuous improvement rather than domination of price.

At present, specialisation is less concerned with service differentiation (Porter 1985) and more concerned with strategic market position for contractors. As maturing takes hold and the return on profits stabilises, contractors will rely heavily on circulating their capital quickly, effectively to yield a high ROCE rather than the profit margin emphasis of most other sectors (Ive & Gruneberg 2000). As this occurs, then the only way to secure competitive advantage would appear to be by investment in process, and hence performance improvement. However, at the present a great deal of investment has gone into restructuring, more than has gone into processes for improving construction performance. Continuous improvement, however, is a process, and it is to this the analysis must turn.

Competition and process

The *process* stream provides concepts and models which are used to understand how the main corporate actors or players operate in a competitive market. One conceptual approach, the *core competency* model (Hamel & Prahalad 1994) is located in the *resource based view* (Wernerfelt 1984; Rumelt 1984) which identifies resources controlled by an enterprise that are hard or costly to emulate. The core competency model will be used to illustrate the process approach.

The *process* approach looks at the way things are done. The approach provides concepts and models to understand how the main corporate actors operate in a competitive market. Skills and expertise, technology, and systems to manage people and technologies are all aspects of this. Competitive advantage is claimed by being best in an area. One main trend has been to elevate formerly tactical processes of business operation to a strategic level, allocating resources to capture advantage in prescribed areas. Such advantage is sustained until competitors catch up. Technological advantage can be copied or bought relatively quickly. Skills can be developed with relative speed, for example, professional skills and expertise. Hamel and Prahalad (1994) realised that other intangibles were much more difficult, sometimes impossible, to emulate, giving greater scope to achieve long-term advantage. These intangibles are called competencies, hence the *core competency model* (Prahalad & Hamel 1990), the example considered to illustrate a process approach.

Core competencies of competitive advantage are sustainable. While competitors 'play catch-up', the originators allocate resources to continue competency development to new levels. Eventually the law of diminishing returns is experienced for a competency, yet competitors face a barrier to emulating a competency as their yield upon investment will be lower than the initial pioneer or followers. Therefore advantage is protected when diminishing returns kick in. At this stage, enterprises can investigate diversifying their competencies into new areas that will yield advantage.

What are the areas of competency advantage? At a tactical level competencies may come in a number of forms: behaviour that improves team effectiveness, or useful information and knowledge that is developed and confined to the enterprise, a revised system or set of procedures releasing abilities, internally generated software to improve efficiency, giving clients effective attention to improve service experience. These may be assigned to particular people or may be owned by or embedded in organisations. At the tactical level, such competencies can be seen as serendipity and accidental. A strategic elevation requires a conceptual framework in order to invest in particular competencies. There are a range of concepts, three of the most frequently considered areas being:

(1) Organisational learning and knowledge management (OL and KM)
(2) Relationship marketing and management (RM)
(3) Emotional intelligence (EI).

Each of these has a long line of associated literature: for OL and KM see, for example, Argyris and Schön (1978), Nonaka and Takeuchi (1995), Senge (1993); for RM see Gummesson (2001) and Grönroos (2000), and for EI see Goleman (1996, 1998). It has been argued that each potentially offers performance improvement as great as other process strategies, such as lean production and supply chain management, and indeed more if the connections between these can be harnessed on projects (Smyth 2004).

As noted, some tactical issues have been elevated to the strategic, such as just-in-time production and delivery, lean and agile production, supply chain management, plus related tools, for example, alliances and partnering. In essence these emanate from a production tradition (cf. Koskela 1992, 2000; Green & May 2003) and are applied in construction through procurement (Smyth 2005). This type of continuous improvement has been the main thinking in construction (Egan 1998), which essentially is a procurement approach based upon driving costs down. Long-term gains must also be secured from adding value, which means that parallel human systems are required to add value to the product and service quality. This requires the application of conceptually based competencies, such as OL, KM, RM and EI to mobilise the technical and systems potential.

Main contractors face a paradox in relation to competency development. On the one hand acquiring competencies differentiates their services, yielding competitive advantage from related inputs of added value for complex projects. This improves contractor effectiveness and client satisfaction at the project level. On the other hand, the position of main contractors in the marketplace and the position for securing contracts require that overheads and transaction costs are kept low. Competency development requires investment. Investment is needed to yield a return. Mobilising competencies on projects has the effect of increasing total costs. It has been shown that innovation and other development generally comes from the client (Ive 1995). The investment incentive is poor, plus any potential short-term advantage contractors can proactively secure is eroded as leading clients can easily adopt demands to impose on competitors, for example, lean production and supply chain management practices. The motive has been absent, yet the cited restructuring may lead to increased service differentiation.

Does this analysis prevent contractors developing core competencies of competitive advantage? They can, yet it increases risk, adding higher market risk to high levels of project risk. However, the potential returns reflect risk, and the advantage will be sustained for a longer period as competitors would remain cautious for a while, perhaps until clients demand competency change. The clients who experience such service improvement may give contractors repeat business, switching costs (the transaction costs of sourcing a new supplier) having been raised (Smyth 1999). The current risk–profit trade-off is insufficient for main contractors. However, as markets evolve and core business activity yields insufficient profit margin and ROCE, main contractors may be forced to increase investment.

Clients are not prepared to wait for proactive process improvement. A key issue for clients is uncertainty over the quality of service. Clients can have a good experience on one project and a contrasting poor experience on the next project with the same contractor. There are a number of potential reasons why this arises. One reason is that project personnel change. A related reason is the lack of support between the corporate headquarters and the project team delivering the project locally. Because the cost of managing such support is high, enterprises in other sectors tend to put systems in place in order to articulate service standards. Even if the design and specification of each activity is different the process can be systematised. *Systems* do not mean IT, procedures or rules, although all these aspects may form part of a human management system. Where the product is standard, the task is easy. For example, McDonalds' products are so standardised that the service can be standardised to the extent it is franchised. The focus then becomes quality management through the McDonalds' 'university' training and quality control of customer service and product over-the-counter. The uniqueness of complex projects renders this difficult, especially where delivery is remote from the headquarters. The absence of such systems means that the quality of service experience is down to the personality and blame culture that typifies many construction firms and projects (Smyth 2000).

From North America there has been a trend amongst major corporate clients, such as Microsoft, to appoint 'project teams' rather than 'contractors' (Smyth 2000). While the contract is signed with the contractor, the terms require that the same team is kept in place in order to induce greater standardisation and continuity of service (Smyth 2004; Wilkinson, in press). This overcomes the worst consequences of transaction cost management where team members are changed by the contractor during the project life cycle in order for team 'A' to be used to bid and start a subsequent contract and so on until team 'N' completes the project. This contractual tool can also be used on a series of projects, a client investment programme. This point raises a profound issue of competition. The competition around contracts is essentially about securing work rather than delivering a service. In other words, uncertainty has been used to obscure quality issues. Although, in theory, competition delivers the appropriate quality for the price, this has not been the case for large projects.

The literature review concerning process has highlighted some of the key issues, especially concerning core competencies. Competency development has been poorly implemented or ignored in construction (Smyth 2004). The irony is clearly that clients and government policies have tried to encourage continuous improvement, yet the necessary investment has not been forthcoming. The law of diminishing returns on waste and cost reduction kicks in early, added value being guided or managed by large clients, such as BAA, with considerable investment programmes. Yet market maturity may demand differentiation in the future, hence opening opportunities for continuous improvement by adding value to the service.

Competition and added value

Linking the structural and process analyses, there is some *market* shift towards differentiation, namely through the restructuring process. A shift towards differentiation at the level of the *firm*, based upon strategic investment, has yet to occur. Client demands for service improvement will ensure expectations are increased in the realm of construction projects. Clients wish to receive *added value*, driving continuous improvement.

Value added represents the minimum value to justify the minimum market price – client *needs* – which in projects would be represented by the iron triangle time–cost–quality/scope (in Fig. 2.5 the area bracketed as 'Needs'). *Added value* increases the benefits received – client *desires*. Delivering added value is a source of competitive advantage which can be secured through structural or process improvements. Both require investment, for although short-term value can be added by squeezing

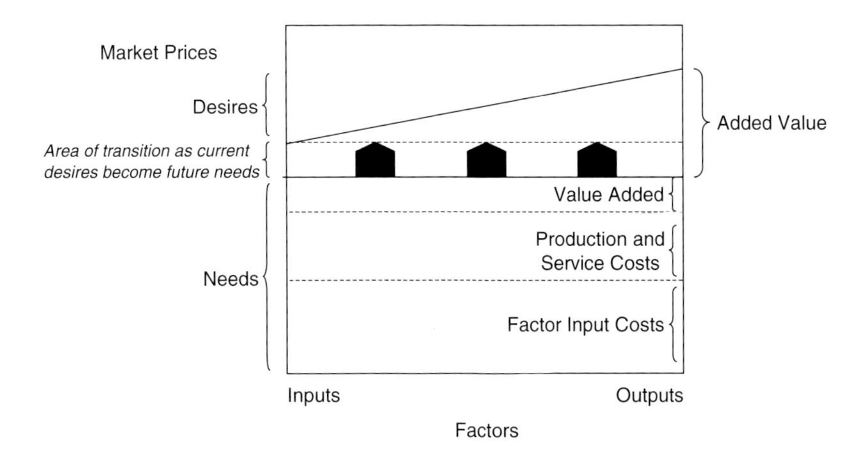

Figure 2.5 Expectations and value dynamics over time (source: Smyth 2005).

existing capacity and performance, long-term rewards are yielded by expanding or enhancing performance capacity.

As many advantages can be copied, current added value is displaced over time to value added as clients come to expect such additions as standard fare (see the 'area of transition' in Fig. 2.5 and the three block arrows indicating the trend for today's added value to become tomorrow's value added). Client desires, as they are routinely supplied, become needs, and expectations are therefore raised to new, higher level desires. Hence, there is need to drive continuous improvement.

Contractors investing to add value are vulnerable to being undercut through price competition until new market norms concerning value are established. Also, new investment to provide *added value* becomes *value added* over time. This dynamic competition carries inherent risk for contractors pursuing service differentiation. Thus improvement will be tentative at best, only becoming established in the long term, and only where the domination of transaction cost management is overcome. The drive must continue to come from clients, as direct competition between firms will not induce such a market change.

The prognosis is reasonably good. Projects are part of client corporate strategies, so their success is important, sometimes critical (Morris 1994; Morris & Pinto 2004). Contractors could be forced to invest and successful ones will dominate as investment will provide higher entry barriers to both market divisions and size and complexity of projects. Some contractors will be shaken out because of the risk elements, and capital concentration will result in all-project markets. The current restructuring amongst contractors may provide the springboard for this to occur.

Conclusion

Competition in complex project markets follows the mainstream management models of structure and process at a very general level. The five forces of competition and the core competency models were used to illustrate these aspects, respectively.

It has been argued that the management literature soon reaches limitations in contracting. Contractors emphasise overhead and transaction cost minimisation in strategic commercial management, with the result that many performance improvement agendas are subsumed under low or zero investment requirements.

Main contractors were specifically shown to address marketing issues through structural solutions, setting up separate divisions for each procurement route, yet are in the process of restructuring the market in the search for higher margins and ROCE into asset-based, income-based and mega-subcontracting market divisions.

Clients were specifically shown as initiators of the drive for innovation and improvement. There is also a trend to appoint 'teams' rather than 'contractors' in order to manage service quality, where contractors have failed to invest to meet client requirements. The main outcome is that the current market trends will continue to mature. Even if contracting firms remain passive recipients of market forces, clients will drive investment in order to achieve strategic advantage for their businesses. Yet contractors who take and survive the risks of proactive investment stand to yield above average profits on a sustainable basis. Contractors face strategic choices, and those that take advantage of current restructuring and respond to client demands will be most favourably placed to become more dominant within their markets. Such transition carries increased market risk on top of high project risk. The outcome of this process cannot be predicted, yet a time of change is unfolding at the level of the sector and firm.

This sets a research context for the next ten years. There are a number of specific agendas, including:

(1) Analysis of competitive processes in relation to *size* of main contractors and complex projects has yet to be conducted, especially concerning the dynamics of moving from one level to another in the market. The research would need to define the levels and thresholds more precisely and identify the barriers of entry at each level
(2) Investment analysis of structural solutions into market divisions based upon procurement option or equivalent, breaking down the amount spent by contractors into structuring and restructuring, including national and international comparisons
(3) Analysis of outworking of market restructuring of firms into asset, income and mega-subcontracting, including comparisons of the features and experiences across different project industries
(4) Performance improvement has primarily focused upon outputs – key performance indicators (KPIs) – rather than investment. Investment by leading contractors into areas that impact all as well as single projects. Investment analysis is also required at more detailed levels in relation to service/product differentiation, management of leanness, agility and supply chains, and management of other competencies such as OL/KM, RM and EI, quantifying each area.
(5) Analysis of continuous improvement areas needs to be conducted to ascertain which yields greatest advantage for the contractor and which yields greatest client satisfaction, particularly in relation to concepts such as OL, KM, RM and EI. This analysis should be supplemented by comparisons with existing tactical-strategic processes of improvements, including lean and agile production, just-in-time delivery and assembly, and supply chain management.
(6) Research needs to be conducted into the neglected area of human systems for continuous improvement within and across projects, covering the main office–project interface and the client–contractor interface. Investment in these areas needs to be monitored and assessed in terms of benefits to clients and contractors, especially competitive benefits. Related investment in people is needed within contracting organisations, and also in terms of the quality of recruits into project enterprises, providing career structures and rewards which relate to long-term strategic requirements rather than short-term incentives that fail to address corporate competition. Research also needs to be conducted into the dimensions to change the image of complex project sectors as the working environment in order to improve capacity to recruit in the future.
(7) Research into the dimensions to change the culture of complex projects from task orientated based upon personalities and blame to one based upon investment in systems, competencies and technologies for performance improvement. The

research will need to embrace the nature and strength of culture necessary to overcome transaction cost forces and will also need to monitor success in this area.

Such research embodies both analyses of what is current and recent practices and trends, what is unfolding, and normative issues. There are many detailed issues within each of these main areas, some of which may be elevated from tactical to strategic ones over the next decade, so awareness will be needed to adjust research to emergent trends. The ability of main contractors to respond to industry and policy agendas, whilst pursuing others to protect their market positions is nothing new. The strategic concern is to be as risk averse as possible, and so, structural and process shifts recommended from research output must embrace this reality, yet demonstrate clearly the necessity and inevitability of considerable change. The current economic and political climate is reasonably sympathetic in most countries and so the next ten years may prove to be as good a window for change as there has been for many decades.

References

Argyris, C. & Schön, D.A. (1978) *Organizational Learning: a Theory of Action Perspective*. Addison-Wesley, Reading, MA.

Boston Big Dig (2004) http://www.bigdig.com (accessed 31 August 2004).

Campbell, N. (1985) An interaction approach to organisational buying behaviour. *Journal of Business Research*, **13**, 35–48.

Cox, A. & Townsend, M. (1998) *Strategic Procurement in Construction*. Thomas Telford, London.

Douma, S. & Schreuder, H. (1991) *Economic Approaches to Organisations*. Prentice Hall, Hemel Hempstead.

Egan, J. (1998) *Rethinking Construction*. HMSO, London.

Gann, D. (2000) *Building Innovation*. Thomas Telford, London.

Goleman, D. (1996) *Emotional Intelligence*, Bloomsbury, London.

Goleman, D. (1998) *Working with Emotional Intelligence*. Bloomsbury, London.

Green, S.D. & May, S.C. (2003) Re-engineering construction: going against the grain. *Building Research and Information*, **31**(2), 97–106.

Grönroos, C. (2000) *Service Management and Marketing*. John Wiley & Sons, London.

Gruneberg, S.L. & Ive, G.J. (2000) *The Economics of the Modern Construction Firm*. Macmillan, Basingstoke.

Gummesson, E. (2001) *Total Relationship Marketing*. Butterworth-Heinemann, Oxford.

Hamel, G. & Prahalad, C.K. (1994) *Competing for the Future*. Harvard Business Books, Boston, MA.

Hillebrandt, P.M., Cannon, J. & Lansley, P. (1995) *The Construction Company in and out of Recession*. Macmillan, Basingstoke.

Ive, G. (1990) Structures and strategies. *Habitat International*, **14**(2–3), 45–58.

Ive, G. (1995) The client and the construction process: the Latham Report in context. In: *Responding to Latham: The Views of the Construction Team* (ed. S.L. Gruneberg). Chartered Institute of Building, Ascot.

Ive, G.J. & Gruneberg, S.L. (2000) *The Economics of the Modern Construction Sector*. Macmillan, London.

Johnson, G. & Scholes, K. (2002) *Exploring Corporate Strategy* (6th edn). Financial Times–Prentice Hall, London.

Koskela, L. (1992) *Application of the New Production Philosophy to Construction*. Technical Report 72, Centre International de Formation Europèene (CIFE). Stanford University, Stanford.

Koskela, L. (2000) *An Exploration towards a Production Theory and its Application to Construction.* Report 408. Technical Research Centre of Finland (VTT), Espoo.

MPA (2003) Recovery from 9/11 in New York. *Proceedings of the Management of Unexpected Projects, Seminar 108.* Major Projects Association, Templeton College, Oxford.

Morris, P.W.G. (1994) *The Management of Projects.* Thomas Telford, London.

Morris, P.W.G. & Pinto, J.K. (eds) (2004) *The Resource Book of Managing Projects.* John Wiley & Sons, New York.

Nonaka, I. & Takeuchi, H. (1995) *The Knowledge-Creating Company.* Oxford University Press, Oxford.

Porter, M.E. (1980) *Competitive Strategy.* Free Press, New York.

Porter, M.E. (1985) *Competitive Advantage.* Free Press, New York.

Prahalad, C. & Hamel, G. (1990) The core competencies of the organization, *Harvard Business Review*, **63**(3), 79–91.

Rumelt, R.P. (1984) Toward a strategic theory of the firm. In: *Competitive Strategic Management* (ed R. Lamb). Prentice-Hall, NJ; pp. 566–570.

Senge, P. (1993) *The Fifth Discipline.* Random House, London.

Smyth, H.J. (1985) *Property Companies and the Construction Industry in Britain.* Cambridge University Press, Cambridge.

Smyth, H.J. (1999) Partnering: practical problems and conceptual limits to relationship marketing. *International Journal for Construction Marketing,* **1**(2), www.brookes.ac.uk/conmark/IJCM/.

Smyth, H.J. (2000) *Marketing and Selling Construction Services.* Blackwell Science, Oxford.

Smyth, H.J. (2004) Competencies for improving construction performance: theories and practice for developing capacity. *International Journal of Construction Management,* **4**(1), 41–56.

Smyth, H.J. (2005) Procurement push and marketing pull in supply chain management: the conceptual contribution of relationship marketing as a driver of in project financial performance. *Journal of Financial Management of Property and Construction,* **10**(1), 4–18.

Wernerfelt, B. (1984) A resource-based view of the firm. *Strategic Management Journal,* **5**, 171–180.

Williamson, O.E. (1985) *The Economic Institutions of Capitalism.* Free Press, New York.

Wilkinson, S. (in press) Client handling models for continuity of service. In: *The Management of Projects: A Relationship Approach* (eds S. Pryke & H. Smyth). Blackwell, Oxford.

Winch, G.M. (2002) *Managing the Construction Project.* Blackwell Science, Oxford.

3 Culture

Richard Fellows

Introduction

All projects are subject to increasing complexities; these are growing both numerically and in individual scope. They arise through both internal factors and external factors, and may be classified further as technical and human. Whilst there is continuing debate on definitions, classifications and measurement of complexities, this chapter adopts a more qualitative perspective, appropriate for addressing underpinnings of human behaviour.

Context is important, because it is through and dependent upon context that people behave and thereby give expression to and promote development (change) of culture. Thus, a project may be regarded as a performance-orientated activity (established objective(s)) which has defined (or definable) temporal limits – a beginning and an end. In the context of a construction project, those limits may be defined differently depending on the perspective and purpose of the definition: so for a 'traditional' constructor, the beginning is likely to be the contractual date for 'start on site' and the end, the scheduled completion date (subject, of course, to extensions of time and delays).

Especially the human aspects of complexities arising on projects and in project-orientated firms are epitomised in the construction industry. Further, that industry is one of the leading project-based industries, operating on a wide diversity of project types and on a vast range of scales – from individual person firms and jobs to global organisations (often joint-venturing) undertaking huge, multi-million (US) dollar projects. Certainly, over the past half century, many reports in a large number of countries have endeavoured to examine aspects of construction projects and firms to identify and analyse performance and to seek mechanisms for improvements; recently, comparisons with the automotive industry and with aerospace have become popular (largely due to performance improvements by Japanese organisations). So, throughout this chapter, examples are drawn from construction to complement the discussion in the text.

Construction is the assembly of components into a completed (sub-) artefact. Unfortunately, it is common for the objectives to be fuzzy, individually-derived, often poorly communicated and subject to change. Those facets are due to the ways in which the parties relate in the various project procurement arrangements. A construction project, therefore, may be regarded as executed by a temporary multi-organisation (TMO), in which the parties form varying power-based coalitions as the design and construction proceed. Hence, the objectives pursued depend upon the parties involved at the time, the existing power structure and the (irreversible) decisions taken in previous stages of the project realisation process (Fig. 3.1). Recent attention to 'relational contracting', 'partnering' and other procurement approaches

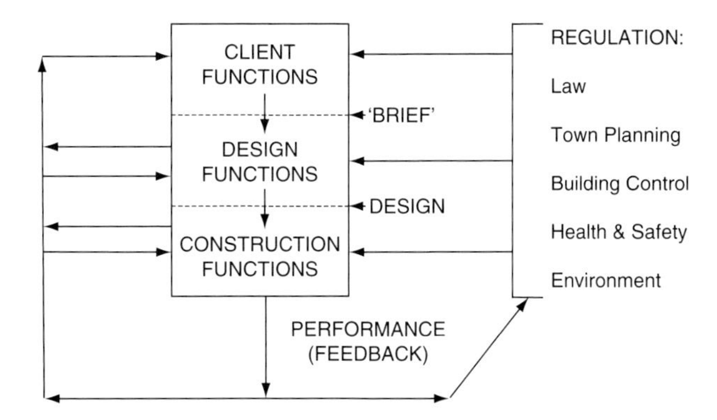

Figure 3.1 The project realisation process. *Notes*: (1) Performance leads to satisfaction of participants and, hence, (perspectives of) project success. (2) Performance–satisfactions–success also produces feedforward in the 'cycling' of project data and information to aid realisations of future projects through participants' perception–memory–recall filtering ('experiences').

which should emphasise co-operation and commonality witness the increasing acknowledgement that process of realisation is vital to product and so both facets of performance must be pursued [as noted in relationship marketing – see, for example, Grönroos (1991)]. However, in practice at least, it may be appropriate to ensure that the parties recognise, on both collective and individual levels, their mutual interdependence for achieving project success.

Management is defined variously, but a pertinent definition is 'the making and implementing of decisions concerning people', thereby lending focus to use of data for prediction in the context of determining the most suitable alternative and carrying out consequent actions through persons as the only 'active' factors. Further, it emphasises people as the object of the decisions and actions – meeting their demands. The context of commerce gives vent to the economic–legal orientation through the effecting of transactions by persons in the production and exchanges of goods and services.

Two dichotomous trends may be observed in many aspects of global activities; both convergence and divergence appear to be occurring coincidentally. People, organisations and rules intermingle, with artefacts and symbols (languages, buildings) becoming increasingly alike. Individuals become aware of uniqueness and seek to preserve individuality (e.g. bespoke headquarters buildings) and heritage (e.g. clothing, dances, dialects). Cuisine, quite obviously, is subject to both trends – fusion food and ethnic dishes sometimes may be obtained in the same restaurant!

'Transparency' in business activities, especially for public sector organisations, is widely regarded as vital to ensure that appropriate behaviour occurs. While transparency (or visibility) is only (accurate) communication, its practice is commonly regarded as embodying value assessments of correctness of behaviour – that is the domain of ethics which has its foundation in culture! Ethics and corruption are such issues that international league tables are produced annually (e.g. Transparency International – http://www.transparency.org/).

Examination of cultures – notably, to determine differences (rather than similarities) – is a rapidly extending field of academic study and of commercial consultancy.

Awareness (understanding) and accommodation of culturally based differences are recognised as important for commercial performance success (as judged by culturally-determined measures). Some regard culture as a tool which may be employed to effect changes to further performance against (often) pre- and (sometimes) distantly-determined criteria and targets; such an approach may lead to confusion between effecting cultural change (which is long term) and behaviour modification (short term and easily reversed): here it is important to recall that people are risk averse and so 'revert to type', i.e. return to well-known ways of doing things. Others regard culture as a medium to and through which adaptation must occur, but in which 'creep' may take place yielding long term change.

Culture, then, has major importance for the environments – the external milieu in which organisations operate. However, culture occurs within organisations too – organisational culture, which often overlaps significantly with organisational climate. If culture is usually defined by reference to commonalities (values, beliefs, behaviour, etc.) then issues of culture are bifaceted: intracultural (what constitutes a culture) and intercultural (similarities and differences between cultures – boundary considerations). Cameron and Quinn (1999), for example, suggest that culture acts as a 'glue' which binds people into social groups. Intracultural awareness is a precursor to intercultural sensitivity, understanding and thence management.

The suggestion that management is all about change leads immediately to people, as people make (prevent, modify) change happen. That is culture in operation... but, then, so is continuity of activity without change!

This chapter endeavours to provide a summary of the basic knowledge blocks of what constitutes culture, techniques which may be employed to further awareness and understanding of culture, and appreciation of the issues involved in managing projects involving cultural complexities. Pragmatically, some attention is devoted to examination of ethics and their variability interpersonally, interorganisationally and internationally. The chapter ends by reviewing current research agendas concerning culture and ethics, with particular reference to the construction industry, as a reflection of the outstanding issues of significant concern for major complex projects.

Definitions

Especially in the realm of social sciences, the concept of definition is of particular import due to the arrays of alternative definitions, and hence the different understandings, dimensions and measures, etc., which may arise. The definitions are therefore offered as being appropriate in the context of major complex projects especially, and are accompanied with brief surrounding explanation and discussion of primary alternatives.

Culture

Hofstede (1994a) defined culture as:

> *the collective programming of the mind which distinguishes one category of people from another*

This definition indicates that culture is learned, rather than something which is innate in the person or inherited genetically; it is, of course, inherited behaviourally through replicating and responding to the behaviour of other people. Further, culture is a collective construct, and so categorisation of people may be by ethnic origin, political nation, or other appropriate grouping – the important aspect relates to the categorisation process in that (as for any meaningful grouping) 'within category' variability is much less than 'between category' variability.

Barthorpe (2002) provided a review of definitions of culture, including the 164 definitions noted by Kroeber and Kluckhohn (1952), and examined how they may be classified. In practice, a 'working description' of culture could be simply expressed as: *'how we do things here'*. That perspective can be somewhat problematic, in that it may encourage confusion between change in culture (long term) and behavioural modification (short term). Further, it masks the cultural issues of the objectives of action and inaction, what is done, who is involved, when, etc. There are, of course, manifestations of culture in addition to behaviour – language, heroes, artefacts (not least of which are buildings), etc.

Perhaps the most informative definition is that of Kroeber and Kluckhohn (1952):

> *...patterns, explicit and implicit of and for human behaviour acquired and transmitted by symbols, constituting the distinctive achievements of human groups, including their embodiment in artefacts; the essential core of culture consists of traditional (i.e. historically derived and selected) ideas and, especially, their attached values; culture systems may, on the one hand, be considered as products of action, on the other as conditioning elements of future action.*

If we follow this definition, it seems that more enduring societies enjoy advantages of cultural development and stability. This perspective accords with the notion of indexicality in sociology (see, for example, Clegg 1992) in which interpretation of messages and thence comprehension of meaning are recognised to be dependent upon socialisation, including education and training.

The cultural dynamics model proposed by Hatch (1993) successfully encapsulates the cyclical processes of manifestation, realisation, symbolisation and interpretation to provide a framework within which to understand the dynamism of organisational cultures. The dynamism comes from the continual construction and reconstruction of culture as contexts for setting goals, taking action, making meaning, constructing images and forming identities.

Organisational culture and organisational climate

Hofstede (1994b, p. 180) defined organisational culture as:

> *the collective programming of the mind which distinguishes the members of one organization from another*

In constructing the definition, Hofstede (1994b, pp. 180–181) asserted that organisational culture is:

> *...holistic,...historically determined,...related to the things anthropologists study,... socially constructed,...soft,...difficult to change....*

A common difficulty is to differentiate organisational culture and organisational climate. Mullins (2002, p. 906) defined organisational climate as:

> *Relating to the prevailing atmosphere surrounding the organisation, to the level of morale, and to the strength of feelings of belonging, care and goodwill among members. Organisational climate is based on the perceptions of members towards the organisation.*

Thus, organisational climate is a subjective, collective construct of the organisation as an entity, thereby lending credence to the view that an organisation has an identity and behaviour beyond the sum of its members; certainly, from the perspective of individual members, the view of the organisation as an independent organism with its own values and behaviour is real. (That is reflected in the legal–institutional view of corporate business units including limited liability.) Further, it appears that organisational climate has strong linkage to ethics – the moral interactions of the organisation and its stakeholders. Indeed, that does raise the question of the extent of membership of an organisation as a subset of stakeholders, where membership commonly comprises owners and employees, but stakeholders includes several other groups associated with the organisation and its activities – customers, suppliers, etc.

Tagiuri and Litwin (1968, p. 27) noted that:

> *Organizational Climate is a relatively enduring quality of the internal environment of an organization that (a) is experienced by its members, (b) influences their behaviour, and (c) can be described in terms of the values of a particular set of characteristics (or attributes) of the organization.*

However, a note of caution was raised by Moran and Volkwein (1992, p. 43) concerning desires to effect changes in organisational climate as a cognitive tool of management:

> [there are] *implications for managerial practice in two respects. First,... since climate operates at a more accessible level than culture, it is the more malleable and, hence, the more appropriate level at which to target short-term interventions aimed at producing positive organizational change. ... Second, the conceptualization... suggests that interventions to change climate must consider the deeper patterns embedded in an organization's culture.*

Cameron and Quinn (1999) emphasised the necessity of distinguishing between organisational culture and organisational climate. They stated that:

> *Culture is an enduring, slow to change attribute of organizations; refers to implicit, often indiscernible aspects of organizations; includes core values and consensual interpretations of how things are...*
>
> *Climate... refers to more temporary attitudes, feelings and perceptions of individuals... can change quickly and dramatically... refers to more overt, observable attributes of organizations... includes individualistic perspectives that are modified frequently as situations change and new information is encountered.*

In project-based industries like construction, organisational boundaries are much more problematic to identify and analyse due to the operation of projects as temporary multi-organisations (TMOs) with diverse formal and informal organisational linkages.

Ethics

The *Oxford English Dictionary* (Swannell 1992) provides a variety of definitions of 'ethics', all of which relate to morals, e.g. 'the whole field of moral science'. Perhaps the definition which is most pertinent is:

> *The moral principles or system of a particular leader or school of thought; the moral principles by which any particular person is guided; the rules of conduct recognized in a particular profession or area of human life.*

Rosenthal and Rosnow (1991, p. 231) noted:

> *...ethics [refers] to the system of moral values by which the rights and wrongs of behavior...are judged.*

Not only should ethics refer to values, but to secure operation, reference must be made to principles and standards regarding behaviour.

Thus, ethics constitutes a vital behavioural link between culture, climate and behaviour of members of an organisation within both the contextual norms of the organisation and also the norms of the wider society (societies), including the behaviour standards required by the prevailing law.

Dimensions of culture

It is evident from the definitions that the constructs are complex and multi-faceted; hence, it is essential to determine the dimensions along which the constructs may be measured and analysed. Although many researchers have, variously, determined dimensions, probably the widest known dimensions of both national and organisational cultures remain those of Geert Hofstede.

Culture is manifested through facets of behaviour. Behaviour is dependent upon values and beliefs, whether any behaviour is determined by conscious thought/evaluation or is 'instinctive'. In the latter case, common survival mechanisms are likely to govern, and so are relatively common amongst humans, whilst in the former case cultural influences will be stronger. That leads to models of culture with physiological instincts and beliefs at the core (survival imperatives; religion, morality etc.), values as the intermediate layer (the hierarchical ordering of aspects of beliefs, perhaps with visions of trade-offs) and behaviour at the outer layer (as in language, symbols, heroes, practices, artefacts, etc.) (Fig. 3.2).

It is, then, the observable outer layer which must be employed to secure measurements indicative of culture through, first, identifying and defining suitable

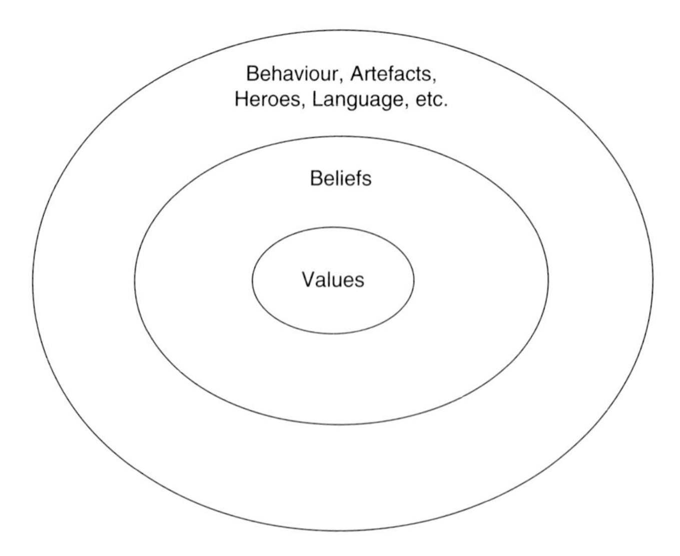

Figure 3.2 Layers of culture.

dimensions – exhaustive in scope and exclusive in content. In studying national cultures, Hofstede (1980) determined four dimensions:

(1) *Power distance*: 'the extent to which the less powerful members of institutions and organizations within a country expect and accept that power is distributed unequally' (Hofstede 1994b, p. 28)
(2) *Collectivism/individualism*: 'Individualism pertains to societies in which the ties between individuals are loose: everyone is expected to look after himself or herself and his or her immediate family. Collectivism as its opposite pertains to societies in which people from birth onwards are integrated into strong, cohesive ingroups, which throughout people's lifetime continue to protect them in exchange for unquestioning loyalty' (Hofstede 1994b, p. 51)
(3) *Masculinity/femininity*: 'masculinity pertains to societies in which gender roles are clearly distinct (i.e., men are supposed to be assertive, tough, and focussed on material success whereas women are supposed to be more modest, tender, and concerned with the quality of life); femininity pertains to those societies in which social gender roles overlap (i.e., both men and women are supposed to be modest, tender, and concerned with the quality of life)' (Hofstede 1994b, pp. 82–83)
(4) *Uncertainty avoidance*: 'the extent to which the members of a culture feel threatened by uncertain or unknown situations' (Hofstede 1994b, p. 113).

Hofstede later added a fifth dimension of long-termism

> *the fostering of virtues orientated towards future rewards, in particular perseverance and thrift* (Hofstede 1994b, p. 261)

and short-termism

the fostering of virtues related to the past and present, in particular respect for tradition, preservation of 'face', and fulfilling social obligations (Hofstede, 1994b, pp. 262–263)

following studies in Asia which detected important impacts of 'Confucian dynamism' (Chinese Culture Connection 1987).[i]

Trompenaars and Hampden-Turner (1997) advanced five value-oriented dimensions of culture which, they suggest, 'greatly influence our ways of doing business and managing as well as our responses in the face of oral dilemmas'. Those dimensions are:

- Universalism–particularism (rules–relationships)
- Collectivism–individualism (group–individual)
- Neutral–emotional (feelings expressed)
- Diffuse–specific (degree of involvement)
- Achievement–ascription (method of giving status).

Hofstede (1994b) proposed six dimensions for analysis of organisational cultures:

- Process–results orientation [technical and bureaucratic routines (can be diverse)–outcomes (tend to be homogeneous)]
- Job–employee orientation (derives from societal culture as well as influences of founders, managers)
- Professional–parochial (educated personnel identify with profession(s) – people identify with employing organisation)
- Open–closed system (ease of admitting new people, styles of internal and external communications)
- Tight–loose control (degrees of formality, punctuality, etc., may depend on technology and rate of change)
- Pragmatic–normative (how to relate to the environment, N.B. customers; pragmatism encourages flexibility).

Scrutiny of the various dimensions used to analyse both national and organisational cultures, essentially, indicates considerable conceptual commonality. Further, Hofstede's dimensions of organisational culture align with the human–task schools of management thought [such as theory X and theory Y of Herzberg *et al.* (1967)].

Schools of ethics

Hinman (1997) distinguished morals and ethics by regarding morals as first order beliefs and practices about what is good and what is bad which guide behaviour and ethics as second order, reflective consideration of moral beliefs and practices. Such issues of definition and perspective, both theoretical and operational, have generated four primary paradigms for ethical analysis (Leary 1991, pp. 261–262). In deontology, a universal moral code is held to apply. In skepticism (alternatively, relativism; subjectivism), ethical rules are arbitrary and relative to culture and to time; that view is extended in ethical egotism where ethics become a matter of the conscience of the individual such that the individual is the arbiter of ethics (what is right and what is wrong). Thus, egotism concerns pursuit of self-interest (hedonism), and so can be related to common business criteria (notably, profit maximisation). Teleology constitutes a utilitarian approach (utilitarianism; consequentialism) where ethics are dependent upon the (anticipated) consequences: that suggests a cost–benefit view, perhaps invoking the judgmental criterion of 'the greatest good for the greatest number' which, itself, is likely to necessitate subjectively determined weightings. Objectivism asserts that there are definitions of what is right and what is wrong which are accepted generally (either universally or more locally).

Perspectives on ethics are, to some degree at least, culturally dependent. The notions of Confucius, who built on earlier Daoist concepts, influence Chinese perspectives on ethics. Here, the emphasis is different from Western ethical/moral ideas of what is good (and bad) by focusing on how to become good. In one sense, *dao* may be used to signify the proper way of life. *De* is a concept similar to the Aristotlean concept of virtue but incorporates kindness and self-sacrifice and emphasises how to achieve virtue. *Jen* concerns love, benevolence and humanism and is actioned as proper motivation; thus, *jen* has an inward focus to guide behaviour. The external-focus equivalent of *jen* is *i* and is a result of socialisation. Hence, an example of *jen* is to love one's parents; an example of *i* is to pay respect to an older person. A well known result of the adoption of Confucian ethics is the hierarchy of respectful (harmonious) relationships: sovereign–subject, father–son, husband–wife, brothers, friends (the male emphasis is clearly evident!).

Ethical dilemmas

Ethical dilemmas arise for a number of major reasons. First, there is no universal theory (set) of ethics. Second, as theories of ethics differ, an individual is often obliged to select actions from mutually exclusive alternatives and has conflicting theoretical guidance. Third, although it is apparent that at the primary (basic) level a set of universal 'goods' exists, at the secondary (culturally-shaped definitions and codes of conduct) and tertiary (specific codes of behaviour) levels notable differences apply. Thus, dilemmas arise due to differences in ethical manifestations which are culturally dependent.

Consistency is a significant ethical consideration such that views and behaviour should be applied across people and time. This is, of course, problematic as situations and individuals are rarely identical to previous occurrences, so the questions become ones concerning degrees of difference – issues of judgement. A further complication arises as environments and human views evolve, including attempts to 'rectify past errors'. Consistency has a cultural dimension too, which is manifest particularly in Trompenaars and Hampden-Turner's (1997) dimension of particularism–universalism.

Much of ethics is concerned with means – the behavioural considerations of what to do and how to do it. However, there are ends dimensions also. What is known, when and by whom, as well as considerations of the anticipated consequences of alternative behaviour, merit examination. Especially where environmental inputs are involved (natural resources such as ground, seed, water, sunlight) the decision involves not only the labour power expended by the individual, but also issues of ownership (forms) of the non-labour resources.

Professional ethics

In the context of professionals, where a profession may be regarded as an organisation of people possessing and practising using a body of unique, expert knowledge, Bayles (1988) identified professional ethics as a system of behavioural norms. Such norms relate to the employment of the particular knowledge, and thus concern the relationship between experts and 'lay' persons. Generally, then, the behavioural rules seek to ensure that (unfair) advantage is not taken by the expert over the lay person due to the knowledge differential – the application of 'customer protection' through (professional) self-regulation. Commonly, the issue of disclosure of interests arises: the professional must disclose any self interest in the subject matter of the relationship to the client; then, continuity of the relationship is on the basis of 'informed consent' on the part of the client.

Particularly in 'Western' societies, the post-industrial revolution period has witnessed an enormous increase in the use of contracts to govern relationships, notably in business circumstances. What has occurred is the accentuation of legally encapsulated rights, duties and remedies, seemingly at the expense of relational duties and reciprocation (in the sense of 'consideration' in contract law). The growth of 'legalism' is based around the cultural dimension of individualism. However, other factors appear to be important too – notably, the total amount of a good (a 'desirable') and the distribution of that good: enter the ethical notion of 'greed'!

In the legal(istic) context of governance and regulation, the domain of ethical rules relates to means primarily, such as in codes of conduct of professional institutions. It may be argued that a system of common law (or religious law) has a strong ethical and/or relational base, but that may not be reliable in terms of the current state of development. In the UK, for instance, the doctrine of 'equity' was developed to address inequalities in legal practices; the 'statute of interpretation' was instigated to assist legal decision makers to follow the intent of legislation (especially when confronted with conflict in the statutes, for example).

Professional codes of practice and/or conduct are contracts entered into by members of an organisation (usually a professional institution) which form the legally-enforceable requirement for the behaviour of members. Stewart (1995, p. 11) noted that such codes, 'do not teach morality, ethics or values: they lay down rules for conduct and, unless they are used in a positive manner as a basis for teaching principles, they will in daily practice be no more than guidelines for action'. Thus, the issue of enforcement is important. Enforcement concerns not only checks and detection of transgressors but imposition of consequences upon such transgressors. If detection is unlikely, consequences are inadequately negative (from transgressors' perspectives), or both, then transgression is far more probable in an opportunistic environment.

Principles have been developed by which approriateness of behaviour is judged. One of the best known of such principles is used in the UK to determine the required *reasonableness* of behaviour of a person in 'everyday' situations – this refers to the behaviour expected of the 'man on the top (deck) of the Clapham omnibus'. However, in special circumstances, other tests are used to determine what behaviour is appropriate to fulfill the requirement of reasonableness. For professionals and experts (and those 'holding themselves out' to have special skills), the behaviour required is that of an ordinary practitioner – *Bolam* v. *Friern Barnet Hospital* (1957); other situations give rise to higher levels of skill being required [to avoid liability for (professional) negligence].

Leisinger (1995, p. 184) quoted Lay (1993, p. 9) in that,

the morals we have do not regulate how institutions should treat people, only how people should treat people

In addition, ethical problems may occur incrementally – through individual steps of negligible ethical detriment that accumulate to an ethical problem. Thus, although it may be easy and tempting to dismiss a marginally unethical (or immoral) action as being insignificant, at least in its own consequences, that is a dangerous perspective due not only to accumulation but also to its possible impact on the person's perspective on what is of ethical significance in the future – a possible progressive change or erosion of ethical standards.

Langlois and Schlegelmilch (1990) found that many large companies in Europe had documented codes of ethics. However, they also found that the codes addressed only parts of business ethics, usually concerning personnel and reliability matters, and that although national differences exist, large companies tend to employ standardised codes, and thereby are likely to be ignoring significant cultural differences.

Organisational ethics

Cohen (1993) defined ethical climate as employees' prevailing perceptions of organisational signals regarding norms in making decisions which have a moral component. Treviño (1986) developed a model of organisational ethical culture including the organisation's normative structure (norms of appropriate behaviour), referent to others' behaviour and expectations concerning obedience to legitimate authority – which encourage people to take responsibility for the consequences of their decisions and actions. Hartman (1996), for example, noted that organisational culture may be employed by management as a tool to change behaviour, implement decisions, etc., often through the application of incentives, rewards and punishment systems. Others regard culture as a situational 'given' within which managers and others must operate. (Irrespective, the question arises of the ethics of any attempt to change the behaviour of others.)

Normally, ethical codes occur as sets of written rules, procedures and/or components of mission statements. Employees and/or members of a profession regard them as a formal basis for self-regulation. De George (1995) contended that ethical codes are helpful in resolving specific issues faced by individuals. Weller (1988) argued that ethical beliefs and behaviour can be improved by the existence and enforcement of corporate policies. Reeck (1982) noted that ethical codes provide guidance for professionals in their determining appropriate action – they aid consistency and stability in making decisions about moral issues. However, Henry (1995) cautioned that ethical codes do not, themselves, solve moral dilemmas but do help to raise levels of awareness and so encourage ethical practice.

Thus, Etheridge (1996) raised the imperative for effectiveness of ethical codes of implementation via appropriate promulgation and training for the changing of attitudes and behaviour, finding in a study of ethics in businesses in Hong Kong, that the existence of ethical codes assisted managers in recognising moral parameters.

Culture profiles

Given that it is important to differentiate organisational culture and organisational climate, as noted by Moran and Volkwein (1992), it is also important to recognise that culture can be observed only 'at the surface' of behaviour, etc. – the level of climate (Figs 3.3, 3.4). Thus, some researchers, notably Schein (1990), regard culture as a pattern of basic assumptions which constitute communal values and which, as such, are taken for granted (so often are not expressed in documents – mission statements, etc.). For Schein, cultures arise, first, through the formation of norms of behaviour relating to critical incidents (commonly in the form of lessons learned from significant mistakes) which are commonly communicated through stories passed on between members of the community (national and/or organisational histories) and, second, through identification with leaders and what they scrutinise, measure and control.

Organisational cultures usually derive from the founders of the organisation and others who have had major impact on the organisation's development (e.g. Henry Ford, Alfred Sloan). Such people, through influence over hiring of staff, have acted to 'shape' the values and behaviour of members of the organisation and to develop the organisation's identity, both internally and externally. Thus, organisational cultures (and climates) tend to be self-perpetuating – persons who 'fit' are hired, 'fit' because they are hired; errors of 'fit' are subject to resignation or dismissal. Thus, organisational cultures develop through the necessity of maintaining effective and efficient working relationships amongst stakeholders (both permanent and temporary). Pressure for cultural change commonly arises

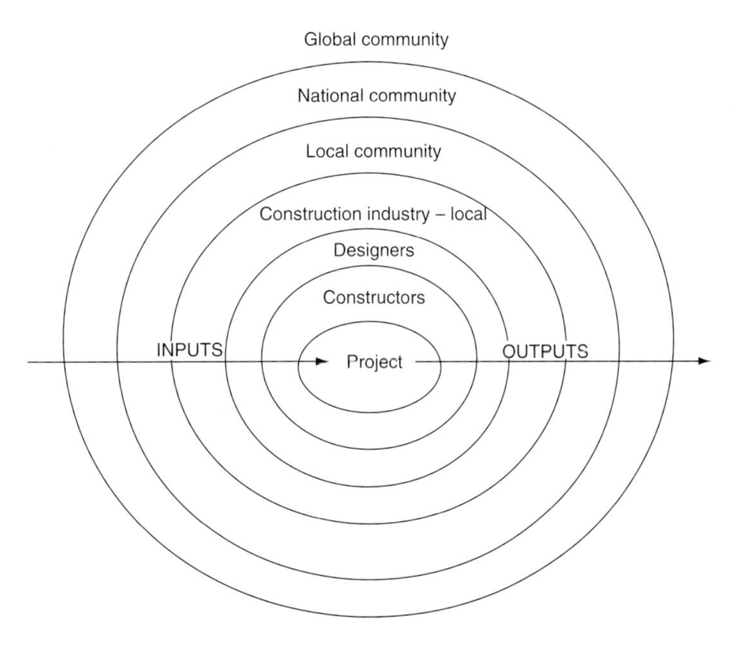

Figure 3.3 Project cultural ('onion') layers. *Note*: Boundaries are likely to vary in permeability.

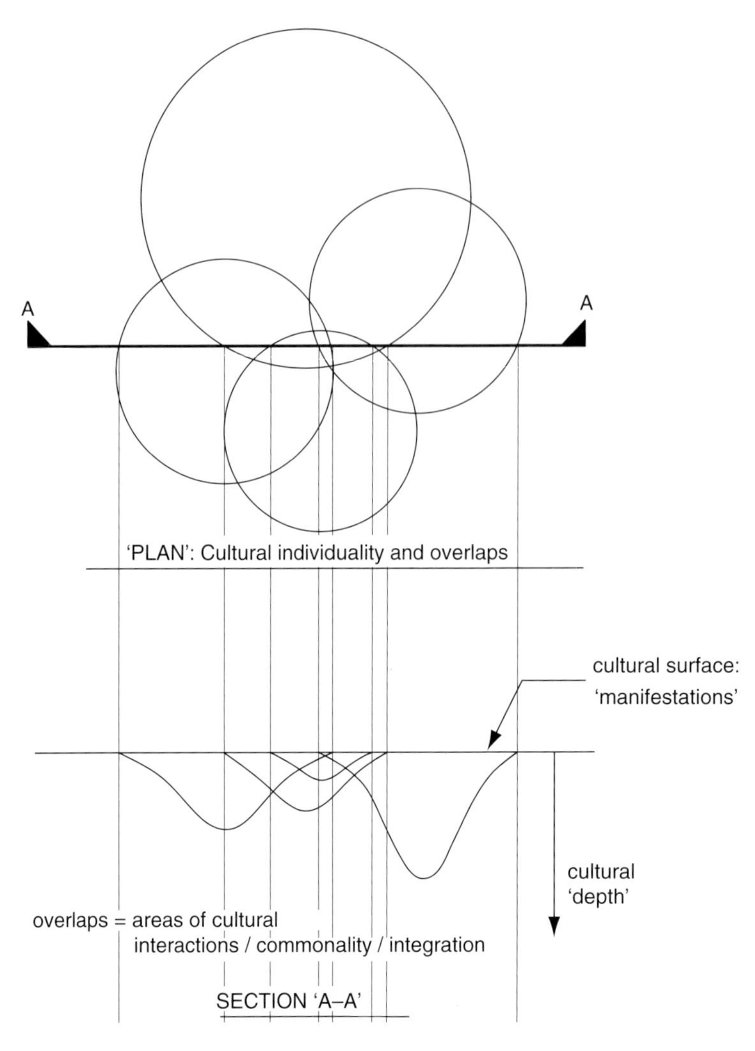

Figure 3.4 Culture interaction profile (developed following Trompenaars & Hampden-Turner 1997).

from external parties, particularly in situations of environmental turbulence and attempts to enter new markets.

There is considerable support for the notion that leadership style should be compatible with the prevailing culture to engender 'success', as noted by Black and Porter (1991). However, such a relationship may occur because behaviour necessary to yield successful performance varies between cultures and/or that which constitutes successful performance is culturally dependent (see, for example, Turpin 1993).

Eldridge and Crombie (1974) noted three concerns for cultures – depth (values and commitment), breadth (coordination of the contributors) and progression (coordination, development and/or change, over time). Cultures are, of course, influenced by general perspectives; much recent interest lies in the general characteristics attributed to Eastern and to Western people. Eastern people are often regarded as 'fatalistic' (or, at least, 'flexible' and/or 'adaptive') whilst Westerners are characterised as (would-be) controllers. Those categorisations arise because Eastern people tend to regard themselves as subservient to natural forces (they desire to be 'in harmony' with nature); change and hence uncertainty, as well as time, are accepted as inevitable; truth is determined by spiritual (religious and philosophical) principles. Western people, however, tend to regard themselves as largely in control of nature and so can harness many of its forces to improve human society; thus, time, change and uncertainty can be managed; truth has a scientific base, and so is determined via measurements and facts.

As ever, but perhaps increasingly clearly, there is evidence of Eastern people seeking to control their environment – notably, the Three Gorges mega-project in the People's Republic of China. In Western societies especially, the issues of environmental protection and sustainability are ascendant; however, the means of seeking to achieve the objectives, and indeed the objectives themselves, can be founded on various arguments based both in the 'host' culture and in 'foreign' cultures! (The Three Gorges project may be designed to maintain harmony for the population in the long term; environmental sustainability may be sought through measures to control both human activities and environmental events, such as 'seeding' clouds to generate rainfall.)

Although age is venerated in most societies, for industrial positions, significant culturally-determined differences arise. In Western organisations, promotion is determined through achievement – educational/training qualifications and (quantified) performance of work activities. In Eastern societies, the tradition is for promotion to be via ascription – employment is for (working) life and promotion accords with age (and experience); 'retired' eminent persons remain venerated and continue to be able to wield high levels of referent power (in such circumstances, the most influential form of power) (Fang 2002). However, both 'traditions' are subject to some breakdown – in Eastern societies, redundancies are being invoked and more recognition is being given to achievements; in Western societies, continuity of employment is recognised as valuable, and permanence of relationships plus stability and wisdom of age are more valued.

Such differences are clearly apparent in languages, which are potent manifestations of cultures. Western languages tend to have a high content of words which are precise in meaning and assembled to convey exact messages. Eastern languages tend to have a high content of words of which meanings are often vague and determined by context in the message; further, it is common for messages to be constructed to be inexact in themselves and require interpretation through knowledge of the prevailing culture to reveal their meaning.

Schein (1984) suggested two primary types of organisational culture: 'free flowing' – an unbounded, egalitarian organisation without (much) formal structure, thereby encouraging debate and (some) internal competition; 'structured' – a bounded, rigid organisation with clear rules and requirements [analogous to the organic-mechanistic analysis of Burns and Stalker (1961)]. That perspective is strong in the discussion of the operation of construction projects – formal systems are in place (organisation charts, contractual procedures, etc.) which are used 'only in the last resort', i.e. when things go wrong, but projects operate through the network of a system of informal relationships which emphasises 'doing the pragmatic' to achieve progress. The belief is that through strict adherence to the formal system, the project would quickly grind to a halt as there are so many potentially delaying bottlenecks (as in the procedures regarding oral variations). However, the risks involved must be understood (which suggests that a low level of risk aversion and/or uncertainty avoidance is appropriate).

Handy (1985) suggested that there are four primary forms of organisational culture:

(1) *Power*: This is configured as a web with the primary power at the centre; emphasis is on control over both subordinates and external factors (suppliers, etc. and nature)
(2) *Role*: Involves functions/professions which provide the structural pillars for the support of the overarching top management (analogous to a Greek temple); emphasis is on rules, hierarchy and status through legality, legitimacy and responsibility (as in contractual rights, duties and recourse)
(3) *Task*: Jobs or projects are a primary focus yielding an organisational net (as in a matrix organisation); structures, functions, activities, etc., are evaluated in terms of contribution to the organisation's objectives ('hire and fire' is common)
(4) *Person*: People interact and cluster relatively freely; emphasis is on serving the needs of members of the organisation with a focus on consensus.

Handy suggests that the main factors which influence organisational culture are: history and ownership, size, technology, goals and objectives, environment and people.

Deal and Kennedy (1982) said they believed that organisational cultures represent widely held philosophies in the (business) environment. As such, the cultures comprise shared values, particular rites and/or rituals and means of informal communication; the formal communications are part of the institutional framework. They proceeded to suggest four categories in their taxonomy of organisational cultures:

(1) *Tough-guy macho*: high risk-taking, individualistic, short term, poor learning (common in construction)
(2) *Work hard–play hard*: low risk-taking, social, persistent, quantity rather than quality (common in real estate)
(3) *Bet your company*: high risk, slow and ponderous, ideas develop (common in architecture)
(4) *Process*: bureaucracy, protectiveness, cautious and low risk-taking, procedures are important (common in utilities; government departments).

Graves (1986) developed a taxonomy of organisational culture comprising 'barbarian', 'presidential', 'monarchical', and 'pharaonic', whilst Williams *et al.* (1989) employ the categories of 'power', 'people', 'task', and 'role'. The categories (sequentially) correspond quite closely and relate to those noted by Handy (1985).

Schein (1990) investigated the process of socialising persons into organisations. Commonly, such processes begin formally with a short 'induction course' but, more significantly, the informal processes are more widespread and enduring. The purpose of both is to influence the new personnel to think and behave in accordance with the accepted norms of those in (that part of) the organisation. Schein determined seven dimensions for such socialisation:

(1) *Group–individual*: The extent to which new personnel are 'inducted' in groups or individually
(2) *Formal–informal*: The extent to which 'induction' is done formally
(3) *Self-destructive and reconstructing–self-enhancing*: The extent to which 'induction programmes' endeavour to destroy aspects of individuals and replace them with 'preferred' behaviours, etc., or seek to enhance and develop the existing facets of the personnel
(4) *Serial–random*: The extent to which mentors, etc., are provided or whether the individuals are left to fend (and discover) for themselves
(5) *Sequential–disjunctive*: The extent to which new personnel are informed and guided through a series of familiarising/inducting activities or whether the process is totally open/unprogrammed
(6) *Fixed–variable*: The extent to which induction/training is programmed according to time or is left open for 'progress when ready'
(7) *Tournament–contest*: The extent to which each part of induction/training is geared to 'pass/fail' (fail is likely to result in dismissal) or where the persons construct portfolios of achievements (to assist in promotion exercises, etc.).

Although there is a tendency for larger organisations (almost of necessity) to be more formalised, the critical variable in Schein's examination concerns control, based around required and enforced compliance. Further, different sections and/or levels of organisations may act differently and do so towards different types of recruits; even in highly structured organisations in which control is essential, there are inevitably occasions where intuition or 'thinking outside the box' is vital (as in 'emergencies').

Cooke and Lafferty (1989) developed an organisational culture inventory (OCI) which has been employed to enable individuals to examine their experiences in relation to specific cultural styles, based on aggregated perceptions of co-workers and in relation, therefore, to norms of the organisation. They identified twelve types of organisational culture:

(1) *Humanistic*: Helpful, person-centred management, facilitating growth and development of people
(2) *Affiliative*: Being friendly, sharing feelings, etc., emphasising constructive relationships
(3) *Approval*: Avoiding conflicts, 'go along' with others
(4) *Conventional*: Follow procedures (bureaucracy)
(5) *Dependent*: Please persons in authority, hierarchical and non-participative
(6) *Avoidance*: Punish errors, negative rewards so avoid responsibility – pass responsibility to others
(7) *Oppositional*: Confrontational, criticise others, make 'safe' decisions only
(8) *Power*: Non-participative use of position power, build power base
(9) *Competitive*: Reward winners, avoid (appearing to be) losing
(10) *Competence/perfectionist*: Hard work, persistence and perfectionism are valued; mistakes must be avoided
(11) *Achievement*: Value persons who set and achieve (difficult) goals, pursue excellence with overt enthusiasm
(12) *Self-actualisation*: Value creativity and quality, task accomplishment and development of personnel.

The types may be categorised into 'satisfactory' (types 1–4), 'security' (types 5–8), and 'dependent' (types 9–12).

Cooke and Rousseau (1988) found that chief executives characterised organisations as best for implementing successful strategies which had satisfaction cultures [see Cooke and Lafferty (1989), above]. Satisfaction cultures tend to be more risk seeking, whilst security cultures tend to be risk avoiding and, in consequence, employ sanctions to foster desired behaviour (thereby becoming inhibiting). The security cultures were people/security (approval, conventional, dependent, avoidance), emphasising control of relationships, and task/security (oppositional, power, competitive, perfectionist), emphasising control over tasks.

Cameron and Quinn (1999) developed an organisational culture assessment instrument (OCAI) as a questionnaire, administered to members of an organisation under study to rate aspects on six dimensions of culture: dominant characteristics, organisational leadership, management of employees, organisation glue, strategic emphasis, criteria of success. Each dimension contains four questions, A–D, and respondents distribute 100 'marks' for each of the six dimensions between the four constituent questions. The questions represent the four major types of organisational cultures identified by Cameron and Quinn (1999) in their research on over 40 000 managers in more than 1000 organisations. The questionnaire is completed twice by each respondent, once regarding the current organisational culture and the other concerning changes to yield the preferred culture. The scores for each question set, A–D, are averaged and diagrammatic profiles are drawn. The resultant profiles denote the current culture of the organisation and the culture which members would like to be in place approximately five years into the future (the preferred culture). Profiles of organisations may be used to examine 'fit' of the organisation to its operating environment and for comparison with other organisations undertaking similar activities: the authors note its use both in effecting understanding of culture and as a valuable aid to effecting change.

Cameron and Quinn (1999) proceeded by employing a 'competing values' model in which 'flexibility and discretion' is juxtaposed to 'stability and control' on one dimension; the other dimension juxtaposes 'internal focus and integration' and 'external focus and differentiation'. The resultant model, shown as Fig. 3.5, yields four quadrants, each denoting a type of organisational culture.

The four major types of organisational culture defined by Cameron and Quinn (1999) are:

(1) *Hierarchy*: The organizational culture compatible with this form is characterised by a formalised and structured place to work. Procedures govern what people do. Effective leaders are good coordinators and organisers. Maintaining a smooth-running organisation is important. The long-term concerns of the organisation are stability, predictability, and efficiency. Formal rules and policy hold the organisation together

(2) *Market*: The major focus of markets is to conduct transactions with other constituencies to create competitive advantage. Profitability, bottom line results, strength in market niches, stretch targets, and secure customer bases are primary objectives for the organisation. Not surprisingly, the core values that dominate market type organisations are competitiveness and productivity

(3) *Clan*: Some basic assumptions in a clan culture are that the environment can be best managed through teamwork and employee development, customers are best thought of as partners, the organisation is in the business of developing a humane work environment, and the major task of management is to empower employees and facilitate their participation, commitment and loyalty
(4) *Adhocracy*: A major goal of an adhocracy is to foster adaptability, flexibility and creativity, where uncertainty, ambiguity and/or information-overload are typical. Effective leadership is visionary, innovative and risk-orientated. The emphasis is on being at the leading edge of new knowledge, products and/or services. Readiness for change and meeting new challenges are important.

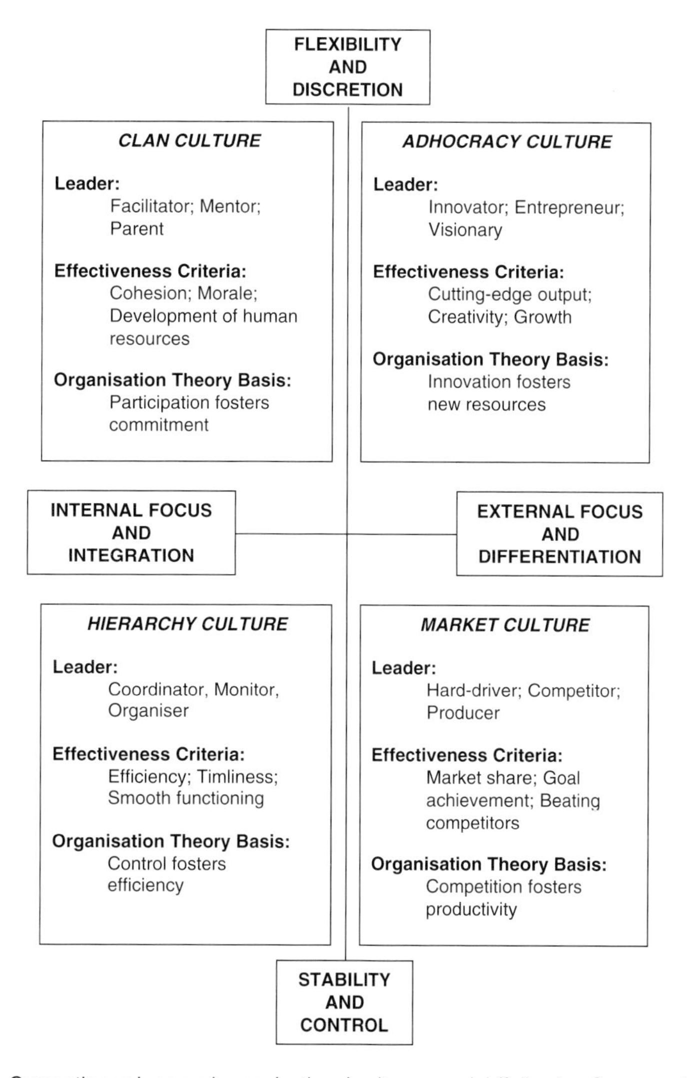

Figure 3.5 Competing values and organisational cultures model (following Cameron & Quinn 1999).

Increasingly for major projects, cultures are brought into contact in ever more diverse ways. Such contacts occur vertically between organisations, as in the value chain of providing a project through 'traditional' (design–tender–construct/production) procurement; others occur horizontally, as between subcontractors or contractors coming together for a joint venture. Both forms take place internationally and multinationally and have to interface much more deeply with more distant functional forms of organisations for highly diverse purposes (e.g. consortia of international banks in build–own–operate–transfer (BOOT) joint venture financing; trading and merchandising organisations in seeking to dispose of payments for project made in used goods, such as old car tyres).

Managers of joint ventures face major difficulties in constructing realities shared by the participants which engender enough coherence in the joint venture unit to facilitate effective operation. Such difficulties, magnified if the project is international, are due to the joint venturers having disparate views of the purpose(s), capabilities and prospects (often concerning profitability, in particular) of the joint venture organisation and/or project, despite the execution of detailed, prior, formal agreement.

Encounters between (inevitably) somewhat differing cultures generate clashes and/or conflict. Research into organisational 'mergers' indicates that:

> *Usually the corporate culture of the most powerful economically successful company dominates.* (Furnham 1997, p. 570)

Hence, 'merger' really means 'take-over'; so in joint ventures, whether formal or informal, one participant is likely to dominate if only transiently – the shifting multi-goal coalition, common in construction project TMOs.

Alvesson (1994) studied an advertising agency and found a tendency amongst the professionals employed there to 'downgrade' the judgements and abilities of clients. Such a tendency is apparent also amongst design professionals and/or consultants on construction projects (see, for example, Fellows *et al.* 2004). This approach to clients' inputs occurs despite the necessity for clients and professionals to work together closely to achieve and enhance good performance, the widely expressed prime (perhaps only) objective of participants in construction projects being to satisfy the client!

Cultures' consequences

Culture is not manifested just as 'how we do things here' but includes 'what', 'why', expectations of consequences and (socio-) structural contexts. It generates the framework of norms and the acceptance limits of behaviour, as well as embodying concepts of justice and suitable repercussions for any transgressors (who are caught). Notably, it gives rise to value-judgements, including 'guilt' or innocence. Laurent (1983) found that middle and senior managers in organisations of various nationalities have different preferences for the structures of their organisations. Structures, institutions, etc., whether physical or socio–politico–legal, which are not culturally acceptable are not sustainable, at least not in the long term.

Thus, culture – however defined, using whatever dimensions and at whichever level – is fundamental to the 'whats', 'hows' and consequences of processes and the resultant products. Further, culture may be seen as dynamic and, in consequence, somewhat fuzzy. Whilst culture shapes behaviour, so behaviour, in turn, shapes culture (change) [Figs 3.5 (culture profile) and 3.6 (cultural adaptation)].

Cultures exert strong but subtle pressures on people to think and behave in particular ways. Obviously, those pressures apply to persons who are members of the society (nation, organisation), but they also apply to those who are 'visitors'. A short, temporary visit may lead to an individual insulating their behaviour from the local culture whilst, for more enduring visits which may become permanent, the internal and external pressures to 'acculturate' are likely to be significantly stronger. Such 'pressures' are reflected in education and training programmes, including learning language, which are commonly undertaken in preparation for spending a period in a different society.

Thus, it is at the cultural boundaries and interfaces where the majority of significant problems are likely to occur, problems which may be more prevalent and/or magnified if the cultural differences are not anticipated (such as can occur in discourse between consultant advisers and contractors from the same locality). Otherwise, allowances are likely to be in place to increase tolerance through flexibility, and hence more effort is made to ensure understanding is correct.

The combination of cultural manifestations – language and behaviour – have a major impact on whether a deal is struck, with whom, within what formal and informal frameworks, how it is executed and with what consequences (see, for example, Trompenaars and Hampden-Turner 1997). Cultures and cultural changes

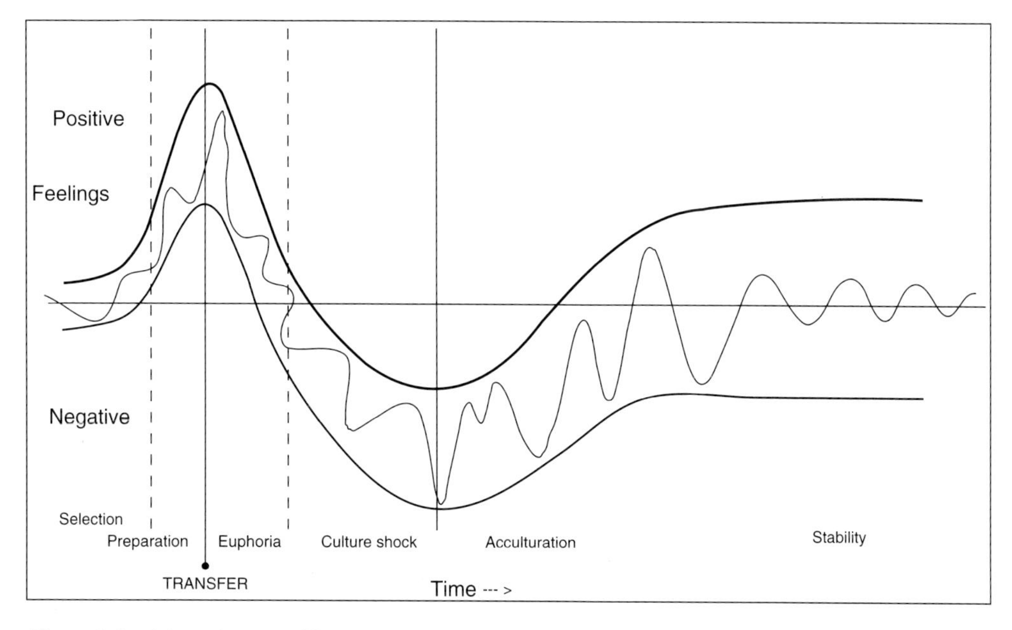

Figure 3.6 Adaptation to a different culture: acculturation (developed following Hofstede 1994b).

impact on performance, including what aspects of performance are regarded as important! Importation of alien methods, etc., commonly generates rapid rejection, whether overt (such as refusal to comply) or implicit (such as in low levels of performance). However, the situation is not straightforward because many subcultures exist contemporaneously; indeed, Baron and Greenberg (1990) consider that organisations may not exhibit a single, unitary culture, and that measures used for strengths of cultures are subject to significant question. In the context of change, it is helpful to recall the preference most people exhibit for the familiar and thus the tendency to resist change; even if change does occur, a rather strong tendency to revert to prior norms is common [as in private finance initiative (PFI) scheme financing; see, for example, Ezulike *et al.* (1997)].

Cultures are manifested through norms of behaviour and, as such, also reflect morals through ethics of practices. Whilst norms imply averages, they also imply variability. There are no imposed 'upper limits' of behaviour, but bottom limits of acceptance are specified in laws, codes of conduct, etc. Of course, issues of detection, enforcement and sanction then arise. A further complexity is due to social dynamism and the essential for appropriate changes to be incorporated to accord with modifications in societies' value structures. Globalisation might suggest that social differences are reducing, that convergence is occurring. However, studies (see, for example, World Bank 2001) indicate that, on various measures, evidence for divergence is just as robust as evidence for convergence.

Cameron and Ettington (1988) stated:

the effectiveness of organisations is more closely associated with the type of culture present than with the congruence or strength of that culture

The strength of a culture is a measure of its power to impact on what happens within an organisation, including processes employed. Cultural congruence concerns the generality of a culture throughout an organisation – the more generally a culture applies, the greater the congruence. Cultural type relates to the kind of culture of an organisation, for example, innovative, risk taking. Kotter and Heskett (1992) carried out a matched comparative study between high performing and lower performing organisations and concluded that cultural type, strength and congruence are the major factors differentiating the performances achieved.

In practice, economic considerations impose upper limits of behaviour in businesses operating in market economies. In command economies, the norms and parameters are prescribed by state organisations in their regulation of economic activity; in the public sector of mixed economies (which tend to have been much reduced over recent years), although the state determines what will be purchased, the purchasing is done from the private sector and so is dependent upon market forces. In construction, public sector activities have not only declined (commonly, through privatisation) but also have changed in their nature – to reduce capital expenditure by government by use of concession-based procurement. That has placed significantly different investment and financing requirements on construction organisations, amending the traditional role of construction being a cash generating activity, the speed and effective management of which leads to generation of profit. Understandably, construction businesses are exploring mechanisms to adapt concession arrangements to align more closely with their more familiar financing of activities.

Within an existing paradigm of business behaviour, the positioning of a firm is dependent on its influential personnel through their determination of the organisational culture. Clearly, such positioning varies, but within limits is determined by society. The greater the tolerance of the society in which an organisation operates and the more self-contained the organisation's activities are, the more diverse will be the spectrum of organisational cultures and behaviour.

Fragmentation (manifested in the separation of design and construction and/or production; functional specialisation and independence amongst designers; widespread subcontracting in construction and/or production), a widespread and long standing criticism of construction (really occasioned by the ensuing communication, co-ordination and co-operation requirements and problems rather than by the division of labour and/or specialisation aspects), means that at any stage of a project there is a distinct, if transient, power hierarchy. The independence of individual specialist organisations, the inputs from which are vital for the realisation of the project, facilitate that situation. It should be unsurprising, therefore, that coordination proves to be so problematic and that real and extensive co-operation is rare. There are always others to blame for faults, and room for manoeuvre is extensive. Thus, claims are widespread and frequently constitute a significant profit-earning opportunity in competitive pricing (Rooke *et al.* 2003), litigation is common, innovation is low and performance suffers. Those problems appear to be perpetuated and proliferated through the price-based competitive tendering mechanisms which remain by far the most commonly used methods to award projects. (It is common for such project awarding mechanisms to remain in use even if projects are, ostensibly, operating under 'collaborative arrangements' such as partnering. Price-based competitive awarding methods are a usual requirement in the project awarding regulations of public sector organisations.)

There seem to be two polar approaches to cultural awareness: one which focuses on similarities and the other which focuses on differences. Whilst cognisance of both similarities and differences is important for understanding, appreciation and adaptation of behaviour, i.e. to achieve compatibility, there are potential negative aspects also. Focus on similarities may induce complacency and lack of sensitivity to important differences, whilst focus on differences may encourage incompatibility through ethno-centrism, 'jokes' about differences, etc., thereby enhancing alienation. Unfortunately, the negative aspects of differences focus is ascendant in construction, leading to participants' behaviour generating a culture of mistrust and disrespect in interpersonal and/or interorganisational dealings to compound the other negative perspectives which abound. To overcome such problems, change initiatives must be enormously robust and be high in valence of expected benefits with evidence of realisation. More commonly, periodic power of shifting market forces leads to (temporary) changes.

Culture is the constant practice of daily existence. We all 'do culture' continuously, which of course yields the dynamics, norms and variability of such human group measures. From a managerial perspective, culture can be regarded as a process or as a tool. Certainly, culture concerns communication of meaning, and so is subject to distortion, noise, etc., resulting in lack of and mis-understanding. Thus, we may regard manifestations of cultures as 'signifying systems' through which collections of signals are given which the sender believes will invoke understanding by the recipient(s), and thence lead to particular behaviour such as an element of project performance.

Problems of communicating meaning (messages) are well known and addressed in detail in the social sciences through conversation analysis (e.g. Potter 1997), discourse analysis (e.g. Heritage 1997), etc. The construction industry has long known of its extensive communication problems (e.g. Higgin & Jessop 1963), but little seems to have been achieved towards their resolution. Dismissive statements abrogating responsibility to act ('it's the culture of the industry') remain common. We respond to communications in relation to their contents and contexts, the presence and importance of each of those elements depending on the nature of the society and the language used (the cultural context). Sensitivity to relevant others influences our communications, interpretations of meaning and consequent action within a power hierarchical context of who those relevant others are (in a business context, at least). Thus, we accept culture as a contextual variable, but we may also believe that we can mould culture for change; the latter is much more likely to constitute behavioural modification in the short term and/or initially at least.

Re-engineering is defined by (Hammer & Champy 2001, p. 35) as:

> *the fundamental rethinking and radical redesign of business processes to achieve dramatic improvements in critical, contemporary measures of performance, such as cost, quality, service and speed*

Whilst such apparently customer-oriented performance measures seem laudable, they are also subject to potential abuse and to the criticisms cited by Green (1999). Further, important issues arise concerning indentification of who the customers are – not just the party commissioning the project directly (signatory of the contract), but also financier(s), current and future owner(s), tenant(s), users, passers-by – all those functionaries impacting on the demand side of the project. Given a perspective of culture as a tool of management, whereby behaviour and thence performance may be altered and shaped according to the manager's desire, the human is reduced to another category of passive business resource rather than being the active participant in the business processes.

A number of 're-engineering' initiatives (partnering, just-in-time (JIT), lean construction, etc.) have been attempted in the construction industry over recent years. Others, such as the new engineering contract (NEC) have overtly sought to change 'the' culture of the industry; in the case of NEC, beyond the domestic national borders – an endeavour of no small bravado! Such initiatives serve to indicate the deep and widespread view that culture is a tool which managers can manipulate readily at their will to obtain changes in performance. Such a view belies the fundamental and complex nature of human cultures, and so, necessarily, those initiatives are doomed. Only insofar as the people affected recognise and accept merits in the attempted change will such changes be incorporated into those cultures.

Thus, it appears that, in re-engineering, Adam Smith's 'hidden hand' has far from decreased in its sinister apparitions. The desires of and opportunities for the (transiently) powerful to manipulate the less powerful may, in fact, have become enhanced!

Re-engineering organisations – essentially, revisiting organisational design, structuring, etc., is heavily value (and thus ethically) laden. Whose values are pursued and to what degrees of vehemence, with what effects on others, are essential issues in which opportunism may play a significant role. On the international level, such a struggle is articulated by Hutton (2002) in his analysis of the global role of the USA.

So, re-engineering constitutes radical organisational change – but how, what is involved and with what consequences? In adopting re-engineering and endeavouring to apply it to people, the perspective of culture as a tool which may be manipulated by management at will is apparent. That perspective raises value judgements – for whose benefit and at whose cost?

Research agenda

The agenda for researching culture and its effects remains extensive and diverse, reflecting the all-pervasive nature of the subject and the inherent dynamism of people. At a fundamental level, research into culture remains differentiated into two primary approaches; this leads to varying perspectives on potential for effecting cultural changes (including mechanisms, duration, etc.). The anthropological approach considers that organisations (human groups) are cultures, whilst the sociological approach considers that organisations have cultures. Such fundamental differences have consequences for measurement and dimensioning of cultures.

In construction, the International Council for Research and Innovation in Building and Construction (CIB) has established a task group (TG-23) to study culture in the industry around the world which set its research agenda for 2003–2005 to be:

- To extend the studies of TG-23 to research cultures existing in the construction industry worldwide from societal and organisational perspectives
- To stimulate and facilitate international exchange and collaboration for studies into re-engineering processes adopted by construction organisations to examine their culturally-dependent consequences
- To research cultural facets of the construction industry to determine feasibilities of changes to improve performance
- To assemble a 'cultural inventory' to assist organisations in adapting to new cultural situations across construction industry sectors and countries.

The cultural inventory comprises two facets – reviewing the current applicability of research profiling national cultures and by using the common instrument of the OCAI questionnaire, researching cultural profiles of construction organisations and sectors in various countries; in each country, the research is co-ordinated by a 'national champion'.

In addition to the research of TG-23, many cultural issues remain within construction, many of which have been well-known for a long time! However, at this juncture, it may be opportune to re-examine them by questioning what has long been 'received wisdom' – the assertions and/or assumptions on which they are based. A widespread assertion (excuse?) is that construction is different, even unique, regarding its outputs and processes. Is that really so, and are the differences between construction and other industries of such significance? Often, construction is said to operate in a highly volatile and/or turbulent environment – but what about the media, fashion goods, electronics, for example? From examining work flow processes, price and cost data, it is evident that many aspects of construction are readily predictable over considerable periods of time; bidding may be fairly random, but price competition alone is tending to decline in importance for work allocation.

Fragmentation could otherwise be described as specialisation and/or division of labour, and as such be regarded as conducive to improving performance. Following Lawrence and Lorsch (1967), fragmentation is problematic, not because of functional separation and/or diversity, but due to the inadequacy of communication, co-ordination and integration, all of which are major cultural concerns.

The culture of the industry is variously described in context not only depicting the nature of the industry, but also alluding to several widespread problems. Common cultural descriptors include 'low technology', 'macho', 'opportunistic' and 'claims orientated', and 'blame'. Low technology reflects the relative level of technology in the output (it's not 'rocket science'), and despite changes (intelligent and/or responsive buildings and components, e.g. glazing), organisations in the industry remain inhibited through risk aversion and lack of investment in research and development. Macho reflects the traditionally (inherent? necessarily?) dangerous, dirty nature of construction activities yielding the 'male domain' of being strong, tough, brave, etc., to overcome the difficulties, employing loud voices, 'particular language' and muscular power – hardly essential now for the operations on highly effective and efficient (often Japanese) project sites!

Opportunistic and claims orientated reflect the approaches commonly adopted to enhance revenue (and hence profits) (Rooke *et al.* 2003, 2005), but also can be seen as opposites – claims under a contract can constitute legitimate recourse (recompense) for default and/or opportunistic behaviour of the other party to a contract. Anecdotally, in some countries it is very difficult to persuade local employees to submit and pursue contractual claims as such behaviour is totally alien to their (local) culture and, hence, means of doing good (harmonious) business. More generally, one may suggest that the opportunism and/or claims orientation is (largely) caused by the orientation of work allocation processes to price competition pertaining to initial contract sum.

Blame reflects reluctance to take responsibility, and, coupled with the division and diversity of inputs (and hence responsibility) with consequent fuzziness, the ready opportunities to assert that others hold the responsibility. To an extent, it may also reflect the necessity of taking only proven responsibility in low profit activities. Further, in a 'Western' context, avoiding blame may be seen as face saving. Taking responsibility alternatively may be regarded as demonstrating trustworthiness, but that adopts a longer term perspective.

A major concern is to generate successful projects, but that involves defining success which is likely to involve measures of performance made at particular time(s), from specific participant(s) and/or perspective(s). A widely-articulated objective is to 'satisfy the client', usually the party who commissions the project (the 'immediate paymaster'). However, given a joint-venturing, mutual interdependence relationship between project participants, and thence performance, it is more appropriate to seek performance which yields (reasonable) satisfaction of all participants to constitute a successful project.

Globalisation is, perhaps, the epitomy of and greatest challenge for, cultural integration: that is not to say convergence to a lowest common norm (as in many shopping malls), but to recognise the variables involved in cultural differences, sensitivities, tolerance and change, be they integrative or divergent. Both geographically and functionally, increasing cultural interactions are inevitable. The challenge is to deal with those effectively and to the enduring benefit of all.

All too frequently, commercial management is judged on the basis of short-term returns. Whilst compatibility of an organisation's culture and its operational environment have been demonstrated to be essential for success, it must be remembered that what success means is, itself, variable! Measures of success often lead to conflict between short-term and long-term perspectives in which efficiency and effectiveness may also be seen as conflicting. Business cultures and climates are commonly manifested through the ethics demonstrated inwardly [employees, etc. – terms and conditions of employment and organisational citizenship behaviour (OCB)] and outwardly [as in corporate social responsibility (CSR)] to examine how people treat people and the consequences of, through reactions to, the treatments adopted. Is it really the case (as asserted by Milton Friedman and others) that the only ethic of business is to maximise the wealth of owners, or should effort be devoted beyond that horizon to consider the welfare of all stakeholders? Those major decisions, indeed all decisions, are based on structures of human values: the values which are employed and with what consequences are a matter of culture, usually most clearly manifested in the practices and results of commercial management.

References

Alvesson, M. (1994) Talking in organizations: managing identity and impressions in an advertising agency. *Organization Studies*, **15**, 535–563.

Baron, R.A. & Greenberg, J. (1990) *Behaviour in organizations: understanding and managing the human side of work* (3rd edn). Allyn and Bacon, Boston, MA.

Barthorpe, S. (2002) The origins and organisational perspectives of culture. In: *Perspectives on Culture in Construction* (eds R.F. Fellows & D.E. Seymour). CIB Publication 275, International Council for Research and Innovation in Building and Construction, Rotterdam; pp. 7–24.

Bayles, M.D. (1988) The professional–client relationship. In: *Ethical Issues in Professional Life* (ed. J.C. Callahan), Oxford University Press, Oxford; pp. 113–119.

Black, J.S. & Porter, L.W. (1991) Managerial behaviours and job performance: a successful manager in Los Angeles may not succeed in Hong Kong. *Journal of International Business Studies*, **22**, 99–113.

Bolam v. *Friern Barnet Hospital* (1957) 1 WLR 582.

Burns, T. & Stalker, G.M. (1961) *The Management of Innovation* (2nd edn). Tavistock Institute, London.

Cameron, K.S. & Ettington, D.R. (1988) The conceptual foundations of organizational culture. *Higher Education: Handbook of Theory and Research*. Agathon, New York; pp. 356–396.

Cameron, K.S. & Quinn, R.E. (1999) *Diagnosing and Changing Organizational Culture*. Addison-Wesley Longman, Reading, MA.

Chinese Culture Connection (a team of 24 researchers) (1987) Chinese values and the search for culture-free dimensions of culture. *Journal of Cross-Cultural Psychology*, **18**(2), 143–164.

Clegg, S.R. (1992) Contracts cause conflicts. In: *Construction Conflict Management and Resolution* (eds P. Fenn & R. Gameson). E. & F.N. Spon, London; pp. 128–144.

Cohen, D. (1993) Creating and maintaining ethical work climates: anomie in the workplace and implications for managing change. *Business Ethics Quarterly*, **3**(4), 343–358.

Cooke, R. & Lafferty, J. (1989) *Organizational Culture Inventory*. Human Synergistic, Plymouth, MI.

Cooke, R. & Rousseau, D. (1988) Behavioural norms and expectations: a quantitative approach to the assessment of organizational culture. *Group and Organizational Studies*, **13**, 245–273.

De George, R.T. (1995) *Business Ethics* (4th edn). Prentice Hall, Englewood Cliffs, N.J.

Deal, T. & Kennedy, A. (1982) *Corporate Cultures*. Addison-Wesley, Reading, MA.

Eldridge, J. & Crombie, A. (1974) *A Sociology of Organizations*. Allen & Unwin, London.

Etheridge, J.M. (1996) *The Influence of Corporate Codes of Ethical Conduct on Ethical Standards in Hong Kong Companies*. Business Research Centre, School of Business, Hong Kong Baptist University, Hong Kong.

Ezulike, E.I., Perry, J.G. & Hawwash, K. (1997) The barriers to entry into the PFI market. *Engineering, Construction and Architectural Management*, **4**(3), 179–194.

Fang, Z.Y. (2002) *Behavioural Analysis of Project Team Performance in China*. PhD thesis (unpublished). Department of Real Estate and Construction, University of Hong Kong.

Fellows, R.F., Liu, A.M.M. & Storey, C. (2004) Ethics in construction project briefing. *Science and Engineering Ethics*, **10**(2), 289–302.

Furnham, A. (1997) *The Psychology of Behaviour at Work: the Individual in the Organization*. Psychology Press, Hove.

Graves, D. (1986) *Corporate Culture – Diagnosis and Change: Auditing and Changing the Culture of Organizations*. Pinter, London.

Green, S.D. (1999) Partnering: the propaganda of corporatism? In: *Profitable Partnering in Construction Procurement* (ed. S.O. Ogunlana). E. & F.N. Spon, London; pp. 3–14.

Grönroos, C. (1991) The marketing strategy continuum: towards a marketing concept for the 1990s. *Management Decision*, **29**(1), 7–13.

Hammer, M. & Champy, J. (2001) *Reengineering the Corporation* (new edn). Nicholas Brealey, London.

Handy, C.B. (1985) *Understanding Organisations* (3rd edn). Penguin, Harmondsworth.

Hartman, E. (1996) *Organizational Ethics and the Good Life*. Oxford University Press, New York.

Hatch, M.J. (1993) The dynamics of organisational culture. *Academy of Management Review*, **18**(4), 657–693.

Henry, C. (1995) Introduction to professional ethics for health care professionals. In: *Professional Ethics and Organizational Change in Education and Health* (eds C. Henry & P. Jane), Edward Arnold, London; p. 13.

Heritage, J. (1997) Conversation analysis and institutional talk. In: *Qualitative Research: Theory, Method and Practice* (ed. D. Silverman), Sage, London; pp. 161–182.

Herzberg, F., Mausner, B. & Bloch Snyderman, B. (1967) *The Motivation to Work* (2nd edn). Wiley, New York.

Higgin, G. & Jessop, N. (1963) *Communications in the Building Industry*. Tavistock Institute, London.
Hinman, L.M. (1997) *Ethics: A Pluralistic Approach to Moral Theory*. Harcourt Brace Jovanovich, Orlando, FL.
Hofstede, G.H. (1980) *Culture's Consequences: International Differences in Work-related Values*. Sage, Beverley Hills, CA.
Hofstede, G. (1994a) The business of international business is culture. *International Business Review*, **3**(1), 1–14.
Hofstede, G.H. (1994b) *Cultures and Organizations: Software of the Mind*. Harper Collins, London.
Hutton, W. (2002) *The World We're In*. Little Brown, London.
Kotter, J.P. & Heskett, J.L. (1992) *Corporate Culture and Performance*. Free Press, New York.
Kroeber, A.L. & Kluckhohn, C. (1952) Culture: a critical review of concepts and definitions. In: *Papers of the Peabody Museum of American Archaeology and Ethnology*, Vol. 47, Harvard University Press, Cambridge, MA.
Langlois, C.C. & Schlegelmilch, B.B. (1990) Do corporate codes of ethics reflect national character? Evidence from Europe and the United States. *Journal of International Business Studies*, **4**, 519–539.
Laurent, A. (1983) The culture diversity of Western conceptions of management. *International Studies of Management of Organization*, **8**, 75–96.
Lawrence, P.R. & Lorsch, J.W. (1967) *Organization and Environment: Managing Differentiation and Integration*. Division of Research, Graduate School of Business Administration, Harvard University, Boston.
Lay, R. (1993) Dis Macht der Moral. *Unternehmenserfolg durch ethisches Management [Success in Business through Ethical Management]*. Econ Taschenbuch Verlag, Düsseldorf.
Leary, M.R. (1991) *Introduction to Behavioral Research Methods*. Wadworth, Belmont, CA.
Leisinger, K.M. (1995) Corporate ethics and international business: some basic issues. In: *Whose Business Values* (eds S. Stewart & G. Donleavy), Hong Kong University Press, Hong Kong; pp. 165–202.
Moran, E.T. & Volkwein, J.F. (1992) The cultural approach to the formation of organizational climate. *Human Relations*, **45**(1), 19–48.
Mullins, L.J. (2002) *Management and Organisational Behaviour* (6th edn). Prentice Hall, Harlow.
Potter, J. (1997) Discourse analysis as a way of analysing naturally occurring talk. In: *Qualitative Research: Theory, Method and Practice* (ed. D. Silverman). Sage, London; pp. 144–160.
Reeck, D. (1982) *Ethics for Professions: A Christian Perspective*. Augsbury, Minneapolis, MN.
Rooke, J., Seymour, D.E. & Fellows, R.F. (2003) The claims culture: a taxonomy of attitudes in the industry. *Construction Management and Economics*, **21**(2), 167–174.
Rooke, J., Seymour, D.E. & Fellows, R.F. (2005) Planning for claims: an ethnography of industry culture. *Construction Management and Economics*, **22**(6), 655–662.
Rosenthal, R. & Rosnow, R.L. (1991) *Essentials of Behavioral Research: Methods and Data Anlaysis* (2nd edn). McGraw-Hill, Boston, MA.
Schein, E.H. (1984) Coming to an awareness of organisational culture. *Sloan Management Review*, **25**(2), 3–16.
Schein, E.H. (1990) Organisational culture, *American Psychologist*, **45**, 109–119.
Silverman, D. (ed.) *Qualitative Research: Theory, Method and Practice*. Sage, London; pp. 144–160.
Stewart, S. (1995) The ethics of values and the value of ethics: should we be studying business values in Hong Kong? In: *Whose Business Values* (eds S. Stewart & G. Donleavy), Hong Kong University Press, Hong Kong; pp. 1–18.
Swannell, J. (ed.) (1992) *The Oxford Modern English Dictionary*. Clarendon Press, Oxford.
Tagiuri, R. & Litwin, G.H. (eds) (1968) *Organizational Climate*. Graduate School of Business Administration, Harvard University, Harvard, MA.

Treviño, L.K. (1986) Ethical decision-making in organizations: a person–situation interactionist model. *Academy of Management Review*, **11**, 601–617.
Trompenaars, F. & Hampden-Turner, C. (1997) *Riding the Waves of Culture: Understanding Cultural Diversity in Business* (2nd edn). Nicholas Brealey, London.
Turpin, D. (1993) Strategic alliances with Japanese firms: myths and realities. *Long Range Planning*, **26**(4), 11–15.
Weller, S. (1988) The effectiveness of corporate codes of ethics. *Journal of Business Ethics*, **7**, 389–395.
Williams, A., Dobson, P. & Walters, M. (1989) *Changing Culture: New Organizational Approaches*. Institute of Personnel Management, London.
World Bank (2001) *Global Economic Prospects and the Developing Countries 2001*. www.worldbank.org/prospects/gep2001/sumeng.htm (accessed 17 May 2001).

Appendix

Societal norms and workplace norms under Hofstede's dimensions of national cultures. [Derived from Hofstede (1994b)]

Small power distance	**Large power distance**
Inequalities amongst persons should be minimised	Inequalities between persons are expected and desired
There should be, and is to some degree, interdependence between more powerful and less powerful persons	Less powerful persons should be dependent on the more powerful persons; less powerful persons tend to be polarised between dependence and counterdependence
Organisational hierarchies mean inequalities of roles; they have been established for convenience	Hierarchies in organisations reflect the existing inequalities between persons
Decentralisation is popular	Centralisation is popular
There are small differences in remuneration packages through the organisation	There are wide ranges in remuneration between persons at the top and the bottom of organisational hierarchies
Subordinates expect and want to be consulted	Subordinates expect and want to be told what to do
The ideal boss is a resourceful democrat	The ideal boss is a benevolent autocrat (or a 'good father')
Privileges and status symbols are unpopular and shunned	Privileges and status symbols for more senior persons are expected and popular
Collectivist	**Individualist**
Persons are born into extended families, or analogous groups, which protect them in exchange for loyalty	Persons grow up to look after themselves and their immediate family only

Identity is based on the social network to which a person belongs	Identity resides in the individual person
Employer–employee relationships are important and perceived in moral terms – as in a family	Employer–employee relationships are set out in contracts and supposed to be based upon mutual gain
Hiring and promotion of employees take the employee's group into account	Hiring and promotion of employees should be based upon skills, qualifications and achievements within pre-determined rules
Management relates to groups of persons	Management relates to individuals
Relationships take precedence over tasks	Tasks take precedence over relationships
Feminine	**Masculine**
Caring for others and preservation are dominant social values	Material success and progress are dominant social values
People and good relationships are important	Material wealth is important and shows social status
Everyone should be modest	Males should be assertive, ambitious and tough
All persons should be tender and concerned with establishing and fostering good relationships	Females should be tender and caring for relationships
Work is to facilitate (better) living standards	Living is for work
Managers use intuition and try to achieve consensus	Managers should be, and are expected to be, decisive and assertive
Equality, solidarity and quality of working life are emphasised	Equity, competition and performance are emphasised
Conflicts are resolved through compromise and negotiation	Conflicts are resolved through 'fighting'
Weak uncertainty avoidance	**Strong uncertainty avoidance**
Uncertainties are normal in life – so, each day is accepted for what it brings	Inherent uncertainties in life are regarded as threats which must be 'fought'
Low stress; subjective feeling of well-being	High stress; subjective feeling of anxiety
Aggression and other emotions should not be shown	Aggression and other emotions may be vented at appropriate times and in suitable places
Persons feel comfortable in ambiguous situations and when faced with unfamiliar tasks	Familiar tasks are accepted; ambiguous situations and unfamiliar risks are feared
There should be no more rules than is absolutely necessary	Rules are required extensively
Time is a framework for orientation	'Time is money' is operated

(Continued)

Persons feel comfortable when being 'lazy'; persons work hard only when necessary	Persons require to feel, and to appear to be, busy; emotional need to (be seen to) work hard
Precision and punctuality have to be learned (and persons tend to be 'relaxed' about them – synchronous activities are usual)	Precision and punctuality are engendered and so, come naturally (activities are executed serially)
Deviant and innovative ideas and behaviour are tolerated readily	Deviant ideas and behaviour are suppressed; innovation tends to be resisted
Motivation works through achievement, esteem and belongingness	Security, esteem and belongingness are motivators
Short term orientation	**Long term orientation**
Traditions are respected	Traditions are adapted to modern contexts
Social and status obligations are respected regardless of cost	Social and status obligations are important and respected but within limits
Great social pressure to (be seen to) maintain and advance status, even is it causes spending beyond income	Thrift and economy in resource usage are important
Savings are secondary; little funding generated for investment	Savings are emphasised – making a pool of funds available for investment
Results are required quickly	If results are slow to emerge, persons persevere
Persons are concerned with 'face' (personal image by self and others)	Persons are willing to subordinate themselves for a purpose/common good
Persons emphasise finding the truth	Virtue is important and respected

Endnote

[i] For expansion of the dimensions of culture and brief discussion of their manifestations for societal and workplace norms, according to Hofstede, see the Appendix to this chapter.

4 Procurement in the Context of Commercial Management

David Langford and Mike Murray

Introduction

Clients who procure the services of external organisations to complete projects are challenged to find the optimum combination of resources that will fulfil their objectives. Typically, the planning, management and control of a project are often viewed through the lifecycle of an engineering project. However, if we consider other industries such as banking, retail and fashion, it is clear that, at some point in time, most organisations, whatever the industry sector, will embark on a project. Such projects will involve the procurement of finance, labour, materials and machinery, and in this respect they have similar requirements. This is particularly so if we consider the generic skills required by project managers. However, this chapter describes the importance of procurement to commercial management by the example of construction and engineering. For more than a century, these industries have played a key role in the UK's relative economic wealth through the employment of labour and the construction of the nation's infrastructure and buildings.

We may ask ourselves, was building ever simple, or has it always been complex? Did the ancient Egyptian, Aztec, Greek or Roman clients consult over the optimum procurement route, or did they simply 'design and build'? It is unlikely that the Egyptian pharaohs were concerned with time and budget, given the longevity of their projects, although no doubt they expected their burial tomb to be ready when required! It is, however, known that the Great Pyramid of Khufu involved the builders setting over 2 300 000 blocks of stone, weighing on average 2.5 tonnes, to a height of 146.5 metres. Clearly this required some form of project management. Indeed, a quotation from the BBC television series *Secrets of the Lost Empires* emphasises the need for construction management expertise:

> *When I look at the Great Pyramid, the marvel there for me is not the stonework as much as the level of organization that the ancient Egyptians had in getting society* [project team] *to pull together in such a way that they not only had blocks of stone, but bread on the table.* (Barnes *et al.* 1996)

The procurement of labour for such a massive project also offers an interesting insight for today's commercial managers. Skilled builders were thought to be 'directly employed' on a permanent basis whilst peasant farmers were organized into competing gangs. The gangs were subdivided into progressively smaller units with the lowest level (divisions) named with single hieroglyphs denoting 'life', 'endurance' and 'perfection' (Barnes *et al.* 1996).

The procurement of modern projects involves a great number of specialists. The demise of the engineer–architect and increasing fragmentation is perhaps a focal point for a postscript analysis of evolution within the construction industry. After all, a project is only established when people engage in activities that constitute the project process. It is the sequencing of these activities and what they represent that manifests itself as a project, and thus is often described as a 'procurement system'. However, this term has much wider connotations than is often understood within the construction supply chain. Indeed, it is the fragmented nature of most project supply chains that lends credence to confusion when we talk of selecting a procurement route. Such misunderstanding often presents itself as a kaleidoscope whereby uncertainty has become a byword for too many construction projects.

Traditional procurement and design and build: it's all the same!

It could be argued that both the supply and demand sides of the construction industry have colluded in an untruth regarding the selection of appropriate procurement systems. The supply side, in particular, has sought to configure the options available to clients as 'packaged' in a manner that suggests great differences in risk transfer and accountability. Granted, the 'standard' contracts penned have been diverse and we can see distinct differences between 'traditional' and 'design and build' projects, particularly in the appointment of risk, prestige and power. This has suited the industry, built on fragmentation, very well. Clients are expected to select a procurement 'option' as if choosing a new bank account, even though, in contrast to the financial sector, the supply side of construction has not been able to offer performance guarantees. Clients are to be 'enticed' with new 'solutions' and this market is now thriving due to dissatisfaction with the status quo. Thus, as construction lawyers continue to argue the fine points of contract law, the pragmatists wish only to design and build.

If we choose to examine the history of 'procurement selection' it is evident that the demand side has played its part with equal enthusiasm, although evidence from experienced clients may reveal a different picture. Ward *et al.* (1991) noted that client expectations of procurement methods are based on client experience with previous projects and associated procurement methods. This would suggest that clients approach this issue from a pragmatic perspective and this gives validity to the study by Ireland (1989) who compared and contrasted six procurement methods in Australia. He argued that the titles used did not indicate what was occurring in the management of projects. Moreover, he suggested that the procurement methods reviewed had 'virtually meaningless distinctions'.

Clients have been seduced by the comfort of an 'options' solution, and this is evident in research undertaken by Hibberd and Djebarni (1996) who found that many clients and consultants were unhappy with the procurement system that they had used most recently. Those questioned in this study believed that a change in procurement route would resolve the procurement process problems. Thus, it is argued that 'commercial gain' has been a significant factor in promoting the delusion of the 'options' available, and construction clients have not benefitted from this. It is akin to supermarkets offering three options of an 'equivalent' product; a brand named product (e.g. Kellogg's Rice Crispies); supermarket equivalent (say, Tesco Ricesnaps) and supermarket 'value' option. Each product is offered for sale on the basis of where a purchaser attributes 'value' or, as we know only too well in construction, 'lowest price'. It would seem that the supply side of construction has a legacy of offering novice clients the 'value' option at 'highest price' (i.e. claims, delays, quality problems) despite an illusion of choice. Procurement 'options' do not intersect this culture and are, it is argued, an illusion, a smokescreen used by the supply side who are unable to 'design and build' as their forefathers did in Victorian Britain. As Tookey *et al.* (2001) note, 'post modern procurement' in the construction industry is chaotic rather than ordered. Offering a 'choice' to clients commissioning construction projects is no longer a simplistic solution constrained by 'procurement'. The actions taken for productivity improvement within the process now dictate the shape and form of these 'procurement systems' and these in themselves are continuously altered by the evolution in the demand and supply sides of the industry.

The rise and fall of management contracting and construction management

The introduction of management contracting (MC) to the UK in 1969 and its close cousin construction management (CM) in the 1980s led to various research groups examining these procurement systems. Ward *et al.* (1991) observed four types of management contracting systems in operation, MC; CM; design and manage; and design, manage and construct. They concluded that these systems would only work if all project stakeholders involved understood their roles and responsibilities within each system. Projects themselves were under the spotlight. Some were considered more successful than others. The Broadgate office development in London was considered by most project stakeholders to have been a huge success. It was examined by the Construction Management Forum (1991) and, although not explicitly mentioned in their paper, by Dodd and Langford (1990). The latter research explored the organisational and personnel issues emerging from the relationships between the CM and trade contractors. However, the New British Library project at St Pancras, London, a project procured partly using CM, was described by Sleep (1997) as a 'comedy of errors'. She also cites a report by the Commons Public Accounts Committee (1996) who described the project as 'a model of how not to manage a construction project'.

It is interesting that, during the early 1990s, the trade contractors were rebranded as 'specialist trade contractors' and this title remains in various forms today, that is, 'specialists', 'package contractors'. Indeed, given the volume of subcontracting that now exists in both traditional contracting and design and build (D&B), these terms have more general use. However, despite the widespread use of package contractors within the industry, CM and MC have fundamental differences to the other procurement systems. In particular, Minogue (1993) noted that CM exposes the management of package contractors to the employer, the professional team and, indeed, the employer's construction lawyers, rather than being hidden behind the veil of a main (traditional) contract. Hughes (1997) explored the contract law and practice issues concerning CM and concluded that CM is more a management philosophy than a contract structure, whilst Langford and Murray (2004) argue that advantages of the system are really drawn from Adam Smith's (1776) classical economic theory. Specialisation produces economic benefits due to lower costs and more efficient production. The use of specialist trade contractors enables designs and installations to be rolled out as the project progresses, thus ensuring a fast track approach. Moreover, the specialist packages can be tendered and so have the dubious benefit of work being let for the lowest cost. This philosophy may be waning in the face of 'best value' rather than lowest priced procurement strategies.

The concept of 'value' is much in evidence in the contemporary procurement environment. Largely, the 'best value' initiative promulgated by government has become part of the recovery of the procurement practice. Langford *et al.* (2003a) explored the way in which best value may be delivered to and from the parties in the construction process. Different perspectives of value may be perceived. Value can be read from an economic, marketing and business perspective. Contemporary theory sees 'value' in a procurement sense, as a social construct; different firms in the procurement supply chain, dependent upon their position and power, will interpret value in different ways. To some, value in procurement will be the traditional 'low price'; to others it will encompass a wider set of selection criteria such as soft values of compatibility, social chemistry, shared vision and co-operation. Hence, value in procurement is a social and political construct as well as being bounded by economic selectivity,

It is clear that, despite over three decades of use, management procurement systems are not well understood by all parties who adopt them. Cox and Thompson (1998) emphasise that the CM method requires strong hands-on project management and the ability to direct and make swift concise decisions as projects progress. However, the inexperience of public clients in particular has been exposed. The problems encountered on the British Library project (National Audit Office 1996) were repeated in the design and construction of the New Scottish Parliament in Edinburgh (Black 2000, 2004; Spencely 2000; Fraser 2004). As Clark (2001) observes, CM is not a suitable procurement system for central government! Moreover, Hill's (2000) retrospective look at MC concludes by suggesting that contractors frequently fail to understand their role in management projects and often behave as though they are operating under a traditional approach. Despite its misuse by public clients, private clients who make repeated and regular commissions (e.g. BAA, Tesco and Stanhope plc) continue to use this system. As Minogue (1996) points out, these types of clients tend to have substantial and expert in-house project management teams. She also comments that because of this level of expertise, it could be argued that there is little reliance on the skill of the construction manager at strategic level!

Trust, commitment, partnering and divorce

In Sir Michael Latham's report, *Constructing the Team* (Latham 1994), the construction industry was advised to adopt a less adversarial cultural by embracing more co-operative and trusting relationships. Clients, contractors and consultants were encouraged to adopt 'partnering' as a solution to resolving a legacy of mistrust and antagonism. The call to adopt a partnering approach left the industry perplexed given that it was acknowledged to be an ideology rather than a new contractual form.

However, a plethora of reports and guidance (Bennett & Jayes 1995, 1998; Barlow *et al.* 1997; Construction Industry Board 1997) was offered by academics, and the conference circuit was boosted by a proliferation of one-day seminars on this topic. The industry was, to some extent, on a 'road to Damascus'. However, if Latham (1994) equates to a journey by road, then the subsequent report by Sir John Egan, *Rethinking Construction* (Egan 1998), was travel by 'superhighway'. A new lexicon penetrated the industry and a new set of buzzwords reinforced an ideology that was often referred to as 'Eganomics'. Partnering would be redefined as 'supply chain management' (SCM) and long-term alliances would be preferable to one-off project partnering. The industry had gone 'strategic' in every sense. Radical reform was called for, and the UK construction industry was challenged to provide evidence of 'world class' performance.

The call for less confrontational behaviour between project participants appeared to acknowledge the difficulty in changing human behaviour, but may have underestimated the challenge in unfreezing an industry culture. Critics of partnering included Green (1999) who, in seeking to challenge the international academic community into debate, linked partnering to the propaganda of corporatism. In particular, he was critical of the dominant power that regular-spend clients (e.g. supermarkets) exerted over their project supply chains. Bresnen and Marshall (2000) argued that partnering is an imprecise and inclusive concept, capturing within it a wide range of behaviour, attitudes, values, practices, tools and techniques. Today, partnering has evolved with formal contractual guidance on 'how to partner' (Mosey 2000; Bennett & Baird 2001) now available. However, despite a gallant attempt by the Movement for Innovation (M4I) to publicise the benefits of partnering, vis-à-vis its 'demonstration projects' programme[i] the ideology has not been wholeheartedly accepted by all stakeholders within the construction industry. Anecdotal evidence would suggest that the novice client may be particularly exposed to professional guidance that does not acknowledge the potential benefits of Latham's so-called 'win–win' culture. Finally, as Bower (2003) notes, partnering does not eliminate problems by mere acceptance of its use; she concluded that partnering has many advantages, but there are also barriers to be surmounted.

Public procurement

In the UK, successive governments have attempted to regulate the performance of the construction industry. Murray and Langford (2003) note that various reports, commissioned by the UK Government or quasi-governmental bodies, have sought to improve the performance of the demand and supply sides of the construction industry. In post World War II Britain, the role of the Government, as client, was paramount in rebuilding the country's housing stock. This contrasts with current ideology where the majority of rented housing is commissioned by housing agencies, and public buildings and infrastructure are procured under the private finance initiative (PFI). Thus, the transfer of risk, from public to private, can be seen to be a core issue in the procurement of public buildings and services. This evolution has taken place in parallel with the UK Government's desire to be a best practice client. However, the developments taking place within central government bodies (i.e. Ministry of

Defence Prime Contracting[ii] and NHS ProCure 21[iii] initiatives) are in contrast to many local authorities who appear not to have fully embraced the 'movement for innovation'. For example, research undertaken by the Specialist Engineering Contractors Group (SEC) revealed that 90 per cent of local authorities continue to use the Joint Contracts Tribunal (JCT) contract forms that impose the use of retentions. The survey of 420 councils found that only a handful were using the Association of Consultant Architects' Project Partnering Contract PPC 2000 (Anon. 2004a). It may be that too few civil servants employed within the works departments are aware of the new ethos of partnerships. However, it may be that finance directors commission too many projects with an eye on capital cost alone, and thus do not request guidance from experienced construction professionals who may be employed in junior local authority positions.

Holt and Rowe (2000) examined a related issue in their research into UK Government departments as clients commissioning civil construction projects. Their findings revealed that project and construction managers believed government clients were unwilling to compromise for fear of being accountable with regard to strict spending targets. Moreover, the public clients were often perceived as being overly bureaucratic, manifesting itself in attentiveness to the details of rule. Such concerns are often cited regarding PFI projects; Reijniers (1994), in The Netherlands, noted that public and private partners often have conflicting interests. This study found that the public client tends to be concerned with: legislation; political opinion and political influence; democratic decision-making influence; the minimisation of risks and the realisation of a social goal. However, the private supplier has an eye on achieving returns on: invested funds; taking business risks; anticipating market and competitive developments and realising a corporate goal. As noted earlier, the UK has a legacy of public projects that have gone awry. Reports on such projects as the new British Library (National Audit Office 1996), Portcullis House (National Audit Office 2002a) and Devonport Dockyards (National Audit Office 2002b) provide ample evidence that too many projects have been overly exposed to financial risk, ultimately paid for by the taxpayer. The 'daddy' of them all may just be the construction of the New Scottish Parliament in Edinburgh: an inquiry was established to find out why this project escalated in cost ten-fold and was over 20 months late in delivery, the report from which concluded that the procurement route selected bore a heavy burden on the project's problems:

> *It verges on the embarrassing to conclude, as I do, that virtually none of the key questions about construction management were asked. Similarly none of the disadvantages of construction management were asked. This, in my view, is the point when the wheels began to fall off the wagon.* (Fraser 2004, p. 243)

The Auditor General's report (Black 2004) also confirmed that lessons from the New British Library project were not taken on board and the Parliament project was commissioned using a construction management method. The quick start on site was followed by numerous variations to packages that sent the provisional sums soaring, with associated design construction and programming complications. It was said that as much as £100m may be attributable to prolongation costs. Why construction management was chosen to procure the project is a compelling question.

Indeed, Fraser was critical of the Project Sponsor and the project's first Project Manager:

> *The selection of Construction Management was the single factor to which most of the misfortunes that have befallen the Project can be attributed. Against that background I am highly critical of the failure of Mr Armstrong and Mrs Doig to ensure that there was an appropriate evaluation of the highly risky contract strategy that was adopted, particularly in view of the choice of architect.* (Fraser 2004, p. 244)

It has been argued that the procurement selection process was a political decision; the late Donald Dewar, anxious to promote the construction of a landmark building for the Parliament, alighted on the idea of CM to compress the time to the start on site. Once ground had been broken, decisions to cancel the project would become more difficult. Furthermore, Fraser (2004) criticised the appointment process that led to the employment of Bovis as Construction Manager. Despite being the highest of the four tenderers, Bovis were excluded and thereafter readmitted to the process. Apparently, this decision was taken by the Project Sponsor after considering 'informal evidence' on Bovis' performance on a previous contract – the Museum of Scotland project. However, as Fraser notes: '[the Project Sponsor] was unable to provide me with any satisfactory reason for her selection of Bovis to be readmitted to the process'. In January 2005, Sir Robert McAlpine (an unsuccessful tenderer) issued a £4.31 million writ against the Scottish Parliament, claiming that the appointment of Bovis contravened EU procurement regulations.

The Scottish Parliament project would appear to have been subjected to 'optimism bias'. This term was coined in a study of 50 major building and civil engineering projects, undertaken by Mott MacDonald (2002). It revealed that too many public projects were beset by this problem. Optimism bias is described as the tendency for a project's cost and duration to be underestimated and/or its benefits to be overestimated. This issue has also been explored by Davidson and Huot (1989) who argued that incentives to produce optimistic estimates of viability are very strong and the disincentives weak. Accountability is often low because politicians who underestimate costs in order to have projects approved are rarely in office when actual viability can be calculated, if it ever is. Moreover, contractors and others with special interests in major projects are also eager to have their proposals accepted because contractual penalties for producing overoptimistic tenders are often relatively small; thus, major project costs are often underestimated, even during tendering (cited in Flyvbjerg *et al.* 2003).

Private procurement

It could be said that private clients who commission projects on a regular basis have adopted a smörgåsbord approach to procurement. A mix and match of various processes that sit within each procurement system can be combined to fit the client's objectives. Indeed, parts of the process have undergone severe 'plastic surgery', and 'intelligent' private clients (e.g. Stanhope plc) are now reconfiguring the construction supply chain within their projects. Of course, the private sector construction industry has various subsectors, be it house building, commercial, industrial or retail, and it is recognised that within each of these subsectors progress in achieving productivity improvements is variable. Indeed, if we consider the private house building sector for a moment, it is evident that spectacular margins are achievable on the back of 1970s site management procedures, that is, the profit comes from intelligent land bank acquisition and shortage of supply in a high demand climate rather than 'lean' process improvement at production level. For example, 'snagging'[iv] continues to be a problem on all projects, despite investigations into design, workmanship and material failures. Moreover, skips continue to be filled with usable materials designated for landfill sites, despite the 'landfill tax' legislation. This scenario would appear to be evident in too many projects that are characterised by 'fragmentation and wealth'. Each party to the construction process (be it designer, constructor or self-employed worker) has the opportunity to gain disproportional direct financial benefit from projects that do not satisfy individual clients, end-users, or contribute to overall industry improvement. This predicament is a manifestation of an industry that is perhaps, 'revolving' rather than evolving, and although predominantly driven by regular-spend clients who seek improvements, is constrained within an industry framework shackled by a legacy of fragmentation.

Guaranteed maximum price (GMP)

The use of a guaranteed maximum price (GMP) contract is one example that redefines the timing and contribution of each party's contribution to the overall process. This approach transfers the financial risk to the contractor and can be lethal if finalised before design detailing is all but complete. GMP is not a form of contract and has been applied to traditional, D&B and management procurement systems. It tends to be arranged in the second stage of a two-stage tender whereby the contractor will price the job as a lump sum, taking on the risk for cost- and time-overruns. However, this risk is often passed downstream to package contractors who may be the least able to manage it. It is also common for this arrangement to incorporate an 'open-book' policy whereby any savings on the contract budget are split between contractor and client (Chevin 1996). An example of GMP is the recently completed new Bullring shopping centre in Birmingham. The £247 m shell and core work was completed by Sir Robert McAlpine under a design and manage contract, incorporating a GMP. The project involved between 40 and 50 major subcontract packages, with a total of 200 packages over the project (Anon. 2001). The ability to offer a GMP with confidence relies on a careful analysis of a project's cost plan, and this requires the assistance of the subpackage contractors and suppliers. However, this appears to be a 'catch-22' predicament given that on one hand, they need to secure work and thus assist in establishing firm costs, and then accept the transfer of risk during the construction period.

Cascading of risk

It could be argued that the separation and delineation of the procurement systems used in the construction industry is a relatively pointless exercise, particularly if we consider the evolution over the past thirty years or so. The transfer of risk downstream in the supply chain can be seen in all procurement systems. In particular, contractors who previously built projects with direct labour, now employ 'skeleton' managerial staff on projects. The bulk of the work is subcontracted to package contractors, 'subbies' and specialists. The term is relatively unimportant, but belies a great weakness in the UK model. It could be argued that many of these subcontractors are the least able to carry the financial risk passed onto them. Perhaps more importantly, Pearce (2003a) has argued that this fragmented structure precludes the exploitation of economies of scale, and hence cost reductions.

The industry model described above frequently results in quality and functional problems, and this has led Rowlinson and Kvan (2004) to describe the system as 'vicious circles in the procurement process'. Thus, if the industry is truly seeking radical improvement in its performance, then it would be reasonable to suggest that these small and medium enterprises (SMEs) must be offered assistance and support to improve. After all, the true benefits that may be derived from lean construction, SCM and concurrent construction require the employment of competent SMEs who should benefit equally in the spoils. Indeed, Male (2003) cites research undertaken by the Warwick Manufacturing Group[v] who examined SCM in the construction industry and concluded that this industry faces major problems with its suppliers that are fundamental to profitability. However, many SMEs are overly burdened by day-to-day business concerns; recent problems include high insurance and pension costs that cannot be offset by the 'creative accounting' undertaken by the large contractors.[vi] Avoiding insolvency by maintaining a positive cash flow is perhaps the key business objective. Wholeheartedly embracing a new industry culture may take a little longer. Furthermore, given the current agenda for construction to adopt Corporate Social Responsibity (CSR) credentials, Pearce (2003b) paints a worrying scenario. He suggests that the costs inherent in 'triple bottom line' objectives (i.e. profitability, environment and social responsibility) will be less affordable to small firms working on low margins. Thus, it would seem that the upstream supply chain (clients, consultants and the 'skeleton' contractors) will continue to obtain financial benefits at the expense of the downstream SMEs who fight for the 'ends to meet'.

The cost of different procurement approaches

The little evidence available suggests that procurement decisions do influence cost and time performance. Rowlinson (1986) studied 12 projects, six by D&B and six by traditional methods, and found small performance advantages in time (7%) and cost (3%) when D&B was used, although the time saving was primarily in the inception–completion cycle rather than the on-site time.

Naoum's (1990) follow-on study considered the performance of management contracting and traditional procurement in communal buildings. Again, performance advantages were observed in a wide range of indicators from client satisfaction with time, out-turn cost, quality, cost and time certainty. The study was undertaken during the mid 1980s, almost two decades before the key performance indicator

(KPI) regime of *Rethinking Construction* (Egan 1998), and provided a backcloth to current developments.

More recently, Langford *et al.* (2003b) considered a comparison of traditional and lump sum bids submitted under a scheme entitled the Alternative Tendering Initiative (ATI). In a comparison of 11 projects in a stretch of motorway, with identical specifications, the tender and out-turn costs were compared. In short, the ATI projects were procured at over £120,000 per km cheaper than traditional methods. Moreover, the ATI delivered greater cost certainty to the project. However, all of these results must be read with care. Issues such as transaction costs and internal client costs were not taken into account.

Corruption in procurement negotiations

A report by Transparency International (Stansbury 2003) revealed that corruption in the construction and engineering industry is a serious problem in both developing and developed worlds. Moreover, a European economic crime survey, undertaken by accountancy firm PricewaterhouseCoopers, found that 20 per cent of construction firms in Europe reported being the target of serious economic crime in 1999 and 2000 (Construction News 2002). Corruption within the construction industry is one of the core issues that falls under a call by the UK Government for businesses to demonstrate their corporate social responsibility (CSR) credentials. CSR, also known as 'corporate citizenship', has been largely ignored by chief executive officers (CEOs) within the construction industry. Indeed, a survey on corporate sustainability sampled the top 100 companies in each of the 19 developed countries and found that only 9% of construction and building materials firms had submitted corporate reports on environment, social and sustainability issues (KMPG 2002). This contrasted with 50% of utilities companies and 45% of chemicals and synthetics firms (Spring 2003). Further evidence from another KPMG survey attracted the following comment from a Friends of the Earth (FoE) spokesperson:

> *The building industry would be wise to take this survey* [KPMG survey of 201 construction industry executives] *as a wake up call on how far behind it is. If firms are not going to take their responsibilities voluntarily, the only way to do it is through regulations.* (Bennett 2003)

It would therefore appear that the construction industry has not addressed the challenges set out in the report *Business and Society: Corporate Social Responsibility* (Department of Trade and Industry 2002a). The DTI challenges private, voluntary and public sector organisations to take account of their economic, social and environmental impacts. Moreover, they should take complementary actions to address key challenges based on their core competences – locally, regionally, nationally and internationally. This is perhaps a case of 'do as I say and not what I do' if we consider past events concerning the UK Government's support of construction interests through its Export Credit Guarantee scheme. The 1994 Pergau Dam 'affair' is an example where the procurement of UK construction contractors and UK arms manufacturers have been inextricably linked together through the offer of 'tied aid' from the Overseas Development Administration's (ODA) budget. In this case the Malaysian government were committed to purchasing over £1000 m worth of British military equipment in return for the UK funding Pergau to a sum of £234 m. A High Court ruling in 1996 found that aid for Pergau was in violation of the Overseas Aid Act (HMSO 1966), which forbids aid money being used for the purchase of arms (United Kingdom Parliament 2001). The involvement of a contractor in the alleged 'bribery' is exposed by Palast (2000), a journalist with *The Observer* newspaper, who interviewed barrister Jeremy Carver, an adviser to Transparency International[vii]. The extract below is taken from the newspaper and is also cited in United Kingdom Parliament (2001):

> *'I went to a DTI reception. I was introduced to someone who identified himself as the chairman of a company and we were talking about corruption. He announced with enormous pride that he personally had handed over the cheque to the government minister for the Pergau Dam "bribe" in Malaysia.' The corporate honcho, the chairman of Balfour Beatty, was not confessing, but boasting about the payment, which he may have considered not a bribe but just the cost of doing business Malaysian-style. Carver noted that the then Tory Trade Minister, learning of the pay-off, publicly congratulated Balfour Beatty on its 'patriotic competitiveness'.*

The extract above suggests an underground world involving spies and intrigue, where international construction projects are a by-product of political decision-making. However, returning to the domestic scene, the UK Government is seeking to impose a cultural change through legislation vis-à-vis a white paper *Modernising Company Law* (Department of Trade and Industry 2002b). This paper advises companies to submit formal reports on their social, environmental and economic impacts. These initiatives are intended to encourage more transparency within business transactions, something that is the antithesis of corrupt practices. Corrupt practices can include bribery, deception and fraud. Within the construction industry, these corrupt practices are often related to the misuse and distortion of information concerning claims for delay and disruption of a contractor's master programme. A 'bung in a brown envelope' may also help with securing a tender bid and, as noted above, this can involve international projects and governments. Clearly the problem is serious for Transparency International who have proposed an anti-corruption initiative for the construction and engineering industry. Hence Transparency International see the construction industries, particularly in developing countries, as being vulnerable to corruption of the procurement process. To combat corrosive practices they have prepared a Business Integrity Management System to be used in consultancy and similar business transactions.

Pre-qualification procedures for contractor and/or consultant selection

It is evident that a 'lowest price wins' culture has dominated the construction industry for decades and that this issue alone has thwarted the adoption of rigorous pre-qualification procedures, intended to assist clients in achieving best value. The problems associated with this practice are well known: lack of transparency and trust in relationships; contractor variations and claims to make profit; performance difficulties in programme and cost; quality control problems and end user dissatisfaction. As John Ruskin pointed out in 1849:

> *It's unwise to pay too much, but it's worse to pay too little. When you pay too much, you lose a little money – that is all. When you pay too little, you sometimes lose everything, because the thing you bought was incapable of doing the thing it was bought to do. The common law of business balance prohibits paying a little and getting a lot – it can't be done. If you deal with the lowest bidder, it is well to add something for the risk you run. And if you do that, you will have enough to pay for something better.*

This quote has been used in various construction texts including (*Accelerating Change* Strategic Forum for Construction 2002). This view is echoed by a construction client representative from pharmaceutical company, Glaxo Wellcome:

> *...we have got the industry we deserve because it's the industry we asked for. If we want a Rolls Royce, we have got to ask for it and be prepared to pay for it. We can't buy a Lada and spend the next two years trying to turn it into a Rolls Royce for the same price.* (Contract Journal 2000).

Renewed interest in pre-qualification procedures has been encouraged by the introduction of SCM ideology within the construction industry. Indeed, the rationalisation of suppliers by both clients and main contractors has left many subcontractors out in the cold. In particular, the use of framework deals, where smaller contracts are packaged together into one, has left civil engineering SMEs without work (Anon. 2002). Suppliers are, however, fighting back; in a case concerning Network Rail's (NR) selection process, a group of subcontractors alleged that the procedures were unfair and threatened to take legal action if it could be proven that the process was outside the terms within the Competition Act 1998. A spokesperson for NR disputed the allegation by explaining that suppliers are selected using a series of weighted criteria to ensure that the best 'fit for purpose' contractor is selected. Selection criteria include safety, management, training and commercial considerations (Anon. 2004b). However, even where continuity of work is guaranteed within a supply chain alliance, a legacy of adversarial practices within the industry has left too many subcontractors with a fundamental mistrust and scepticism about forming closer relationships with main contractors (Larson & Drexler 1997; Dainty *et al.* 2001).

The scepticism noted by Dainty *et al.* (2001) finds support from a study undertaken by Langford and Hardcastle (2004). Main contractors and small subcontractors were asked about adherence to the cultural change advocated by *Rethinking Construction* (Egan 1998). The study found that whilst main contractors sought to secure framework agreements with clients, they continued to procure subcontractors at the lowest price. The culture of low bidding in commercial management is pervasive. However, it would be a mistake to see this as clients driving down their own costs. Some supply side contractors see benefits in winning work at the lowest price and have set their value proposition to clients to be, in Treacy and Wiersema (1997) terms, 'operationally excellent' – that is, be better at supplying a service or good at a lower price than their competitors.

It is clear from the NR case noted above that those clients who commission projects regularly are becoming more sophisticated in their choice of suppliers. Less sophisticated clients and irregular spend clients may, however, be adopting a 'better the devil you know' approach whereby they are either unwilling or are unable to apply more intelligent sourcing procedures. However, contractors and consultants are increasingly being requested to demonstrate a diverse range of competences: financial; managerial; safety; environmental; technical; creativeness, etc., that supplement their ability to complete their role in projects successfully. These pre-qualification checks are designed to whittle out the incompetent, incapable, immoral and inept! They represent a move away from subjective and intuitive decision-making towards a more rigorous and transparent methodology, and this is being supported through academic research (Sonmez *et al.* 2001). The industry itself has also made progress, and individual clients have developed new procedures (e.g. The Highways Agency Capability Assessment Toolkit), whilst national improvements have been made through the use of Constructionline (2004), a central pre-qualification hub that allows the supply side of the industry to present its capabilities and communicates a profile of individual clients.

Financial penalties for 'mis-selection'

Pre-qualification can often involve what may be termed 'mis-selection' and often results in heavy financial penalties for clients. The two examples that follow have striking similarities in that they both involve the employment of US 'cladding' suppliers for the construction of new parliamentary buildings in the UK. The construction of Portcullis House in London left the UK Government with an additional £10 m bill after it was successfully sued by an unsuccessful tenderer (Harmon) for unfair treatment and contravention of procurement regulations in relation to the contract for the fenestration (prefabricated wall and window units). Harmon, a US contractor, had submitted a bid that was £2 m cheaper than Seele Alvis, the UK–German joint venture contractors who were awarded the contract on the basis that their bid was technically superior and would offer better value for money. In 1996, Harmon issued a writ against the UK Government, but subsequently went into liquidation; however, the liquidator continued legal action. The High Court found that the client (Corporate Office of the House) had failed to give an adequate statement of the criteria on which the contract would be awarded and had been guilty of 'misfeasance' (dishonest abuse of powers given to a public officer) in that it had been obvious to officials when awarding the contract that to do so would not comply with European Union Law and the Public Works regulations (National Audit Office 2002).

On the Scottish Parliament project, the supplier for the cladding and windows package was Flour City Architectural Metals (UK) Ltd, a UK registered member of a group of companies owned by Flour City International Inc. registered in the USA. The company had no track record as a supplier in its own right, and its capability to perform the contract was based on the reputation and record of its overseas holding company. The contract started in December 2000, but by August 2001 there were doubts about the ability of Flour City to perform the works. By October that year

Flour City (UK) underwent compulsory winding-up for insolvency. An Audit Scotland (2002) report into the fiasco revealed that the termination of Flour City's contract and subsequent retendering of the work would cost the client (approximately) an additional £3.9m. The report revealed that the appraisal of Flour City's financial standing was not sufficiently thorough and that even before the execution of the full trade contract, warning signs were evident. Flour City had requested advance payments for works, had not been paying its own suppliers and was aggressive in its pursuit of early payment of valuations. The report concluded that the award of the contract to Flour City was not improperly made, but that there were deficiencies in the selection, award and management procedures for the contract.

E-tendering and reverse auctions

The construction industry has a legacy of 'cut-throat' tendering procedures where lowest-bid has been a prerequisite to securing work. However, the increase in strategic partnering between the demand and supply sides, and within the supply side network, has resulted in an increase in 'open-book' tendering where costs are transparent throughout a project's life. The benefits derived from this new project culture (delivering best value) may be under attack from the use of Internet auctions. Indeed, in a review of one reverse auction[viii] conducted by Stein *et al.* (2003), the suppliers' level of distrust increased, despite a massive cost saving made by the client. Critics of e-tendering[viii] (auctions) within the supply side of the industry have accused clients of returning to a lowest-bid culture where tender costs are struck down over an open forum. However, clients argue that this method will eliminate the potential for bribery and force the supply side to examine its operating costs. However, Stein *et al.* cite research by Emiliani and Stec (2001) who correctly argue that too many suppliers are 'electronically coerced' into participation. Despite agreeing on a bottom line tender figure, the ability to watch competitor's bids in real time may influence suppliers to buy work knowing that any profit may be difficult to obtain.

Public clients in particular are being driven towards reverse auctions due to an impending EC Consolidate Public Procurement Directive that directs them to select on lowest price. Both the DTI and the Office of Government Commerce (OGC) are supporting the use of reverse auctions in construction 'where appropriate' (Anon. 2004c). However, despite fierce opposition from the Specialist Engineering Contractors Group, there are supporters within the supply side of the industry. Le Marechal (2004), a managing director of a mechanical and electrical business, argues that clients should employ strict pre-qualification checks on their suppliers before allowing them to compete in an on-line forum. Moreover, he argues that the productivity challenge set for the industry, in *Rethinking Construction* (Egan 1998) can be achieved or even exceeded using this model. Thus, it would seem that a fresh review of the tendering procedures used in the construction industry might be upon us, but with a time old analysis of lowest cost, value and trust. Latham and Egan need not be shocked; after all, these issues have been the mainstay for successive commentaries of the industry since The Simon Report, *The Placing and Management of Contracts* (Ministry of Works 1944).

Conclusions

The issue of procurement has been at the forefront of commercial management in many industries. Since the late 1940s, the way construction work is procured has been the focus of practically every government report on the construction industry (Murray and Langford 2003). In other industries where large capital expenditures are made, such as defence, the issues are similar. How do we procure large capital projects on time, on cost and with an appropriate level of performance? Over time has come the realisation that traditional methods of procurement are not working. The sequence of low cost bidding (contract administration, progress chasing the supplier, monitoring performance, etc.) are becoming obsolete. The separated and sequential process has been replaced with integrated logistics involving key concepts of supply chain management. Understanding the supply chain process is one of the key competences in the commercial management of most project-based industries.

The most sophisticated companies seek to procure work in which value is optimised rather than price minimised. This value enhancement is said to be neutral in its impact upon all in the supply chain, be they customers, suppliers, designers and assemblers, with all parts of the supply chain being beneficiaries. The quest for integration brings different firms into closer proximity and this changes the relationships between the firms and key actors within the firms. Hence procurement is less likely to be a simple transaction-based exchange, but a more complex, socially and politically alert relational process where time and value are prized over lowest out turn cost. These changes have taken place against a backcloth of technical advances in the building process. As projects become technically more sophisticated then the number of suppliers for specialist trades becomes smaller and so framework agreements which bond specialists and major contractors into a lasting embrace (or restrictive bear-hug, depending upon your point of view) are welcomed by contractors and specialists. The contractor has ensured a supply line for a critical resource and the specialist has a secure workload whilst a framework exists. The symbiotic relationship of framework agreements is extended when it happens between client and contractor. As clients become more actively involved in major projects, from whatever industry, they will seek greater control of the process, and dissatisfaction with traditional methods will drive experimentation with new forms of procurement.

Politics inevitably plays a part in shaping procurement routes – a management idea or indeed a procurement system has to be consonant with the dominant political idea of the day. Langford and Murray (2004) use Morgan's (1986) idea of metaphors to explain the movement from traditional procurement to partnering and performance based contracts. The features of the business environment and the procurement system are importantly connected. The traditional methods of procurement were, as Murray and Langford (2003) have pointed out, defective. In short they emphasised 'contractual' rather than 'trusting' relationships. The report *Trust and Money* (Latham 1993) suggested that construction contracts delivered a paucity of trust and too little money. In Morgan's (1986) terms the relationships were played out in a business environment which was characterised as the machine metaphor. Roles and relationships were shaped by inflexible professional duties, each role being carefully defined and supported by legal obligation through the formal contract and extended by case law.

Much work is proceduralised and the defined roles enable easy replacement in the way that a defective part of an engine can be replaced.

The machine metaphor starts to crumble when projects become more complex in technical and organisational terms, and the system cannot contain the objectives of the participants. Clients grumble about delays and cost overruns, contractors protest about design changes, designers complain that design quality is being compromised. Managing a multi-organisational and multi-contractual environment using traditional means is irksome. From here we move to Morgan's (1986) organic metaphor. Contractors who laid bricks and mortar add design services to ensure that they are part of the new wave of 'single point responsibility' contractors. They have grown organically from 'just' being builders, into design and build contractors and in some cases, service providers (e.g. Amec, Amey, Jarvis, Interserve). The client's needs are served by greater simplicity in the process and the contractors benefit by reducing the risk of delivering projects which they have designed. The organic metaphor has been created by shifts in client expectations of better project performance in terms of time and cost.

A third era may be seen in the movement from design and build to the management routes. Langford and Murray (2004, pp. 654–677) chart this transition as follows:

> *US Construction Management...came into the UK environment as an alternate* [sic] *to management contracting and stimulated other forms of construction procurement, not least project management. Clients and constructors no longer polished the mirror to 'tell it like it is'. The challenge for such innovative organisations was to 'tell it like it may become' and to unseat conventional assumptions about roles and relationships in the procurement process.*

Visionaries abound; the metaphor here is the 'political'. Disputes extend from being based on sites and are widened to encompass ideologies of how the best way forward is to be charted. In this epoch, new roles and responsibilities start to emerge. Quantity surveyors recreate themselves as project managers, and new professions emerge: value engineers/managers, programme managers, key client managers, to name but a few. Architects withdraw from the project managerial battlefield to focus on being designers. In this epoch, disputes break out and Government is moved to appoint Sir Michael Latham to report on the state of the industry. His recommendations in *Trust and Money* (Latham 1993) are accepted by Government overnight; legislation is introduced to make arbitration mandatory in an attempt to cease hostilities, and law regarding payment systems is introduced in an effort to calm the waters.

The 'political' era is over and movement is to the cultural metaphor. Culture is about learned behaviour and the central plank of this era is that norms of behaviour, performance, expectations, etc., are uniformly accepted across a supply chain for a project. Institutions to oversee the introduction of the new culture (Construction Forum, Constructing Excellence and other Government-sponsored organisations) are put in place. Rhetoric abounds about teamwork: 'singing from the same song sheet' 'pulling together' and 'pain and gain sharing' are the cries de coeur of the cultural metaphor. What triggered this? *Rethinking Construction* (Egan 1998) saw the industry as backward and at the heart of this backwardness was the procurement process: if the building site could only be run like a factory then performance would improve.

The ideology of 'customer focus' was brought out, with performance being measured through key performance indicators used on demonstration projects.

The cultural metaphor is about transforming the culture of an industry, and the introduction of supply chain methodologies which align the technical performance but also harmonise the organisational culture of the participants in a project is evident. The central question is, what now? How will procurement practice of the future be experienced? As has been seen, the procurement process has flowed through a set of transactional mechanisms where price was the key determinant, through to relational procurement where partnering and alliances determine procurement route. Whilst all is speculation, the growing concern with corporate social responsibility (CSR) and environmental audits gives a signal that such issues will be part of procurement decisions of the future. Taking CSR, it will surely be necessary for clients to request the CSR record of companies within pre-qualification checks. Only those companies who abide by stringent CSR standards would be admissible.

Environmental audits would also become part of the scenery. Research at Strathclyde University (Ferguson 2004) considered six construction companies to see whether a formulated environmental strategy improved corporate performance. It was found that evidence of an environmental strategy did not directly improve financial results but enhanced the environmental respect of the firms which had them. Moreover, it enabled some differentiation from contractors who were not as advanced in developing and implementing an environmental policy. The secondary outcomes were an increased flow of opportunities for projects, with the prospect of improved profitability. Consequently, environmental audits are likely to be evident as frameworks for procurement policies of clients.

A further development in the procurement process concerns organisational learning (OL) and knowledge management (KM). Few project teams spend time on post-project reviews, and the harvesting of knowledge to facilitate continuous project-to-project learning has been largely untapped, despite an interest in OL and KM at corporate level. Perhaps the adoption of OL and KM at company level is less contentious than its application in a multi-organisational project environment. However, even where fewer adversarial project coalitions are formed Walker and Walker (1999) contend that the adoption of partnering and strategic alliances has not facilitated organisational learning. This may be the catalyst for a new professional within the construction project team, that of a 'knowledge manager' (see Kamara *et al.* 2005) who will be responsible for managing and collating the 'project story' as a means to improve the procurement process in future projects.

The chapter has emphasised the problems and opportunities of procurement management. It has shown that different procurement routes suit different types of projects and different types of clients. These points have been illustrated by numerous examples of good and bad practice. Moreover the social and political forces which move the industry from one procurement method to another have been speculated upon.

As Murray and Langford (2003) have shown, procurement is an enduring issue for commercial managers of whatever industry. Churchill (1943), in a speech to the House of Lords when requesting that the bombed out Parliament be rebuilt exactly as before, said, 'We shape our buildings and afterwards they shape us'. To the modern commercial manager the apposite quote may be, 'We shape our procurement practices and then they shape our projects'.

References

Anon. (2000) Clients are to blame for the state of the industry. *Contract Journal*, 18 October, 2.

Anon. (2001) The BullRing. *Contract Journal*, 31 October, 16–17.

Anon. (2002) Small Firms face framework ruin. *Construction News*, 9 May, 1.

Anon. (2004a) 90% of authorities still adversarial. *Contract Journal*, 15 April, 3.

Anon. (2004b) Subbies slam NR's selection process. *Contract Journal*, 31 March, 2.

Anon. (2004c) DTI backs e-auction. *Contract Journal*, 26 May, 1.

Audit Scotland (2002) *The 2001/02 Audit of the Scottish Parliamentary Corporate Body: Additional report on the Flour City Contract*. Audit Scotland, Edinburgh.

Barlow, J., Cohen, M., Jashapara, A. & Simpson, Y. (1997) *Towards Positive Partnering: Organisational and managerial Challenges for the UK Construction Industry*. Policy Press, University of Bristol, Bristol.

Barnes, M., Brightwell, R., Lehner, M., Page, C. & von Hagen, A. (1996) *Secrets of Lost Empires: Reconstructing the Glories of Ages Past*. BBC Books, London.

Bennett, J. & Jayes, S. (1995) *Trusting the Team: The Best Practice Guide to Partnering in Construction*. Centre for Strategic Studies in Construction, University of Reading, Reading.

Bennett, J. & Jayes, S. (1998) *The Seven Pillars of Partnering: a Guide to Second Generation Partnering*. Thomas Telford, London.

Bennett, J. & Baird, A. (2001) *NEC and Partnering: The Guide to Building Winning Teams*. Thomas Telford, London.

Bennett, C. (2003) Cited in Talking 'bout evolution. *Building Journal*, 3rd October, 47–48.

Black, R.W. (2000) *The New Scottish Parliament Building: An Examination of the Management of the Holyrood Project*. Prepared with the assistance of Audit Scotland. Laid before the Scottish parliament by the Auditor General for Scotland in pursuance of the Public Finance and Accountability (Scotland) Act 2000, AGS/2000/2. Audit Scotland, Edinburgh.

Black, R.W. (2004) *Management of the Holyrood Building Project, Audit Scotland, Edinburgh*. AGS/2004/7. Audit Scotland, Edinburgh.

Bower, D. (2003) Effective partnering. In: *Management of Procurement, Construction Management Series* (ed. D. Bower). Thomas Telford, London; pp. 96–114.

Bresnen, M. & Marshall, N. (2000) Partnering in construction: a critical review of issues, problems and dilemmas. *Construction Management and Economics*, **1**(18), 229–237.

Chevin, D. (1996) The max factor. *Building Journal*, 17 May, 40–42.

Churchill, W. (1943) Speech to House of Lords, 28 October. *Hansard*.

Clark, P. (2001) Whatever happened to construction management? *Building*, 11 May, 36–37.

Construction Management Forum (1991) *Report and Guidance*. Centre for Strategic Studies in Construction, University of Reading, Reading.

Constructing Excellence (2004) *Demonstrating Excellence: An Evolution of the Programme of Demonstrations*. Report, London. www.constructingexcellence.org.uk/pdf/bpknowledge/demonstrating_exc.pdf (accessed 24 July 2005).

Construction Industry Board (1997) *Partnering in the Team*. A Report by Working Group 12 of the CIB. Thomas Telford, London.

Constructionline (2004) www.constructionline.co.uk (accessed 9 June 2004).

Cox, A. & Thompson, I. (1998) *Contracting for Business Success*. Thomas Telford, London.

Dainty, A.R.J., Briscoe, G.H. & Millett, S.J. (2001) Subcontractors' perspectives on supply chain alliances. *Construction Management and Economics*, **19**, 841–848.

Davidson, F.P. & Huot, J.-C. (1989) Management trends for major projects. *Project Appraisal*, **4**(3), 137.

Department of Trade and Industry (2002a) *Business and Society: Corporate Social Responsibility*, Report. www.societyandbusiness.gov.uk/pdf/2002_report.pdf (accessed 10 October 2004).

Department of Trade and Industry (2002b) *Modernising Company Law: A White Paper*. www.dti.gov.uk/companiesbill/prelims.pdf (accessed 24 July 2005).

Dodd, J. & Langford, D.A. (1990) Construction management on one large project in London: a case study. *Construction Management and Economics*, **8**, 385–398.

Egan, J. (1998) *Rethinking Construction*. HMSO, London.

Emiliani, M.L. & Stec, D.J. (2001) Squaring online reverse auctions: key issues for purchasing process improvement. *Supply Chain Management*, **5**(4), 32–34.

Ferguson, H. (2004) *Environmental strategy as competitive advantage in construction*. PhD work in progress, University of Strathclyde, Glasgow.

Flyvbjerg, B., Bruzelius, N. & Rothengatter, W. (2003) *Megaprojects and Risk: An Anatomy of Ambition*. Cambridge University Press, Cambridge; p. 44.

Fraser, P. (2004) *The Holyrood Inquiry: A Report by the Rt Hon Lord Fraser of Carmyllie QC*. SP Paper No.205, Session 2, 15 September. Scottish Parliament Corporate Body, Edinburgh.

Green, S.D. (1999) Partnering: the propaganda of corporation. In: *Profitable Partnering in Construction Procurement* (ed. S.O. Ogunlana). E. & F.N. Spon, London; pp. 3–13.

Hibberd, P.R. & Djebarni, R. (1996) Criteria of choice for procurement systems. In: *Proceedings of the first RICS Construction and Building Research Conference (COBRA)*, University of West of England, 19–20 September. Royal Institution of Chartered Surveyors, London.

Hill, C. (2000) A matter of life and death. *Building Journal*, 21 January, p. 64.

HMSO (1966) *The Overseas Aid Act*. HMSO, London.

HMSO (1998) *The Competition Act*. HMSO, London.

Holt, R. & Rowe, D. (2000) Total quality, public management and critical leadership in civil construction projects. *International Journal of Quality and Reliability Management*, **17**(4/5), 541–553.

Holti, R., Nicolini, D. & Smalley, M. (2000) *The Building Down Barriers Handbook of Supply Chain Management – The Essentials*. Construction Industry Research and Information Association (CIRIA), London.

Hughes, W. (1997) Construction management contracts: law and practice. *Engineering, Construction and Architectural Management*, **4**(1), 59–79.

Hughes, W., Hillebrandt, P. & Murdoch, J. (1998) *Financial Protection in the UK Building Industry*. E. & F.N. Spon, London.

Ireland, V. (1989) Virtually meaningless distinctions between nominally different procurement methods. *Proceedings of CIB W65 4th International Symposium of Organization and Management in Construction*, Vol. 1. International Council for Research and Innovation in Building and Construction, Rotterdam; pp. 203–212.

Kamara, J.M., Anumba, C.J. & Carrillo, M. (2005) Cross-project knowledge management. In: *Knowledge Management in Construction* (eds C.J. Anumba, C. Egbu and P. Carrillo), Blackwell Publishing, Oxford; pp. 103–120.

KPMG (2002) *International Survey of Corporate Sustainability Reporting 2002*. www.kpmg.com/search/index.asp?cid=594 (accessed 24 July 2005).

Langford, D.A. & Hardcastle, C. (2004) *Changing the Culture of Construction*. Final Report to DETR. The Stationery Office, London.

Langford, D. & Murray, M. (2004) Procurement systems. In: *The Wiley Guide to Managing Projects* (eds P.W.G. Morris and J.K. Pinto), John Wiley & Sons, Hoboken, NJ; pp. 654–677.

Langford, D., Martinez, V. & Bititci, U. (2003a) Best value in construction: towards an interpretation of value from a client and contractor perspective. *Journal of Construction Procurement*, **9**(1), 56–67.

Langford, D.A., Kennedy, P., Conlin, J. & McKenzie, N. (2003b) Comparison of construction costs on motorway projects using measure and value alternative tendering initiative contractual arrangements. *Construction Management and Economics*, **21**, 831–840.

Larson, E. & Drexler, J.A. (1997) Barriers to project partnering: report from the firing line. *Project Management Journal*, **28**(1), 46–52.

Latham, M. (1993) *Trust and Money: Joint Review of Procurement and Contractual Arrangements in the United Kingdom Construction Industry*, Interim Report. HMSO, London.

Latham, M (1994) *Constructing the Team: Final Report of the Government/Industry Review of Procurement and Contractual Arrangements in the UK Construction Industry*. HMSO, London.

Le Marechal, N. (2004) E-auctions are good for contractors: viewpoint. *Construction News*, 27 May, p. 13.

Male, S. (2003) Future trends in construction procurement: procuring and managing demand supply chains in construction. In: *Management of Procurement* (ed. D. Bower). Construction Management Series. Thomas Telford, London; pp. 228–249.

Ministry of Works (1944) *Report of the Committee on the Placing and Management of Building Contracts (The Simon Report)*. HMSO, London.

Minogue, A. (1993) CM's 'problems' are the industry's. *Building*, 23 July, 26–27.

Minogue, A. (1996) Managing the future. *Building*, 12 June, 36.

Morgan, G. (1986) *Images of Organisations*. Sage, London.

Mosey, D. (2000) *Guide to ACA Project Partnering Contracts PPC 2000 and SPC 2000*. Association of Consultant Architects, Bromley.

Mott Macdonald (2002) *Review of Large Public Procurement in the UK*. A report commissioned by HM Treasury, London.

Murray, M. & Langford D.A. (eds) (2003) *Construction Reports 1944–1998: How the Government Has Shaped the Post-War Construction Industry*. Blackwell, Oxford.

Naoum, S. (1990) *A comparison of management contracting and traditional methods of procurement of commercial buildings*. PhD Thesis (unpublished), Brunel University.

National Audit Office (1996) *Progress in Completing the New British Library*. Report by the Comptroller and Auditor General, HC362 Session 1995–1996. HMSO, London.

National Audit Office (2002a) *Construction of Portcullis House, the New Parliamentary Building*. Report by the Comptroller and Auditor General, HC 750 Session 2001–2002. HMSO, London.

National Audit Office (2002b) *The Construction of Nuclear Submarine Facilities at Devonport*. Report by the Comptroller and Auditor General, HC 90, Session 2002–2003. The Stationery Office, London.

Nicolini, D., Tomkins, C., Holti, R., Oldman, A. & Smalley, M. (2000) Can target costing and whole life costing be applied in the construction industry? Evidence from two case studies. *British Journal of Management*, **11**(4), 303–324.

Nicolini, D., Holti, R. & Smalley, M. (2001) Integrating project activities: the theory and practice of managing the supply chain through clusters. *Construction Management and Economics*, **19**, 37–47.

Palast, G. (2000) War on corruption/not quite minister. *The Observer*, 9 July, Business, p. 5.

Pearce, D. (2003a) *The Social and Economic Value of Construction: the Construction Industry's Contribution to Sustainable Development*. New Construction Research and Innovation Strategy Panel (nCrisp) Report, London.

Pearce, D. (2003b) Environment and business: socially responsible but privately profitable? In: *The Challenge of Change: Fifty Years of Business Economics* (ed. J. Hirst). Profile Books, London; pp. 54–65.

Reijniers, J.J.A.M. (1994) Organization of public–private partnership projects: the timely prevention of pitfalls. *International Journal of Project Management*, **12**(3), 137–142.

Rowlinson, S. (1986) *Comparisons between traditional and design build contracts for industrial buildings*. PhD Thesis (unpublished), Brunel University.

Rowlinson, S. & Kvan, T. (2004) Procurement systems. In: *Architect's Handbook of Construction Project Management* (eds M. Murray & D. Langford), RIBA Enterprises, London; pp. 141–157.

Ruskin, J. (1849) *The Seven Lamps of Architecture*. Smith Elder, London.

Sleep, K (1997) New British Library: a comedy of errors. *Construction News*, **6**(March), 48–49.

Smith, A. (1776) *An Enquiry into the Nature and Causes of the Wealth of Nations*. W. Strathan and T. Cadell, London.

Sonmez, M., Yang, J.B. & Holt, G.D. (2001) Addressing the contractor selection problem using an evidential reasoning approach. *Engineering Construction and Architectural Management*, **8**(3), 198–210.

Spencely, J.D. (2000) *Report on the Holyrood Project to the Scottish Parliamentary Corporate Body*. March 2000, SP Paper 99, Session 1, Annexe 1, Edinburgh. www.scottish.parliament.uk/business/businessBulletin/bb-00/holy-04.htm (accessed 1 June 2000).

Spring, M. (2003) In business: talking 'bout evolution. *Building*, 3 October, 47–48.

Stansbury, N. (2003) *Anti-corruption Initiative in the Construction and Engineering Industry, Introductory Report*. Transparency International (UK). www.transparency.org/integrity_pact/dnld/anti_corr_initiative_constr_engineer.tiuk.pdf (accessed 24 June 2005).

Stein, A., Hawking, P. & Wyld, D.C. (2003) The 20% solution? A case study on the efficacy of reverse auctions. *Management Research News*, **26**(5), 1–20.

Strategic Forum for Construction (2002) *Accelerating Change*. Rethinking Construction & The Construction Industry Council, London.

Tookey, J., Murray, M., Hardcastle, C. & Langford, D.A. (2001) Construction procurement routes: redefining the contours of construction procurement. *Journal of Engineering, Construction and Management*, **8**(1), 20–30.

Treacy, M. & Wiersema, F. (1997) *The Discipline of Market Leaders: Choose your Customers, Narrow your Focus, Dominate your Market*. Harper Collins, London.

United Kingdom Parliament (2001) *Select Committee on International Development Appendices to the Minutes of Evidence, Appendix 1 Recent cases of Corruption Involving UK Companies and UK-Backed International Financial Institutions*. www.parliament.the-stationeryoffice.co.uk/pa/cm200001/cmselect/cmintdev/39/39ap06.htm (accessed 12 July 2004).

Walker, D.H.T. & Walker, B.M. (1999) Organisational learning as a vehicle for improved building procurement. In: *Procurement Systems: A Guide to Best Practice in Construction* (eds S. Rowlinson & P. McDermott). Routledge E. & F.N. Spon, London; pp. 99–137.

Ward, S.C., Curtis, B. & Chapman, C.B. (1991) Advantages of management contracting – critical analysis. *Journal of Construction Engineering and Management*, **117**(2), 195–211.

Endnotes

[i] Data from the report *Demonstrating Excellence* (Constructing Excellence 2004) shows that from the 134 active projects at that time, 72 per cent involved better improvements through some form of partnering. Partnering has been the main process innovation within the demonstrations programme.

[ii] Prime contracting is a procurement method piloted by the Ministry of Defence (MoD) Estates Division. Its aim was to establish an integrated supply chain under the single point responsibility known as the prime contractor. The two pilot projects involved Laing and Amec and went under the guise of 'Building Down Barriers' (see Holti *et al.* 2000; Nicolini *et al.* 2000, 2001) and www.defence-estate.mod.uk.

[iii] The aim of ProCure 21 is to promote better capital procurement in the National Health Service. This will be achieved through delivering better quality healthcare buildings and improved value for money by invoking a major cultural change 'vis-à-vis' the adoption of *Rethinking Construction* ideology (Egan 1998). The 'framework' system involves three types of partners: principal supply chain partners (PSCPs); primary supply chain members (PSCM) and supply chain members (SCM) (see www.nhsestates.gov.uk).

[iv] 'Snagging' is a term commonly used throughout the construction industry and is understood to mean post-completion defect rectification. From a constructor's perspective, these defects are often considered to be inconsequential. However, the demand side of the industry continues to impose on itself a 'defects liability' culture and its associated 'retentions', and this appears to condone the existence of snagging. Paradoxically, Hughes *et al.* (1998) point out that as 50% of the retention fund is released at practical completion, it seems unlikely that its sole purpose is to prevent snagging.

[v] The Warwick Manufacturing Group (WMG) is a group of professionals working as an integral part of the University of Warwick, but with independence of operation. Their mission is to enable companies to realise their full potential through the application of 'state of the art' business processes, methodologies and technology. This is to be achieved through education (full and part-time), training, research and consultancy operating on a worldwide basis (www.wmg.warwick.ac.uk/index.php).

[vi] For example, PFI contractor Amey did not put bid costs down as a loss in its accounts. It tucked them away on the balance sheet; thus the revenue position for the business could not be seen. Bid costs should have been written off as an expense (Accountancy Standards Board). Amey had to admit it was spending £1m per month and this turned profits of £55m into £18m losses.

[vii] Transparency International (TI) was formed in 1993; it is a non-profit making, independent, non-government organisation dedicated both to increasing awareness of the economic and human damage that results from corruption, and to curbing corruption. It works in a non-confrontational way with government, business and civil society (see www.transparency.org.uk).

[viii] Reverse auctions and e-tendering are terms used to describe a tendering process that does not involve paper and envelopes. Companies willing to offer their service link on to the Internet site where a 'live' competitive auction takes place. However, this fundamentally differs from a traditional auction hall at, for example, Sotheby's. Instead of seeking the highest price for a product, the true objective of this mechanism is to drive the price down.

Part 2
Corporate Milieu

Introduction

All commercial organisations exist for a purpose (usually to generate a profit); therefore, they will have a governance structure and a strategy (even if this strategy is to have no strategy at all) to achieve this aim. This part covers the interfaces between the organisation at a corporate level and both the projects milieu (the environment within which individual projects collectively reside) and the external milieu addressed in the previous part. In this context, a major influence of the external environment is in the imposition of legal and regulatory frameworks, requiring the establishment of a corporate governance system. Likewise, in order to fulfil the organisation's objectives in the light of external factors such as markets, purchasers, competitors and suppliers, there is a requirement for the development of an organisational strategy, again, which is linked to its structure and governance. Within these inter- and intraorganisational relationships the concept of trust is important. The areas of corporate governance, strategy, marketing, trust and outsourcing have been selected for further explanation.

Corporate governance

A corporate governance system is the combination of internal, external and regulatory mechanisms that ensures that the organisation is run for the benefit of one or several stakeholders, such as shareholders, creditors, suppliers, purchasers (clients), employees and other parties with whom the firm conducts its business. Optimally, the corporate governance system ensures sufficient returns on investment and an increase in corporate value. In Chapter 5, Marc Goergen and Luc Renneboog provide an all encompassing account of the available mechanisms that help to ensure that management run an organisation in the best interest of its principles. Taking a cross-European perspective they show how differences in legislation and regulatory frameworks, culture and tradition influence corporate governance systems, and hence how an organisation operates.

Strategy

The concept of strategy could be construed as a set of commitments made by an organisation that define and rationalise its objectives and fashion the way in which it seeks to fulfil them. An organisation's strategy tends to define the desired organisational structure in a general way; thus the chosen strategy will influence how the organisation is structured and governed, and how decisions are made and implemented, i.e. it largely determines what it does. In Chapter 6, Andrew Davies and Michael Hobday consider a strong emerging trend, that of firms changing the focus of their business strategies from being producers of physical products to providers of services. Drawing upon in-depth case-study research, they illustrate how suppliers of capital goods have developed strategies for integrated solutions by changing their position in the industry value stream, thereby transforming their organisational structures thereafter.

Marketing

An integral part of any business strategy is to ensure future work for the enterprise. Most firms therefore will, in one way or another, try to market themselves towards present and future purchasers (clients). These activities can be undertaken in a multitude of ways; however, organisations need to ensure that marketing effort is in their best interest and generating a return for the company. In Chapter 7, Christopher Preece, Krisen Moodley and Michael Brown examine marketing concepts and techniques within a business-to-business environment. Using the construction industry as an example, they illustrate how firms have adopted various strategies in an attempt to generate business. They then proceed to discuss how these strategies can be evaluated and measured.

Trust

Trust is a complex phenomenon. The term is highly ambiguous, not least because of its association with people as well as to circumstances and context. Accordingly, there are many ways of viewing trust, yet few would refute its importance for inter-, or indeed, intraorganisational relationships. In Chapter 8, Will Swan, Peter McDermott and Malik Khalfan deal with issues surrounding trust within a business-to-business environment. In particular, they draw upon the findings from an exploratory study into the role of trust within construction teams. They highlight several key factors influencing trust and, with the aid of four case studies, illustrate how different project arrangements influence the level of trust on a project and how trust impacts on commercial managers.

Outsourcing

The decision to outsource an activity presents managers within an organisation with complex choices. Any significant change in the number of activities that are outsourced will have an impact on the structure of the affected divisions/departments and most probably the organisation as a whole. Assessing the merits and consequences of outsourcing a service is, therefore, far from a trivial matter. In Chapter 9, Jan Bröchner provides a comprehensive and thorough overview of studies carried out on outsourcing. Covering a wealth of academic work in the area, he critically examines the origins of the concept, several of its applications, and the theories and methods used in researching outsourcing.

The contributors' profiles

Marc Goergen

Professor Marc Goergen has held lectureships at the Manchester School of Management (UMIST), the Manchester School of Accounting and Finance (University of Manchester) and the ISMA Centre at the University of Reading. In 2005, he took up a chair in finance at the University of Sheffield School of Management. His research interests are corporate governance, corporate ownership and control, initial public offerings, mergers and acquisitions, corporate investment models and dividend policy. He has published papers in *European Financial Management*, the *Journal of Corporate Finance*, the *Journal of Business Finance and Accounting*, the *Journal of Corporate Law Studies* and the *Journal of Law, Economics and Organization*. He has written a book on *Corporate Governance and Financial Performance* (Edward Elgar) and another on *Dividend Policy and Corporate Governance* (Oxford University Press). He has also contributed chapters to several edited books. Marc is a fellow of the International Institute of Corporate Governance and Accountability at the George Washington University and a Research Associate of the European Corporate Governance Institute.

Luc Renneboog

Dr Luc Renneboog is an associate professor at the Department of Finance of Tilburg University and research fellow at the CentER for Economic Research (Tilburg) and the European Corporate Governance Institute (ECGI) in Brussels. He held appointments at the Catholic University of Leuven and Oxford University, and visiting appointments at European University Institute (Florence), London Business School, HEC (Paris), Venice University, and Centro Universitario de Estudios Financieros (Madrid). He graduated from the Catholic University of Leuven with degrees in management engineering (BSc/MSc) and in philosophy (BA), from the University of Chicago with an MBA in finance and from the London Business School with a PhD in Financial Economics. He has published widely on corporate finance in the *Journal of Financial Intermediation*, *Journal of Corporate Finance*, *Journal of Law, Economics and Organization*, *Journal of Banking and Finance*, *Cambridge Journal of Economics*, and others. Luc co-authored the book *Dividend Policy and Corporate Governance* (2003) and co-edited the books *Corporate Governance Regimes: Convergence and Diversity* (2002) and *Venture Capital Contracting and the Valuation of High Technology Projects* (2003), all published with Oxford University Press. His research interests are corporate finance, corporate governance, dividend policy, insider trading, corporate restructuring, financial distress and the economics of art.

Andrew Davies

Dr Andrew Davies is a principal research fellow at the Innovation Studies Centre, Tanaka Business School, Imperial College, London. Previously, he was Senior Lecturer and Senior Research Fellow at the Science and Technology Policy Research Unit (SPRU), Sussex University, and Deputy Director of the ESRC Complex Product Systems (CoPS) Innovation Centre at the Universities of Sussex and Brighton. He is co-author of *The Business of Projects: Managing Innovation in Complex Products and Systems*, to be published by Cambridge University Press in 2005, co-editor of *The Business of Systems Integration*, published in 2003 by Oxford University Press, and author of *Telecommunications and Politics* published by Pinter in 1994. He holds a BA (First Class) in economic geography and an MA (Distinction) and DPhil in political economy from the University of Sussex. He was previously a post-doctoral fellow at the University of Amsterdam (1991–1994), a research fellow at the Rotterdam School of Management, Erasmus University (1998), and a visiting research fellow at the Technology and Innovation Management Centre, Queensland University in Australia (2003). His academic research has been published in journals such as *Research Policy, Industrial and Corporate Change* and *Organization Studies*. His policy research has been published in high-profile journals such as *Demos Quarterly, New Economy, ITPS Report* and *Telecommunications Policy*.

Michael Hobday

Professor Michael Hobday is Co-Director of the ESRC Complex Product Systems (CoPS) Innovation Centre at SPRU, University of Sussex. He is also Professor of Innovation at CENTRIM, University of Brighton. Michael has worked at SPRU since 1984, focusing on innovation management and East and South Asian innovation studies. He is co-author of *The Business of Projects: Managing Innovation in Complex Products and Systems*, to be published by Cambridge University Press in 2005, co-editor of *The Business of Systems Integration* published in 2003 by Oxford University Press, and author of *Innovation in East Asia*, published in 1995 by Edward Elgar. This book provided the first comprehensive analysis of East Asian firm innovation strategies, including detailed case studies of companies from Korea, Taiwan, Singapore and Hong Kong. He holds a BA (First Class), an MA in economics and a DPhil in telecommunications policy from the University of Sussex. Over the past decade he has been the Principal Investigator on three major CoPS projects funded by the UK Engineering and Physical Sciences Research Council. He has produced more than 140 publications, including five books, numerous journal articles on technology management, and policy reports for governments.

Christopher Preece

Dr Christopher N. Preece is leading research within the School of Civil Engineering at the University of Leeds, in the fields of business and marketing management in the construction industry. In collaboration with a number of leading edge civil engineering and building companies and consultancies, the focus of these studies is on increasing client satisfaction and improving the business culture of the construction industry through more effective and competitive strategies. His interests include procurement and the pre-qualification process, corporate communications, human resource management and construction education. He has published widely, delivered keynote addresses, chaired and organised conferences, nationally and internationally. His professional involvement includes national committee work for the Chartered Institute of Building. He is also a Member of the Chartered Institute of Marketing.

Krisen Moodley

Krisen Moodley, BSc, MSc, is programme leader for MSc taught courses in Construction Management, in the School of Civil Engineering, University of Leeds. After graduating from the University of Natal, his initial employment was as a quantity surveyor with Davis Langdon Farrow Laing in southern Africa, before his first academic appointment at Heriot-Watt University. He spent four years at Heriot-Watt before joining Leeds in 1994. His research interests are concerned with the strategic business relationships between organisations and their projects. Other specialist research interests include procurement, project management and corporate responsibility. Krisen has published and presented papers on these subjects both nationally and internationally. His recent books include: *Corporate Communications in Construction* and *Construction Business Development: Meeting New Challenges, Seeking Opportunity*, and chapter contributions to *Construction Reports 1944–1998*, and *Engineering Project Management*.

Michael Brown

Michael Brown graduated with a first class honours MEng degree in civil engineering with construction management in the School of Civil Engineering, University of Leeds. The culmination of Michael's studies was a research project into marketing strategies and spending within the construction industry. This aimed to identify current trends and philosophies within a range of construction related organisations pertaining to marketing spending and strategies. Since graduating, Michael has worked as an assistant project manager for White Young Green in Leeds and has recently transferred to the structural engineering department to gain a variety of experience within a civil engineering consultancy.

Will Swan

Dr William Swan is currently working as a Senior Project Manager with the Centre for Construction Innovation (North-West). In addition to completing his doctorate in knowledge management in 2002, he has been the Senior Research Fellow on projects concerning collaborative environments in construction, and latterly, on the Trust in Construction project that is reported here.

Peter McDermott

Dr Peter McDermott is a Management Board Member of the Salford Centre for Research and Innovation (SCRI). He is the Principal Investigator for the *Trust in Construction* and *Integrating the Supply Chain* research projects that have informed the development of the chapter published here. Through Salford he is also managing funded research projects against the *Rethinking Construction: People and Culture*, and doctoral studies including design quality indicators, a critique of rethinking construction, lean construction, public sector infrastructure projects, and procurement as a driver for change in construction. He has recently taken on the Editorship of the *International Journal of Construction Procurement*.

Malik Khalfan

Dr Malik M.A. Khalfan graduated with a first class degree in civil engineering from NED University, Karachi, Pakistan, in 1998. He worked briefly as a site engineer before undertaking his postgraduate degree in construction project management, followed by research on concurrent engineering in construction, both at Loughborough University, UK. On completion of his PhD in 2001, he joined Loughborough University and was involved in an EPSRC funded project as a Research Associate. Currently, he is working with Salford Centre for Research and Innovation at the University of Salford as a Research Fellow. Dr Khalfan's research interests include concurrent engineering (CE), readiness assessment of CE in construction, sustainable construction, knowledge management in construction, supply chain management, and partnering and strategic alliances within the construction industry. He was the winner of the CIOB Innovation Award 2000 in the best research paper competition.

Jan Bröchner

Professor Jan Bröchner was born in Stockholm, Sweden, in 1948. After graduating with an MSc (CE) from the Royal Institute of Technology (KTH) and receiving a BA from Stockholm University, he completed his PhD at KTH with a thesis on economic aspects of housing rehabilitation. During the 1980s, he combined teaching and research with work for a large Swedish construction contractor, developing specialist contracts related to environmental and occupational hazards. He has also been a research manager for the Development Fund of the Swedish Construction Industry (SBUF). In 1998, he was appointed to the Chair of Organization of Construction, at Chalmers University of Technology in Göteborg, Sweden. Jan Bröchner has published a number of articles in international journals and has contributed chapters to several books. He serves as a regional editor for *Facilities* and is a member of the editorial board for *Construction Management and Economics*. A key theme for his research is how technical uncertainty, technical development and human expertise are handled in business relationships that affect durable, fixed and complex physical resources.

5 Corporate Governance and Shareholder Value[i]

Marc Goergen and Luc Renneboog

Introduction

A corporate governance system is the combination of mechanisms which ensure that the management (the agent) runs the firm for the benefit of one or several stakeholders (principals). Such stakeholders may cover shareholders, creditors, suppliers, clients, employees and other parties with whom the firm conducts its business. Becht *et al.* (2003) considered that corporate governance rules are the outcome of the contracting process between the management and the principals. Agency problems arise when management violates contracts with the principals by running the firm against the interests of the principals. The efficiency criterion varies with the corporate governance system. Whereas in many Continental European countries (e.g. Germany) the definition of corporate governance explicitly mentions stakeholder value maximisation, the UK and US view – expressed by, for example, Shleifer and Vishny (1997) – focuses on generating a fair return for the providers of finance. Contrary to Shleifer and Vishny, Jensen and Meckling (1976) showed that the maximisation of shareholder value is equivalent to economic efficiency, provided that the following conditions are fulfilled: (i) the firm is considered as a nexus of complete contracts with creditors, employees, customers, suppliers and other relevant parties, (ii) the shareholders can claim the residual returns after all other contractual obligations have been met and (iii) there are no agency problems.

There are numerous corporate governance devices to ensure economic efficiency. They include the market for corporate control, large shareholder monitoring, changes in blockholdings, creditor monitoring, internal control mechanisms such as the board of directors and various committees, executive compensation contracts, and the corporate law regime and stock exchange rules. Goergen *et al.* (2005) have argued that dividend policy is another important corporate governance device: a high payout policy pre-commits managers to generate sufficient cash flows to pay the required dividend to the shareholders. As such, the dividend payout policy can be a substitute to the above-listed corporate governance mechanisms. Conversely, some of the corporate governance devices may be complementary to a high dividend payout policy. For example, one needs the presence of a large shareholder or a strong board of directors to impose and enforce such a dividend policy. Similarly, firms have used high dividend payout policies to fend off hostile takeovers.

This chapter has five sections. The second section discusses the differences in ownership and control concentration across countries. Not only does concentration differ across countries, but so does the concentration by type of shareholder. It is important to discuss such differences, as the distribution of ownership and control has important implications on the range of possible agency problems. The third section shows that control is not necessarily the same as ownership, as there are several mechanisms which cause deviations from the one-share–one-vote principle. Such deviations may cause or correct certain types of agency problems. The fourth section reviews the theoretical models on the various alternative corporate governance mechanisms and the empirical evidence on their efficiency. The fifth section concludes and presents an agenda for further research in this area.

The patterns of ownership and control

Differences in control concentration across countries

There is now ample empirical evidence that corporate control differs substantially across countries. The European Corporate Governance Network (ECGN),[ii] led by Barca and Becht (2001), was the first to carry out an exhaustive cross-country study on control patterns. Their focus is on control[iii] which may be distinct from ownership for two reasons. First, not all shares grant voting rights as, for example, non-voting shares do not confer any control rights to their owners. Second, as control is not necessarily exercised directly, i.e. at the first tier, Barca and Becht (2001) take not only direct stakes into account, but also indirect ones, as they accumulate all voting stakes controlled directly or indirectly by the same ultimate shareholder or investor group.[iv] In the remainder of this section, we discuss control rather than ownership.

In most Continental European companies, the majority of the voting equity is held by one shareholder or a shareholder group (Fig. 5.1), and in the majority of firms

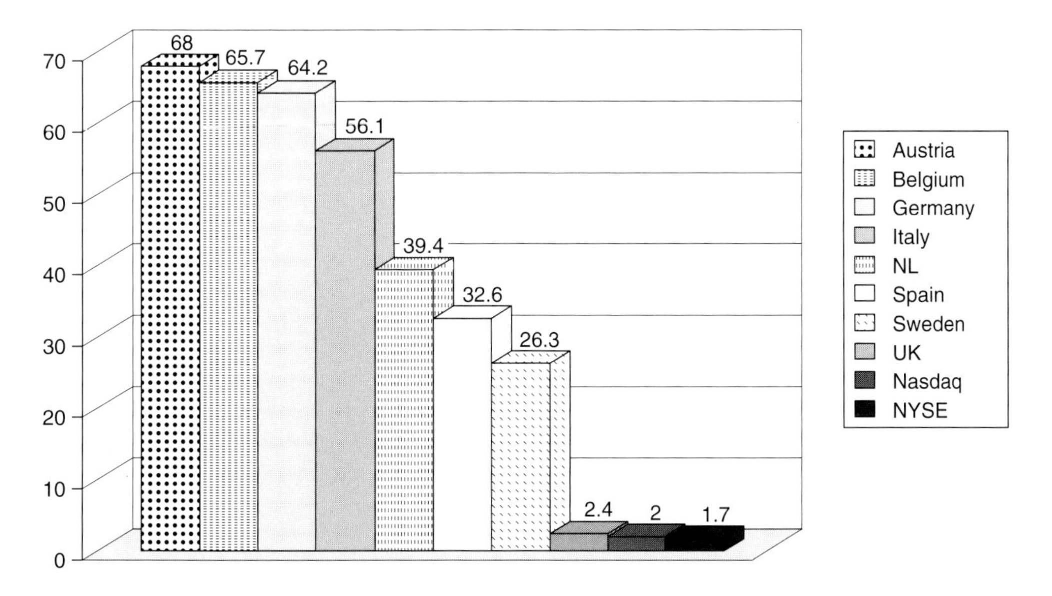

Figure 5.1 Percentage of listed companies under majority control (source: Barca & Becht 2001).

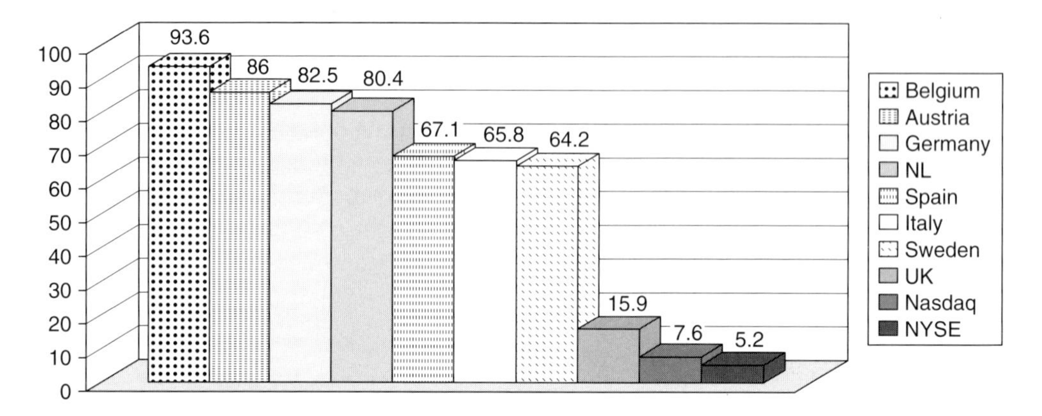

Figure 5.2 Percentage of companies with a blocking minority of at least 25% (source: Barca & Becht 2001).

there is a shareholder who possesses at least a blocking minority, i.e. 25 per cent (Fig. 5.2). Conversely, the UK and USA have dispersed equity and broad discretion on the part of management to run the business.

Table 5.1 confirms the high degree of control concentration in Continental Europe. In Austrian, Belgian, French, German and Italian companies, a single shareholder (or a group) owns an absolute majority of the voting shares. This is in sharp contrast to the UK where the largest shareholder controls on average only 14 per cent. In the USA, the dilution of ownership and control is even more pronounced. Whereas in

Table 5.1 Distribution of largest shareholders in Europe and the USA.[a]

				Shareholdings			
	Year	**Sample**[b]	**Note**[c]	**Largest**	**2nd largest**	**3rd**	**4–10th**
Austria	1996	600	(1)	82.2	9.5	1.9	6.5
France	1996	403	(1)	56.0	16.0	6.0	5.0
Italy	1996	214	(1)	52.3	7.7	3.5	5.1
Netherlands	1996	137	(2)	28.2	9.2	4.3	7.1
Spain	1996	394	(1)	38.3	11.5	7.7	10.3
UK	1993	248	(2)	14.0	8.3	6.1	9.2
USA (NYSE)	1996	1309	(2)	8.5	3.7	1.8	0.9
USA (NASDAQ)	1996	2831	(2)	13.0	5.7	3.0	1.6
				Largest	**2nd + 3rd**	**4 + 5th**	**6 – 10th**
Belgium	1994	135	(1)	55.8	6.9	0.6	0.2
Germany	1996	402	(1)	59.7	8.6	2.6	0.3

[a] This table gives the average size of the largest control stakes (stakes of voting equity) for European countries and the USA.
[b] The sample companies in all countries are listed; the Austrian sample consists of both listed and non-listed companies.
[c] (1) Both direct and indirect shareholdings are considered; (2) only direct shareholdings.
Source: research by European Corporate Governance Institute; Barca and Becht (2001).

Continental Europe a coalition of the three largest shareholders votes for more than 60 per cent, in the UK and USA a similarly-sized coalition owns a mere 30 per cent of the votes.

The question arises as to why the levels of control are so substantially different from country to country. Goergen and Renneboog (2003) have provided some answers to this question. They studied how control evolves subsequent to the initial public offering (IPO) of firms from two countries with substantially different corporate governance regimes: Germany (a system with strong control, complex control structures, few listed companies and weak shareholder protection) and the UK (a market-based regime with many listed firms, diffuse ownership and strong shareholder rights). Probably the primary reason why shareholders hold larger voting stakes in German firms is the different legal system and regulation. Analysing the German and UK stock exchange regulations, rules on minority shareholder protection, informational transparency and the takeover codes, Goergen and Renneboog (2003) concluded that shareholders are less protected in Germany. The casting of votes at annual meetings, the structure of the board of directors and fiduciary duties of directors further weaken the rights of German shareholders. Hence, this makes control more valuable to shareholders of German firms who want to avoid expropriation of their investments and/or to take advantage of private benefits of control. Furthermore, holding large control stakes is less expensive in Germany than in the UK, because ownership pyramids, the use of non-voting shares and the appointment of one's representatives to the board of directors make it possible to have control with relatively low levels of cash flow rights.

Although the characteristics of the legal system may explain the higher levels of control in Germany, they do not necessarily explain how the differences in control concentration come about. Given that firms from different systems are all exposed to increasing rates of competition at the global level, it is surprising that corporate governance regulation is still so different (McCahery *et al.* 2002). Goergen and Renneboog (2003, 2005) explained the lack of convergence by economic characteristics which determines control retention by the initial shareholders after the IPO, the dispersion of control and control transfers. They distinguished between control transfers to bidders that are ultimately widely held and those that are closely held. For a sample of UK firms and a sample of German firms, which are floated on the stock exchange with initially high levels of family control, they demonstrated that the initial shareholders in the average German company not only had much larger stakes, but also kept majority control over the five years after the IPO. In contrast, the initial shareholders of UK companies had already lost majority control two years after going public. They found strong evidence that firm characteristics explain differences in the evolution of control, not only across firms from the same country but also across the two countries. They showed that firm size is an important determinant of control concentration in the UK but not in Germany. Whereas new shareholders of large German firms hold large voting stakes, large UK companies end up being widely held. The reason for this difference is that wealth constraints become binding for UK shareholders, whereas German shareholders can leverage control via ownership pyramids. If the founder of a German firm is still a shareholder at the IPO and if there are non-voting shares outstanding, control is likely to remain tightly in the hands of the initial shareholders. This is not surprising, as founding families often extract private benefits of control and non-voting shares help them obtain additional financing whilst keeping control at the same time. Whereas growth has no impact on the control of UK firms, high-growth German firms experience transfers of control to new large shareholders.

Distribution of control across categories of shareholders

As the mandatory disclosure threshold is 5 per cent of the voting equity in European countries, except the UK (3%) and Italy (2%), this subsection only considers holdings of at least 5 per cent. Table 5.2 shows that not only is control different across countries, but so are the main categories of shareholders. The main shareholders are grouped into (i) institutional investors (banks, insurance companies, investment and pension funds), (ii) individuals (excluding directors) or families, (iii) directors and their families and trusts, (iv) industrial and holding companies, and (v) the federal or regional governments. We add up the voting blocks (of 5 per cent or more) held directly (by holding shares directly in a target firm) and indirectly (via other companies) by each of these categories.

Industrial and holding companies

In Belgium and France, industrial and holding companies control on average 38 per cent and 35 per cent of the shares, respectively (Table 5.2). They control an average stake of 21 per cent in listed German firms. Similarly, the corporate sector holds high levels of control in Austrian, Italian and Spanish firms. In contrast, the industrial and commercial sector of the UK and The Netherlands controls only 6 per cent and 11 per cent, respectively, of listed companies.

Banks and other institutional investors

Banks generally hold only small stakes in Continental European firms. On average, banks control between 0.4 per cent and 7.2 per cent in Belgium, Italy, The Netherlands and Spain. In France, although direct shareholdings amount to only about 2.7 per cent, overall bank control is strongest with an average of 16 per cent of the votes (Table 5.2). In Germany, only 5.8 per cent of the large voting stakes of 5 per cent and more are held (directly as well as indirectly) by banks, resulting in an average of 1.2 per cent of the votes (Table 5.2). However, the influence of banks in Germany is higher than these numbers suggest as banks are normally granted the right to exercise the votes on shares deposited by the banks' customers. Most German companies also tend to have a housebank (*Hausbank*) which often exerts a major influence on the firm (via, for example, board membership). In Anglo-American countries, in contrast, bank control is almost negligible.

Goergen and Renneboog (2001) argued that the aversion to and avoidance of potential conflicts of interest is the main reason why bank holdings in listed companies are relatively unimportant in most countries. Banks owning shares in

Table 5.2 Distribution of largest shareholders in Europe.[a]

	Sample[b]	Year	Control[c]	Individuals and families	Banks	Insurance companies	Investment funds	Holding and industrial companies	State	Directors
Austria	600	1996	(1)	38.6	5.6	0.0	0.0	33.9	11.7	0.0
Belgium	155	1994	(1)	15.6	0.4	1.0	3.8	37.5	0.3	0.0
France	402	1996	(1)	15.5	16.0	3.5	0.0	34.5	1.0	0.0
Germany	402	1996	(1)	7.4	1.2	0.2	0.0	21.0	0.7	0.0
Italy	214[b]	1996	(1)	68.6	7.2	0.0	0.0	24.2	0.0[d]	0.0
Netherlands	137	1996	(2)	10.8	7.2	2.4	16.1	10.9	1.3	0.0
Spain	394	1996	(1)	21.8	6.6	8.8	0.0	32.6	0.0	0.0
UK	248	1993	(2)	2.4	1.1	4.7	11.0	5.9	0.0	11.3

[a] This table gives the total large shareholdings of voting rights (over 5%) held by different investors classes.
[b] Numbers for Italy refer to both listed and non-listed companies; for other countries only listed companies are taken.
[c] (1) Both direct and indirect shareholdings are considered; (2) only direct shareholdings.
[d] Of the listed Italian companies about 25% are directly and indirectly controlled by state holdings; this is classified in the table under 'Holding and industrial companies'.
Source: research by the European Corporate Governance Institute; Barca and Becht (2001).

firms frequently also hold debt in these firms. However, each type of claim may require a different optimal decision process in the wake of financial distress. When the firm is in financial distress and the bank faces a refinancing demand by the firm, creditor claims may encourage the bank to push the firm file into liquidation, whereas the equity claims may lead the bank to revoke its loans.[v]

The main shareholders in Dutch and UK firms are investment trusts, pension funds and insurance companies. As a category, they hold on average 19 per cent and 16 per cent, respectively (Table 5.2).[vi] Given the lack of institutional blockholders in most Continental European countries, as compared to the UK and USA, one should have expected little 'shareholder activism' from institutional investors.[vii] Even in the UK and USA, evidence of monitoring by financial institutions is scarce because institutional investors avoid monitoring firms and gathering non-public information. Stapledon (1996) and Stapledon and Bates (2002) argued that, if they were to obtain such information, they would be prevented from trading by insider trading legislation and, as a result, the liquidity of their portfolio would be reduced. Monitoring the many firms making up their portfolios may also be prohibitively expensive.

Families and individuals

Table 5.2 shows that individuals and families are one of the main shareholder categories in Continental Europe.[viii] Family control is especially strong in the largest German firms (Franks & Mayer 2001). Becht and Boehmer (2001) have corroborated this pattern: in 37 per cent of their sample, individuals or families controlled, on average, 20 per cent of the votes. A particular category of individuals controlling share stakes is that of the directors who are insiders and therefore possess superior information on the firm's prospects. Very little is known about directors' control for the case of most Continental European countries as: (i) the stakes of most directors do not exceed the disclosure thresholds, (ii) although large family blockholders frequently nominate board members (which can be family members), the origin of board representation is not publicly disclosed, and (iii) the use of intermediate investment companies further obscures the picture of directors' control. In contrast, directors of UK firms are, along with institutional investors, the most important category of owners. Franks *et al.* (2001) showed that, in the few cases where executive directors held large stakes, their voting rights could lead to managerial entrenchment and resistance to disciplinary actions undertaken by other shareholders.

To summarise: (i) control concentration is substantially higher in Continental Europe than in the UK or USA; (ii) pyramids and other complex control structures in Continental Europe enable shareholders to retain control over long periods; (iii) the Continental European corporate sector owns large stakes in listed companies; (iv) in all the countries, banks – unless they are part of a financial group – hold in general only small stakes; (v) institutional investors and directors are the main shareholders in the UK, but are relatively unimportant in Continental Europe; and (vi) directors' control is comparatively high in the UK and can lead to managerial entrenchment.

Why and when is control not necessarily identical to ownership?

This section discusses the differences between control and ownership, and reviews the different devices that cause deviations from the one-share–one-vote rule. Whether or not the one-share–one-vote principle is upheld has important implications for the range of potential agency problems. Diffuse ownership combined with weak shareholder control may cause serious agency conflicts between management and shareholders because of the lack of monitoring. This situation may be especially problematic in countries with weak shareholder protection. Monitoring management may be prohibitively expensive for small shareholders as the monitor bears all the costs related to their control efforts but only benefits in proportion to their shareholding (Grossman & Hart 1980, 1988; Demsetz 1983). Hence, only a large holding provides sufficient incentives to monitor the management. Increased share liquidity and exposure of the company to the disciplining role of the market for corporate control are the main advantages of dispersed control. Conversely, strong ownership and voting power come with low liquidity, but the presence of a large controlling shareholder reduces the likelihood that managers will intentionally not meet the firm's objectives.

Panels A (dispersed ownership and control) and D (concentrated ownership and control) of Table 5.3 are the two basic cases where ownership and control coincide. One would expect these two cases to represent, respectively, most Anglo-American companies (dispersed ownership and control) and most Continental European and Japanese firms (concentrated ownership and control). However, in practice, panel D does not apply to most Continental European firms as the one-share–one-vote rule is not necessarily endorsed. There are several mechanisms which cause violations of the rule, such as ownership pyramids, shares with multiple voting rights and voting restrictions. Panel C is the case where the concentration of control is lower than that of ownership: this occurs because voting caps prevent large shareholders from having control. The use of voting caps improves the protection of small shareholders against expropriation by large shareholders, and can also be used when a company is fending off a hostile bidder. Franks and Mayer (1998) reported that in each of the three hostile takeover battles in Germany since World War II such voting limits were used to reduce the control held by several large shareholders from, for instance, 30 per cent to 5 per cent. In two out of three cases, the voting caps contributed to the failure of the takeover bid. The issuing of multiple voting shares was outlawed in Germany,[ix] Italy, Spain and the UK as of May 1998 (Faccio & Lang 2002; Goergen & Renneboog 2003).

Panel B of Table 5.3 shows the case of dispersed ownership with concentrated voting power. Whereas such a situation combines the advantage of large-shareholder monitoring with that of wealth diversification, it is also possible that the concentrated control will be misused to extract private benefits from minority shareholders. Most Continental European law systems provide a number of mechanisms that allow controlling shareholders to earn a return on their investments that exceeds the financial return by the private benefits of control. The most frequently used mechanism to exercise control with a limited investment is ownership pyramids which enables shareholders to maintain control throughout multiple layers of ownership

Table 5.3 Ownership and voting power.

	Control	
Ownership	**Weak**	**Strong**
Dispersed	Panel A	Panel B
Concentrated	Panel C	Panel D

Panel A: Dispersed ownership and weak voting power
Where: USA, UK
Advantages: (a) high potential for portfolio diversification and high liquidity; (b) existence of a takeover market
Disadvantages: insufficient monitoring; free-riding problem
Agency conflicts: management vs shareholders

Panel B: Dispersed ownership and strong voting power
Where: countries where a stakeholder can collect proxy votes, where shareholder coalitions are allowed, where non-voting shares are issued and where shareholding pyramids exist, e.g. in Continental Europe
Advantages: (a) monitoring of management; (b) portfolio diversification and liquidity
Disadvantages: (a) violation of one-share–one-vote; (b) reduced takeover possibility
Agency conflicts: controlling blockholders vs small shareholders

Panel C: Concentrated ownership and weak voting power
Where: any company with voting right restrictions, e.g. some German firms
Advantages: protection of minority rights
Disadvantages: (a) violation of one-share–one-vote; (b) low monitoring incentives; (c) low portfolio diversification possibilities and low liquidity; (d) high cost of capital
Agency conflicts: management vs shareholders

Panel D: Concentrated ownership and strong voting power
Where: Continental Europe, Japan, in any company after a takeover, in recently floated companies
Advantages: high monitoring incentives
Disadvantages: (a) low portfolio diversification possibilities and low liquidity; (b) reduced takeover possibilities
Agency conflicts: controlling blockholders vs small shareholders

whilst sharing the cash flow rights with other shareholders at each intermediate ownership tier (Fig. 5.3). Consequently, ownership pyramids reduce the liquidity constraints that large shareholders face while allowing them to retain substantial voting power. If shareholder C owns 51 per cent of the voting equity of firm B, which in turn owns 51 per cent of the voting equity of firm A, there is an uninterrupted control chain which gives shareholder C uncontested control at each tier of the pyramid. At the same time, the cash flow rights of shareholder C in firm A are a modest 26 per cent. This discrepancy between ownership and control is a characteristic of countries that allow the exploitation of private benefits of control.

Proxy votes is a second mechanism that ensures control with a limited amount of cash flow rights (panel B). As mentioned above, German banks are allowed to cast

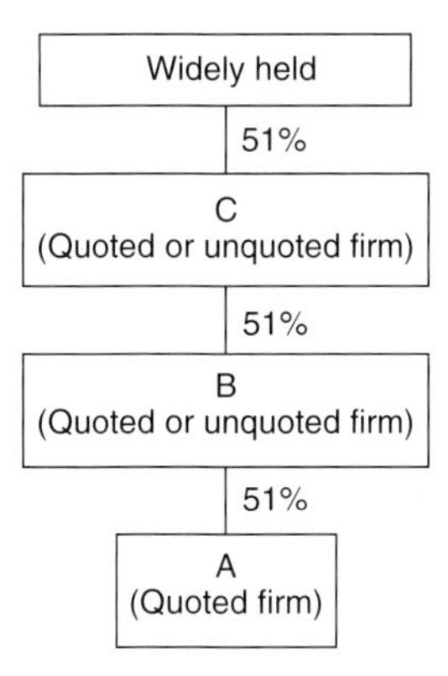

Figure 5.3 Example of a 'pyramid' structure. Firm A is one of the firms in our sample. At the first-tier of the pyramid, it is controlled by firm B (which holds 51 per cent of its voting shares), which can be a quoted or unquoted firm. At the next layer, we find that firm B is controlled by firm C (which holds 51 per cent of its voting shares), which again can be quoted or unquoted, but is in this case widely held. *Ultimately*, firm C controls firm A.

the voting rights of the shares deposited with them by the firm's shareholders. The exercise of the votes is conditional on the bank announcing its voting intentions regarding specific resolutions at the general meeting and on the absence of receiving alternative instructions from the depositors. Another example of proxy voting can be found in the USA where the management can make proposals to be voted on at the general meeting and solicit proxy votes for their support.

Starting from a situation characterised by both diffuse ownership and control (panel A), voting pacts is a third mechanism to achieve the case of panel B. Such pacts enable small shareholders to exercise a much higher degree of control jointly than individually. There is not much evidence of such shareholder coalitions, given their potentially substantial costs. For example, in the UK, a long-term shareholder coalition is treated as a single shareholder by the regulator and, as a consequence, has to comply with all the rules on mandatory tender offer, disclosure of strategic intent, etc. Hence, a coalition controlling at least 30 per cent of the votes would be compelled to make a bid for all the other shares. As a result, most coalitions are formed on an *ad hoc* basis with a specific aim (e.g. the removal of badly performing management). Crespi and Renneboog (2002) studied the power of shareholder coalitions in the UK by determining the relative voting power of each blockholder coalitions using Shapley values, which measure the extent to which individual shareholders (or groups of shareholders) are pivotal in potential winning voting coalitions. They concluded that such coalitions play a monitoring role in poorly performing UK firms.

A fourth mechanism separating ownership from control is dual class shares. According to Grossman and Hart (1988), a large stake generates benefits of control, which are made of private benefits and security benefits. The latter cover benefits of ownership and control concentration (for example, the positive effects from monitoring) that accrue to all the shareholders. However, large stakes have their own potential agency costs as the private benefits usually come at the expense of other shareholders or stakeholders. An example of such an agency cost is the 'squeezing-out' of minority shareholders at a price below the value of their shares in a tender offer, and the diversion of resources from shareholders to a blockholder (Zwiebel 1995; Pagano & Röell 1998; Johnson *et al.* 2000). The private benefits of control are usually non-transferable benefits beyond the financial return on investment. For example, a supermarket chain, by owning a large stake in a brewery, may have an important strategic advantage. As a large shareholder it will typically be represented on the brewery's board and will thus be able to obtain private information on the firm's cost structure or on supply contracts with competing supermarket chains. For example, after acquiring such strategic information, the large shareholder may renew negotiations about the price charged by the brewery for its products. Consequently, such transactions can lead to the creation of another kind of agency conflict, namely the oppression of minority shareholder rights. Take, for example, a shareholder who owns 51 per cent of the voting shares in firm A and also owns 100 per cent of the equity of another firm, firm B. If firm A is a supplier to firm B, the controlling shareholder may be tempted to reduce the transfer price of goods sold to firm B. Profits are then maximised at the level of firm B which the shareholder controls and all of whose cash flow rights he owns. At the same time, the minority shareholders of firm A lose out as its profits are not maximised. Renneboog (2000, p. 1991) gave the example of a blockholder of a holding company. He argued that the blockholder will most likely focus on the interests of the holding company rather than on those of the minority shareholders in the group's subsidiaries, for instance, by dividing the markets across the group's subsidiaries to prevent competition between them. For example, the French holding company Suez may divide the international utilities markets up between its subsidiaries, the French Lyonnaise des Eaux and the Belgian firm Tractebel. Such a strategy would reduce the investment value of the minority shareholders of these subsidiaries. Johnson *et al.* (2000) provided other examples of minority shareholder expropriation in the context of transition economies.

Correia da Silva *et al.*, (2004) argued that control over the dividend payout policy may also yield the large shareholder a private benefit of control, for example, by setting a dividend payout ratio which is optimal to their own specific tax situation but not to those of the majority of the smaller shareholders. Furthermore, a large shareholder, who has other ways of extracting funds from the company (such as in the above examples), may be tempted to set a low payout ratio. As a result, the monitoring or pre-commitment role of dividend policy is discarded at the expense of the minority shareholders.

European firms frequently have dual or multiple class shares. One class (B-shares) has fewer voting rights than the other class (A-shares). They are used by firms in Germany, Italy, Scandinavia and Switzerland, but are rarely used in the UK (Goergen & Renneboog 2001). Faccio and Lang (2002) showed that in Denmark, Finland, Italy, Sweden and Switzerland, the percentage of firms with dual class shares ranged from about 35 per cent to 65 per cent. For Austria, Germany, Ireland, Norway and the UK, the range varied from 13 per cent to 24 per cent (Table 5.4). In France, Portugal and Spain, the proportions are almost negligible. Since the 1980s, when stock exchanges liberalised the originally restrictive policy on multiple and

Table 5.4 Dual class shares in Europe.

Country	Number of firms	Number of firms with dual class shares	Percentage of firms with dual class shares
Austria	99	23	23.23
Belgium	130	0	0.00
Denmark	210	70	33.33
Finland	129	47	36.43
France	607	16	2.64
Germany	704	124	17.61
Ireland	69	16	23.19
Italy	208	86	41.35
Norway	155	20	12.90
Portugal	87	0	0.00
Spain	632	1	0.16
Sweden	334	185	55.39
Switzerland	214	109	50.93
UK	1953	467	23.91
Total	5531	1164	21.05

Source: based on Faccio and Lang (2002) and Bennedsen and Nielsen (2002).

dual class shares, dual class shares have become an increasingly important concern to investors in the USA. There is considerable diversity across Europe in terms of the specification of dual class shares. For example, in Sweden, each B-share has normally one tenth of the vote of an A-share. At the extreme, there are some countries where B-class shares have no voting rights. Moreover, firms from certain other countries (for example, Germany) can issue preference shares which are risk-bearing capital without votes, but with special dividend rights. A special case of a multiple voting share is a so-called 'golden share' which gives one or more shareholders (e.g. the government) a veto right in certain clearly defined situations.

Barclay and Holderness (1989) were the first who attempted to quantify the private benefits of control. Since their study, several other studies have attempted to do the same, using more or less sophisticated techniques. Most studies measure the extent of the private benefits by the premium paid for control blocks in publicly traded companies. Nicodano and Sembenelli (2000) and Trojanowski (2003) have argued that this methodology is inappropriate, since it neglects the firm's overall ownership and control structure. Along the lines of Zwiebel (1995), they argued that the fraction of control rights transferred in a block trade should be measured by changes in the relative strategic importance of shareholders. The relative strength of a shareholder is calculated by measuring the degree to which the shareholder is pivotal in potential voting coalitions, as proxied by changes in Shapley values. Rydqvist (1987) and Zingales (1994) analysed samples of companies with dual class stocks and measured control using power indices. Price comparisons of shares carrying different control rights allowed them to make inferences about the value of private benefits of control.

Dyck and Zingales (2004) measured the private benefits of control based on 412 control transactions in 39 countries during 1990–2000 and found that the value of control, measured by the difference between the price paid for the block and the share price two days after the announcement of the block transaction relative to the share price, ranged from –4 to +65 per cent with an average of 14 per cent. In some European countries, extremely large private benefits of control are evident (Table 5.5); for example, the mean premiums in Austria, Italy and Portugal are 38 per cent, 37 per cent and 20 per cent, respectively. In contrast, in most other European countries, the mean premium is below 10 per cent.

However, measuring private benefits of control is tricky, as the control premium depends on a number of factors, such as the level of competition in the market for corporate control (Dyck & Zingales 2004), the size of the block sold (Barclay & Holderness 1989; Zingales 1994, 1995), the dispersion of shares in the target firm

Table 5.5 Block premium as a percentage of firm equity.[a]

Country	Mean	Median	S.d.	Min.	Max.	Number of observations	Number of observations positive
Australia	0.02	0.01	0.04	–0.03	0.11	13	9
Austria	0.38	0.38	0.19	0.25	0.52	2	2
Canada	0.01	0.01	0.04	–0.02	0.06	4	2
Denmark	0.08	0.04	0.11	–0.01	0.26	5	3
Finland	0.02	0.01	0.06	–0.07	0.13	14	9
France	0.02	0.01	0.10	–0.10	0.17	5	3
Germany	0.10	0.10	0.13	–0.24	0.32	18	15
Israel	0.27	0.21	0.32	–0.01	0.89	9	8
Italy	0.37	0.16	0.57	–0.09	1.64	8	7
Japan	–0.04	–0.01	0.09	–0.34	0.09	21	5
Netherlands	0.02	0.03	0.05	–0.07	0.06	5	4
New Zealand	0.03	0.03	0.09	–0.17	0.18	19	14
Norway	0.01	0.01	0.05	–0.05	0.13	14	9
Portugal	0.20	0.20	0.14	0.11	0.30	2	2
Singapore	0.03	0.03	0.03	–0.01	0.06	4	3
Spain	0.04	0.02	0.06	–0.03	0.13	5	4
Sweden	0.06	0.02	0.08	–0.01	0.22	13	12
Switzerland	0.06	0.07	0.04	0.01	0.15	8	8
UK	0.02	0.01	0.05	–0.06	0.17	43	23
USA	0.02	0.02	0.10	–0.20	0.40	47	28
All the above	0.14	0.11	0.18	–0.04	0.48	412	300
Europe and the USA only	0.13	0.04	0.17	–0.03	0.44	200	140

[a] Block premiums are computed by taking the difference between the price per share paid for the control block and its market price two days after the announcement of the control transaction, and by dividing it by the market price of the block two days after the announcement and multiplying the ratio by the proportion of cash flow rights represented in the controlling block.

Source: based on Dyck and Zingales (2004).

(Barclay & Holderness 1992), the distribution of voting power across shareholders (Zwiebel 1995; Nicodano & Sembenelli 2000), the nationality of the buyer (Nenova 2003), and the firm's financial situation (Berglöf & Burkart 2002). In addition, the control premium is lower in the Anglo-American market-based corporate governance system characterised by dispersed ownership and good shareholder protection.

Corporate governance mechanisms

This section reviews the theoretical models on corporate governance mechanisms and summarises the main conclusions from empirical research. We focus on internal governance mechanisms, such as large blockholder monitoring and the structure of the board of directors, as well as external mechanisms, such as the market for corporate control, changes in control concentration and creditor monitoring, managerial pay-for-performance schemes and dividend policy.[x]

Internal mechanisms

Blockholders

The key question is whether blockholders enhance firm value. The increased monitoring of management by large blockholders is expected to create shareholder value (see, amongst others, Admati *et al.* 1994; Maug 1998; Kahn & Winton 1998). Furthermore, if directors own blockholdings, this should align their interests with those of the other shareholders. Morck *et al.* (1988) measured firm value by Tobin's Q, proxied by the market-to-book ratio of assets, and came up with a non-linear relation between Q and managerial ownership. Firm value increased as managerial ownership increased up to 5 per cent; it then decreased as managerial ownership increased from 5 to 25 per cent, and increased again above 25 per cent managerial ownership. Hence, at low levels, an increase in managerial ownership seems to bring about additional incentives for the managers to maximise shareholder value; at levels between 5 and 25 per cent, additional ownership leads to entrenchment; the entrenchment effect disappears if managerial ownership exceeds 25 per cent. McConnell and Servaes (1990) found a different relationship: Q increased in an inverse U-shaped fashion until it reached its maximum at 40–50 per cent of managerial ownership and then declined with higher levels of ownership. In addition, they reported a positive linear relation between firm value and the fraction of shares owned by institutional investors. McConnell and Servaes (1995) investigated the link between a firm's ownership structure, its growth potential and leverage. They showed that the distribution of equity ownership among insiders, institutions, corporate blockholders and atomistic shareholders was of marginally greater significance in low-growth than in high-growth firms.

Mehran (1995) showed that blockholders also have an impact on managerial remuneration, equity-based compensation being used less extensively in US firms with stronger outside blockholders, implying that blockholder monitoring acts as a substitute for equity-based compensation contracts. Conversely, in the case of Spanish companies, Crespi *et al.* (2002) reported that managerial remuneration contracts related to share price performance were complements to outside blockholdings, suggesting that in Spain a strong blockholder is needed to impose such contracts. Companies without outside blockholders have their managerial compensation contracts based on accounting performance which is subject to managerial control and may be subject to window-dressing. Renneboog and Trojanowski (2003) reached a similar conclusion for the UK.

However, concentrated ownership may also generate its own agency costs. Trojanowski (2003) criticised the above studies for failing to control these costs. There are at least four types of costs. First, control by a large shareholder may result in reduced risk sharing (Demsetz & Lehn 1985; Admati *et al.* 1994). Second, as shown above, ownership concentration reduces the liquidity of all the shares (Bolton & Thadden 1998). Third, the large blockholder of highly leveraged firms, especially those performing poorly and with high bankruptcy costs, may force management to take excessive risks; this may eventually lead to the expropriation of the debtholders (Jensen & Meckling 1976; Coffee 1991). Fourth, even when tight control by shareholders is efficient *ex post, ex ante* it constitutes an expropriation threat that reduces managerial incentives to exert effort and undertake value-maximising strategies (Burkart *et al.* 1997; Pagano & Röell 1998).

The above studies examine the link between ownership or control and financial performance. Another strand of the literature focuses on the question whether blockholders take appropriate actions, for example, in the case of poor corporate performance or financial distress. The answer depends not only on the concentration of ownership or control but also on its nature, as particular shareholder categories put a different value on control (Jensen & Meckling 1976). This argument has empirical support from the USA (Demsetz & Lehn 1985; Barclay & Holderness 1989, 1991; Holderness & Sheehan 1988). In the case of the UK, Franks *et al.* (2001) investigated whether the presence of blockholders in poorly performing companies increased the likelihood of board restructuring but found no such evidence. The only consistent and significantly negative relationship was between managerial ownership and executive board turnover. Franks *et al.* (2001) argued that this was clear evidence of managerial entrenchment as managers who had control were able successfully to fend off any attempts to remove them. Banerjee *et al.* (1997) reported no evidence that French holding companies (the main shareholder group in France) assumed a governance role; on the contrary, holding companies as major shareholders seemed to reduce firm value. Renneboog (2000) examined whether different types of blockholders monitored listed Belgian companies. The only blockholders having an influence on executive turnover were industrial and commercial companies which initiated board restructuring when the firm's accounting and stock performance decreased. To summarise, there is little empirical evidence on the corporate governance role assumed by large blockholders.

The above studies assumed that the current owners, and their corporate governance actions, drove future performance. However, Goergen (1998) argued that any conclusions regarding the direction of causality between the two are premature; Kole (1996) provided empirical support for this argument. Using the same sample as Morck *et al.* (1988) for each year between 1977 and 1985, Kole (1996) regressed the

Tobin's Q of each year on the same three board ownership variables measured in 1980 by Morck *et al.* There was a significant relationship between Q and 1980 ownership for 1977–1980, but no relationship between Tobin's Q for 1981–1985 and 1980 managerial ownership, suggesting that performance determined a firm's future ownership structure and not *vice versa*. Köke (2004) also favoured a reversal of causality. Himmelberg *et al.* (1999) gave the following example: suppose that there are two firms, A and B; firm A has a higher degree of market power than firm B. If competition has a disciplining effect on managerial decisions, then firm B requires less monitoring, whereas firm A requires more monitoring. Hence, firm A's management may receive more equity to align their interests with those of the other shareholders. However, due to its higher market power, firm A also has higher profit rates than firm B. Empirically, there would then be a positive correlation between managerial ownership and profitability. But the estimated coefficient of managerial ownership will reflect only spurious correlation, not a causal relationship. Therefore, the results of the empirical papers described in this section should be interpreted with caution.

Boards of directors

Research concerning the board of directors focuses on the following questions:

- Does board composition influence performance and growth?
- Does board composition have an impact on the likelihood of corporate governance actions? For example, are boards dominated by outsiders more likely to remove bad managers?
- What determines the board's composition? For example, does good performance lead to the appointment of more insiders to the board, and *vice versa*; does poor performance lead to the appointment of more independent directors?

Board composition and corporate performance

Numerous studies concentrate on the link between board independence – proxied by the number of outside directors – and performance. Few studies have found a positive relationship between performance (measured by accounting benchmarks and Tobin's Q) and the proportion of outside directors (see, for example, MacAvoy *et al.* 1983; Hermalin & Weisbach 1991; Mehran 1995; Klein 1998; Bhagat & Black 2000). Still, the lack of a cross-sectional result may be due to the endogeneity of board composition.[xi]

Another board characteristic is board size. Large boards may be less effective because some directors may free-ride on the efforts of others (Lipton & Lorsch 1992). Yermack (1996) showed support for this for a sample of US firms. An interesting study concerning the value added by the board is that by Gertner and Kaplan (1996) on reverse leveraged buy-outs (these are firms which, after having been taken private, are going back to the stock market): on examining the boards of such firms, they concluded that these tended to be smaller than in otherwise similar firms.

Interlocked directorships may also influence the board's corporate governance actions. There is interlocking of directorships if directors sit on each other's boards.

Hallock (1997, 1999) reported that interlocked chief executive officers (CEOs) have higher pay than otherwise similar CEOs.

The board of directors and its corporate governance actions

Managerial disciplining (in particular that of the CEO) has received considerable attention in the empirical literature, as such disciplining is one of the few easily observable actions by the board. Several studies document a negative relationship between CEO or executive board turnover and performance.[xii] Furthermore, boards dominated by outside directors are more likely to remove the CEO when stock and accounting performance are poor. In contrast, CEO turnover on insider-dominated boards is not performance-driven. Another measure of independence, the separation of the jobs of CEO and (non-executive) chairman, appears to be an important condition leading to improved corporate governance in the UK (Franks *et al.* 2001).

However, Correia *et al.* (2004) have argued that good corporate governance is not just equivalent to the dismissal of bad managers. First, current poor performance may be due to past poor corporate governance; as such, the dismissal of the management may be too late. The results of Franks *et al.* (2001) suggest that managerial disciplining only occurs when firms are in the lowest quintile of stock performance and are suffering losses. Second, the success of the governance action of removing underperforming management should be considered along with the managerial alternative. Dherment and Renneboog (2002), who studied the French stock market's reaction to the appointment of CEOs with different backgrounds, found that voluntary resignations did not cause price reactions. However, the appointment of an external manager following the performance-related forced resignation of a CEO caused a significant increase in abnormal returns of more than 2 per cent. If an internal candidate was promoted to the post of CEO in a poorly performing firm the abnormal return was negative (1 per cent on the day of the announcement). This negative market reaction presumably occurred because the internal candidate was held responsible, at least partially, for poor past performance. Finally, Denis and Denis (1995) documented performance increases following forced CEO turnover in the USA, but Renneboog (2000) and Franks *et al.* (2001) did not find evidence, for Belgian and British firms respectively, of a significant improvement over the two-year period following the CEO's replacement.

Factors influencing the structure of the board

Hermalin and Weisbach (1988) reported three factors that caused board changes. First, poor performance increased the likelihood that inside directors would be replaced by outside directors. Second, the CEO succession process was intertwined with the board selection process, as firms tended to appoint inside directors to the board when the CEO was close to retirement age. In addition, inside directors tended to leave the board just after a CEO replacement, as they considered themselves to be the losing candidates in the line of succession. Third, after a firm had left a product market, inside directors tended to leave the firm and outside directors tended to join the board.

Kaplan and Minton (1994) and Morck and Nakamura (1999) analysed whether membership of a *keinetsu* group, an industrial group which usually also includes a bank, influenced the board composition of Japanese firms; they showed that banks took a more active role and appointed their representatives to the firm's board when a firm was performing poorly.

In a study of the board-selection process in US IPOs, Baker and Gompers (2003) analysed whether the CEO's bargaining power, proxied by the Shapley value of their voting stake, influenced the selection of board members, and concluded that there was a positive relation, especially when the CEO's tenure and bargaining power were high.

Performance-related remuneration contracts

One way to align the objectives of managers and shareholders is to introduce managerial compensation contracts which relate remuneration to corporate performance. Obviously, the performance criterion should capture value creation as measured by share prices (rather than by accounting performance). The theoretical blueprint of pay-for-performance remuneration was laid down by the principal-agent models of Jensen and Meckling (1976), Holmström (1979) and Grossman and Hart (1983). Gibbons and Murphy (1992) and Brickley *et al.* (1999) argued that new career opportunities may even form incentives for CEOs on the verge of retirement; indeed, they reported that such CEOs, if they had been performing well over their past career, were more likely to be awarded non-executive directorships after their retirement. Holmström (1982) and Gibbons and Murphy (1990) have pointed out that relative, rather than absolute, performance should be taken as the benchmark for CEO remuneration. Indeed, one should not reward CEOs who create value but whose performance lags behind the firm's industry peers as much as CEOs responsible for value creation substantially above the industry mean.

Performance-sensitivity of managerial compensation is well documented for US firms (e.g. Coughlan & Schmidt 1985; Jensen & Murphy 1990). Executive pay seems to depend both on past stock returns and past accounting measures as well as on relative measures of performance (Gibbons & Murphy 1990). Still, in many firms, the level of executive compensation depends not only on past performance: also important are company size (Murphy 1985) and CEO age and tenure (Murphy 1986; Conyon and Murphy 2000).[xiii]

The optimal balance of stock and cash-based compensation solves a trade-off between long- and short-term incentives (Narayanan 1996). While cash compensation creates short-term incentives (and therefore mitigates the long-run overinvestment problem), stock-based compensation provides appropriate incentives and reduces the long-term underinvestment problem (Dechow & Sloan 1991). Finally, Kole (1997) has argued that optimality of a given compensation structure crucially depends on the characteristics of the assets managed by a given CEO.

Recent literature criticises the agency approach that considers managerial compensation to be the optimal outcome of the contracting problem. According to the 'skimming model' of executive remuneration, directors themselves are able to set their own (excessive) pay in firms with inferior governance standards (Bertrand &

Mullainathan 2000). Furthermore, there is evidence that current managerial remuneration contracts do not include benchmarks that are able to separate value creation resulting from good management from that resulting from luck (Bertrand & Mullainathan 2001).

Apart from the availability of funds, the only constraint deemed to curb such managerial discretion is the fear of causing outrage among shareholders potentially angered by excessive pay of the company executives (Bebchuk *et al.* 2002a). Bebchuk and Fried (2004a, b) have argued that the pay-for-performance criterion remains an unfulfilled promise.

Dividend policy

Dividends may assume two roles in terms of corporate governance. First, as mentioned above, a high dividend payout policy may be a bonding mechanism precommitting managers to pursue value maximisation. Such a dividend policy may then allow shareholders to reduce their monitoring efforts. Second, as 'dividend cuts are interpreted by the market as powerful signals of bad news both about the current situation and about future prospects' (Marsh 1992, p. 50), dividend policy may also constitute an important signal. Hence, if managers fail to meet the anticipated dividend payout, this may activate other corporate governance mechanisms which are better suited to deal with poor performance or financial distress. Goergen *et al.* (2004) argued that a pre-condition for dividends to be a signal of failing performance and corporate control is dividend stickiness. They reported that managers of Anglo-American companies were reluctant to reduce their dividends. However, they questioned whether the same reluctance applied to Continental European managers whose firms were frequently controlled by a large shareholder.

External mechanisms

This subsection reviews the three main external mechanisms of corporate governance: the market for corporate control, the market for block trades and creditor (especially bank) monitoring.

The market for corporate control

Jensen (1986) has argued that the hostile takeover market of the 1980s was a reaction to managers pursuing their own objectives and the failure of internal corporate governance mechanisms. The inefficiency was most pronounced in firms and industries generating large free cash flows. In principle, free cash flow, i.e. the cash flow in excess of what is required to finance all positive net present value (NPV) projects, should be returned to the shareholders via dividends or share repurchases. As this does not always happen, especially in industries with excess capacity and in need of downsizing, there is a need for disciplinary action via the takeover market.

Several studies have focused on the disciplinary role of hostile takeovers by investigating whether poorly performing firms are more likely to be the target of a hostile takeover. There is little support for this thesis. Poor performance only slightly affects the probability of a takeover, and the main determining factor is firm size [see Morck *et al.* (1988), Martin & McConnell (1991), Comment & Schwert (1995) for the USA, and Franks & Mayer (1996) for the UK]. Franks and Mayer (1996) also questioned the corporate governance role of the market for corporate control, given that the pre-bid financial performances of targets of hostile bids were not substantially different from those of targets of friendly bids. Contrary to what the theory on the market for corporate control predicts, takeovers with restructuring have higher abnormal share price performance over periods of five and two years before the bid. Franks and Mayer (1996) concluded that there was little evidence that takeovers entailing corporate restructuring were the result of poor past performance. In contrast, Franks *et al.* (2001) showed that companies, forced to cut or omit their dividends, were frequently drastically restructured via mergers and acquisitions which led to the replacement of most of the directors.

Generally, the effectiveness of hostile takeovers as a disciplining device of bad managers seems to be in doubt. Still, the market for corporate control may have a rather indirect role. First, the mere threat of a takeover may raise efficiency *ex ante* (Scharfstein 1988; Shleifer & Vishny 1986). Second, companies shielded from the takeover market have lower share prices. The adoption of anti-takeover devices usually results in significant negative abnormal returns of –2 per cent (Karpoff & Malatesta 1989; Ryngaert 1988; Jarrell & Poulsen 1989). Burkart (1999) concluded that although managers shielded from the takeover threat did not behave like empire-builders, they tended to become 'sluggish'. For example, Bertrand and Mullainathan (2003) and Borokhovich *et al.* (1997) showed that, in the case of US firms, increased insulation from takeovers increased managerial salaries and lowered total factor productivity. Garvey and Hanka (1999) have provided evidence that anti-takeover legislation has led to fewer new investments and fewer disinvestments. Hence, it seems that the existence of an active market for corporate control is valuable. As pointed out in the second section of this chapter, there is only limited potential for a more important role of the market for corporate control in Continental Europe given that the vast majority of firms are controlled by a large shareholder. However, for firms with strong control, a market for partial control (a market for share blocks) may still operate (see next subsection).

Block trades (the market for partial control stakes)

Holderness and Sheehan (1988) and Sudarsanam (1996) reported that transfers of control via block sales were usually accompanied by positive abnormal performance. Barclay and Holderness (1989) showed that the price reaction was positive regardless of the price paid for the share block. The main reason for the positive market reaction is that changes in control may improve corporate governance, especially when the firm is performing poorly and is in need of a substantial reorganisation, e.g. board or financial restructuring (Barclay & Holderness 1991). When performance is poor, passive shareholders are expected to sell their shares, while monitoring shareholders may increase their stakes. Consequently, block transactions giving the purchaser control over the firm may trigger a more favourable market reaction than those transactions that do not confer control to the purchaser. Evidence for this conjecture is provided by Holderness and Sheehan (1988); they also found that, for the USA, the market reaction was more favourable when block transfers were accompanied by a tender offer on all outstanding shares. In addition, the market has been observed to react more positively to block transactions in those firms that subsequently experience a full acquisition (Barclay & Holderness 1992). However, Sudarsanam (1996) has documented that, even when there is no subsequent takeover, the benefits of ownership concentration outweigh the costs.

The following studies distinguish between different types of acquirers of blocks. Banerjee *et al.* (1997) showed that block deals did not generally trigger any positive abnormal performance for listed French firms; however, block acquisitions by holding companies reduced firm value. This is similar to what Renneboog (2000) found for Belgian holding companies. Although institutions and holding companies traded actively in share stakes, ownership increases by these categories did not result in changes in board structure. Conversely, acquisitions of substantial share stakes by industrial companies and families were normally accompanied by changes in the management when prior performance was poor, suggesting that there was a market for control stakes: poor performance triggers block trades which are followed by corporate restructuring. For the UK, Franks *et al.* (2001) examined the impact of block transfers by type of acquirer (amongst other disciplinary mechanisms) on managerial turnover in the wake of poor performance, demonstrating that block purchases by individuals or families were associated with significant increases in executive board turnover. If family control increased substantially in a poorly performing company, executive board turnover increased by 8 per cent. Increases in the control stakes held by families or corporations were associated with significant subsequent increases in board turnover but also coincided with decreases in institutional ownership.

Bethel *et al.* (1998) showed that block purchases by strategic investors in the USA did not cause a significant market reaction, while those by activist shareholders caused significantly positive abnormal returns. (Activist shareholders are defined as those who acquire stakes in poorly performing companies, and subsequently attempt to restructure the firm to achieve a considerable improvement in its performance; in contrast, strategic investors invest for the long run and do not interfere with management.) Keim and Madhavan (1996), who distinguished between buyer- and seller-initiated block transactions in the USA, found that the market reaction to buyer-initiated transactions was normally positive, whereas that to seller-initiated transactions was always negative.

Creditor monitoring

Shleifer and Vishny (1997) argued that large creditors fulfil a role similar to large shareholders because both have large amounts of money invested in the firm and have strong incentives to monitor its management. Diamond (1984) showed that delegation of monitoring to banks is efficient because duplication of monitoring by small investors (creditors) can be avoided, provided the bank's lending portfolio is sufficiently diversified.

Rajan and Zingales (2003) argued that relationship-based financing, as in the German system, performs better when markets and firms are smaller, shareholder protection is low, there is little transparency, and innovation is mostly incremental rather than revolutionary. Large creditors often have a variety of control rights and therefore sufficient power to monitor. Hence, bank monitoring may act as a substitute to alternative governance devices. For example, a disciplinary change in control is expected to be less profitable and hence less likely to occur in firms monitored by banks. Köke (2004) has analysed corporate governance in the German bank-based economy and confirmed that non-market monitoring devices play a larger role because hostile control transactions are rare and because other constituencies, such as large creditors, typically have considerable power. The long-term lending relationships give banks considerable power which is often reinforced by bank representation on the supervisory board of the firm. One reason why banks have such a strong influence in Germany is that historically they have acted as so-called housebanks (*Hausbanken*), providing long-term loans to firms (Baums 2000).

The corporate governance regulatory framework

The importance of the above governance mechanisms, together with the interactions between different mechanisms, needs to be studied by taking into account a country's specific regulatory context. For example, strong shareholder protection reduces the danger of expropriation of minority shareholders. La Porta *et al.* (1998, 1999, 2000) argued that the differences in corporate governance systems can be explained by differences in terms of the protection for minority shareholders and the level of capital market development. They concluded that common law systems, such as those of the UK and USA, tend to offer better shareholder protection than civil law systems, such as those of Scandinavia and Germany. Likewise, creditor protection, measured by creditor rights indices which are based on bankruptcy law and the regulation regarding financial distress, is strongest in common law countries and worst in French civil law countries (Table 5.6) with the Scandinavian and German countries somewhere in between.

The level of shareholder protection has also been shown to be inversely linked to the size of the premium over the market price paid for a majority voting block, as higher premiums are paid in countries with weak protection (Zingales 1994). The volume of IPOs has also been shown to be higher in countries with good shareholder protection (Table 5.7). Work by La Porta *et al.* (2000) and Beck *et al.* (2003) found that firms based in countries with strong shareholder protection had a better growth potential.

Better legal institutions also influence equity rates of return and the demand for equity finance by companies (Lombardo & Pagano 2002). There may be two reasons for this relationship. Efficient legal rules and courts, (1) reduce the private benefits of managers, and (2) make contracts between the firm and its customers and suppliers easier to stipulate and enforce. Better corporate law and more efficient courts raise corporate profitability and growth, which in turn increases the availability of financing. Lombardo and Pagano (2002) showed that legal limits on transactions with companies related through ownership pyramids preserved the cash flow rights of minority shareholders and reduced private benefits of control; they concluded that

Table 5.6 Shareholder and creditor protection.

	Shareholder protection[a]	One-share–one-vote[b]	Creditor protection[c]
UK	4	0	4
USA	5	0	1
English origin average[d]	3.39	0.22	3.11
France	2	0	0
Belgium	0	0	2
Italy	0	0	2
Spain	2	0	2
Portugal	2	0	1
Netherlands	2	0	2
French origin average[d]	1.76	0.24	1.58
Germany	1	0	3
Austria	2	0	3
Switzerland	1	0	1
Japan	3	1	2
German origin average[d]	2.00	0.33	2.33
Denmark	3	0	3
Finland	2	0	1
Norway	3	0	2
Sweden	2	0	2
Scandinavian origin average[d]	2.50	0.00	2.00
Overall average	2.44	0.22	2.30

[a] The shareholder protection index is higher if shareholders can mail their proxy votes, are not required to deposit their shares prior to the general meetings, cumulative voting is allowed, minority shareholders are protected and a minimum percentage of share capital allows a shareholder to call for an extraordinary general meeting.
[b] One-share–one-vote is a dummy variable which equals unity if one share carries one vote (no multiple class voting rights).
[c] The creditor rights index is higher if absolute priority is followed in case of financial distress.
[d] Not all countries reviewed by La Porta *et al.* (1998) are listed, but the averages are based on the entire sample.
Source: based on La Porta *et al.* (1998).

the size of these effects on the equilibrium rate of return was increasing with the degree of international segmentation of equity markets.

Gompers *et al.* (2003) computed a corporate governance index for US firms based on a set of corporate governance provisions; they stated that firms with better shareholder protection had higher Tobin's Qs, higher profits, higher sales growth, lower capital expenditures and made fewer acquisitions. Similarly, Drobetz *et al.* (2004) constructed a governance index for German firms based on five categories of corporate governance rules, reporting that better protection of shareholders led to higher shareholder value.

In general, these studies document a positive effect of better corporate governance protection on financial market development. Some argue that the conclusions that can be drawn from these studies are limited because the directions of causality

Table 5.7 External finance and legal origin (source: based on La Porta *et al.* 1997).

	External capital/GDP[a]	Listed Domestic firms/Population	IPOs/ population[b]	Debt/GDP[c]
UK	1.00	35.68	2.01	1.13
USA	0.58	30.11	3.11	0.81
English origin average[d]	0.60	35.45	2.23	0.68
France	0.23	8.05	0.17	0.96
Belgium	0.17	15.50	0.30	0.38
Italy	0.08	3.91	0.31	0.55
Spain	0.17	9.71	0.07	0.75
Portugal	0.08	19.50	0.50	0.64
Netherlands	0.52	21.13	0.66	1.08
French origin average[d]	0.21	10.00	0.19	0.45
Germany	0.13	5.14	0.08	1.12
Austria	0.06	13.87	0.25	0.79
Switzerland	0.62	33.85		
Japan	0.62	17.78	0.26	1.22
German origin average[d]	0.46	16.79	0.12	0.97
Denmark	0.21	50.40	1.80	0.34
Finland	0.25	13.00	0.60	0.75
Norway	0.22	33.00	4.50	0.64
Sweden	0.51	12.66	1.66	0.55
Scandinavian origin average[d]	0.30	27.26	2.14	0.57
Overall average	0.40	21.59	1.02	0.59

[a] External capital is defined as the equity capital held by shareholders other than the largest three shareholders.
[b] Initial public offerings are companies which are brought to the stock exchange.
[c] Debt is defined here as the sum of the issued corporate bonds and the funds provided by banks.
[d] Not all countries reviewed by La Porta *et al.* (1997) are listed, but the averages are based on the entire sample.

between the legal system and financial structure may run in opposition i.e. financial structure prompts the occurrence of transformations in the legal regime (Bolton & von Thadden 1998; Bebchuk & Roe 2000).

Conclusions

This chapter shows that corporate governance is an intricate amalgam of internal, external and regulatory mechanisms that should, in an optimal scenario, lead to superior returns and to increases in corporate value. A first set of governance devices are designed by the firm itself: the size of the board of directors, the degree of independence of the non-executive directors (or of the supervisory board), the presence of audit, remuneration and promotion committees consisting of non-executive directors, the separation of the functions of CEO and chairman, managerial remuneration contracts based on pay-for-(share price) performance, and dividend policy. These mechanisms should decrease managerial discretion and reduce potential agency conflicts between shareholders and managers. The empirical evidence on internal mechanisms is mixed. While board structure has a beneficial impact on corporate performance and facilitates managerial disciplining, there is ample evidence that remuneration contracts play little in the way of a governance role given the benchmarks upon which they are based.

Agency conflicts are also reduced by external monitoring by blockholders and creditors. While it is costly for shareholders to hold substantial share stakes, there are many mechanisms which allow shareholders to maintain strong control while reducing their direct investment. Ownership pyramids, proxy votes, shareholder voting coalitions and dual class shares erode the one-share–one-vote principle. However, there is little evidence of shareholder monitoring. In addition, in countries like the UK and the USA, directors with strong voting power are entrenched; they are able to defend (the private benefits of) their positions even in the wake of poor performance.

The market of corporate control (the takeover market and the market of controlling share blocks) plays a more powerful corporate governance role. A proportion of takeover targets are poorly performing organisations, and a takeover is virtually always followed by board, financial or corporate restructuring. In addition to the disciplining role of the equity markets, bond markets are also important as the bond ratings agencies largely determine at what interest rate a firm can maintain or attract additional leverage. Direct creditor monitoring is only evidenced in cases of very poor performance and financial distress, when servicing of the debt is threatened. This is especially the case in a bank-based economy (like Germany) where the loan portfolios of the firm are concentrated with the large universal banks (*Hausbanken*).

However, future research has to take into account that the efficiency of these corporate governance mechanisms should not be studied in isolation, as some mechanisms are complements or substitutes to one another. Moreover, future research also needs to bear in mind the regulatory framework (in terms of corporate law, stock exchange regulation, codes of corporate governance) as well as the enforceability of existing rules because they determine the effectiveness of corporations in generating shareholder value.

References

Admati, R.A., Pfleiderer, P. & Zechner, J. (1994) Large shareholder activism, risk sharing, and financial markets equilibrium. *Journal of Political Economy*, **102**, 1097–1130.

Agrawal, A. & Knoeber, C.R. (1998) Managerial compensation and the threat of takeover. *Journal of Financial Economics*, **47**, 219–239.

Baker, M. & Gompers, P. (2003) The determinants of board structure and function in entrepreneurial firms. *Journal of Law and Economics*, **46**, 569–598.

Banerjee, S., Leleux, B. & Vermaelen, T. (1997) Large shareholders and corporate control: an analysis of stake purchases by French holding companies. *European Financial Management*, **3**, 23–43.

Barca, F. & Becht, M. (2001) *The Control of Corporate Europe*. Oxford University Press, Oxford.

Barclay, M.J. & Holderness, C.G. (1989) Private benefits from control of public corporations. *Journal of Financial Economics*, **25**, 371–395.

Barclay, M.J. & Holderness, C.G. (1991) Negotiated block trades and corporate control. *Journal of Finance*, **46**, 861–878.

Barclay, M.J. & Holderness, C.G. (1992) The law and large-block trades. *Journal of Law and Economics*, **35**, 265–294.

Barro, J. & Barro R. (1990) Pay, performance and turnover of bank CEOs. *Journal of Labor Economics*, **8**, 448–481.

Baums, T. (2000) *Corporate governance in Germany – system and current developments*. Working paper, University of Osnabrück.

Bebchuk, L. & Fried, J. (2004a) *Pay Without Performance: The Unfulfilled Promise of Executive Compensation, Part I: The Official View and its Limits*. Harvard University Press, Cambridge, MA.

Bebchuk, L. & Fried, J. (2004b) *Pay Without Performance: The Unfulfilled Promise of Executive Compensation, Part II: Power and Pay*. Harvard University Press, Cambridge, MA.

Bebchuk, L. & Roe, M. (2000) A theory of path dependence of corporate ownership and governance. *Stanford Law Review*, **52**, 775–808.

Bebchuk, L.A., Fried, J.M. & Walker, D.I. (2002a) Managerial power and rent extraction in the design of executive compensation. *University of Chicago Law Review*, **69**, 751–846.

Bebchuk, L., Coates IV, J.C. & Subramian, G. (2002b) The powerful anti-takeover force of staggered boards: theory, evidence, and policy. *Stanford Law Review*, **54**, 887–924.

Becht, M. & Boehmer, E. (2001) Ownership and voting power in Germany. In: *The Control of Corporate Europe* (eds F. Barca & M. Becht), Oxford University Press, Oxford; pp. 128–153.

Becht, M., Bolton, P. & Röell, A. (2003) Corporate governance and control. In: *Handbook of the Economics of Finance*, Vol. 1, Part 1 (eds G. Constantinides, M. Harris & R. Stulz), North-Holland, Amsterdam; pp. 1–109.

Beck, T., Demirgüç-Kunt, A. & Levine, R. (2003) Law and finance: why does legal origin matter? *Journal of Comparative Economics*, **31**, 653–675.

Bennedsen, M. & Nielsen, C. (2002) *The impact of the break-through rule on European firms*. Working paper, Copenhagen Business School.

Berglöf, E. & Burkart, M. (2002) European takeover regulation? *Economic Policy*, **36**, 171–208.

Bertrand, M. & Mullainathan, S. (2000) Agents with and without principals. *American Economic Review*, **90**, 203–208.

Bertrand, M. & Mullainathan, S. (2001) Are CEOs rewarded for luck? The ones without principals are. *Quarterly Journal of Economics*, **116**, 901–932.

Bertrand, M. & Mullainathan, S. (2003) Enjoying the quiet life? Managerial behavior following anti-takeover legislation. *Journal of Political Economy*, **111**, 1043–75.

Bethel, J.E., Liebeskind, J.P. & Opler, T. (1998) Block share purchases and corporate performance. *Journal of Finance*, **53**, 605–634.

Bhagat, S. & Black, B. (2002) The non-correlation between board independence and long-term firm performance. *Journal of Corporation Law*, **27**, 231–273.

Blackwell, D., Brickley, J. & Weisbach, M. (1994) Accounting information and internal performance evaluation: evidence from Texas banks. *Journal of Accounting and Economics*, **17**, 331–358.

Bolton, P. & von Thadden, E.-L. (1998) Blocks, liquidity and corporate control. *Journal of Finance*, **53**, 1–25.

Borokhovich, K., Brunarski, K. & Parrino, R. (1997) CEO contracting and anti-takeover amendments. *Journal of Finance*, **52**, 1495–1517.

Börsch-Supan, A. & Köke, J. (2002) An applied econometrician's view of empirical corporate finance studies. *German Economic Review*, **3**, 295–326.

Bratton, W. & McCahery, J.A. (1999) Comparative corporate governance and the theory of the firm: the case against global cross reference. *Columbia Journal of Transnational Law*, **38**, 213–297.

Brickley, J. & Van Horn, R.L. (2002) Incentives in nonprofit organizations: evidence from hospitals. *Journal of Law and Economics*, **45**, 227–249.

Brickley, J.A., Link, J.S. & Coles, J.L. (1999) What happens to CEOs after they retire? New evidence on career concerns, horizon problems, and CEO incentives. *Journal of Financial Economics*, **52**, 341–377.

Burkart, M. (1999) *The economics of takeover regulation*. SITE Working Paper 99/06, Stockholm School of Economics.

Burkart, M., Gromb, D. & Panunzi, F. (1997) Large shareholders, monitoring, and the value of the firm. *Quarterly Journal of Economics*, **112**, 693–728.

Coffee, J.C. (1991) Liquidity versus control: the institutional investor as corporate monitor. *Columbia Law Review*, **91**, 1277–1368.

Comment, R. & Schwert, G.W. (1995) Poison pill or placebo? Evidence on the deterrence and wealth effects of modern anti-takeover measures. *Journal of Financial Economics*, **39**, 3–43.

Conyon, M.J. & Murphy, K. J. (2000) The prince and the pauper? CEO pay in the United States and the United Kingdom. *Economic Journal*, **110**, 640–671.

Core, J., Holthausen, R. & Larcker, D. (1999) Corporate governance, chief executive officer compensation, and firm performance. *Journal of Financial Economics*, **51**, 371–406.

Core, J., Guay, W. & Larcker, D. (2003) Executive equity compensation and incentives: a survey. *Federal Reserve Bank of New York Economic Policy Review*, **9**, 27–50.

Correia da Silva, L., Goergen, M. & Renneboog, L. (2004) *Dividend Policy and Corporate Governance*. Oxford University Press, Oxford.

Coughlan, A. & Schmidt, R. (1985) Executive compensation, managerial turnover, and firm performance: an empirical investigation. *Journal of Accounting and Economics*, **7**, 43–66.

Crespi, R. & Renneboog, L. (2002) *Coalition formation and shareholder monitoring in the UK*. Discussion Paper. CentER, Tilburg University.

Crespi, R., Gispert, C. & Renneboog, L. (2002) Cash-based executive compensation in Spain and UK. In: *Corporate Governance Regimes: Convergence and Diversity* (eds J. McCahery, P. Moerland, T. Raaijmakers & L. Renneboog). Oxford University Press, Oxford; pp. 647–667.

Dechow, P. & Sloan, R. (1991) Executive incentives and the horizon problem. *Journal of Accounting and Economics*, **14**, 51–89.

Demsetz, H. (1983) The structure of ownership and the theory of the firm. *Journal of Law and Economics*, **26**, 375–390.

Demsetz, H. & Lehn, K. (1985) The structure of corporate ownership: causes and consequences. *Journal of Political Economy*, **93**, 1155–1177.

Denis, D. & Denis, D. (1995) Performance changes following top management dismissals. *Journal of Finance*, **50**, 1029–1055.

Dherment, I. & Renneboog, L (2002) Share price reactions to CEO resignations and large shareholder monitoring in listed French companies. In: *Corporate Governance Regimes: Convergence and Diversity* (eds J. McCahery, P. Moerland, T. Raaijmakers & L. Renneboog). Oxford University Press, Oxford; pp. 297–324.

Diamond, D.W. (1984) Financial intermediation and delegated monitoring. *Review of Economic Studies*, **51**, 393–414.

Drobetz, W., Schillhofer, A. & Zimmermann, H. (2004) Corporate governance and expected stock returns: evidence from Germany. *European Financial Management*, **10**, 267–293.

Dyck, A. & Zingales, L. (2004) Private benefits of control: an international comparison. *Journal of Finance*, **59**, 537–600.

Faccio, M. & Lang, L. (2002) The ultimative ownership of Western European companies. *Journal of Financial Economics*, **65**, 365–395.

Franks, J. & Mayer, C. (1996) Hostile takeovers and the correction of managerial failure. *Journal of Financial Economics*, **40**, 163–181.

Franks, J. & Mayer, C. (1998) Bank control, takeovers and corporate governance in Germany. *Journal of Banking and Finance*, **22**, 1385–1403.

Franks, J. & Mayer, C. (2001) Ownership and control in Germany. *Review of Financial Studies*, **14**, 943–977.

Franks, J., Mayer, C. & Renneboog, L. (2001) Who disciplines management of poorly performing companies? *Journal of Financial Intermediation*, **10**, 209–248.

Garvey, G. & Hanka, G. (1999) Capital structure and corporate control: the effect of anti-takeover statutes on firm leverage. *Journal of Finance*, **54**, 519–546.

Gertner, R. & Kaplan, S. (1996) *The value-maximising board.* Working paper, University of Chicago.

Gibbons, R. & Murphy, K.J. (1990) Relative performance evaluation for chief executive officers. *Industrial and Labor Relations Review*, **43**, 30S–51S.

Gibbons, R. & Murphy, K.J. (1992) Optimal incentive contracts in the presence of career concerns: theory and evidence. *Journal of Political Economy*, **100**, 468–505.

Girma, S., Thompson, S. & Wright, P. (2002) *Merger activity and executive pay*. CEPR Discussion Paper, no. 3255. Centre for Economic Policy Research, London.

Goergen, M. (1998) *Corporate Governance and Financial Performance – A Study of German and UK Initial Public Offerings*. Edward Elgar, Cheltenham.

Goergen, M. & Renneboog, L. (2001) Strong managers and passive institutional investors in the UK. In: *The Control of Corporate Europe* (eds F. Barca & M. Becht). Oxford University Press, Oxford; pp. 258–284.

Goergen, M. & Renneboog, L. (2003) Why are the levels of control (so) different in German and UK companies? Evidence from initial public offerings. *Journal of Law, Economics and Organization*, **19**, 141–175.

Goergen, M. & Renneboog, L. (2004) Shareholder wealth effects of European domestic and cross-border wealth effects. *European Financial Management*, **10**, 9–45.

Goergen, M. & Renneboog, L. (2005) Insider retention and long-run performance in German and UK IPOs. In: *The Corporate Governance Life-Cycle* (eds I. Filatotchev & M. Wright). Edward Elgar, Cheltenham.

Goergen, M., Renneboog, L. & Correia da Silva, L. (2004) *Dividend policy of German firms. A dynamic panel data analysis of partial adjustment models*. Working paper.

Goergen, M., Renneboog, L. & Correia da Silva, L. (2005) When do German firms change their dividends? *Journal of Corporate Finance*, **11**, 375–399.

Gompers, P., Ishii, J. & Metrick, A. (2003) Corporate governance and equity prices. *Quarterly Journal of Economics*, **118**, 107–155.

Grossman, S. & Hart, O. (1980) Disclosure laws and takeover bids. *Journal of Finance*, **35**, 323–334.

Grossman, S.J. & Hart, O. (1983) An analysis of the principal-agent problem. *Econometrica*, **51**, 7–45.

Grossman, S. & Hart, O. (1988) One share–one vote and the market for corporate control. *Journal of Financial Economics*, **20**, 175–202.

Hallock, K. (1997) Reciprocally interlocking boards of directors and executive compensation. *Journal of Financial and Quantitative Analysis*, **32**, 331–334.

Hallock, K. (1999) Dual agency: corporate boards with reciprocally interlocking relationships. In: *Executive Compensation and Shareholder Value* (eds J. Carpenter & D. Yermack), Kluwer Academic, Amsterdam; pp. 55–75.

Hermalin, B. & Weisbach, M. (1988) The determinants of board composition. *RAND Journal of Economics*, **19**, 589–606.

Hermalin, B. & Weisbach, M. (1991) The effects of board composition and direct incentives on firm performance. *Financial Management*, **20**, 101–112.

Hermalin, B. & Weisbach, M. (1998) Endogenously chosen boards of directors and their monitoring of the CEO. *American Economic Review*, **88**, 96–118.
Himmelberg, C, Hubbard, R. & Palia, D. (1999) Understanding the determinants of managerial ownership and the link between ownership and performance. *Journal of Financial Economics*, **53**, 353–384.
Holderness, C.G. & Sheehan, D.P. (1988) The role of majority shareholders in publicly held corporations: an explanatory analysis. *Journal of Financial Economics*, **20**, 317–346.
Holmström, B. (1979) Moral hazard and observability. *Bell Journal of Economics*, **10**, 74–91.
Holmström, B. (1982) Moral hazard in teams. *Bell Journal of Economics*, **13**, 324–340.
Huson, M., Parrino, R. & Starks, L. (2001) Internal monitoring and CEO turnover: a long-term perspective. *Journal of Finance*, **56**, 2265–2297.
Jarrell, G. & Poulsen, A. (1989) The returns to acquiring firms in tender offers: evidence from three decades. *Financial Management*, **18**, 12–19.
Jensen, M.C. (1986) Agency costs of free cash flow, corporate finance and takeovers. *American Economic Review*, **76**, 323–329.
Jensen, M. & Meckling, W.H. (1976) Theory of the firm: managerial behaviour, agency costs, and ownership structure. *Journal of Financial Economics*, **3**, 305–360.
Jensen, M. & Murphy, K. (1990) Performance pay and top-management incentives. *Journal of Political Economy*, **98**, 225–264.
Johnson S., La Porta, R., Lopez-de-Silanes, F. & Shleifer, A. (2000) Tunelling. *American Economic Review*, **90**, 22–27.
Kahn, C. & Winton, A. (1998) Ownership structure, speculation and shareholder intervention. *Journal of Finance*, **53**, 99–129.
Kaplan, S. (1994) Top executive rewards and firm performance: a comparison of Japan and the U.S. *Journal of Political Economy*, **102**, 510–546.
Kaplan, S. & Minton, B. (1994) Appointments of outsiders to Japanese boards: determinants and implications for managers. *Journal of Financial Economics*, **36**, 225–258.
Karpoff, J. & Malatesta, P. (1989) The wealth effect of second generation state takeover legislation. *Journal of Financial Economics*, **25**, 291–322.
Keim, D.B. & Madhavan, A. (1996) Large block transactions: analysis and measurement of price effects. *Review of Financial Studies*, **9**, 1–36.
Klein, A. (1998) Firm performance and board committee structure. *Journal of Law and Economics*, **41**, 275–299.
Köke, J. (2004) The market for corporate control in a bank-based economy: a governance device? *Journal of Corporate Finance*, **10**, 53–80.
Köke, J. & Renneboog, L. (2005) Do corporate control and product market competition lead to stronger productivity growth? Evidence from market-oriented and blockholder-based governance regimes. *Journal of Law and Economics*, **48**, 1–42.
Kole, S. (1996) Managerial ownership and firm performance: incentives or rewards? In: *Advances in Financial Economics*, Vol. 2 (eds M. Marr & M. Hirschey). JAI Press, London; pp. 119–149.
Kole, S. (1997) The complexity of compensation contracts. *Journal of Financial Economics*, **43**, 79–104.
La Porta, R., Lopez de Silanes, F., Shliefer, A. & Vishny, R. (1997) Legal determinants of external finance. *Journal of Finance*, **52**, 1131–1150.
La Porta, R., Lopez de Silanes, F., Shleifer, A. & Vishny, R. (1998) Law and finance. *Journal of Political Economy*, **106**, 1113–1155.
La Porta, R., Lopez de Silanes, F., Shleifer, A. & Vishny, R. (1999) Corporate ownership around the world. *Journal of Finance*, **54**, 471–517.
La Porta, R., Lopez de Silanes, F., Shleifer, A. & Vishny, R. (2000) Investor protection and corporate governance. *Journal of Financial Economics*, **58**, 3–27.

Lewellen, W., Loderer, C. & Martin, K. (1987) Executive compensation and executive incentive problems: an empirical analysis. *Journal of Accounting and Economics*, **9**, 287–310.

Lipton, M. & Lorsch, J. (1992) A modest proposal for improved corporate governance. *Business Lawyer*, **48**, 59–77.

Lombardo, D. & Pagano, M. (2002) Law and equity market: a simple model. In: *Corporate Governance Regimes: Convergence and Diversity* (eds J. McCahery, P. Moerland, T. Raaijmakers & L. Renneboog). Oxford University Press, Oxford; pp. 343–362.

MacAvoy, P., Cantor, S., Dana, J. & Peck, S. (1983) ALI proposals for increased control of the corporation by the board of directors: an economic analysis. In: *Statement of the Business Roundtable on the American Law Institute's Proposed Principles of Corporate Governance and Structure: Restatement and Recommendation*. Business Roundtable, New York.

McCahery, J., Moerland, P., Raaijmakers, T. & Renneboog, L. (2002) *Corporate Governance Regimes: Convergence and Diversity*. Oxford University Press, Oxford.

McConnell, J. & Servaes, H. (1990) Additional evidence on equity ownership and corporate value. *Journal of Financial Economics,* **27**, 595–612.

McConnell, J. & Servaes, H. (1995) Equity ownership and the two faces of debt. *Journal of Financial Economics*, **39**, 131–157.

Marsh, P. (1992) *'Dividend announcements and stock price performance', mimeo*. London Business School, London.

Martin, K. & McConnell, J. (1991) Corporate performance, corporate takeovers and management turnover. *Journal of Finance*, **46**, 671–687.

Martynova, M. & Renneboog, L. (2005a) *Mergers and acquisitions in Europe*. Working paper, European Corporate Governance Institute.

Martynova, M. & Renneboog, L. (2005b) *Takeover waves: triggers, performance and motives*. Working paper, European Corporate Governance Institute.

Maug, E. (1998) Large shareholders as monitors: is there a trade-off between liquidity and control? *Journal of Finance*, **53**, 65–98.

Mehran, H. (1995) Executive compensation structure, ownership, and firm performance. *Journal of Financial Economics*, **38**, 163–184.

Morck, R. & Nakamura, M. (1999) Banks and corporate control in Japan. *Journal of Finance*, **54**, 319–340.

Morck, R., Shleifer, A. & Vishny, R. (1988) Managerial ownership and market valuation: an empirical analysis. *Journal of Financial Economics,* **20**, 293–315.

Murphy, K. (1985) Corporate performance and managerial remuneration: an empirical analysis. *Journal of Accounting and Economics*, **7**, 11–42.

Murphy, K. (1986) Incentives, learning, and compensation: a theoretical and empirical investigation of managerial labor contracts. *RAND Journal of Economics*, **17**, 59–76.

Narayanan, M.P. (1996) Form of compensation and managerial decision horizon. *Journal of Financial and Quantitative Analysis*, **31**, 467–491.

Nenova, T. (2003) The value of corporate votes and control benefits: a cross-country analysis. *Journal of Financial Economics*, **68**, 325–351.

Nicodano, G. & Sembenelli, A. (2000) Private benefits, block transaction premia and ownership structure. Working paper, University of Torino.

Pagano, M. & Röell, A. (1998) The choice of stock ownership structure: agency costs, monitoring and the decision to go public. *Quarterly Journal of Economics*, **113**, 187–225.

Rajan, R. & Zingales, L. (2003) Banks and markets: the changing character of European finance. Working paper, University of Chicago.

Renneboog, L. (2000) Ownership, managerial control and the governance of poorly performing companies listed on the Brussels stock exchange. *Journal of Banking and Finance*, **24**, 1959–1995.

Renneboog, L. & Trojanowski, G. (2003) *The managerial labor market and the governance role of shareholder control structures in the UK*. Working paper, European Corporate Governance Institute.
Rosenstein, S. & Wyatt, J. (1990) Outside directors, board independence, and shareholder wealth. *Journal of Financial Economics*, **26**, 175–184.
Rydqvist, K. (1987) *Empirical evidence of the voting premium*. Working Paper 35, Northwestern University.
Ryngaert, M. (1988) The effect of poison pill securities on shareholder wealth. *Journal of Financial Economics*, **20**, 377–417.
Scharfstein, D. (1988) The disciplinary role of takeover. *Review of Economic Studies*, **55**, 185–199.
Shleifer, A. & Vishny, R. (1986) Large shareholders and corporate control. *Journal of Political Economy*, **94**, 461–488.
Shleifer, A. & Vishny, R. (1997) A survey of corporate governance. *Journal of Finance*, **52**, 737–784.
Stapledon, G. (1996) *Institutional Shareholders and Corporate Governance*. Clarendon Press, Oxford.
Stapledon, G. & Bates, J. (2002) Unpacking the 'interest-holders' in a share: making voting easier for institutional shareholders. In: *Corporate Governance Regimes: Convergence and Diversity* (eds J. McCahery, P. Moerland, T. Raaijmakers & L. Renneboog). Oxford University Press, Oxford; pp. 567–602.
Sudarsanam, S. (1996) Large shareholders, takeovers and target valuation. *Journal of Business Finance and Accounting*, **23**, 295–314.
Trojanowski, G. (2003) *Equity block transfers in transition economies: evidence from Poland*. Working paper, Tilburg University.
Warner, J., Watts, R. & Wruck, K. (1988) Stock prices and top-management changes. *Journal of Financial Economics*, **20**, 461–492.
Weisbach, M. (1988) Outside directors and CEO turnover. *Journal of Financial Economics*, **20**, 431–460.
Yermack, D. (1996) Higher valuation of companies with a small board of directors. *Journal of Financial Economics*, **40**, 185–212.
Zingales, L. (1994) The value of the voting right: a study of the Milan Stock Exchange. *Review of Financial Studies*, **7**, 125–148.
Zingales, L. (1995) What determines the value of corporate votes? *Quarterly Journal of Economics*, **110**, 1047–1073.
Zwiebel, J. (1995) Block investment and partial benefits of corporate control. *Review of Economic Studies*, **62**, 161–185.

Endnotes

[i] Acknowledgements: Parts of this chapter are related to Correia da Silva *et al.* (2004). We are grateful to L. Correia da Silva and B.R. Leffe for stimulating discussions. We acknowledge financial support from the Netherlands Organization for Scientific Research.

[ii] The successor of the ECGN is the European Corporate Governance Institute (ECGI).

[iii] The reason for this is that the EU Large Holdings Directive (88/627/EEC), which governs disclosure of shareholders in the EU member states, requires the disclosure of voting stakes, rather than ownership stakes, exceeding certain thresholds. Hence, information on ownership is not normally available for most EU countries.

[iv] For a more detailed account on the use, occurrence and control consequences of ownership pyramids or cascades, see the third section of this chapter.

[v] Such conflicts of interest may even be exacerbated by the fact that, in countries such as Belgium, France and Italy, intricate ownership-based networks exist which may also comprise banks. For instance, a bank's equity may be controlled by a large industrial holding

company that also owns large share stakes in other subsidiaries in which that bank also owns equity or debt (Renneboog 2000).

[vi] The true control by institutional investors is much higher as most of their holdings are below the disclosure threshold of 5 per cent.

[vii] Bratton and McCahery (1999) question whether Anglo-American style institutional shareholder activism would lead to improved corporate results in Continental Europe because, in their opinion, a minimum level of takeover activity is a precondition of relational engagement between institutional shareholders and managers.

[viii] The higher importance of family control in Austria and Italy can be explained by the fact that the sample consists of both listed and non-listed companies. Still, even after excluding the non-listed Italian firms, a majority of the listed Italian companies are family-controlled.

[ix] For German firms a grandfather clause was put in place in 1998. Before 1998, German firms could be authorised to issue shares with multiple voting rights by the Ministry of Economics of the Land of their headquarters (Goergen 1998, p. 71).

[x] For an overview of the European mergers and acquisitions market, see Goergen and Renneboog (2004) and Martynova and Renneboog (2005a, b).

[xi] If poor performance causes increased board independence, the cross-sectional result on the potential relationship between the degree of board independence (independent variable) and corporate performance (dependent variable) may be underestimated. The reason is that more independent directors will be on boards of firms with historically poor performance (Hermalin & Weisbach 1998; Börsch-Supan and Köke 2002). However, attempts to correct for this problem using simultaneous equations (Hermalin & Weisbach 1991; Bhagat & Black 2000) have not yielded a significant relationship. An alternative approach is event studies focusing on whether changes in board composition are followed by changes in performance and firm value. The stock price reaction to the announcement of the appointment of outside directors in the USA triggers a significant abnormal return of 0.2% (Rosenstein & Wyatt 1990).

[xii] These studies include, amongst others: Coughlan & Schmidt (1985), Warner *et al.* (1988), Weisbach (1988), Jensen & Murphy (1990), Barro & Barro (1990), Blackwell *et al.* (1994), Huson *et al.* (2001), Brickley & Van Horn (2002) for the USA; Kaplan (1994) and Kaplan & Minton (1994) for Japan; Franks & Mayer (2001) and Köke (2004) for Germany; Renneboog (2000) for Belgium; Dherment & Renneboog (2002) for France; Franks *et al.* (2001) for the UK, and Köke and Renneboog (2005) for Germany and the UK.

[xiii] Furthermore, the following characteristics also explain part of the changes in remuneration: ownership structure (Core *et al.* 1999), board composition (Hallock 1997), threat of takeover (Agrawal & Knoeber 1998), merger and acquisition policy (Girma *et al.* 2002), company risk, growth opportunities, dividend policy (Lewellen *et al.* 1987), and the country where the company is operating (Conyon & Murphy 2000).

6 Strategies for Solutions

Andrew Davies and Michael Hobday

Introduction

Since the early 1990s, a growing number of firms across all types of industries have been changing the focus of their business strategy from making physical products to providing high-value added services, such as distribution, finance and maintenance (Quinn 1992; Oliva & Kallenberg 2003). In this chapter we concentrate on one important strategy, to move into services, which is being adopted widely in high-value capital goods – sometimes called complex products and systems – that are produced on a project basis as one-offs or in small tailored batches (Hobday 1998).[i] In a shift away from their traditional product or service offerings, Alstom, the train manufacturer, now offers solutions for 'train availability', and Cable and Wireless, the telecom operator, provides 'global outsourcing' solutions for its corporate network customers.

These firms are developing new business models to provide tailored bundles of products and services as high-value 'integrated solutions' from a single source that address each customer's business or operational needs (Slywotzky 1996; Slywotzky & Morrison 1998; Wise & Baumgartner 1999; Shepherd & Ahmed 2000; Davies *et al.* 2001, 2003; Davies 2003, 2004; Sandberg & Werr 2003). Integrated solutions refer to a variety of new types of large capital goods projects to specify, design and integrate systems and provide services to finance, maintain and operate a product during its life cycle, such as contracts for turnkey, outsourcing, design–build–operate, and public–private partnerships. Providers of integrated solutions undertake many of the core activities, such as systems integration and maintenance, previously performed internally by large business or government customers, such as corporate IT users, airlines and telecom operators.

Drawing upon in-depth case-study research, the chapter argues that suppliers of capital goods have developed strategies for integrated solutions by changing their positions in the industry value stream. Rather than simply moving 'downstream' from manufacturing to services, as some authors suggest (Wise & Baumgartner 1999), our research reveals that suppliers are moving into integrated solutions from upstream and downstream positions to occupy the high-value space situated 'between' manufacturing and services. Previous studies have neglected to identify the core 'systems integration' capabilities required to design and 'integrate' internally or externally developed components – product hardware, software and services – into a functioning system, while coordinating the activities of internal or external component, subsystem or product manufacturers (Prencipe *et al.* 2003).

An in-depth knowledge of their customers' operational needs and of the system they have designed, places systems integrators in a strong position to provide services (including consultancy advice and finance) to operate and maintain a product during its life cycle. These systems integration and service capabilities can be developed in-house, but often firms have to develop partnerships with external suppliers of key components of a solution. Whether originally based in manufacturing or services, our empirical research shows that the provision of integrated solutions also calls for far-reaching transformations in the way firms are organised to win bids and execute projects. Four of our case study firms are following IBM's example by creating new types of customer-facing organisations. Flexible and responsive project organisations are being established to manage all front-end activities with the customer from initial engagement, through bid preparation and project execution, to the provision of services during the operational life of the product or system. Back-end units organised along functional or business process lines are being established to provide the front-end units with the standardised and replicable product platforms and service portfolios they need to provide customers with tailored solutions. This new model of customer-centric organisation represents a major change in the industrial landscape of major capital projects.

The chapter is based on the findings of a three-year collaborative research project[ii] which examined recent changes in the strategies of five international firms:

- Alstom Transport – rolling stock and signalling systems
- Ericsson Mobile Systems – mobile phone networks
- Thales Training and Simulation – flight simulators
- WS Atkins – infrastructure and the built environment
- Cable and Wireless (C&W) Global Markets – corporate telecom networks.

A case study method was chosen to analyse strategic decisions and motivations to move into the provision of integrated solutions. In-depth interviews with up to ten senior project managers and directors in each of the five firms were conducted in 2000. This provided an opportunity to study the key motivations and drivers encouraging the firms to move into integrated solutions. The managers were asked to describe and explain strategic changes in the focus of each firm's activities between 1995 and 2000. This was followed, during the period 2001–2003 by in-depth case studies of the business organisation and two major integrated solutions projects in each of the firms to verify the extent and nature of moves in the value stream. The cross-sectoral sample of firms was designed to examine the differences and similarities in firm strategies across manufacturing and services sectors. However, in a dynamically changing technology and market environment, firm strategies are always changing, so that lessons about integrated solutions provision that may be true for the past decade will not necessarily be true for the next.

The chapter is divided into three sections. The first outlines the main features of the integrated solutions business model. It develops the value stream approach which is used in the second section to examine the strategies our case study firms have been developing to move into integrated solutions. The third section examines the capabilities and organisational structures that these firms have been establishing to implement their integrated solutions strategies.

Integrated solutions: a value stream approach

In a brief review of the literature, this section outlines the main features of an integrated solutions business model. A value stream framework is developed, based on our case study findings, to analyse the ways in which firms are occupying new positions in their industry supply chains to provide integrated solutions.

The integrated business model

The attraction of high-value services is encouraging firms to rethink the focus of their manufacturing strategies. Several authors claim that the traditional sources of competitive advantage in manufacturing – backwards integration, developing superior products, and scale economies – are no longer sufficient to guarantee competitive success in many industries (Slwotzky & Morrison 1998, p. 249; Wise & Baumgartner 1999; Oliva & Kallenberg 2003). They argue that firms are increasingly competing by building on their 'core manufacturing capabilities' and 'moving downstream' (Wise & Baumgartner 1999) or 'transitioning' (Oliva & Kallenberg 2003) into the provision of high-value services that address each customer's needs.[iii] Services are attractive because they provide continuous streams of revenue, tend to have higher margins and require fewer assets than product manufacturing. For many large manufacturing firms, services represent an increasing proportion of their total revenues. In 2001, for example, the revenues IBM obtained from services (43 per cent) overtook hardware and technology (42 per cent) for the first time in the firm's history (Gerstner 2002, p. 363).

The strategies developed by firms to move into services vary widely across industries. In consumer goods, cars, domestic appliances and soft drinks, firms have concentrated on moving into highly profitable distribution and after-sales services, offering final consumers a range of comprehensive services, such as consumer credit, maintenance contracts and short-term warranties. In capital goods, the focus of this chapter, some of the world's leading capital goods manufacturers – IBM, General Electric (GE) and ASEA Brown Boveri (ABB) – pioneered the move into services in the early 1990s and were among the first firms to create the integrated solutions business models that other suppliers are now emulating. Integrated solutions add value by creating unique benefits for each customer.

Beyond bundling and customer-centric

Integrated solutions can be distinguished from traditional bundling strategies. A product bundle comprises entirely standardised components, at set prices, and is offered on the condition that the customer purchases the full line of internally developed products, irrespective of the differences in customer needs or capabilities (Porter 1985, p. 425). In the early 1980s IBM, for example, sold low-cost, standardised bundles of internally developed personal computer hardware, software and service support. Integrated solutions, by contrast, are provided as a bundle of products and services that are customised and priced according to a specific customer's needs (Hax & Wilde 1999, p. 13). By the 1990s, firms like IBM recognised that providers of integrated solutions must be prepared to tailor individual components to each customer's unique requirements and willing to specify a competitor's product if this provides the best solution to a customer's needs.

Providing solutions that address each customer's needs means that firms have to understand how value is created 'through the eyes of the customer' (Slywotzky & Morrison 1998, p. 18; Wise & Baumgartner 1999, p. 135; Galbraith 2002a). Under the traditional 'product-centric' approach to value creation, firms have focused their efforts on making, selling and delivering products. In many cases, managers focused their efforts on meeting traditional measures of project success (i.e. within cost, on schedule and to the required specifications) rather than attempting to add value by creating unique benefits for the customer, beyond basic technical support and short-term warranties, after the product was 'handed over the wall' on completion of the project and the customer took over responsibility for operating, maintaining and financing it during its lifespan.

Adopting customer-centric thinking involves a 'backwards-thought process' (Womack & Jones 1996, p. 18). Engaging in a close dialogue with their customers, suppliers must first identify their customer's business needs and then develop the capabilities to offer products and services that link uniquely well to a customer's priorities. For example, in a move away from its traditional design and build activities, Alstom Transport now offers to provide its customers – UK train operating companies – with complete transport solutions for train availability during the life cycle of the product. Its first project of this kind, awarded in 1995, was to renew the train fleet on London Underground's Northern Line. Rather than specify the size of the total fleet, the contract only required that 96 trains be available for service each day for the duration of a 20-year contract. To achieve the customer's targets for train availability, Alstom built 106 trains and set up a maintenance organisation to service them.

Repositioning in the value stream

One way of identifying and classifying the capabilities required to offer integrated solutions is to identify the activities they perform in the 'value stream' (Davies 2003, 2004), flowing from raw materials to the final consumer (Womack & Jones 1996, p. 19). A value stream framework, which can be used to analyse, plot and compare strategic requirements, moves into integrated solutions provision.

Until the mid-1990s, a traditional capital goods industry – such as railways and telecommunications – typically consisted of two main vertically-integrated value-adding stages: equipment manufacture and operations. In recent years, these sectors have developed more elaborate divisions of labour, as firms specialise in performing an increasingly narrow range of activities in vertically-disintegrated value streams. To account for this increase in the type and range of activities performed, previous

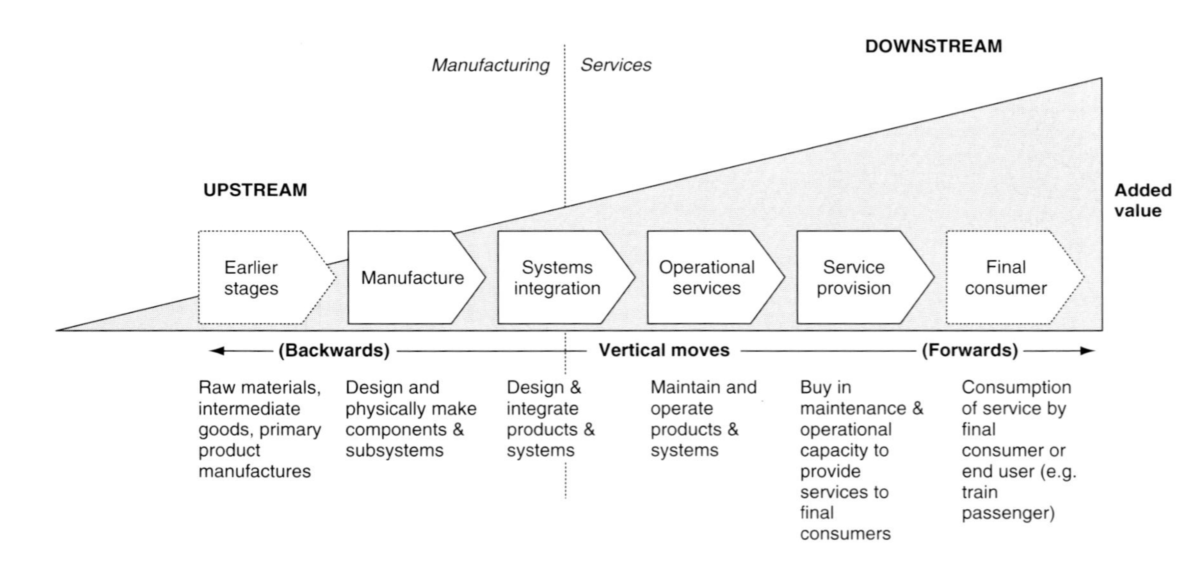

Figure 6.1 Moving base in the capital goods value stream.

research has identified four main value stream stages in a typical capital goods industry, as depicted in Fig. 6.1 (Davies *et al.* 2001; Davies 2003). The stages include:

- *Manufacture*: The first stage is responsible for taking raw materials and sub-assemblies and transforming them into physical components and subsystems that are manufactured to meet an overall system design.
- *Systems integration*: The second stage adds value through the design and integration of physical components – product hardware, software and embedded services – that have to work together as a whole in a finished product.
- *Operational services*: In the next stage, an operator or business user runs and maintains a system to provide services, such as a corporate telecom network, baggage handling, flight simulation training, and train services.
- *Service provision*: In some industries, services are provided to the final consumer through intermediary organisations called service providers. These firms buy in the system capacity they require from external operators and concentrate on brand, marketing, distribution and customer care activities.

The four stages can be illustrated by the example of mobile communications. Since the mid-1990s, suppliers like Ericsson and Siemens have been concentrating on becoming systems integrators and outsourcing a growing proportion of their non-core manufacturing activities to specialised contract manufacturers, such as Flextronics. In addition to traditional operators, a new type of service provider has also entered the market. For example, as a so-called 'mobile virtual network provider', Virgin Mobile specialises in developing its subscriber base through brand image, advertising and customer care activities while buying in network capacity from another operator to carry its radio communications traffic.

Since the mid-1990s, the traditional boundary between upstream suppliers and downstream customers has continued to be redrawn. Buyers of capital goods (operators, services providers and business users) are focusing on the provision of their core products and services and outsourcing non-core activities. To meet the demand for outsourcing, suppliers are undertaking systems integration and operational activities previously performed as part of their customers' businesses. In full outsourcing solutions, this includes the transfer of assets and staff to supplier firms. Buyers of capital goods are entering into long-term partnerships with their suppliers to ensure that providers of solutions share the responsibility and risks of performing outsourced activities.[iv]

The process of customer outsourcing in capital goods has been accelerated by the liberalisation and privatisation of former state-controlled sectors, such as telecoms and railways. For example, Ericsson recognises that its integrated solutions offerings must address the needs of the variety of different customers operating in more competitive mobile communication markets. Whereas experienced and vertically-integrated mobile operators, such as Vodafone, often want to perform a broader range of activities in-house, less sophisticated and more specialised customers with limited in-house capabilities, such as Virgin Mobile, tend to rely on suppliers for complete solutions to their needs.

Suppliers in the UK, such as Thales, Alstom and WS Atkins, are also being encouraged to move downstream by increasing use of private finance in public sector projects. Under the private finance initiative (PFI), introduced in 1992, private sector companies 'design, build, finance and operate' public sector projects, ranging from schools to complex weapons systems. Under the policy of public–private partnership (PPP), adopted in 1997, public projects are financed partly by private firms, while the state shares some of the risk. PFI and PPP suppliers perform activities along the value stream from systems integration to services provision, as well as offering financing and business consultancy services as vital inputs to each stage in the value stream.

Integrated solutions strategies

Building on Penrose's (1959) analysis of the resource base of the firm, Richardson (1972) argued that a firm achieves competitive advantage by developing distinctive capabilities (knowledge, experience and skills) to perform particular activities in its industry. When a firm changes position in the value stream – by moving upstream or downstream – it must develop new capabilities and create new organisational structures. The notion of a base suggests that there is a 'centre of gravity' in the value stream arising from a firm's initial success in the industry in which it grew up (Galbraith 1983, p. 316). Firms develop their capabilities and create organisations that fit their particular industries and value-adding stage. They can be in the same industry but have different capabilities and organisations because of their different starting positions, experiences and initial successes. A firm's centre of gravity 'establishes a base from which subsequent strategic changes take place' (Galbraith 1983, p. 319). Slywotzky and Morrison (1998, p. 19) argue that as a firm succeeds by learning to provide integrated solutions that address customer needs, its centre of gravity moves closer to the customer. But a strategic centre-of-gravity move of this kind is difficult to accomplish without rethinking the traditional focus of the firm in the value stream.

By proposing that firms are simply moving in one direction only – downstream from manufacturing to services – previous studies have failed to recognise the key role of systems integration in the delivery of integrated solutions. All five of the case study firms are moving from upstream and downstream positions to strengthen their capabilities as systems integrators. From their new base in the value stream, each firm is developing additional capabilities – in operational services, business consultancy and financing – to provide complete solutions to its customers' needs. Table 6.1 summarises strategies of the case study firms.

Strategic moves

For the product-based firms, integrated solutions does entail a move downstream from a traditional base in manufacturing. As their traditional centre-of-gravity stage in the value stream – making physical products – has become less profitable, Alstom, Ericsson and Thales have outsourced a growing proportion of their manufacturing activities and are focusing on becoming systems integrators and providers of services to operate, maintain and finance products.

Alstom illustrates the shift in strategies of these product-based firms. Within the diversified Alstom energy and transportation, the Alstom Transport division was responsible for train and signalling system design and manufacture. Since the mid-1990s, Alstom has developed a strategy to evolve from a 'seller of goods to a system and

Table 6.1 The shift to integrated solutions.

Company	Traditional product or service focus (1995)	Integrated solutions (2003)
Alstom Transport – railways	Products: • subsystems (e.g. propulsion, traction, drive, electronic information systems) • rolling stock • signalling and train control systems	Transport solutions (e.g. 'train availability'): • systems integrator – turnkey solutions for project management, fixed infrastructure, and finance • services for maintenance, renovation, parts replacement and service products – 'Total Train-Life Management'©
Ericsson – mobile communications systems	Products: • mobile handsets • mobile system • subsystem products: radio base stations, base station controllers, mobile switches, operating systems, customer databases	Turnkey solutions to design, build and operate mobile phone networks: • mobile systems – complete supplier, systems integrator and partner • global services – services and business consulting to support a customer's network operations
Thales Training & Simulation – flight simulation	Products: • standalone flight simulators for commercial and military aircraft	Training solutions (e.g. 'pay as you train'): • systems integration • training services – networked training, independent training centres for training services, synthetic training environments
WS Atkins – infrastructure and the built environment	Engineering consultancy, project management and technical services for infrastructure projects	Integrated solutions for the built environment: • design, build, finance and operation of infrastructure across industrial sectors • Total Solutions for Industry (TS4i) provides one-stop-shop for design, construction, maintenance and finance
Cable & Wireless Global Markets – corporate networks	Provides 'managed network services' for multinational corporations: • network design • supply telecom infrastructure and applications • network management	Global outsourcing solutions for a multinational corporation's entire telecom and IT needs on a global basis: • network design • network management • supplies telecom infrastructure and applications • ownership of the network • network operation • business process applications • service level agreements

service provider' (Owen 1997). In 1998, a Service Business was created as a result of a strategic review of Alstom's global activities, which recognised the huge growth in the market for rolling stock services, such as maintenance, technical support, product upgrades and renovation. The Systems Business was set up to carry out major systems integration, turnkey and PFI contracts for complete bundles of train, signalling and infrastructure. By 2000, the Passenger Business, traditionally responsible for the design and manufacture of critical subsystems, such as traction systems, was outsourcing the manufacture of up to 90 per cent of components integrated in its rolling stock products and systems.

For the service-based firms, providing integrated solutions requires the development of upstream capabilities as well as moving further downstream into operational service provision. Cable and Wireless (C&W) and WS Atkins are strengthening their capabilities as integrators of systems using components sourced from external manufacturers. WS Atkins, for example, started out as a provider of design engineering consultancy services as a specialised input to the systems integration stage across several industries as diverse as transport, property management, defence and public health. Under a strategic review in 1998, WS Atkins was reorganised to meet customer demands for long-term contracts that entail the provision of 'an increasing range of services' (WS Atkins 1999, p. 6). The firm's strategic vision is to be a customer-focused, service-based organisation which is 'the world's first choice supplier for technical services and integrated solutions for the built environment' (WS Atkins 1999, p. 4). In April 1999, the group was reorganised again to focus on the provision of integrated solutions across three consolidated UK national business streams – called Property, Transport, and Management and Industry.

Strategic imperative and choices

To move successfully into the provision of integrated solutions, our research shows that a firm must gain control of the channel to market. Channel control is imperative in capital goods sectors where suppliers sell directly to a few large customers in oligopolistic markets, because any attempts to move downstream may be blocked if a firm sells its products and services through independent firms that control the channel to the customer. For example, Thales's attempts into flight simulation training solutions for commercial airlines have been prevented by the specialised independent training schools which purchase simulators and control channels to market. Performing a role similar to distributors or retailers of consumer goods, these training schools have resisted attempts by Thales and other producers to enter the training market. As major airlines have outsourced training, it has been the training schools, rather than simulator producers, which have taken on the training tasks, despite the efforts of Thales and other producers to move into training services.

Such channel conflicts can be overcome if the firm can develop partnerships with channel controllers or buy the channel. For example, C&W's initial attempts to move into the high-value end of corporate networks markets were hampered by the presence of global business consultancy organisations, such as Accenture and PriceWaterhouseCoopers, which use their scale and global reputation to exploit and control channels to the business user. To gain a channel to market, C&W actively pursued and finally formed a strategic partnership with Accenture. C&W now provides global data and Internet protocol (IP) components of outsourcing solutions provided to Accenture's clients.

Alstom and Ericsson, by contrast, have not faced such channel conflicts as they move downstream because they already have direct relationships with their customers – railway and mobile phone operators. They may, however, face conflicts with their customers if they move too aggressively into the customers' territory, or move without prior agreement.

Once channel control has been established, a firm faces two strategic choices about where to position itself in the value stream. The first choice concerns the scope of systems integration. Firms providing solutions are developing the systems integration capabilities to provide single-vendor and multi-vendor systems. Single-vendor systems are 'internally' developed technology, components and subsystems developed by vertically-integrated manufacturers, such as Thales's proprietary flight simulation products.

Multi-vendor systems are assembled or integrated from 'externally' developed components. WS Atkins, for example, designs and project manages the integration of systems supplied by external manufacturers across diverse sectors. Its railway division, WS Atkins Rail, buys and integrates equipment from Alstom, Bombardier and Siemens as well as more specialised suppliers. Alstom and Ericsson are developing the capabilities to integrate both types of systems. Alstom, for example, designs, manufactures and builds rolling stock and offers to integrate equipment supplied by its competitors, Bombardier and Siemens as part of its turnkey solutions offerings.

The second choice relates to the spread of a firm's industrial activities, vertically within an industry and horizontally across different industry segments. Vertically-integrated strategies focus on the provision of integrated solutions to customers within a specific industry, such as C&W's development of global outsourcing solutions for multinational customers. Horizontally-integrated strategies focus on the provision of integrated solutions to customers across different industries.

Using our case study findings, we can begin to distinguish between different types of integrated solutions strategies, represented as a matrix – shown in Fig. 6.2 – along two different dimensions: the scope of systems integration; and the vertical or horizontal spread of industrial activities (Davies 2003). The matrix reveals that there is a great deal of variety in the strategies pursued by the case study firms. It can be used to analyse the strategy of a division within a diversified firm, such as Thales Training and Simulation (which provides single-vendor solutions within a single industry). Some of these firms, like WS Atkins, specialise in multi-vendor solutions and sell their services to horizontal industry segments. Others, like Ericsson, provide single- and multi-vendor solutions within a single industry. As a diversified firm, Alstom provides single- and multi-vendor solutions to customers across a number of energy and transportation sectors.

Implementing the strategy

In this section, we examine how firms have been implementing their strategies for integrated solutions. We argue that a firm's original base in the value stream shapes

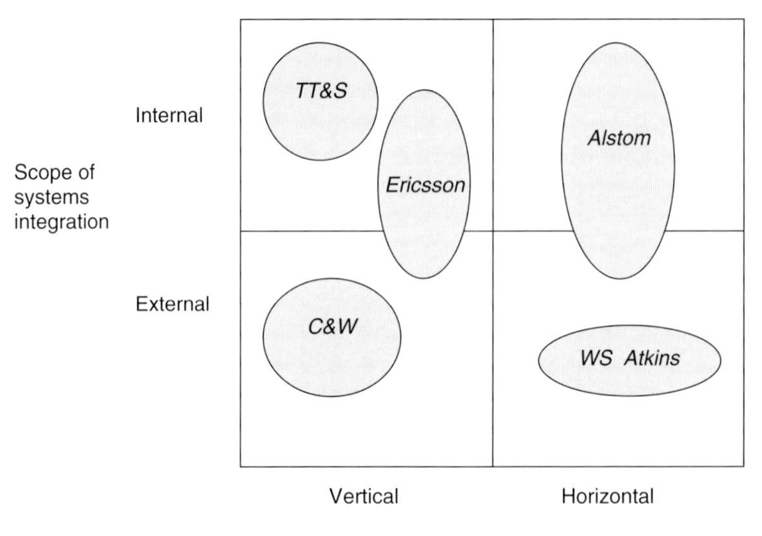

Figure 6.2 Positioning integrated solutions providers.

the types of capabilities and organisational structures it must establish when it moves position.

Building integrated solutions capabilities

The case study findings provide an opportunity to consider the types of capabilities required to compete successfully in the provision of integrated solutions (Davies *et al.* 2001; Davies 2003), as summarised in Table 6.2. This section develops the hypothesis that firms moving into the provision of integrated solutions need to build a core capability in systems integration. But to offer complete solutions to their customers' needs throughout the life cycle of a product or system, firms are offering combinations of operational services to support life-cycle solutions, and financing and business consulting services.

Systems integration

To provide customers with physical products that can easily be deployed with services as part of a solution to a customer's needs, the five case study firms are developing their systems integration capabilities. Traditionally, the product-based firms designed, manufactured and integrated systems using in-house developed components. By the late 1990s, Alstom, Ericsson and Thales were developing strategies to focus on the systems integration function, while outsourcing a growing proportion of their manufacturing activities.

As service-based firms with no in-house manufacturing capabilities, C&W and WS Atkins are specialised systems integrators. Performing little or no technology development in-house, these firms specialise in providing systems integration services

Table 6.2 Capabilities of integrated solutions providers (2003).

Company	Systems integration	Operational services	Business consulting	Financing
Alstom Transport	Design and build trains and signalling systems, using equipment developed in-house or externally. Act as prime contractor in large turnkey projects	Maintain, upgrade and operate trains	Consultancy-based approach to meet customer needs	Vendor financing and asset management
Ericsson	Design, manufacture and integrate mobile phone systems, using equipment developed in-house or externally, e.g. 'multi-vendor' systems	Maintain, support, upgrade and operate mobile networks	Two business consultancy organisations to meet needs of Ericsson and external customers	Considering, but not yet offering, vendor financing
Thales Training & Simulation	Design and integration of flight simulators Co-ordination of external contractors for component supply	Provide services to train pilots and manage simulator building facilities Joint venture with GE Capital Training	Consultancy organisation to meet customer needs	Revenue sharing agreement for simulators, e.g. split between TT&S and United Airlines
WS Atkins	Design and integrate external manufacturers' equipment across diverse sectors, such as railway and baggage handling systems Co-ordination of external contractors for component supply	Maintain, operate and provide services to end-users, e.g. setting up independent service provider to design, build, finance and operate baggage handling services	Consultancy-based approach to meet customer needs	Created joint-venture company, TS4i, with Royal Bank of Scotland to provide integrated solutions for design, construction, maintenance and finance
Cable & Wireless Global Markets	Design and integrate networks using externally supplied equipment Developing capability to integrate internet and IT systems Co-ordination of external contractors for component supply	Design, build, operate and manage a global customer's IT and telecom needs	Consultancy-based approach to meet customer needs	Sometimes takes on responsibility for ownership of networks for duration of contract

using components sourced from external manufacturers. C&W, for example, is developing partnerships with new 'best-of-breed' IP suppliers such as Nortel and Cisco Systems to provide corporate customers with systems that C&W designs, installs, maintains and supports.

The range of activities performed by systems integrators is increasing as a result of customer demands for turnkey solutions. Under contracts for turnkey solutions, the supplier is responsible for the entire set of activities involved in the design, integration, construction, testing and delivery of a fully functioning system. All the customer has to do is turn a key. If other products, services or capabilities are required to provide complete solutions to a customer's needs, systems integrators co-operate with partners in joint ventures or consortia to carry out those portions of the work. For example, Alstom Transport has established its Systems business unit to integrate components, subsystems and services developed internally or externally by partners in a consortium. By combining skills in project management, systems integration, financial engineering, fixed infrastructure and civil engineering, the Systems business is able to provide track infrastructure, rolling stock and signalling systems as a single turnkey package.

Operational services

Suppliers are building on their base in systems integration and moving into provision of operational services to maintain, renovate and operate a product or system during its lifespan from sales to decommissioning. Two product-based firms, Alstom and Ericsson, have set up new divisions to provide these services. For example, Alstom's Service business unit offers services – which it calls 'Total Train-Life Management' – to capture value created during all stages in the 30 year operating life cycle of a train, such as maintenance, renovation, spare parts, and asset management.[V]

As they take over operational activities, suppliers have an incentive to design systems from the start that are reliable and easily maintainable. Lessons learnt can be fed back into the design and build of current and future generations of systems. Because manufacturers like Alstom, Ericsson and Thales develop technology as well as integrate systems and perform operational services, they are able to create new feedback loops within different parts of the same firm. This can initiate a cycle of innovative improvements between systems integration and operational service activities, leading to the design of more reliable and efficient systems being built in the future. In the case of Alstom's contract for the Northern Line extension of the London Underground, the managers responsible for maintenance and operational services were deeply involved in the front-end design of the rolling stock. As a result of their recommendations, the train designers made more than 250 modifications to create easy-to-maintain and easy-to-use trains. By contrast, pure systems integrator firms like WS Atkins and C&W that rely on external manufacturers for equipment and technology development are unable to take advantage of these dynamic feedback loops.

Business consultancy

As part of an integrated solutions bundle, firms are offering customers consultancy advice on how to plan, design, build, finance, maintain and operate systems (Sandberg &

Werr 2003). Firms are expanding their business consultancy capabilities by creating joint ventures with other firms that have such capabilities; acquiring firms already operating in this field; or by developing business consultancy skills in-house.

Some of the case study firms, such as Ericsson and WS Atkins, have developed these skills internally by establishing specialist business consultancy organisations. Ericsson Global Services helps the company's market-facing units provide customers with consultancy advice on their strategies for mobile communications, such as how to write business plans, produce network designs, finance and manage their assets, and develop applications for third generation (3G) services. Alstom and Thales have developed a consultancy-based approach within their existing business units. C&W, by contrast, has been seeking a strategic partnership or joint venture with one of the major business consultancy organisations because in global corporate network markets these firms dominate the value stream and control access to the customer.

Financing

The ability to provide finance is the fourth capability being developed by some integrated solutions providers. Financial services play a vital role in the negotiation phase when customers require assistance with financing the purchase of high-cost products. Firms can offer value-sharing contracts which lower the purchase price of a product in return for a proportion of the future value generated during the operational phase. The supplier is paid in part by sharing in the operational efficiency that it creates for the customer. Value-sharing contracts provide an opportunity to engage in strategic discussions with customers during the negotiation stage and to open doors to many other projects that might otherwise have been unavailable to the firm.

The growing importance of private finance in the UK is generally associated with large public sector PFI and PPP projects. But our case studies suggest that it has also grown in importance in recent years as an industry-led initiative to provide vendor financing and asset management services in capital-intensive telecom, railway and other large infrastructure systems. Vendor financing is driven by the high costs of constructing new systems. In 3G mobile phone markets, for example, vendor financing is being offered to help mobile operators with limited funds to build 3G mobile phone networks on expectation of payment at a later date. But suppliers vary in their approach to vendor financing: whereas Nokia has used vendor financing to gain market share, Ericsson has been less willing to be financially exposed in this way.

Asset management is also of growing importance as a service for customers, such as train operating companies, seeking to reduce the costs and extend the operating life of an installed base of products. In 2000, for example, WS Atkins created 'Total Solutions for Industry' – a joint venture with the Royal Bank of Scotland – to provide customers with a one-stop-shop source of integrated solutions for finance together with design, construction and maintenance. Serving contracts with an asset value of between £5m and £20m, the joint venture offers to manage assets for customers such as mobile telephone base stations, baggage handling systems and power stations. The bank supplies the finance (and specific financial services such as equity savings) and WS Atkins undertakes design, construction management and asset management.

An attempt to capture the variety of capabilities offered as part an integrated solution is shown in Fig. 6.3. It identifies the new combinations of systems integration and service capabilities that firms are developing in-house or through strategic partnerships to meet the differing needs and capabilities of various customer segments.

Organising to deliver integrated solutions

This section examines some of the new forms of organisation being developed by the five case study firms to implement their integrated solutions strategies. In moving away from their traditional product- or service-based business units, several of the firms have been creating customer-focused organisations to harness and leverage the resources and capabilities – from internal and external suppliers – to provide efficient and repeatable solutions (Davies *et al.* 2003).

Repeatability is a measure of a firm's progress in integrated solutions provision. The high costs of developing the capabilities to provide initial one-off solutions for lead customers have to be recouped by replicating the product and service components of solutions until they become standardised offerings, sold repeatedly to many customers at lower costs (Davies & Brady 2000; Galbraith 2002a).

Previous studies of the customer-focused organisations have focused on the ways in which large manufacturing firms have reorganised their internal activities for the delivery of repeatable solutions (Foote *et al.* 2001; Galbraith 2002a,b). They describe a model of solutions provision implemented by vertically-integrated manufacturing firms like IBM, Sun Microsystems, ABB and Nokia. In a shift away from traditional structures with operational units organised along product, brand or geographical lines, these firms are forming front-end units to develop, package and deliver integrated solutions, refocusing their product-based units as back-end providers of standardised solutions-ready components, and developing strong strategic centres to manage the interfaces between the two types of operational units.

Although useful, the model is applicable only to vertically-integrated manufacturers that are reorganising their internal organisations, because it assumes that back-end units are internal to the firm and that product units may continue to sell their own products to external customers. We offer an alternative model for integrated solutions provision developed as an ideal type, which can be used to analyse and compare the different structures being set by the case study firms (Fig. 6.4). It identifies the operational units in a model of organisation applicable to firms traditionally based in manufacturing and services. The model consists of front-end, customer-facing units (CFUs) set up as project organisations which control the channel to the customer. Back-end units – either divisions within the firm or external suppliers – are responsible for providing CFUs with the products, service portfolios, capabilities and resources they need to provide customers with tailored solutions.

The five case studies are establishing different versions of this model depending on whether the firm started out from a base in manufacturing or services. Alstom and Ericsson, for example, have established new operational units in addition to their product-based business units to provide 'service' components of solutions. C&W and WS Atkins, by contrast, are establishing long-term partnerships with external manufacturers to provide the 'product' component of an integrated solution.

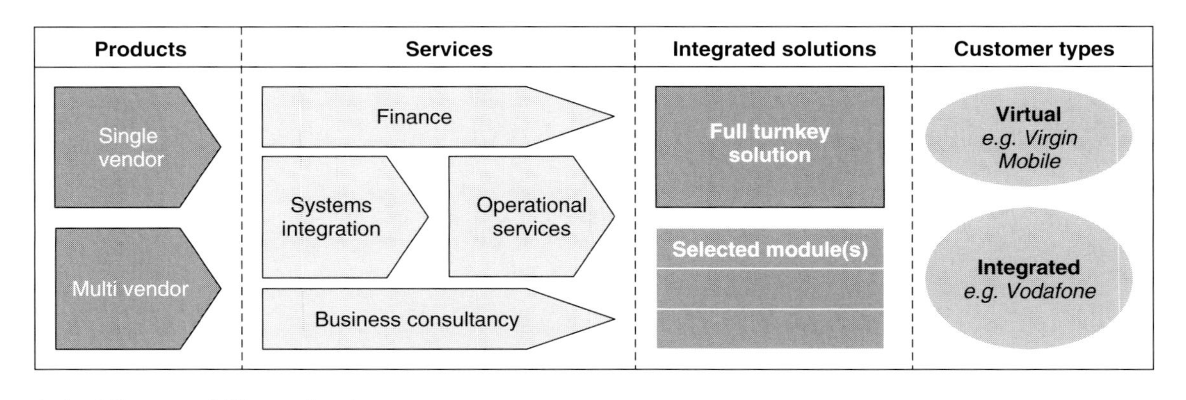

Figure 6.3 Integrated solutions capabilities and customers.

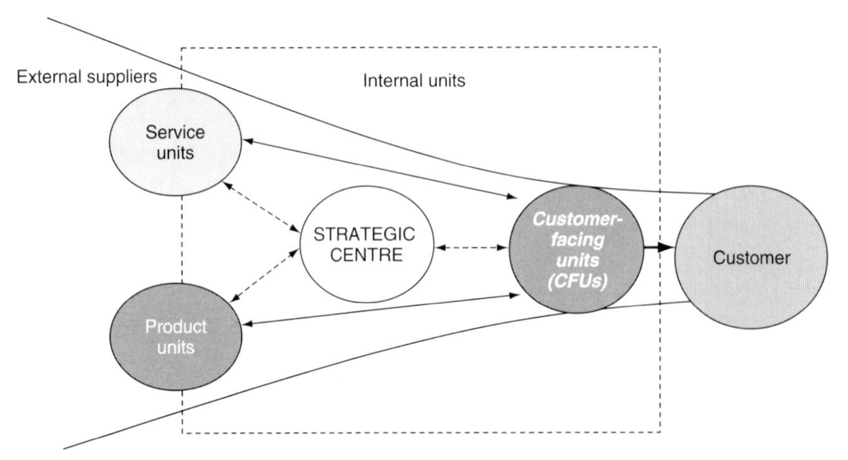

Figure 6.4 Customer-facing organisation.

Front-end units: customer-facing solutions providers

The front-end CFUs are responsible for leveraging the capabilities required to understand and meet a customer's needs for an integrated solution. These units design and integrate the systems and arrange the provision of services during the operational life of a product. Thales Training and Simulation is the only firm in our study which has attempted to mobilise the additional capabilities required for integrated solutions provision without altering its traditional product-based organisation. The other firms have created new types of CFUs to meet the needs of different customer segments, including:

- Industry-based segments
- Customer groups within an industry
- Single customers
- Single projects.

Four of our case study firms have restructured their entire organisations to focus on different customer segments. Alstom, for example, has set up a variety of different customer-facing structures: the Services unit provides rolling stock operational services; the Systems unit offers turnkey solutions for rolling stock and signalling systems. In addition, it establishes special purpose vehicle project organisations, such as the West Coast Train Care Company which Alstom set up to meet Virgin Train's needs for the UK's West Coast Main Line. Ericsson has refocused its entire organisation to create market units for customers in different geographical regions and CFUs for global mobile phone operators, such as its Ericsson Vodafone division which is dedicated to meeting the mobile operator's needs on a global basis.

Service-based firms such as C&W and WS Atkins have created front-end structures able to leverage external sources of 'best-in-class' technology and products. C&W has created partnerships with so-called 'best-of-breed' IP suppliers so that its front-end Global Markets division can meet the needs of different groups of corporate customers with various systems that C&W designs, installs, maintains and supports. WS Atkins has set up a number of front-end units to deliver solutions based on leading-edge technology across different industry-based customer segments. For example, its railway division, WS Atkins Rail, buys and integrates equipment from Alstom, Bombardier and Siemens as well as more specialised suppliers.

The front-end units are organised so that the entire capabilities of the firm and its external partners are channelled through the project implementation team at the point of contact with the customer (Davies *et al.* 2003). CFUs are set up as project-based organisations to respond rapidly and flexibly to different customer needs, acting as 'flexible resource units' for constantly forming projects. However, resources from the firm's back-end units and network of external partners may be required when resources cannot be provided internally by the front-end unit. Project teams, composed of people from the front-end unit, its partners and the customer's organisation, can be easily assembled, disbanded and reassembled around the needs of each customer's solution. Teams operate in ways similar to consultancy organisations, moving routinely from one project to another (Galbraith 2002a, p. 204).

Project teams are set up to engage with solutions customers during an extended project life cycle (Davies *et al.* 2003). The CFU is responsible for engaging with existing or potential customers about their strategic needs and operational priorities, often before the customer has requested formal assistance. In these senior management pre-bid discussions, solutions providers offer consultancy advice on how to improve a customer's operations or develop its business model.

If the customer requests help or issues an invitation to tender, a proposal team is assembled to develop a value proposition (Sharma & Molloy 1999; Cornet *et al.* 2000; Foote *et al.* 2001; Bennett *et al.* 2001). If the customer is a strategic partner, the proposal team makes an 'offer' rather than a bid. The proposal team requires people with skills and expertise in commercial management, technical design and project management. The proposal team is responsible for ensuring that the value created by the solution during systems integration and operational service provision will meet the customer's expectations. It has to work quickly to develop the proposition. The exact degree of customisation specified in the proposal will vary according to the sophistication, size and unique requirements of each customer. Rather than sell solutions from predefined price lists, the team has the authority to decide how to tailor products and services to solve specific customer problems. They must break down an integrated solution into its component parts. Each part is then scheduled and assigned to internal or external product and service suppliers.

Unlike traditional arms-length bargaining over products or services, an integrated solution requires a commercial framework based on co-operation and trust. This helps the proposal team and customer develop a common understanding of how value will be measured in terms of pricing and margins, the volume and mix of products and services, capital costs and distribution of risk.

Back-end units: product platforms and service portfolios

Back-end product and service units are organised to provide resources and capabilities that can be leveraged and integrated by the front-end project teams. Several of our case study firms, including WS Atkins, C&W and Ericsson, have been developing structures which prevent the back-end units from having direct relationships with the customer. In this way, the back-end capabilities of the firm and its network of external suppliers are channelled to the customer through a front-end unit. These units – whether internal or external to the firm – are responsible for developing standardised solutions-ready components.

Product-based units supply the technology and physical components of integrated solutions delivered by the front-end units. In-house product-based organisations often view the creation of front-end solutions providers as a threat to their power base and status within the firm. Under the new organisational arrangements, these internal product units have to become flexible and responsive to the front-end demands for resources and capabilities. They must provide tailored components, products and systems, particularly where the front-end units are developing highly customised solutions based on new generations of technologies for lead customers, such as Ericsson's 3G product portfolio developed for Vodafone. Product-based units are also responsible for developing a platform of standardised or modular products that can be easily configured at low cost as part of a solution to each customer's needs.

Service-based units are responsible for developing a portfolio of simplified, consistent and easy-to-understand services that the front-end project teams use to sell and deliver integrated solutions. Rather than develop customised services for each new project, the front-end solutions providers can select pre-developed service offerings from the portfolio. Back-end service units can also provide the front-end teams with additional resources and support when the required capabilities cannot be found locally. Based on standardised processes, tools, pricing and guarantees for service reliability, the service portfolio is constantly revised to improve the process of selling and delivering solutions.

Services can be provided by internal business units or external service providers. Product-based firms like Alstom and Ericsson have set up stand-alone service divisions to develop service portfolios. Ericsson's Global Services division, for example, was formed in 2000 to develop a portfolio of services that can be used by its front-end units. The division has no direct contact with Ericsson's customers. Instead it provides resources and service portfolios that can be configured as part of each solution delivered to mobile phone operators. Service-based firms may also need to access a broader range of services than they can traditionally provide in-house, such as WS Atkins's joint venture with the Royal Bank of Scotland to provide customers with financial services as a part of an integrated solutions package.

Strategic centre

The strategic centre is responsible for encouraging collaboration between the front- and back-end units and between the firm's internal and external activities. It develops corporate functions, consistent product and service offerings, processes, pricing and brand names to support global delivery of solutions. Unless a strong centre can provide effective management of the interfaces between the front and back units, these organisations may pursue diverging strategies, becoming deadlocked over priorities, and pass accountability for success and failure back and forth (Foote *et al.* 2001, p. 93).

At Ericsson, for example, the product-based units – such as Ericsson Mobile Systems – no longer sell directly to the customers. In a radical restructuring of Ericsson's global activities, they have become internal providers of technology and solutions-ready products to the firm's front-end market units, which serve customers in different regions, and customer-facing units such as Ericsson Vodafone, which are dedicated to the needs of global mobile phone operators. To manage the interfaces between its internal product and service units, Ericsson has created corporate functions to plan and allocate resources, to manage transfer pricing, and to support competence development throughout the group.

When firms create the customer-facing units with sole responsibility for channels to market and customer relationships, they often face strong resistance from the in-house product or service units. These internal units are reluctant to lose control of their own customer accounts and unwilling to accept that the front-end units will propose solutions that incorporate a competitor's technology or products. A senior management team has to be prepared to resolve any conflicts by overcoming any resistance to the new structure from the firm's established units. For example, C&W's strategy to create a global structure for integrated solutions provision met strong resistance from its regional business units (RBUs) who were unwilling to relinquish control of their profitable corporate customers. In response to this internal resistance, as well as an unfavourable market environment due to the collapse of telecom markets in 2001, C&W's senior management team decided to re-establish its traditional structure of delivering solutions through its RBUs. Ericsson, by contrast, continues to fashion an organisation for global integrated solutions composed of front and back units overseen by a strong corporate centre.

Conclusions

This chapter has presented a value stream framework to analyse, plot and compare the strategies of firms moving into the provision of integrated solutions. The framework has helped to show that rather than simply 'going downstream' or 'transitioning' into solutions, integrated solutions providers are developing core capabilities as systems integrators and service providers. To move successfully into the provision of integrated solutions, we have argued that it is imperative that firms gain control of the channel to market. Failure to do so will prevent the firm from developing the close customer relationships required to provide integrated solutions. Once channel control is assured, a firm faces two strategic choices about where it should be positioned in the value stream. First, it can specialise in one particular activity, such as systems integration, or integrate forwards and backwards within a single industry's value stream. In addition to these vertical moves, a firm can provide integrated solutions horizontally to customers operating across different industries.

Second, a firm must decide whether to provide single vendor or multi-vendor products as part of the systems integration services. Some authors argue that the 'acid test of a true solutions provider' is a willingness to propose a solution that incorporates a competitor's product or technology if this provides an optimal solution to a customer's needs (Foote *et al.* 2001, p. 87). Indeed, a key competitive advantage of service-based firms is their ability to specify 'best-in-class' products as multi-vendor solutions, often by forming strategic partnerships with equipment manufacturers. While some product-based firms, like Alstom and Ericsson, now offer multi-vendor systems, others like Thales remain single vendor providers. Firms that remain tied to internally developed proprietary technology have to be confident that they can persuade their customers that their products are best in class. However, an involvement in technology development enables product-based firms to benefit from dynamic feedback loops not available to service-based firms: knowledge of operational performance can be used to make technical improvements in current and future product generations.

In a departure from the traditional product or service-based organisations, several of the case study firms are creating new customer-facing structures to mobilise the capabilities from inside and outside the firm required to deliver integrated solutions. Front-end units act as a distribution channel to the customer for products (single and multi-vendor) and services (systems integration, operations, finance and consultancy) supplied by a firm's in-house units or external partners. Back-end units are responsible for developing solutions-ready components that can be incorporated in the solutions designed and integrated by the front end. A strong 'strategic centre' is required to implement strategies for integrated solutions and to co-ordinate internal interfaces between different units within the organisation and external interfaces with the customer, supplier and other collaborators.

The challenges of moving into integrated solutions provision should not be underestimated. To be able to develop a profitable business, firms have to gain control of the channel to the customer, avoid moving so far downstream that they begin to compete with their customers, and manage the risks associated with financing life-cycle solutions. For many firms, the biggest challenge will be developing the capabilities and organisations required to integrate different pieces of a system provided increasingly by an external network of specialised component suppliers, subcontractors and service providers. This is a challenge which may take many years: IBM, for example, is still engaged in the process of implementing a strategy for integrated solutions initiated by its former chief executive officer, Louis Gerstner, in the early 1990s.

The rise of integrated solutions calls for more in-depth research on the implications of this new activity for project activities and project business organisations. For example, the traditional project life cycle now extends over many years or even decades and includes important pre-bid and post-project phases of activities. Our research points to four main phases in an integrated solutions life cycle: (1) engaging with the customer in high-level strategic negotiations, often before an invitation to tender has been issued; (2) working closely with the customer to develop a value proposition during a bid or offer phase; (3) project managing the systems integration process; and (4) operating a product or system during a specified contractual period. Future research is required to analyse and explain how suppliers and customers are working together in strategic partnerships to identify, capture and share in the added value created over such an extended life cycle.

References

Bennett, J., Sharma, D. & Tipping, A. (2001) Customer solutions: building a strategically aligned business model. In: *Insights: Organization and Strategic Leadership Practice*. Booz Allen and Hamilton, Boston, MA; pp. 1–5.

Cornet, E., Katz, R., Molloy, R., Schädler, J., Sharma, D. & Tipping, A. (2000) Customer solutions: from pilots to profits. In: *Viewpoint*. Booz Allen and Hamilton, Boston, MA; pp. 1–15.

Davies, A. (2003) Integrated solutions: the changing business of systems integration. In: *The Business of Systems Integration* (eds A. Prencipe, A. Davies & M. Hobday), Oxford University Press, Oxford; pp. 333–368.

Davies, A. (2004) Moving base into high-value integrated solutions: a value stream approach. *Industrial and Corporate Change*, **13**(5), 727–756.

Davies, A. & Brady, T. (2000) Organizational capabilities and learning in complex product systems: towards repeatable solutions. *Research Policy*, **29**, 931–953.

Davies, A. & Hobday, M. (2005) *The Business of Projects: Managing Innovation in Complex Products and Systems*. Cambridge University Press, Cambridge.

Davies, A., Tang, P., Hobday, M., Brady, T., Rush, H. & Gann, D. (2001) *Integrated Solutions: The New Economy between Manufacturing and Services*. SPRU-CENTRIM, Brighton; pp. 1–43.

Davies, A., Brady, T. & Tang, P. (2003) *Delivering Integrated Solutions*, SPRU-CENTRIM, Brighton; pp. 1–34.

Department of Trade and Industry (DTI) (2002) *The Value Added Scoreboard*. DTI Business, Finance and Investment Unit, London.

Domberger, S. (1998) *The Contracting Organization: A Strategic Guide to Outsourcing*. Oxford University Press, Oxford.

Foote, N.W., Galbraith, J.R., Hope, Q. & Miller, D. (2001) Making solutions the answer. *McKinsey Quarterly*, **3**, 84–93.

Galbraith, J.R. (1983) Strategy and organization planning. *Human Resource Management*. Reprinted in: H. Mintzberg & J.B. Quinn (eds) (1991) *The Strategy Process*, 2nd edn, Prentice Hall, Upper Saddle River, NJ; pp. 315–324.

Galbraith, J.R. (2002a) Organizing to deliver solutions. *Organizational Dynamics*, **31**(2), 194–207.

Galbraith, J.R. (2002b) *Designing Organizations: An Executive Guide to Strategy, Structure, and Process*. Jossey-Bass, Wiley, San Francisco, CA.

Gerstner, L.V. (2002) *Who Said Elephants Can't Dance? Inside IBM's Historic Turnaround*. Harper Collins, London.

Hax, A.C. & Wilde, D.L. (1999) The delta model: adaptive management for a changing world. *MIT Sloan Management Review*, Winter, 11–28.

Hobday, M. (1998) Product complexity, innovation and industrial organisation. *Research Policy*, **26**, 689–710.

Kay, J. (1993) *Foundations of Corporate Success*. Oxford University Press, Oxford.

Oliva, R. & Kallenberg, R. (2003) Managing the transition from products to services. *International Journal of Service Industry Management*, **14**(2), 160–172.

Owen, D. (1997) GEC Alstom in career discussions. *Financial Times*, 19 November.

Penrose, E. (1959) *The Theory of the Growth of the Firm* (3rd edn 1995). Oxford University Press, Oxford.

Porter, M.E. (1985) *Competitive Advantage: Creating and Sustaining Superior Performance*. Free Press, London.

Prencipe, A., Davies, A. & Hobday, M. (eds) (2003) *The Business of Systems Integration*. Oxford University Press, Oxford.

Quinn, J.B. (1992) *Intelligent Enterprise: A Knowledge and Service Based Paradigm for Industry*. Free Press, New York.

Richardson, G.B. (1972) The organization of industry. *Economic Journal*, **83**, 883–896.

Rosenberg, N. (1982) Learning by using. In: *Inside the Black Box: Technology and Economics* (ed. N. Rosenberg), Cambridge University Press, Cambridge; pp. 120–140.

Sandberg, R. & Werr, A. (2003) The three challenges of corporate consulting. *MIT Sloan Management Review*, **44**(3), 59–66.

Sharma, D. & Molloy, R. (1999) The truth about customer solutions. In: *Viewpoint*. Booz Allen and Hamilton, Boston, MA; pp. 1–13.

Shepherd, C. & Ahmed, P.K. (2000) From product innovation to solutions innovation: a new paradigm for competitive advantage. *European Journal of Innovation Management*, **2**, 100–106.

Slywotzky, A.J. (1996). *Value Migration: How to Think Several Moves Ahead of the Competition*. Harvard Business School Press, Boston, MA.

Slywotzky, A. & Morrison, D.J. (1998) *The Profit Zone: How Strategic Business Design Will Lead You to Tomorrow's Profits*. John Wiley & Sons, Chichester.

Wise, R. & Baumgartner, P. (1999) Go downstream: the new profit imperative in manufacturing. *Harvard Business Review*, September-October, 133–141.

Womack, J.P. & Jones, D.T. (1996) *Lean Thinking: Banish Waste and Create Wealth in Your Corporation*. Simon & Schuster, New York.

WS Atkins (1999) *Annual Review*.

Endnotes

[i] This chapter is based on Chapter 8 in a forthcoming book by the authors (Davies & Hobday 2005).

[ii] Mastering Service Capabilities in Complex Product Systems: A Key Systems Integration Challenge, funded by the UK's Engineering and Physical Sciences Research Council (EPSRC) Systems Integration Initiative (Grant no. GR/59403).

[iii] Regarded as the key measure of firm performance, added value is the difference between the market value of a firm's output and the costs of its inputs (Porter 1985, p. 38; Kay 1993, p. 23; Department of Trade and Industry 2002).

[iv] For the outsourcing customer, a capital good no longer represents a fixed cost incurred on an intermittent basis, but a variable cost paid for in regular instalments for the duration of a service-based contract. A supplier of capital goods can achieve efficiency gains by spreading the costs of providing solutions over a larger number of customers. These economies of scale benefit customers that cannot achieve the same cost savings when designing and implementing one-off solutions solely for internal requirements. A supplier can achieve lower unit costs by strengthening its capabilities, because these capabilities represent a fixed set-up cost, which the supplier does not have to reinvest in each time it serves a new customer (Domberger 1998, p. 78).

[v] The typical life cycle extends over 30 years – two years to design, build and manufacture rolling stock and 28 years to provide services. So, for example, whereas the cost of building a fleet of 70 diesel trains is around £65m, the service life generates revenues worth £200m.

7 The Effectiveness of Marketing Spend

Christopher Preece, Krisen Moodley and Michael Brown

Introduction

This chapter illustrates the application of business-to-business and services marketing concepts and techniques using the construction sector as a model. The dynamic nature of the industry requires companies to have effective, flexible marketing strategies. In recent years, private finance initiatives (PFI) have been increasingly used to procure public sector work. Also, with an increasing number of international projects available and the added competition of international companies bidding for work within the UK, the industry has become exposed to even more diverse conditions and business environments. It is vital, therefore, that companies are able to adapt to the changes within the industry.

Companies operate in an age where value and customers are the key drivers. Marketing makes a critical contribution in the process of finding new and keeping old customers, and ultimately achieving long-term business success. Evaluating marketing performance should be a key task for management. Tim Ambler, in his work on marketing and the bottom line, suggests that the businesses that concern themselves with a rigorous evaluation of marketing results are in the minority (Ambler 2000). This chapter examines marketing strategies and the manner in which to measure them.

Marketing strategies in construction

Construction companies have adopted various different strategies over the years to improve marketing communications with clients, the public and within the company itself. The strategies vary from the traditional, to more advanced, high-tech options. By creating successful marketing strategies, organisations obtain an advantage when competing in the marketplace. Promoting the company's values and skills through a successful marketing plan helps to start the long and challenging process of generating business in today's construction industry.

The Chartered Institute of Marketing has adopted this definition of marketing:

> *Marketing is the management process responsible for identifying, anticipating and satisfying customer requirements profitably.* (Pearce 1992, p. 35)

Although quite a general description, the Chartered Institute of Marketing's definition summarises the aims of marketing within a business. The main aim of any business is to make a profit, and this can only be done if the company is marketed correctly and customers are getting what they want. The definition of marketing can be applied to both business-to-business marketing and marketing within consumer industries.

The main concern for any marketing strategy is to generate repeat business and to ensure that clients and customers continue to buy from the business. There are, however, different ways of generating this future business and the methods employed may vary to an extent depending on the industry and area of business. The customer–client relationship in a consumer market and a construction market is completely different and this accounts for the difference in successful marketing strategies across the two types of industry. Pettinger (1998) recognises these differences as being key to the different marketing approaches taken in various industries:

> *. . . with the consumer industry, the customer is distant, there is a product-orientated relationship and the customer has no loyalty to the company but to the product, whereas in the construction industry, there are extensive professional and personal relationships between organisations and the client bases, the commission on professional knowledge rather than perceived quality of work.* (Pettinger 1998, p. 48)

For these reasons, the marketing approach has to be different. Whereas Cadbury can market brands of different chocolate bars, a construction company's worth is in the reputation of the company as a whole. Cadbury can make a name for itself by successfully marketing a good product while maybe also having one or two not so good ones; a construction company cannot afford such luxury. One bad project that has the company's name attached to it is enough to make marketing any other services that the company successfully provides virtually impossible. Clients want to be associated with companies that have proven records of delivering quality projects over and over again. Due to the nature of word of mouth contact within the industry, information will soon get around if a company is regularly underperforming, and no amount of marketing strategy can make up for that. The adage that 'all publicity is good publicity' definitely does not apply to construction companies.

Strategies available

A marketing strategy aims to fulfil the ideals of the definition of marketing. It is up to the marketing teams to come up with ways to identify, anticipate and satisfy customer requirements profitably. There are a number of tried and tested methods to do this, and this section aims to give an outline of these methods and their importance and significance.

Branding

The American Marketing Association states that:

> *. . . a brand is a name, term, sign, symbol or design – or a combination of these – intended to identify the products, goods or services of a seller or group of sellers and to differentiate from those of competitors.* (Pettinger 1998, p. 57)

In business-to-business terms and in particular the construction industry, it is hard to identify brand names that command an equal impact or are as well known as some consumer brands. Branding is not something which construction companies have typically looked at, but has a potentially high value within a marketing strategy (Pettinger 1998). The importance of a strong brand name in a business-to-business climate is not as high as in a consumer market, although many companies have tried to re-brand in the hope that their image within the business sector and with clients is improved. A typical example of the re-branding of a company within a business-to-business environment is that of British Steel or Corus, as it is now known. The re-branding process is incredibly expensive, not only due to changing stationery, website design and company literature, but also the cost associated with launching the new name into the marketplace. Re-branding is supposed to be a way for the company to re-assess and assert its goals and ideals, but is often 'perceived as a cynical attempt by a company to boost its share price or shrug off the past' (TCM Archives 2001).

Branding is one of the keys to a successful business. By associating with a brand, clients can expect a certain standard of product or service. People, in general, associate certain brands with a certain type of product, and the construction industry is no different. Coca-Cola and Microsoft are the most valuable brands in the world, and the success of their business is both a cause and an effect of the quality associated with the brand name. The construction industry is no different. With the importance of the tradition of quality services outlined in the background to marketing, if clients can associate a traditionally good service with a name, then that business will see more work than a company with a weaker brand name.

The re-branding of a company to try and achieve the benefits of a brand is, however, sometimes ineffective. It can be an expensive process and is sometimes seen as an attempt to rid the company of a bad reputation associated with the previous name. A strong, developed brand is extremely effective in the external and internal marketing of the company. Not only do clients perceive the brand and company in a good way, but also employees will prefer to work for a company with a strong history, tradition and brand value.

Advertising

The role of advertising in the construction industry is not the same as the role that it plays in other industries, such as retailing and the food industry. The construction company has to advertise itself to ensure that people are aware of its existence, but much less emphasis is placed on traditional advertising strategies for business generation.

Pettinger (1998) looks at the importance of advertising within construction companies who have to target the consumer when the consumer is the direct purchaser. This is especially true with regard to housing extensions, home improvements and refurbishment. Periods in history such as property booms lead to increased spending and money available for advertising campaigns for construction companies. Double-glazing, conservatory construction companies and other similar companies invest heavily in consumer marketing and advertising, especially in newspapers, on television and with leaflets.

Construction companies in a business-to-business environment do not rely as heavily on marketing in the form of leaflets, brochures and general advertising. In fact, Smyth (2000) suggests that advertising is under-used in construction, especially given the symbolic importance of some key buildings and projects. Smyth (2000) also recognises that measuring an advertising campaign is important by looking at recognition and the attributes of the company. Reports show that the construction industry could benefit from paying more attention to its image (Smyth 2000).

Business-to-business advertisements are usually written to audiences seeking problem solving or profit-improving information. Batra *et al.* (1996) state that business-to-business advertisements 'should be informative and offer specifics (but not [be] boring), and (ideally) offer case histories of how the advertised brand helped someone else in a similar situation'. This is perfectly matched to the construction industry, as many of the advertisements that are produced by construction companies, in magazines, or the more tailored advertisements to clients, state relevant case studies and are generally informative.

Preece *et al.* (1998) view advertising and sponsorship as being very similar ways of communicating a marketing message. Advertising is seen as: 'paying a media institution in order to display a chosen message'. Sponsorship, although similar is when 'an organisation's name is linked to some team, personality or event'. Sponsorship can be an expensive way of getting a company's name into the market, but can be quite effective if sponsorship of a renowned event always uses the company's name.

The most important reason to advertise is to increase awareness of the company. Without creating awareness of the company, marketing and business development is made virtually impossible. Advertising must be carefully placed to ensure that the right people see the right advertisements in the right places. Television advertising is very effective in most other industries, but is limited in marketing the business-to-business services offered by the construction industry; it is also an expensive advertising option and must therefore be carefully considered by construction companies before adoption. Television is effective where a product is more visible. House builders have used television to build brands and to sell their products and lifestyle options. This strategy is appealing to a mass market rather than a niche. Much more effective advertising includes adverts in sector magazines and advertisements in the industry publications. The value of an advert lies in its eye-catching nature and its ability to demonstrate, quickly and effectively, the expertise, experience, successes and values of the company.

Sponsorship is another method of advertising that is suited to both business-to-business environments and mass markets. Companies can sponsor in a variety of ways ranging from sports events and teams, orchestras, animals and shows, to television programmes. The higher the profile of the activity being sponsored, the higher the cost of sponsorship and the higher the public awareness of the sponsor. Global activities such as the Olympic Games or the Football World Cup raise the profile of sponsors to a worldwide audience. This approach does not specifically target a niche but simply raises profile through association with the event. A more targeted approach is where companies sponsor activities at relevant industry and sector events in order to try and make the name of the company more visible. Sponsorship is targeted rather than generic and may be more suitable for business-to-industry approaches. Sponsorship can be an expensive way to raise company profile, but can be used to develop and work alongside other marketing approaches.

Public relations

The Institute of Public Relations (IPR) in the UK defines public relations as follows:

> *Public relations is about reputation – the result of what you do, what you say and what others say about you. Public relations practice is the discipline which looks after reputation – with the aim of earning understanding and support, and influencing opinion and behaviour. It is the planned and sustained effort to establish and maintain goodwill and mutual understanding between an organisation and its publics.* (IPR 1999 – from Harrison 2000)

Public relations management is an integral part of the marketing strategies of any company in any particular industry. Within the construction industry, the reputation of a company is essential to its success as a business. Therefore, maintaining good public relations practice to look after reputations is a vital part of marketing within a construction company. The concept of public relations management has sometimes had cynical connotations, with a view that it is sometimes used for 'putting a gloss on bad news' (Harrison 2000).

Public relations is an integral part of the marketing process and relates to all aspects of marketing. The process of public relations is also more than just maintaining a good relationship with the general public. Good communications must be practised throughout the company to maintain relationships with all stakeholders of the project. These include the client, the media, financial institutions, the government and the community as a whole. By ensuring that everyone in the company understands the importance of communications with all stakeholders of the project, the reputation of the company can be preserved. By improving communications, public relations strategies can be an effective way of marketing the company in terms of increasing awareness. By demonstrating the values, standards and qualities of the company, the reputation of the company and the industry as a whole can be improved.

Customer relationship management

Customer relationship management (CRM) uses modern methodologies and technologies, along with e-commerce, to improve customer relations and ultimately to improve sales and business development. Companies require a CRM strategy that will ensure continued good working relationships with their previous clients. An effective CRM strategy will look at what a company can do to ensure that they are considered for future work with these clients.

Once the objectives of the CRM strategy have been identified, management can start looking at the important factors to include in a CRM strategy and the ways in which this can be successfully implemented. The purpose of this section is to suggest ways in which this can be done. The budget of the CRM is important because the company must be realistic about the cost of the CRM and the balance between the cost and the potential future rewards. It will be difficult to justify spending a significant part of their budget on the strategy, but the view must be to the future and the repeat business that can be generated by ensuring continued excellent relationships with clients.

The infrastructure of the company will also affect the CRM strategy. Management will need to look at the structure of the company and whether it is conducive to the implementation of the CRM strategy. Some of the questions that need to be addressed in terms of infrastructure are as follows: Is there sufficient hardware, software and everything in place to run an effective database system? Are there any corporate standards that may hinder the implementation of the CRM? Are the necessary resources available? The process of implementation and the importance of the people involved in the CRM strategy are also noted by Stone *et al.* (2001) who also raise the point of measuring the effect of the strategy as being important to its successful implementation.

CRM is a relatively new way by which companies can ensure that their client base is fully researched and the marketing department has a systematic tool to enable them to target clients methodically and with relevant information. CRM provides a database with which the company can view the type of organisation that they are dealing with, the previous projects or services that have been provided for them and any future work that could be offered. This type of database is very flexible and can be used to good effect by the marketing team.

Corporate hospitality

The use of corporate entertainment within the construction industry has always been seen as a good method of generating business, retaining clients and a generally useful marketing strategy. Up until a few years ago spending on corporate hospitality in the UK had been increasing rapidly. Between 1994 and 1999, the amount spent on corporate hospitality rose from £300 m to £700 m (Benady 2002). Corporate hospitality is not a cheap marketing option for companies; the British Grand Prix last year provided packages ranging from £300 to £1300 a person, and spending the day at Twickenham to watch an England rugby match cost a company on average £500 per person (Flack 2002). In terms of spending on corporate hospitality, there is a significant trend to spending around the same amount of money, but putting on fewer events and making sure that they are done well.

Although there is still a lot of money spent on the 'big five' corporate hospitality events (Royal Ascot, Wimbledon, British Open Golf tournament, Henley Regatta and the Chelsea Flower Show) overall, there is a reduction in corporate entertainment, in line with a general downturn in marketing expenditure.

The most important aspect of corporate hospitality events is the feedback and the information that can be gained from staging them. Companies must identify clear objectives for their corporate hospitality events. The right people must be targeted and research must have been done to ensure that the attendees are interested in the event. It is important to understand fully the objectives of the event and who will be involved. It is also important not to try and sell too much at a corporate event. The people at the entertainment event are there primarily to take advantage of the free event and will not be receptive to a hard sell or promotional presentation. There is the opportunity to follow up with an attempt to generate business, but the idea of the event is to attract the right people and make them more aware of the existence of the company.

Corporate hospitality is not as popular now with businesses as it was in the late 1980s and early 1990s. It is, however, seen by a lot of companies as an effective way of maintaining relationships. The only way to ensure that corporate hospitality is used as an effective tool is for the company to develop a mixed guest list which will benefit everyone who is invited to attend. Attendees have to benefit from taking part in corporate hospitality events. The majority of companies view networking as the major benefit of corporate hospitality.

The company should not underrate the importance of the type of event, and it is important to think innovatively when planning corporate hospitality. If the company can offer something different to the average day out or corporate hospitality event, then clients are more likely to attend and the event will be more successful. These events can be expensive to plan and stage, so it is important that everything is planned in detail. Achieving a mixed guest list, offering innovative events and following up the events with appropriate marketing activities are keys to success.

Corporate/promotional literature

Preece and Male (1997) concluded that promotional literature in construction companies provides an invaluable marketing tool, but it is important that it 'presents clearly the benefits to the client and his advisers of the contractual service offered'. A good quality promotional document will communicate to the client one or more of the following things:

(1) *Background information*: The company brochure can now be seen as the first link in generating client awareness of the company. A hard hitting, attractive, informative brochure that gives a background to the organisation and points anyone interested in further information to an equally attractive, user friendly and informative website is a way in which promotional literature can really benefit a construction company. The only problem with this strategy is that construction companies at present do not seem to be able to successfully deliver both components
(2) *Organisational values and achievements*: Corporate literature within the construction industry must communicate values, expertise and successes of the company and must be relevant to the sector within which the client operates.

Promotional gifts

The promotional gift industry is ever growing, with its value reaching £1 billion in 2000 (Greaves 2000). Over 90% of exhibitors at trade shows give something away to attendees of the event, indicating the importance that businesses attribute to promotional gifts. But the nature of the gift has to be carefully considered, as with any marketing tool. For the construction company, the importance of the company brand is to generate awareness, which will hopefully lead to the successful generation of work. It is important, therefore, that the corporate image and brand values are upheld in any form of activity, no matter how small that is. This is why many companies produce umbrellas, caps, stress balls, pens, coasters and diaries to keep the name of the company in clients', competitors' and the general public's mind.

The main intentions of a company producing promotional gifts as cited by Daly (1993) are to:

(1) Influence a positive perception of the way a company does business or the image of its product
(2) Reinforce the buying decision
(3) Strengthen relationships
(4) Stimulate interest.

If a positive perception of the business is generated, clients will perhaps assume that the way the company does business is good. However, if given a poorly researched, low quality, insignificant gift, clients may judge this to be a reflection of the way that the company does business in general. If, however, the client is in receipt of a good quality gift that reflects their interests, they are going to be more appreciative and hold the construction company in much higher regard.

The main purpose of the promotional gift is to maintain and strengthen relationships between the client organisation and the construction company. The carefully planned promotional gift can send the message to the recipient that, 'these people think that I really am important' (Daly 1993). The gift has to be targeted to the right people within the organisation, and the culture of the organisation must also be considered. Research must be done into the ethics of the company and to establish whether they are receptive to gifts. Many of these practices are dying out in today's business-to-business environment, although gifts around Christmas time are still an accepted way of maintaining healthy relationships.

The gift can also be used to stimulate interest in the company, and, as discussed previously, by combining this with a good website and readily available further information, the construction company can make effective use of the promotional gift within their marketing strategy.

So far, the positive aspects of giving gifts have been discussed; Robson (1993) looked at the possible negative connotations of gift giving within the industry:

> *The question of what is ethical and what is not is obviously a subject that should always be considered very closely and some companies have quite firm policies concerning gifts. Sometimes, for example, executives are obliged to return such items if it is considered the gifts are of excessive value. But there are some situations, especially when dealing with some overseas companies, when offence might be taken if the offer of gifts is declined.* (Robson 1993)

He brings up two important points that are critical to the practice of giving promotional gifts within the business arena, the first issue being that individual companies should make policies regarding receiving promotional gifts clear and everything should be in the open. The second is that companies must understand the policies of other organisations, especially when dealing with overseas businesses, and appreciate their own individual values and ethical policies.

Promotional gifts have traditionally been a part of the construction industry. Christmas presents and gifts throughout the year are commonplace and, as with corporate hospitality, some people see it as a slightly unethical way of generating business or maintaining relationships with clients and suppliers. The use of promotional gifts can be useful, however, in increasing the awareness of the company. Carefully designed promotional gifts that are not just gimmicks and could be of use to someone can be beneficial in spreading the name of the company. As long as the gift is well thought out and the targets have been selected carefully, the gift will not just be thrown away with the rest of the marketing material that inundates companies on a daily basis.

The question of different cultures is important when considering corporate and promotional gifts. Companies must research cultures within organisations and countries to ensure that giving and receiving or not giving and not receiving gifts cause no offence.

Exhibitions, trade shows and seminars

An exhibition or trade show provides a construction company with the perfect opportunity to target potential clients and increase awareness of their company. Exhibitions are unique in that they are the only medium that brings the whole market together: buyers, sellers and competitors (Smith 2002). At a site equipment demonstration exhibition in May 2002, almost 20 000 people (ABC 2002) attended from the construction industry, including contractors, consultants, plant suppliers and many other businesses within construction. This gave attendees the opportunity to mix with competitors, suppliers and clients to gain knowledge from what others were doing and, therefore, potentially to improve their own business.

The exhibition industry in general is thriving, with £2.04 billion spent in 2001. This marked a 16% increase on the 1999 figures (Exhibition Venues Association 2002). The construction industry's main annual exhibitions are the Interbuild Conference, the Building Exhibition and the 'Civils' exhibition. All three events attract numerous companies from the construction industry over a period of 3–4 days

Defining the event budget is important because there are numerous costs associated with putting on a stand at an exhibition: direct costs, such as stand space, building the stand, graphics, literature, carpets, catering; and indirect costs, such as the time and effort spent by staff in preparation for the event.

Exhibitions, trade shows and seminars are good for increasing the awareness of the construction company. As with all the strategies, they must be carefully planned, and the objectives clearly defined at the start. The attendees of the events must be considered to ensure that the material available and information provided are relevant to their needs and interests. Exhibitions can be used as a vehicle for all the other marketing strategies available. Promotional literature, gifts, advertising and public relations strategies can be implemented with the help of an exhibition or trade show. Companies can also use seminars to increase awareness of the company, current projects or areas in which they excel.

The Internet

The advent of the digital age has introduced a new medium to the marketing arena. The Internet is the fastest growing area in marketing and commercial development.

The importance of the Internet in the developed world is demonstrated by the fact that it is the primary source of information for a variety of stakeholders, including customers. It is also an environment that is changing the transactional nature of business and forcing business models to be re-examined. A good illustration of the impact is on booksellers where Amazon has redefined the nature of traditional bookselling, forcing the sector to change their approach. Spend on the Internet is the biggest growth area in marketing.

The most basic presence on the Internet is to develop a company website. It is cheap and relatively easy to set up a basic information website, provided the company has the basic software, hardware, a domain and an Internet provider. There are three essential dimensions to website design and management (Plant 2000): content, format and access. The web presence has to be part of an overall marketing strategy, and one of the most important areas is the domain name. The domain name is essentially the Internet brand, and where possible should tie closely to the name of the company. For example, Virgin.com is the unmistakeable domain of the Virgin group of companies. The domain name is critical in directing the enquiry to the company website, as a domain name that is not associated with the company can often lead to misdirection. A close relationship between the domain name and the brand name is therefore desirable. Merger, acquisitions or rebranding may have a knock-on impact on the Internet. To summarise, there is a need for consistency between general brand building and brand building on the Internet.

In the process of developing an effective web presence, it is useful to remember that the website is not simply an extension of a generic company brochure. The website may attract a variety of stakeholders, including potential clients, partners, subcontractors, investors, interest groups, potential employees, etc. The flexibility to tailor a website to provide interest to all these audiences is possible; a clear strategy for the website is essential and should tie into the overall marketing plan. The key is to match the user's psychological, demographical and technological sophistication profile with the website's initial and subsequent impact. The broader promotional and informational aspects must not be forgotten, as the website provides substantial opportunity to promote all aspects of the company. There must be a balance between information provision to the customer and information collection by the business on visitors. The aim must be to develop customer–site interaction to use connections to move the customer towards the 'sell'. The content and design of the site must build relationships with customers that lock them into a site, for example, by offering continuous advice. There is a common misconception that a website that is technologically advanced with all the 'bells and whistles' is good. Effective websites are not only technologically good, but are well designed, attractive, speedy and have good content.

A website on its own is not adequate. A good back-up operation to support the website is essential. This does not just involve technical support, but the support to information requests. A well-designed site will have a feedback and information request section. The manner in which this is managed will indicate how seriously a company takes its website. It is better not to have an information request section than a poorly managed operation. Every request represents a future client, employee or supporter. Companies that are not heavily involved in e-commerce often neglect the backroom operations necessary to support a website.

The web allows a company to control its domain. Its use continues to grow and is becoming a significant part of marketing strategy and marketing spend. Efficiently developed, it can provide a useful support tool to other marketing efforts of the organisation. As more companies adopt e-commerce applications for their business operations, the role of the website and Internet presence will expand.

Marketing metrics

A constant conundrum for all marketing activity is the question, 'Does marketing represent value for money?' Accountants often query what real value marketing professionals bring to the company. The concept of measuring marketing has been employed for a long time. Marketing metrics have measured everything from customer preferences to market penetration, market share, brand loyalty, brand awareness, etc. Measurement is relatively straightforward, and some companies have developed sophisticated measuring systems.

Within a consumer industry, increased sales can be almost directly related to an extensive advertising or promotional campaign. Typically, in an area such as financial services, objectives are set for each advertising or marketing campaign, including total budget and the anticipated level of new business in real terms. There is careful targeting or profiling of potential clients, through use of either internal databases or third party agencies. Once a campaign has been initiated a management information and marketing tracking team follow up sales and enquiry data. Both teams require data for different parts of the organisation: the marketing team will look at responses and sales and the extent to which profiling has worked; the management information team will look at marketing cost per enquiry, marketing cost per sale and other defined performance indicators. This information then forms the basis of future campaigns and, from a managerial perspective, influences the decision process on future requests for spending. The financial services sector will track all forms of marketing related activity in a similar manner. All activities, for example, choosing to sponsor one sport over another or deciding where adverts are placed in the print media, are tracked through metrics. This information is then used in all subsequent decisions.

Kokkinaki and Ambler (1999) suggest that marketing activity measures fit into six categories:

- Financial measures (turnover, contribution margin, profitability)
- Competitive market measures (market share, advertising share, promotional share)
- Consumer behaviour measures (customer penetration, customer loyalty, new customers gained)
- Consumer intermediate measures (brand recognition, satisfaction, purchase intention)
- Direct customer measures (distribution level, quality of service)
- Innovativeness measures (new products, revenue from new products)

Financial measures are the most easily used. Market share measures are also fairly prominent in their use, while customer related measures are also commonly applied in service businesses.

An alternative approach is to evaluate the impact of marketing activity on brand equity. Brand equity is the set of associations and behaviours on the part of the brand's customers, channel members and parent company that permit the brand to earn greater volumes or margins than it could without the brand name (Srivastava & Shocker 1991). Brand equity is also the most prized asset for many companies and the principal marketing asset.

These measures are important, but do they provide answers about the impact on long-term sales, profits and financial futures? In general terms, this is important because industry has changed from considering tangible assets to taking into account intangibles, such as brand loyalty and customer focus. Most current metrics provide answers to short-term impacts of marketing expenditure. For metrics to deliver value it has to track through both short- and long-term answers.

Typical contractor spending

The amount of spending on marketing and the returns it generates have always been a source of contention between marketers and other business managers within construction. Marketers believe that marketing spending brings returns to the company, while many other functions of business see much of marketing as uncontrolled expenditure. Marketing spend has to be justified, and the best way to do this is to try to measure how much business is generated by a particular marketing strategy. Typically, construction companies spend on the following activities as part of their marketing remit:

(1) Advertising
- Strategic
- Corporate
- Feature/editorial

(2) Corporate literature
- Brochures
- Brochure inserts

(3) Exhibitions, seminars, conferences
- Attendance
- Exhibiting

(4) Corporate entertainment
- Sporting events
- Trade association/award dinners
- General entertainment

(5) Promotional gifts
- General
- Christmas

(6) Public relations

- Consultants

(7) Mail shots.

These activities form the traditional base of the marketing departments in construction. Activities such as public relations, business development and technology are not normally within the marketing remit. It is therefore appropriate to ask the following questions. Where does the marketing remit start and finish? Who, for example, is responsible for a company website? How is this spend attributed and who controls design? Measurement of marketing is a more complex activity because of the crossover between departments.

Measuring the effectiveness of marketing spending

The problem of using marketing metrics in business-to-business environments is that the generation of business is hard to assign to any particular strategy or method. For example, the construction industry, be it consulting or execution, is a difficult environment in which to measure the effectiveness of marketing. Therefore, mass consumer models are not effective in marketing construction firms. The manner in which work is acquired is also different, in that the buyer is often in a position of power. The needs of the buyers of construction services also vary. Some buyers will consider quality and reputation as important, while others will still consider low cost as the most important aspect to them. While every construction company strives to have long-term buyers of their services, there is also the realisation that these relationships are also capable of being terminated relatively quickly. A constant level of negotiated contracts would be ideal, but very few companies achieve this utopia.

Despite the difficulties facing marketing, it is important to develop a system that is measurable to a certain degree. This can be approached in two stages, the first being to recognise that marketing, like all other activities, is an important part of the business model. Marketing, like all other aspects of the business, has to perform adequately and justify its existence. Performance measurement of this function has to be accepted, and key performance indicators developed. Ultimately, if performance is not achieved, questions have to be asked about the activity.

The second stage to control is the importance of planning within the marketing procedure. Each phase within the strategy has to be planned meticulously to ensure its success. Objectives must be set at each step to ensure that goals are attainable and that everybody involved is working towards these. Due to the nature of marketing, and its history within construction, marketing planning is something which is not seen as being very important. Companies have started to view it from a strategic point of view, but many are not stepping back and looking at the effectiveness of what they are doing. By planning carefully exactly what strategies are being used and, more importantly, why they are being used, companies will have more effective marketing strategies (Kokkinaki & Ambler 1999). Metrics have to be introduced to make strategies more manageable. For example, a construction company can decide that traditional advertising is not a productive strategy for them; they will have evaluated the benefits of advertising and have decided that it is not worth their time and money. Advertising may not reach the correct stakeholders, help access new markets or build relationships with existing clients. Application of metrics should help determine if certain activities are delivering the required responses. If companies could do this, spending could be reduced in areas where the benefits are not as good as they potentially could be in others.

By carefully planning all aspects of the marketing procedure, companies can identify targets, research the best way of contacting the target and determine the precise objectives. The traditional approach of marketing departments to target as many potential clients as possible with as much information that may be useful, has been proven not to be successful, as much of business-to-business marketing is based on relationship building. The main objective for construction companies at the outset of the marketing strategy is still primarily to increase company awareness and by so doing go on to increase their clients' understanding of the company in terms of its expertise, tradition, successes and values.

Although designated marketing departments have most of the control over marketing strategies and techniques, everything that any employee does in the name of the company has an effect on a stakeholder, be it a client, potential client, competitor or any other organisation with an interest in the company or a project in which they are involved. Making everyone aware of their responsibility for productive marketing of the company will ensure that no employee jeopardises future work or relationships. Companies should make a point of training employees to understand their responsibilities and how being aware of the marketing process and how they fit into it can benefit the company as a whole. Application of broader financial metrics work can be used to test the success of the strategies.

Having a varied approach to marketing is essential. By understanding the importance of all the strategies and the benefits that they can offer, construction companies can come up with versatile marketing strategies which can be adapted to different situations. The strategies can be tailored to the target, the stage of the relationship, the objectives of the marketing and the type of sector within which the target organisation works. By understanding what types of strategies work best in certain situations, companies can have a structured approach to all potential situations.

Opportunities for cross-selling are revealed by looking at the services offered across the company. By combining this information with the potential needs of the client, companies should almost be able to offer clients things before they even realise they want them. In doing this, the company is seen by the client as being advanced. This type of strategy will impress client organisations and demonstrate a willingness to really understand their business.

The importance of the company brand should also be considered by construction companies. Due to its nature, many construction companies have a long history in the industry, and this is something that can be used by marketing managers to promote the company. Clients look for history and a traditional quality of work in construction companies, so if the company has a successful history, it can be exploited by marketing departments to the company's benefit. Further, the importance of building a brand should not be overlooked in the marketing strategy. By gaining a reputation within sectors for delivering projects successfully, the company brand will become well known and the organisation considered for more work because the name is already established in the client's mind. The importance of building on the brand should be realised and money spent accordingly. Brand awareness metrics can be applied to ensure that the money spent delivers results. Ultimately, the brand is critical for the creation of brand equity.

Mergers and acquisitions should also be seen as an opportunity to use brand names successfully. If a construction company merges with or buys another company, the value of the brand of the other company should be carefully considered before changing the name. If the company being bought has a strong brand identity in a particular sector, it is important to keep that brand name to ensure clients do not get confused when the name changes

Expansion should also be seen as a way by which companies can continue to market themselves successfully. By demonstrating how the company is expanding and becoming more versatile in its areas of expertise and geographical capabilities, marketing teams can target clients with further examples of previous experience. This means that clients can expect more from the company, and also have the perception that a growing company is a successful one. By being able to offer potential clients different services, or the same services in different areas of the country, the construction organisation will be more attractive to them. Expansion should definitely be addressed more within marketing strategies.

The importance of innovation must also be emphasised to marketing teams within construction. By introducing innovative, interesting and eye-catching marketing strategies, companies can increase awareness within client organisations. The purpose of making marketing material interesting is also to ensure that it stands out from competitors and that potential clients look at it first. It is then up to the other departments to be able to demonstrate company capabilities. As soon as the client is interested and asking questions of the company, the hardest part of the marketing process is over.

Conclusions

In conclusion, companies should include studies of the effectiveness of their marketing techniques as part of their key performance indicators (KPIs). This would mean not only evaluating their profitability, productivity, safety, client satisfaction, product delivery time and cost and all the other KPIs of the industry, but also including an evaluation of marketing efforts throughout the year and how successful the marketing department think they have been. It is vital to the future of marketing in construction that senior employees are made aware of its importance. By being able to look at marketing strategies and how they benefit the company, marketing departments can demonstrate to executives and directors that their work is instrumental in ensuring the future success of the business.

The future direction for construction marketing is for companies to take a more planned approach, being able to identify objectives early and evaluating how effectively marketing strategies generate business for the company. In addition to numerical measurement (if possible), it is important that the evaluations take the form of a general impression from the marketing managers as to the pros and cons of the various marketing strategies. By completely understanding the strategies and how they work, marketing teams can develop more and more advanced strategies. It is important to try to develop marketing within construction, so that companies remain competitive in the national and international marketplace.

Other recommendations for the future are for construction companies to try to model or come up with frameworks to identify the most appropriate marketing tool for any given situation. This situation will depend on the stage of the project, the type of client, the sector in which the client operates and the previous work that may have been carried out, whether for an individual client or within a particular sector. By systematically planning contingencies for all types of situations, a methodical marketing approach can be realised, which is what is missing from the industry today. Marketing is a key business function and requires management attention to monitor performance.

Future research could further investigate the marketing and promotional mixes used by companies. Consideration could be given to the role of marketing in relation to other related functions, such as corporate communications, and their impact on the brand reputation of the firm. Brand development will take on increased importance, especially where firms operate in long-term relationships and in activities such as facilities management, maintenance and concession contracts. The brand will define the reputation of the organisation and be the most visible aspect of the business. Future marketing research will also need to address brand building and associated communication strategies. The research agenda will move towards investigating brand protection and reputation management.

A major criticism of marketing activities relates to their perceived value for money and will continue to be used as a threat to marketing activity. Research into the application of marketing metrics has to be undertaken to answer the question of value: this research has to consider developing a series of performance indicators and the application of benchmarking. These measurement tools should cover a range of marketing activities, from the effectiveness of promotional mixes, to the contribution of marketing to the organisation's profit and turnover. Marketing metrics research will also open up further avenues of investigation into both marketing spend and its effectiveness. Measuring marketing will also address areas such as customer service and client satisfaction. The impact of these areas on marketing and firm performance can also be measured. Marketing metrics, therefore, will be a key theme of future marketing research.

References

ABC (2002) *Certificate of Attendance*. www.abc.org.uk/pdf/certificates/10641378.pdf (accessed May 2002).

Ambler, T. (2000) *Marketing and the Bottom Line. The New Metrics of Corporate Wealth*. Financial Times, London.

Batra, R., Myers, J.G. & Aaker, D.A. (1996) *Advertising Management* (5th edn). Prentice Hall, New York.
Benady, D. (2002) Dinner, deals and dancing. *Marketing Week Magazine*, January 31.
Daly, E. (1993) Gifts? *Manager's Magazine*, **5**, 33–39.
Exhibition Venues Association, (2002) *Review 2002*. www.exhibitionvenues.com/pages/Review/EVA.pdf, 2002.
Flack, J. (2002) It's not the taking part. *Marketing Week Magazine*, August 22.
Greaves, S. (2000) Promotional gifts that make a stand. *Marketing*, **8**, 40–42.
Harrison, S. (2000) *Public Relations: An Introduction* (2nd edn), Thomson Learning, London.
Kokkinaki, F. & Ambler, T. (1999) *Marketing Performance Assessment: An Exploratory Investigation into Current Practice and the Role of Firm Orientation*. Marketing Science Institute, Cambridge, MA.
Pearce, P. (1992) *Construction Marketing: A Professional Approach*. Thomas Telford, London.
Pettinger, R. (1998) *Construction Marketing: Strategies for Success*. Macmillan, Oxford.
Plant, R. (2000) *eCommerce: Formulation of Strategy*. Prentice Hall PTR, Upper Saddle River, NJ.
Preece, C.N. & Male, S. (1997) Promotional literature for competitive advantage in UK construction firms. *Construction Management and Economics*, **15**(1), 38–46.
Preece, C.N., Moodley, K. & Smith, A. (1998) *Corporate Communications in Construction: Public Relations Strategies for Successful Business and Projects*. Blackwell Science, Oxford.
Robson, P. (1993) Promotional gifts. *Purchasing and Supply Management*, (Easton-on-the-Hill), November.
Smith, P.R. (2002) *Marketing Communications An Integrated Approach*. Kogan Page, London.
Smyth, H. (2000) *Marketing and Selling Construction Services*. Blackwell Science, Oxford.
Srivastava, R.K. & Shocker, A.D. (1991) *Brand Equity: A Perspective on its Meaning and Measurement*. Marketing Science, Cambridge, MA.
Stone, M., Woodcock, N. & Starkey, M. (2001) Assessing the quality of CRM, In: *Marketing Business*, **8**, 24–30.
TCM Archives (2001), *Re-branding can be waste of time*. archives.tcm.ie/breakingnews/2001/12/25/story34330.asp (accessed December 2001).

8 Trust and Commercial Managers: Influences and Impacts

Will Swan, Peter McDermott and Malik Khalfan

Introduction

Trust is a multidimensional (Ganesan 1994; McAllister 1995), multifaceted social phenomenon (Fukuyama 1995; Misztal 1996), which is regarded by some as an attitude (Luhmann 1979; Flores & Solomon 1998), and by others as a vital social lubricant (Gambetta 1988; Fukuyama 1995). In spite of the large amount of literature on the subject, Gambetta (1988) still saw trust as an elusive concept, while a few years later Misztal (1996) noted the continuing conceptual confusion that surrounded this social phenomenon. For the purposes of this discussion we define trust as:

> *a willingness to rely on the actions of others, to be dependent upon them, and thus be vulnerable to their actions.... trust... affects the willingness to co-operate* (Wood & McDermott 1999, p. 3)

From the above definition, it could be concluded that trust is an exchange-based concept that is centred on risk, with an element of reliance, an element of goodwill, of vulnerability and an expectation of outcomes.

Within construction delivery networks (Pryke 1999; Swan *et al.* 2001), the role of the commercial manager can be seen to be key, in linking strategic goals to the activities of project delivery teams, specifically at the inter-organisational boundaries. To understand how the commercial manager influences and is influenced by trust drivers and barriers, it is important for us to have an understanding of what trust is. This must then be contextualised within the project arena, looking at how issues such as contractual arrangements, management systems and project relationships can impact the role of the commercial manager.

This chapter illustrates the issues surrounding trust in business-to-business projects by reference to data collected for, and the findings from, four case studies presented in the Trust in Construction Project (Swan *et al.* 2002), which was funded by the Engineering and Physical Sciences Research Council and was designed to be an exploratory study of the role of trust within construction project teams. Further, this chapter investigates how commercial managers function in terms of other construction network members, and examines their role in impacting trust within a project team. Finally, the chapter presents recommendations and conclusions, based on the findings from the Trust in Construction Project, that can also be applied to other industry sectors, for example, car manufacturing, aerospace and shipbuilding, which use multi-disciplinary and multi-organisational teams to deliver projects and provide solutions.

What is trust?

Trust has become a key research area within construction management, as well as in the wider business and management literature. It is seen as pervading all elements of the way that we engage with each other.

> *The importance of trust pervades the most diverse of situations where cooperation is at one and the same time a vital and fragile commodity. . . . but this very pervasiveness seems to have generated less analysis than paralysis: in the social sciences the importance of trust is often acknowledged but seldom examined, and scholars tend to mention it in passing, to allude to it as a fundamental ingredient or lubricant, an unavoidable dimension of social interaction, only to move on and deal with less intractable matters.* (Gambetta 1988, p. viii)

Gambetta highlights one of the clear difficulties within the area, 'how do we define trust?' The literature is varied, reflecting the fact that trust is a crosscutting element of social relations, in economic, social and inter-personal contexts. The definition for construction presented in the previous section, informed an interim study (Swan *et al.* 2002), but it is important to drill down into the issues identified by this definition and unpick them. In the definition given by Wood and McDermott (1999) there are clear issues with regard to reliance, risk and uncertainty (Inkpen & Currall 1998; Wood *et al.* 2002). There is also an element of reliance, an element of goodwill, of vulnerability (Bigley and Pearce 1998) and an expectation of outcomes.

Trust is a constituent element of a relationship. It provides a context, or pragmatic (Liebenau & Backhouse 1990), to borrow the terms of semiotics, in which communication can take place. Decisions are made and communications interpreted on the basis of mental models that individuals hold (Piaget 1961) concerning other individuals (Ring-Smith 1997). In some respects it may be viewed in a similar way to information or knowledge (Wilson and Kennedy 1999). Like all mental models, the concept of trust should be considered dynamic: it is subject to change during the course of events; a specific event may reinforce or damage the level of trust that is held. These issues are entirely dependent on the strength of the mental model, fragile or resilient, and the nature of the event (Ring-Smith 1997).

Elements of trust

During the course of the Trust in Construction project, a tool, or Trust Inventory (Couch & Jones 1997; Cummings & Bromiley 1997), was developed to measure trust in relationships (Swan *et al.* 2001). Further, through the use of a semi-structured questionnaire, 32 interviews were conducted and elements of trust identified. In Table 8.1, the figure in brackets indicates the number of times a specific word was mentioned during these interviews.

These elements were then compiled into a questionnaire comprising 53 items; this was tested using a pilot sample of 187 practitioners. Factor analysis identified four separate but related factors: relationship, communication, commitment and reliability. We hold that these factors are not limited to the construction industry; they are

Table 8.1 Content analysis of trust interviews.

Category	Example vocabulary
Trust (654)	Trust/trusted/trustworthiness (577), mistrust/untrustworthy (24), betray trust (17)
Relationships (124)	Partnering/relating/friendship (57), support/co-operation (26)
Value (76)	Value (76)
Confidence (51)	Confidence (33), faith (18)
Competence (28)	Competence (28)
Professional (21)	Professional (20), unprofessional (1)
Promise keeping (149)	Promise/promise keeping (42), delivery (74), reliability (27)
Fairness/reasonableness (110)	Fairness (58), unfairness (16), reasonable (28)
Mutuality/reciprocity (97)	Mutuality (39), reciprocity (14), expectations/obligations/duty (32)
Honesty/integrity (94)	Honesty (51), integrity (13), truth (16)
Openness/communications (82)	Communications (42), openness/frankness (40)
Values/ethics (72)	Values (39), ethics/morals (19)
Reputation (70)	Reputations (44), respect/valued (26)
Blame culture (21)	Blame culture (21)

generic and applicable to other industries. These factors are now discussed in the following subsections.

Relationship

Relationship is concerned with the perception of an individual's values and motivations. Within this category a number of separate issues are addressed; they have been collected together as they are concerned with perception of the relationship a specific individual has of another with regard to their values, rather than direct communication, such as action or commitment to broader shared goals in respect to the network. This may be viewed as trust based on identification with the trustee (Doney *et al.* 1998). A key element of this is how individuals have shared values with regard to treating one another, which is seen as a key determinant of success in attaining network goals, particularly with respect to building effective teams (Morgan & Hunt 1994; Bresnan 1996). Hertz (1992) notes that shared norms and values are 'the basis of trust in networks'. This issue of 'organisational fit' is seen as a key driver in the early stages of a relationship:

> *organizational similarities between partners help establish trust and enhance the appropriability of knowledge necessary to form the basis of a common frame of reference* (Inkpen & Currall 1998, p. 9)

Another stance, however, is that although shared values are important; it is the understanding and accommodation of differing values that may make a successful project team (Uher 1999). Values and norms are, according to Checkland and Scholes (1992, p. 48), central to any social system; they are implicit behaviours and rules by which social groups operate. While the construction industry as a whole can be perceived to have a specific culture (Barlow 1996; Liu & Fellows 1999), organisations themselves also have specific sets of rules and norms (Bresnan 1996; Nonaka & Konno 1998) in which exchange activity takes place. In addition to this, individuals themselves will establish rules and norms from the many other different social settings within which they may participate, as well as those emanating from their individual psychological make-up (Roy & Dugal 1998). The interview stage of the research project determined that the issue of shared values rather than appreciated differences was important, so this approach was taken as a measure of the level of trust.

Blois (1998) highlights goodwill as a key driver of trust: the issue of trust appears to be more than just a fulfilment of action against an agreed set of specified criteria. Goodwill indicates that the trusted party is willing to go beyond the act of being reliable and looks towards the well-being of the trusting party, an element of trust that is developed through experiential learning (Warwick Manufacturing Group 1999). This is related to the perceived intentionality of goodwill on the part of the 'trustor' (Doney *et al.* 1998). Reciprocity was also stated during the interview as an important area of trust and trust building. The act of give and take is essential in trust building (Tyler & Kramer 1996), while reciprocity highlights the issue of the dynamic nature of trust (Hatush & Skitmore 1997). Furthermore, it is subject to processes, although often tacit, that cause it to increase or decrease. It is held that in stable trusting relationships reciprocity may be viewed in a balanced way by trusting parties (Brower *et al.* 2000).

Fairness (Smyth & Thompson 1999) was also shown to be a vital part of the relationship (Swan *et al.* 2002). It was important for individuals to perceive that the person they were working with would be fair in the face of uncertainty (Smyth & Thompson 1999). The concept of fairness is closely linked to the issue of shared values as to what 'being fair' actually consists of. This view of justice is a central process in building trust (Mishra & Spreitzer 1998). What all these elements point towards is an expectation of behaviours with regard to how individuals treat one another, the ethical framework in which they operate (Bejou *et al.* 1998; Wood *et al.* 2002). These rules of engagement were common and can be seen as powerful elements of trusting relationships.

Communication

This factor is concerned with the characteristics of transmitted communications. This essentially addresses the mode and nature of communication. Within this section four key issues are identified: honesty, timeliness, integrity and openness. Although these reflect values that the individual may hold, the nature of their communication is an explicit representation of this. Interviewees noted that any positive communication could be viewed as a trust building exercise, often making reference to a 'trust bank' or store of trust.

Honesty and integrity of communications can be addressed together as they are closely related concepts (Smyth & Thompson 1999). Honesty is concerned with the accurate passing of information and the clear interpretation of facts (Cummings & Bromiley 1997). In construction, for example, project teams rely on group members to communicate information in an honest fashion, as the production and/or implementation system requires accurate information to respond effectively. The interconnectedness of multiple actors in a construction environment means inaccurate information and the impact this has on future action can create difficulties. Walker and Hampson (2003) highlighted the issue of honesty and how it interconnects with issues of trust.

> *Issues of trust and commitment* [are important] *because it* [sic] *involve teams and individuals feeling safe in measuring performance to learn from their experience. This requires openness and honesty rather than recording idealized or false performance records in order to hide mistakes or to attempt to extract unwarranted credit.* (Walker & Hampson 2003, p. 170)

Openness is closely related to honesty; it is essentially concerned with honesty of omission of information for benefit. Honesty could be considered to be concerned with acts of deliberate lying.

Commitment

Commitment is seen as a fundamental element of the trusting team (Morgan & Hunt 1994) and can be viewed as the adherence to a shared set of external goals and values (Walker & Hampson 2003). The issue of commitment requires a principle to commit, the generation of which may be seen as the main role of leadership (Webber 2002). This working description of commitment allows us to define it within the context of the Trust Inventory (Wood *et al.* 2002), which is commitment to the shared goals and values of the project. Within many groups these are arrived at through a process of continuous exchange (Weick 1995; Checkland & Holwell 1997; Walker & Hampson 2003), or an autopeotic[i] generation of shared meaning (Vicari & Troilo 1998). This assumes that shared meaning is generated by constant exchange in which the goals of the group emerge. It is these goals that create high performing project teams (Viall 1996). Within the research project it was stated as a proposition that trust could be managed through 'interventions'. Examples of trust intervention could be shared risk registers, partnering workshops or the application of problem resolution ladders. These are all formalised systems designed to address key issues of trust building or mitigation of risks to trust being damaged. Many of these interventions are part of the partnering process (Barlow 1996).

These interventions look to identify mutual objectives (Barlow & Cohen 1996) and to make them explicit (Carmichael & Cooper 1999). Tools can then be used to reiterate and support the development of trust (Swan *et al.* 2002). However, this does not discount the role of experiential development of mutual goals and undertaking of the project processes in a more traditional way (Swan *et al.* 2002). What is important is that these goals are communicated and shared as a 'project culture' to ensure that individuals can commit. Without them they may commit to any number of organisational and individual values (Lok & Crawford 1999) rather than working towards the development of a 'mutual partnering culture' (Carmichael & Cooper 1999). It is held that where there is an absence of commitment, team members will look to comply with, rather than commit to, the goals of the project (Walker & Hampson 2003).

Reliability

Reliability on its own could be considered sufficient to give the appearance of trust. Reliability means that an individual has positioned themselves against an expected outcome, but can trust be considered to run deeper? Blois (1999) determines that trust is not about reliability alone, as the mere act of being reliable does not have the underlying intentionality (Palmer 1969; Ricoeur 1991) of goodwill that trust requires.

It is clear from the literature and the interview data that reliability forms a key element of a trusting relationship (Svensson 2001; Swan *et al.* 2002). Without the underlying understanding that individuals will deliver what they need to and when they need to, then the trusting individual is taking a high level of personal risk which is central to building trust (Doney *et al.* 1998). This can be seen to have two elements of trusting: the first is the predictive element of trust, based on past behaviours, and the second, closely related, based on perceived capability (Doney *et al.* 1998).

What is trust for?

While much of the literature, both academic and policy-based (Latham 1993; Comptroller and Auditor General 2001), identifies that trust is important, it is not always clear how it works. Fukayama (1995) notes that 'successful' societies have higher levels of trust than those that do not. There are two ways of looking at this 'glue' (Doney *et al.* 1998). The first is a generic set of rules that society follows, values and norms that allow participants to understand the likely outcomes of their actions (Jefferies *et al.* 1999; Kwan & Ofori 2001). This means that when an individual undertakes a certain course of action, the outcome will be predictable; they have trust in the social system's rules. The issue of cultural values is often difficult to distinguish from the role that institutions play in supporting trusting behaviours (Hagan & Choe 1998).

> *Since each culture's collective programming results in different norms and values, the processes trustors use to decide whether and whom to trust may be heavily dependent on a society's culture.* (Doney *et al.* 1998, p. 602)

Within this national context, the construction industry operates as a subculture with its own rules, values and norms. The construction industry, it is claimed, does not have positive values (Egan 1998; Comptroller and Auditor General 2001). The industry sector's values seem, to some extent, to counter those held in esteem in society as a whole. In addition, at the project level, there are clear examples of trusting behaviour (Swan *et al.* 2002). When we consider that individual values are also different, we are left with a confusing array of signals (Lok & Crawford 1999). It is clear that while social values and norms are important, organisational, project and individual values must be considered when assessing the social structure of an

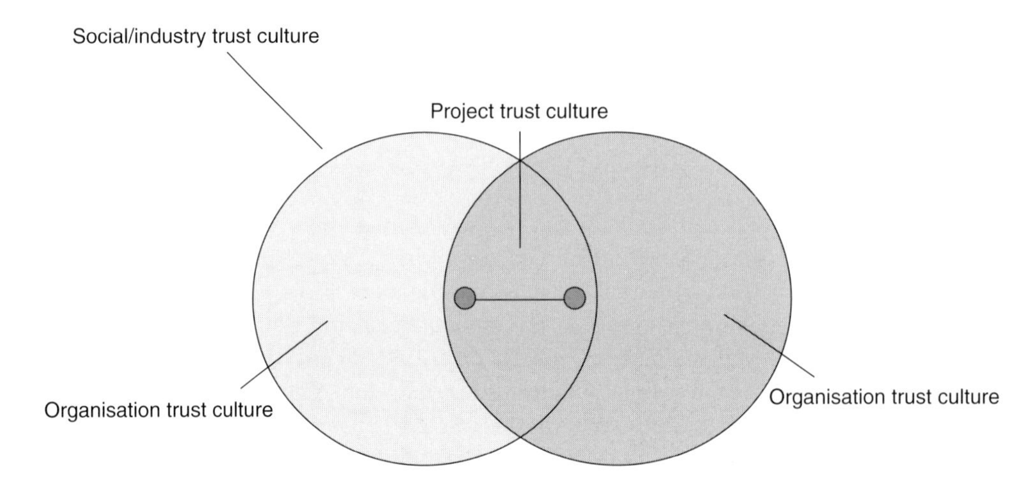

Figure 8.1 Interaction of trust cultures.

organisation. Two organisations with radically different cultures may not be in a situation to create the glue of shared norms and values that Fukayama (1995) discusses.

Figure 8.1 shows this hierarchical interaction of cultures. The societal values, organisational values and project values all interact on the individual in a complex way. It is within these multiple contexts that the relationship is developed. It is these relationships that form the basis of social networks (Ouchi 1980; Scott 1991).

We can also view trust within an informational context. This is allied to the view of trust as a mechanism to deal with uncertainty (Tomkins 2001). When we consider the issues of risk (Edwards 1998), many are concerned with having accurate and timely information (Woodward & Woodward 2001) and issues of partner capabilities (Hatush & Skitmore 1997). The management of risk and, indeed, project management itself, are concerned with the effective management and communication of information.

Risk and uncertainty are key concepts for the effective management of construction projects (Edwards 1998; Carmichael and Cooper 1999). In trusting projects the variables that drive the levels of risk are better understood and allocated, rather than passed to where they may not be well managed (Bachmann 2000). The factor to which this most closely relates is communication. Open flows of information allow risk to be effectively managed between project team members, while a lack of openness or timeliness in information exchange means that the various partners may not be able to manage the various risk factors effectively (Edwards 1998). Trusting someone allows us to behave as if uncertainty had been reduced (Tomkins 2001).

The costs of exchange are lowered with higher levels of trust (Ouchi 1980; Bejou *et al.* 1998) for a number of reasons.

> *Trust is said not only to reduce transactions costs, make possible the sharing of sensitive information, permit joint ventures of various kinds, but also to provide a basis for expanded moral relations in business.* (Brenkert 1998, p. 195)

The primary value is the reduction of control and monitoring mechanisms (Masden 1991) required to manage less trusted partners (Cummings & Bromiley 1997; Wilson & Kennedy 1999). This is especially prevalent in the construction industry; multiple individuals gather the same information, specifically with reference to cost, time and quality, in order to protect the interests of the client from the unscrupulous, and the contractor from not getting paid a fair level for work they have delivered.

Partnering approaches attempt to reduce fragmentation by building trusting relationships and therefore reduce transactions costs (Benheim & Birchall 1999). Trust is used to manage information flows, which is seen as a more effective approach than the use of contract.

> *For the last 70 years, reviews of the construction industry have relentlessly criticized its fragmentation and adversarial attitudes. Commentators have spoken of a lack of integration between design and construction expertise, and the way that problems are tackled in a 'contractual' manner by and between client, designers, main contractors and their suppliers.* (Holti *et al.* 2000, p. 3)

Trust is seen as a key element of building the integrated project team called for by Egan (1998) in *Rethinking Construction* and again in *Accelerating Change* (Egan 2002).

Flexibility (Allen & Cooper 1999) as a response to uncertainty is one of the stated benefits of trusting networks. Many individuals in the interview stage of the project cited a 'willingness to go beyond the contract' (Swan *et al.* 2002) as a key characteristic of those they were willing to trust. This flexibility is concerned with commitment to the shared goals of the project rather than to individual or organisational goals. It should be noted that the ability to be flexible is not just based on the individual managing their own individual risk; there are issues of responsibility (Warwick Manufacturing Group 1999) and authority (Cummings and Bromiley 1997). The individual must be given the ability to act in a flexible way by his or her own organisational structure. It is this ability of networks to respond to environmental indicators that forms a source of competitive advantage (Wilson & Kennedy 1999; Svensson 2001), thus making the right choice of partners important.

Trust and networks

Network theory in its various forms offers a powerful way to regard project teams.

> *While networking can take different forms, all forms are characterised by recurring exchange relationships among a limited number of organisations that retain residual control of their individual resources yet periodically jointly decide over their use.* (Ebers 1997, p. 5)

These networks can be viewed as being made up of nodes and links (Scott 1991) (Fig. 8.2). In the case of the Trust Project these nodes were viewed as individuals. The links were determined as the relationships between those people, of which trust was one dimension. It is through these relationships that information flows. While it is possible for individuals to trust groups, it was revealed from interviews that stronger trust building occurred between individuals. The level of trust towards a company

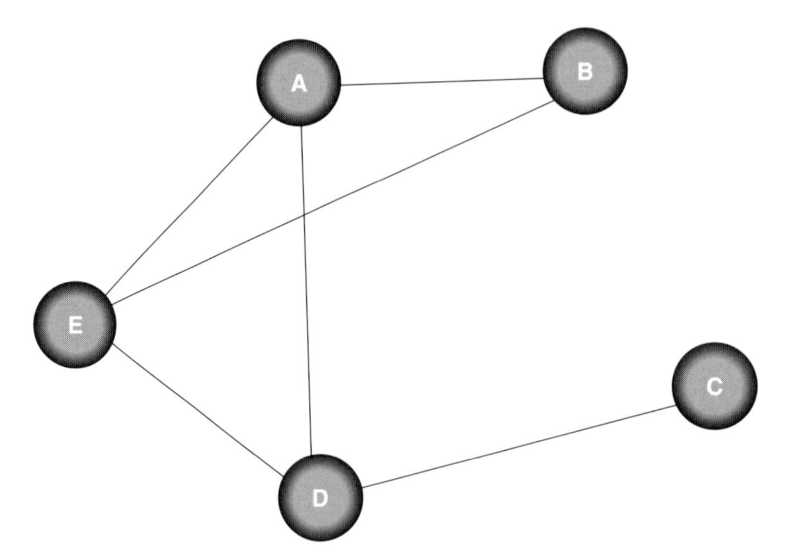

Figure 8.2 Example of a network.

was viewed in a similar way to brand awareness. This was generally formed from past experience with other individuals from the same organisation and the third party experience of other trusted parties. The management of trust in this way could be considered a marketing function. Ultimately, interviewees determined that, although this view formed a base level of trust, their experience with individuals was the stronger driver for building trust.

Commercial manager in construction networks

The research project undertook six case studies, of which we will consider four here: case studies 1–4. The commercial manager role, as defined by the trust project, could be considered as the layer of management below the strategic board level of the organisation, but above the project management role. The role is often spread across several projects engaging in high-level relationship management. The four case studies presented in the next sub-sections will allow us to consider the role of commercial managers between different partner organisations and their contribution to the overall project.

Case Study 1: Large scale infrastructure project

The project was valued at £37 m, had a predicted duration of 18 months and was studied in month 15 of its development. The initial budget cost was £18m; the growth of budget was due to the change of the scope of the works. The timeline for the project was unaffected as the client was very specific with regard to the completion date. This was due to the high levels of capital investment and the requirement to generate an income stream from the project as soon as possible. The project team included many personnel who had worked on a 'world-class' retail development in the same region, as the developer had expressed a preference to work with the same groups of contractors and consultants from the previous project.

> *We have a 'knot' of contractors that know what we expect and how we expect to develop the relationship with the end users.* (Contracts manager of the client organisation)

Many of the individuals within the construction team had been attached to the earlier retail development which had been valued at £300m and had used the same main contractor, architects and structural engineers. There was some retention of key individuals, especially with the architect. The main change was by the contractor who introduced a new project manager to the site. One of the main factors for the selection of the specific team was an attempt to recreate the successful team that had delivered the previous development and thereby transfer learning to the new project. Moreover, the client considered the potential learning between teams would be beneficial.

Key issues

The level of complexity had been determined as high by the individuals involved in this project, with high informational complexity due to the large amount of information flowing between partners. Project management issues were further complicated by operating within a live environment. The participants also considered organisational complexity to be high because of the large number of work packages and specialist contractors. Project complexity issues required commercial managers to address key concerns rather than become hampered with detail.

Project uncertainty was considered to be high due to the use of concurrent design and construction, leading to a large number of design changes that created issues of dependency of tasks and rework. The project was undertaken using management contracting, with the contractor and consultant teams having a direct link to the client, and the contractor managing the works through subcontractors. These works packages were let through the main contractor using standard subcontracts with terms defined by the contractor.

Commercial managers looked to manage the overall strategic relationship, rather than individual project relationships. This case confirms that there are differing perceptions of trust within organisations at different managerial levels (Swan *et al.* 2002). Commercial managers are concerned with the perception of trust in multi-project environments. Although personnel change from project to project, commercial managers tend to function in multi-project environments and keep the trust capital at the strategic level.

Strong messages from the client were reiterated by the commercial managers for all organisations involved within the project. This meant that activity at the site level was aligned to the values of teamwork and flexibility required by the client; therefore, the project delivery team had very clear goals. The client had determined the team approach, replicated from previous projects. The design team was geared towards communicating with the key client representative and addressing their needs. The client felt very strongly about this approach.

Case Study 2: Reconstruction of coastal defences

The project was undertaken over 80 weeks. The initial budget cost for the job was approximately £9 m, but had been revised to include changes to concrete quality and value engineered, which removed about £500 000 from the final cost. This cost was set approximately one-third of the way through the construction phase of the project; it should be noted that the final account was settled nearly seven months before completion of the project and that 14 weeks were added to the project duration due to client changes while the project was delivered against the delivery date of the finalised programme. This meant that, in real terms, the project was delivered 14 weeks early.

Key issues

The main complexity issue was designing working patterns determined by the tides; this proved key to timely delivery of the project. The project itself was considered to be of low to medium complexity by both the client and the main contractor. The commercial managers and directors solved problems at the highest level, the commercial managers being central to the management of conflict, which resolved the issues. Successful management of issues is central to the role of the commercial manager. They have a higher position in the problem resolution hierarchy than the project delivery teams.

When the contract was awarded, the contractor highlighted under-measurement issues by the client, and, at the time of interview, cases were being prepared for adjudication. At this point the regional commercial director proposed a solution to the problem which raised several trust issues that will be dealt with separately within this section.

The proposed solution was to generate a new bill of quantities for the project and so derive a new price. This was then converted to a guaranteed maximum price (GMP) agreement which reflected the original scope of works, while additional items would still be added to the price of the contract. The client then settled the account seven months before practical completion. This was unusual for the contractor who stated that normal settlement periods averaged 6–9 months after the completion of a contract. The client stated that the problem was solved, with the adjudication information both parties had selected having been used to resolve matters to their mutual satisfaction.

Additionally, it was recognised that the contractual arrangements were creating a problem for the client. The contractor and client at commercial manager level resolved to change from a traditional contract to a partnering GMP agreement to resolve potential conflict. This was undertaken with the support of the directors; it could not have been the responsibility of site personnel. Further, the contractor's commercial managers and directors took a clear stance that the project should be non-confrontational: this message was strongly filtered to site staff and proved effective.

Case Study 3: School project

The project was valued at approximately £1.2m and was 40 weeks in construction. There was a long period before development of the project, while the Council team addressed planning and environmental issues. The school would be delivered to the local education authority (LEA) as a 'ready to use' establishment, with the contractor taking responsibility for the internal fittings, such as information technology, the kitchens and furniture. The contractor also assisted with the organisational issues of moving the old school into the new facility. The contractor's commercial manager was central to the management of the project. The project, at the time of data collection, was at the design and construction phase.

Key issues

The project was undertaken as one of two schools within a single partnering project. Considerable delays at both sites due to environmental and site constraints caused the job to be delayed by several months. The construction of the school represented low complexity from a technical perspective. There were only a few key subcontractors.

The relationship uncertainty within the supply chain was also low; however, the client and contractor had never worked together before and this meant their relationship uncertainty was high, so the partnering process was an attempt to mitigate this. The contractor's commercial manager played a vital role in developing and managing the strong relationships with both the client and the rest of the supply chain.

While this was the first job the contractor had undertaken for the client, they had a pre-assembled supply chain of key subcontractors previously used for jobs of this type. Further, there was a clear understanding between the participants with regard to the 'open' partnering approach, with which the contractor and their supply chain were familiar.

> *We believe partnering is the way we want to do business. It may cost more but it reduces risk and creates sustainable business.* (Contractor's commercial manager)

All the participants felt this approach was key: the client organisations had the authority to adopt this approach, and all the key members of the network were willing to engage in the processes. The project team had engaged in a long set-up period, partially imposed by site issues. This allowed the team time to build relationships and was stated by all involved in the process to be a positive and useful experience. The individuals involved in undertaking the works were included in the partnering process from a very early stage, and were encouraged by the commercial managers to work in a way that supported the organisation's partnering philosophy.

The project operated on an open book account; this meant there was transparency of finance. The client representative from the LEA stated that this was the best financial reporting they had experienced on a project and that this approach engendered trust between the parties.

The project was delivered on time and to budget. The technical services project manager stated that he was very happy with the quality and the implementation process was a positive experience, while the end users stated that their strategic brief had been met. These issues had a strong impact on the levels of trust at the data collection stage. The client also appreciated the role played by the commercial managers of the companies involved during the difficult times when conflict arose because of the innovative design and related higher cost per floor area.

Case Study 4: Road development project

The project was valued at £2m and was 32 weeks in construction; it was in the first weeks of construction when the research began. After the final stage of data collection eight weeks of the project remained.

Key issues

The level of complexity had been determined as low as there were few key subcontractors. The organisational complexity was also considered to be low, with many of the key decision-makers having direct access to site meetings, including the commercial managers of all the companies involved. The relationship uncertainty, however, was high, with this job representing the first time that the two main organisations (the contractor and the client) had worked together. Many of the subcontractors were also new to the contractor and the client. However, for this project it was determined that the perspective taken would be for the commercial managers to develop the key relationships between the main contractor and the client, to develop a more focused view of site relationships.

The two teams were from identifiably different approaches to construction. The contractor felt they were strongly aligned with *Rethinking Construction* (Egan 1998) and was comfortable with the tools and techniques of negotiated work. They stated that they looked to gain commercial benefit from the flexibility of these approaches.

Alternatively, the client team stated that they were more comfortable with traditional procurement approaches. They were familiar with the issues, yet stated they felt themselves not entirely able to implement the ideas owing to institutional constraints.

There was a failure to recognise differing values at commercial manager level. Both commercial managers had different approaches; traditional versus relational. This created further problems on site.

It was accepted by both the client and the contractor that the tendered price was insufficient to do the works specified. This created a negative spiral between the two parties. The client accepted that the contractor was not performing 'on a level playing field' and that many of the perceived problems were products of the financial position of the project.

One of the reasons for this was the inability of the commercial managers to resolve the financial issues. This was based partially on a lack of willingness to align approaches between the contractor and client, with options to resolve problems not being taken. The other reason was that the client stated a traditional distrust of contractors, which created a climate of conflict. There were low levels of trust between contractor and client, although the client favoured the contractor more.

Relationships at the senior level were stated to be good. The key project relationship was between the contractor's contracts manager and the technical services manager. There was some feeling between both parties that this was not translated to site. The contractor's regional commercial director stated that they were requiring the site teams to work in a different way, but the site teams were unable to translate this into working practice; the efforts were a complete failure to align goals between the teams.

Cross-case key issues

The commercial manager's role as a person who is creating and managing trust is quite evident from the case studies presented. The key issues emerging clearly from the above case studies are as follows.

Authority and responsibility

Authority and responsibility are important builders of trust if people think they have the ability to make decisions. These authorities and responsibilities must be explicit for a specific project. However, the role of the commercial manager must be to make decisions at the right level, i.e. they must make the decision if the point of issue is contractual, but action at site means that teams cannot resolve their own problems.

Effective management of conflict

The commercial manager is in a strong position to resolve disputes. Their ability to manage conflict effectively contributes towards the success of a project.

Long-term relationships

In most cases, the commercial manager operated in an actual or potential multi-project environment, so they provided a context for project delivery teams. This meant a decision could be made in a wider context which was usually less confrontational. Past relationships were often carried through commercial managers even when project teams disassembled.

Generation of culture

Commercial managers are more visible than director level personnel and therefore provide a conduit to the values of the organisation. They have the ability to drive the values of the shared project organisation.

Process

Commercial managers are usually involved at the contract stage. They have the capacity to engage with the processes that can dictate site behaviours.

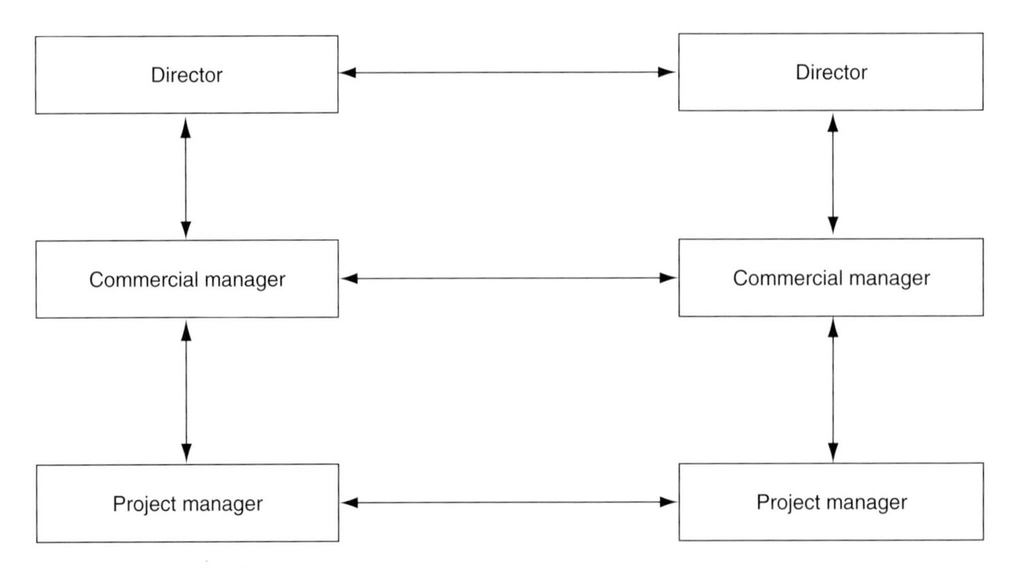

Figure 8.3 Organisational structure.

Common structure

The structure shown in Fig. 8.3 was found in most of the organisations. This ties closely into the concept of structural equivalency (Pryke 1999), meaning that relationships are brokered between organisations at different levels, with each person dealing with their equivalent in their partner organisations. This is a common model used to construct problem resolution ladders.

Other issues

- *Scale and complexity will dictate the role*: Larger projects will give contracts managers a more generic role which will move away from the day-to-day management of the project; this is generally a function of numbers of people involved rather than value.
- *Structure*: The commercial manager is a conduit at a specific level, i.e. above the project. Failure of commercial managers to relate creates problems (as with case 3); this is worse than non-participation.
- *Project stage*: Commercial managers are generally involved at the early stages: this is when they have opportunities to change things. Once the project has been undertaken site issues will take precedence. The commercial manager will address high-level issues, but all activity should be undertaken early if possible.

Conclusions

We can see that there are a number of key elements, both implicit and explicit, acting on the development of trust. Explicit management of trust can be used to manage the risk of relationships breaking down, or not forming in the first place, while the implicit issues may be more complex in terms of management and impact.

Experience is a key driver. While global trust, a basic willingness to trust anybody, varies between individuals, and is possibly associated with risk taking behaviours, it has been noted that experience of individuals with one another is the central driver. The four elements identified: reliability, commitment, relationship and communication, work together to feed the mental models of individuals. Trust is dynamic: both past events and potential future relationships will impact on the current position. It is developed implicitly as part of day-to-day activities of teams working together on projects. High performing projects were generally seen to be more trusting.

This may be viewed as a risk model, with a number of different factors feeding into a mental model. When relationships are new there are high levels of uncertainty that need to be addressed effectively to manage risk out of the system. It is for this reason that we may see the value of intervening directly in the development of trust through the application of team building tools. Looking at trust as a risk factor means we can address the issues of potential breakdown by attempting to manage or mitigate potential conflict. The ability of these interventions to work depends on how robustly these models are held. If interventions are to be applied to project teams, it is important that this is done before the teams have been implicitly formed through experience, otherwise changes may be difficult to effect.

Trust is generally formed between individuals: generally, trust is bilateral in terms of value, i.e. one individual will trust another to about the same extent. When disequilibrium has been observed, it is usually driven by a different interpretation of conflict issues. Trust is usually stronger between individuals of the same organisation and is driven by three key issues: the first is concerned with having a shared set of values driven by the organisation, the second is the longevity of the relationship and the potential consequences of behaving in a distrusting manner, and finally the increased level of experience and lower level of uncertainty in trusting. When considering trust between organisations, trust at the strategic level between structural equivalents, will generally be replicated through lower levels of inter-organisational relationships.

Procurement mechanisms can be key drivers for trust. Formal procurement mechanisms can put in place barriers to effective trust building. Relational procurement mechanisms provide interventions to enable better communications and mitigate potential conflicts by opening information flows, specifically with regard to money. Flexible approaches to procurement allow individuals to address potential conflict and jointly resolve issues. The application of long-term relational models can make untrusting behaviour more risky. If a decision is made to behave in an untrusting manner in a one-off project, the implications are far less damaging than if the relationship is within a multi-project context.

In terms of project drivers, it can be seen that organisational complexity can create issues for the management of trust. Less time can be invested in developing close, trusting relationships with many individuals; this can be mitigated through network trust, where relationships are mediated through a trusted third party. Organisational complexity can also create difficulties in aligning a large number of organisational cultures into a single project culture; this can be mitigated through the application of strong client leadership or through interventions such as partnering workshops. Informational complexity and uncertainty did not appear to impact trust; however, it was felt that higher levels of communication and stronger relationships were used to manage this issue.

We have seen how the role of commercial managers is important for the development of trust. They serve as a conduit between the strategic and the operational and have high levels of authority to address issues that project managers or site supervisors may feel are out of their remit. They support and reinforce the values of the directors, which is critically important when organisations are trying to present themselves as trustworthy. The importance of this link can be seen from a client perspective:

> *We had a contractor that was new to us, and for whatever reason the project did not go well. At the site level they were quite confrontational. The commercial team came back and admitted they were wrong and asked for another chance... We felt that there was commitment from the top, but when it came back to working on site the people that they used either couldn't or wouldn't work in a different way.* (Client commercial director)

From the conclusions presented above, it can be seen that there is a need for more in-depth research to develop more effective trust building relationships within the construction industry as a whole. On the other hand, there is also a strong need to clearly define the roles and responsibilities of commercial managers in the different organisations involved within a construction project to achieve a no-blame, truthful, reliable and responsive culture within the construction industry. An agenda for further research could include investigating:

- How the experience of commercial mangers in interacting with one another can contribute towards better trust building
- How the high levels of uncertainty are addressed in a new relationship in order to manage risk effectively
- How procurement mechanisms can be a key driver of trust within the construction industry
- How organisational complexity can create issues for the management of trust and how they could be resolved.

Last but not least

- Why the role of commercial managers is important in the development of trust.

References

Allen, S. & R. Cooper (1999) *Partnering: Interim Report*. University of Salford, Salford.

Bachmann, R. (2000) Trust, power and control in trans-organisational relations. In: *Inter-Firm Networks: Theory and History*. UMIST, Manchester.

Barlow, J. (1996) Introduction – Partnering in construction. *Partnering in Construction*, University of Salford, Salford; pp. i–vii.

Barlow, J. & Cohen, M. (1996) Implementing partnering: some common red herrings in the literature. *Partnering in Construction*. University of Salford, Salford; pp. 21–35.

Bejou, D., Ennew, C.T. & Palmer, A. (1998) Trust, ethics and relationship satisfaction. *International Journal of Bank Marketing*, **16**(4), 170–175.

Benheim, M. & Birchall, D. (1999) Networks in the construction industry: emerging regulation procedures in the production process. In: *Profitable Partnering in Construction Management*. CIB W92 Harmony and Profit Symposium, Chaing Mai, Thailand (ed. S.O. Ogunlana). E. & F.N. Spon, London; pp. 63–75.

Bigley, G.A. & Pearce, J.L. (1998) Straining for shared meaning in organization science: problems of trust and distrust. *Academy of Management Review*, **23**(3), 405–421.

Blois, K. (1999) Trust in business to business relationships: an evaluation of its status. *Journal of Management Studies*, **36**(2), 197–215.

Brenkert, G. (1998) Trust, business and business ethics: an introduction. *Business Ethics Quarterly*, **8**(2), 195–203.

Bresnan, M. (1996) Cultural change in the construction industry: developing the client's management role to improve project performance. *Partnering in Construction*. University of Salford, Salford; pp. 8–20.

Brower, H.H., Schoorman, D.F. & Tan, H.T. (2000) A model of relational leadership: the integration of trust and leader–member exchange. *Leadership Quarterly*, **11**(2), 227–250.

Carmichael, S. & Cooper, R. (1999) *Adversaries or Partners – Best Practice Guidelines for Partnering in Construction*. University of Salford, Salford.

Checkland, P. & Holwell, S. (1997) *Information, Systems and Information Systems*. John Wiley, Chichester.

Comptroller and Auditor General (2001) *Modernising Construction*. National Audit Office. London.

Couch, L.L. & Jones, W.H. (1997) Measuring levels of trust. *Journal of Research in Personality*, **31**, 319–336.

Cummings, L.L. & Bromiley, P. (1997) The organisational trust inventory (OTI): development and validation. In: *Trust in Organisations: Frontiers of Theory and Research* (eds R.M. Kramer & T.R. Tyler). Sage, Thousand Oaks, CA; pp. 302–330.

Doney, P.M., Cannon, J.P. & Mullen, M.R. (1998) Understanding the influence of national culture on the development of trust. *Academy of Management Review*, **23**(3), 601–620.

Ebers, M. (1997). Explaining inter-organisational network formation. *The Formation of Inter-Organisational Networks*. (ed. M. Ebers). Oxford University Press Oxford; pp. 3–27.

Edwards, P. (1998) Risk management and risk management techniques. *Journal of Construction Procurement*, **4**(2), 103–115.

Egan, J. (1998) *Rethinking Construction*. HMSO, London.

Egan, J. (2002) *Accelerating Change*. (Review of the changes since *Rethinking Construction* and key issues to be addressed). HMSO, London.

Elangovan, A.R. & Shapiro, D.L. (1998) Betrayal of trust in organizations. *Academy of Management Review*, **23**(3), 547–566.

Flores, F. & Solomon, R.C. (1998) Creating trust. *Business Ethics Quarterley*, **8**(2), 205–232.

Fukayama, F. (1995). *Trust: The Social Virtues and the Creation of Prosperity*. Penguin, Harmondsworth.

Gambetta, D. (1988) *Trust: Making and Breaking Cooperative Relations*. Basil Blackwell, New York.

Ganesan, S. (1994) Determinants of long-term orientation in buyer–seller relationships. *Journal of Marketing*, **58**, 1–19.

Hagan, J.M. & Choe, S. (1998) Trust in Japanese interfirm relations: institutional sanctions matter. *Academy of Management Review*, **23**(3), 589–600.

Hatush, R. & Skitmore, M. (1997) Evaluating contractor prequalification data: selection criteria and project success factors. *Construction Management and Economics*, **17**, 129–147.

Hertz, S. (1992) Towards a more integrated industrial system. In: *Industrial Networks: A New View of Reality* (eds B. Axelsson & G. Easton). Routledge, London; pp. 105–123.

Holti, R., Nicolini, D. & Smalley, M. (2000) *Building Down Barriers: The Handbook of Supply Chain Management Essentials*. Tavistock Institute, London.

Inkpen, A.C. & Currall, S.C. (1998) The nature, antecedents and consequences of joint venture trust. *Journal of International Management*, **4**(1), 1–20.

Jefferies, M.C., Chen, S.E. & Mead, J.D. (1999) Project team performance – managing individual goals, shared values and boundary roles. In: *Profitable Partnering in Construction Management*, CIB W92 Harmony and Profit Symposium, Chaing Mai, Thailand (ed. S.O. Ogunlana). E. & F.N. Spon, London; pp. 47–60.

Kwan, A.L. & Ofori, G. (2001) Chinese culture and successful implementation of partnering in Singapore's construction industry. *Construction Management and Economics*, **19**, 619–632.

Latham, S.M. (1993) *Trust and Money* (Interim Report). Department of Transport and Regions, London.

Liebenau, J. & Backhouse, J. (1990) *Understanding Information: An Introduction*. Macmillan, London.

Liu, A. & Fellows, R. (1999) Cultural issues. In: *Procurement Systems: A Guide to Best Practice in Construction* (eds P. McDermott & S. Rowlinson). E. & F.N. Spon, London; pp. 139–162.

Lok, P. & Crawford, J. (1999) The relationship between commitment and organizational culture, subculture, leadership style and job satisfaction in organizational change and development. *Leadership and Organization Development Journal*, **20**(7), 365–373.

Luhmann, M. (1979) *Trust and Power*. John Wiley & Sons, Chichester.

McAllister, D.J. (1995) Affect and cognition based trust as a foundation for interpersonal cooperation in organisations. *Academy of Management Review*, **38**(1), 24–59.

Masden, S.E. (1991) The costs of organization. *Journal of Law, Economics and Organization*, **7**(1), 1–25.

Mishra, A.K. & Spreitzer, G.M. (1998) Explaining how survivors respond to downsizing: the role of trust, empowerment, justice and work redesign. *Academy of Management Review*, **23**(3), 567–588.

Misztal, B.A. (1996) *Trust in Modern Societies*. Polity Press, Cambridge.

Morgan, R.M. & Hunt, S.D. (1994) The commitment trust theory of relationship marketing. *Journal of Marketing*, **58**, 20–38.

Nonaka, I. & Konno, N. (1998) The concept of 'Ba': building a foundation for knowledge creation. *California Management Review*, **40**(3), 40–54.

Ouchi, W.G. (1980) Markets, bureaucracies and clans. *Administrative Science Quarterly*, **25**, 129–141.

Palmer, R. (1969) *Hermeneutics: Interpretation Theory in Schleiermacher, Dilthey, Heidegger and Gadamer – Studies in Phenomenology and Existential Philosophy*. Northwestern University Press, Evanston.

Piaget, J. (1961) The genetic approach to the psychology of thought. *Journal of Educational Psychology*, **52**, 275–281.

Pryke, S.D. (1999) Proposing a social network analysis approach to the evaluation of post Latham initiatives in construction project management and procurement systems. In: *Proceedings of the RICS Construction and Building Research Conference (COBRA)*, 1–2 September, University of Salford (eds D. Baldry & L. Ruddock). RICS Foundation, London; pp. 235–245.

Ricoeur, P. (1991) *A Ricoeur Reader*. Harvester Wheatsheaf, Hemel Hempstead.

Ring-Smith, P. (1997) Processes facilitating trust in inter-organisational networks. In: *The Formation of Inter-organisational Networks* (ed. M. Ebers). Oxford University Press, Oxford; Chapter 5.

Roy, M.H. & Dugal, S.S. (1998) Developing trust: the importance of cognitive flexibility and co-operative contexts. *Management Decision*, **36**(9), 561–567.
Scott, J. (1991) *Social Network Analysis: A Handbook*. Sage, Thousand Oaks, CA.
Smyth, H. & Thompson, N.J. (1999) Partnering and the conditions of trust. In: *Customer Satisfaction: A Focus for Research and Practice* (eds P. Bowen & R. Hindle) CIB W55 and W65 Joint Triennial Symposium 5–10 September, University of Cape Town, Cape Town, South Africa.
Svensson, G. (2001) Perceived trust towards suppliers and customers in supply chains of the Swedish automotive industry. *International Journal of Physical Distribution and Logistics Management*, **31**(9), 647–662.
Swan, W., Cooper, R., McDermott, P. & Wood, G. (2001) A review of social network analysis for the IMI trust in construction project. In: *Proceedings of the 17th Annual ARCOM Conference*, Vol. 1 (ed. A. Akintoye). University of Salford, 5–7 September. Association of Researchers in Construction Management; pp. 57–67.
Swan, W., McDermott, P., Wood, G. & Cooper, R. (2002) *Trust in Construction: Achieving Cultural Change*. Centre for Construction Innovation, Manchester.
Tomkins, C. (2001) Interdependencies, trust and information in relationships, alliances and networks. *Accounting, Organizations and Society*, **26**, 161–191.
Tyler, T.R. & Kramer, R. (1996) Whither trust? In: *Trust in Organisations* (eds R. Kramer and T.R. Tyler). Sage, Thousand Oaks, CA; pp. 1–15.
Uher, T. (1999) Partnering and strategic alliances in Australia. *Journal of Construction Procurement*, **2**(1), 163–167.
Viall, P.B. (1996) The purposing of high performance systems. In: *How Organisations Learned* (ed. K. Starkey). Thompson Business Press, London; pp. 60–81.
Vicari, S. & Troilo, G. (1998) Errors and learning in organisations. In: *Understanding Knowledge in Organisations* (eds G. von Krough, J. Roos & D. Kleine). Sage, London; pp. 204–222.
Walker, D. & Hampson, K. (2003) Developing cross team relationships. In: *Procurement Strategies: A Relationship-Based Approach* (eds D. Walker & K. Hampson). Blackwell Science, Oxford; pp. 169–203.
Warwick Manufacturing Group (1999) *Implementing Supplier Chain Management in Construction*. Warwick University, Warwick.
Webber, S.S. (2002) Leadership and trust facilitating cross-functional team success. *Journal of Management Development*, **21**(3), 201–204.
Weick, K.M. (1995) *Sense Making in Organisations*. Sage, Thousand Oaks, CA.
Wilson, P.N. & Kennedy, A.M. (1999) Trustworthiness as an economic asset. *International Food and Agribusiness Management Review*, **2**(2), 179–193.
Wood, G. & McDermott, P. (1999) Looking for trust in construction: an interim view. In: *Profitable Partnering in Construction Procurement* (ed. S.O. Ogunlana). E. & F.N. Spon, London.
Wood, G., McDermott, P., Wood, G. & Cooper, R. (2002) The ethical benefits of trust-based partnering: the example of the construction industry. *Business Ethics: A European Review*, **11**(1), 4–13.
Woodward, D. & Woodward, T. (2001) The efficacy of action at a distance as a control mechanism in the construction industry when a trust relationship breaks down: an illustrative case study. *British Journal of Management*, **12**, 355–384.

Endnote

[i] Autopeosis is the process of forming shared meaning within a knowledge network.

9 Outsourcing

Jan Bröchner

Introduction

In 1989, when Eastman Kodak decided to transfer their in-house information technology services to IBM and two other external suppliers, it was seen as a major business event. Trade magazines, such as *Computerworld*, covered the developments ensuring that the word *outsourcing* gained currency. Eastman Kodak was soon imitated by many large corporations (Loh & Venkatraman 1992a). The phenomenon, however, was far from new, and manufacturers had been outsourcing for many years without using the term.

In this chapter, we shall look closely at what is really known about outsourcing and how the phenomenon can be approached as an object of investigation. It is a complex phenomenon, and there is a huge volume of related literature. To begin with, we shall consider a narrow but widespread definition of outsourcing. An early definition of outsourcing was 'the purchase of a good or service that was previously provided internally' (Lacity & Hirschheim 1993) and an influential overview of the field defines it in much the same way as 'the process whereby activities traditionally carried out internally are contracted out to external providers' (Domberger 1998, p. 12). However, there is also a broader definition, where outsourcing is seen as what was 'previously termed contracting out', and 'the contracting of any service or activity to a third party' (Kakabadse & Kakabadse 2002). But unless we make the distinction that outsourcing refers to activities that have been performed by in-house employees, we would have to include any type of contractual reliance on external providers, including, for example, traditional subcontracting in the construction sector, and then the concept of outsourcing loses its edge and identity.

Thus, we see outsourcing as a special case of contracting, one that has a project and process nature. Buying what you have once produced yourself is also a special case of the old make-or-buy decision.

Before leaving the issue of definitions, we must note that there is a whole family of terms related to outsourcing. First, there is the root in 'sourcing'. Car manufacturers started to import components from foreign plants at a much earlier date than information technology (IT) outsourcing began. The *Oxford English Dictionary* notes the earliest uses of 'sourcing' in this sense, quoting 1960 articles in *Wall Street Journal* and *Business Week*. 'Selective sourcing' (Lacity *et al.* 1996) refers to 'short-term contracts of less than five years for specific activities' or 'outsourcing a subset of IT activities' (Lacity & Willcocks 2001, p. 4), which makes it almost a synonym for 'out-tasking' as used in the area of facilities management (FM): 'hiring individual, specialized vendors to provide one or more FM functions' (Kleeman 1994). 'Total outsourcing' is given a precise and operational interpretation for the IT field by

Lacity and Willcocks (2001) when they define it as outsourcing greater than 80 per cent of the IT operating budget. Unfortunately, we find a variety of definitions for terms such as 'co-sourcing' where 'the provider takes over responsibility for the roadblocked function or process, but the retained employees remain on the tiger's payroll' [as defined by Linder (2004)]. Or is co-sourcing characterised by 'performance-based contracts, tying supplier payments to business performance' (Lacity & Willcocks 2001)?

Semantic ambiguity debilitates the term 'insourcing'. Frequently, insourcing refers to the in-house reintegration of production that has been previously outsourced. But the term appears to be stretched when Lacity *et al.* (1996) use '*de facto* insourcing' to describe the situation where an organisation uses internal IT departments to provide products and services 'that arise from historical precedent'. And it does not get easier when insourcing is used for the process of a country attracting foreign firms to locate production or suppliers there.

Outsourcing and 'theories of the firm'

When managers are asked about their reasons for outsourcing, the first thing that often springs to their mind is the reduction of operational costs, as numerous surveys testify. Depending on industry sector, type of function to be outsourced, and also country, many respondents tend to point to an ability to improve support by access to new technology or competence. Increasing the level of service quality is also often mentioned. On the other hand, risks are perceived: for example, reduced control and flexibility, dependence on supplier, and service failures. Moreover, a perceived inability to identify and quantify risks can be a barrier to outsourcing. Another line of reasoning springs from the observation that an in-house IT manager might be socially unable to introduce uniform practices in the organisation. It is only the approach of an external supplier that appears to make this standardisation feasible. The same obstacle to standardisation is claimed to be found for business processes such as accounting, finance and human resources management. Therefore, setting up a centre for all customer inquiries may require outsourcing. It is not immediately obvious which body of theory is best suited for analysing this collection of phenomena and perceptions.

The first and still perhaps the most influential attempt at making sense of the 1989 Eastman Kodak decision was a *Harvard Business Review* article that launched the concept of core competencies, intended to explain what could and should be outsourced. In this article, Prahalad and Hamel (1990) presented three tests to identify core competencies in a company. First, a core competence 'provides potential access to a wide variety of markets'; second, it 'should make a significant contribution to the perceived customer benefits of the end product'; finally, it 'should be difficult for competitors to imitate'. The reasoning is clearly affected by theories of what constitutes competitive advantage (Porter 1985). Given that the deduction of these three tests is weak and also that the tests are anything but easy to apply to an actual firm that is considering to outsource part of its activities, the success of this article, which is probably the most widely cited in the whole literature on outsourcing, is remarkable.

Nevertheless, it did break new ground and pointed in an important direction: the role and the management of specialised knowledge in firms. The two authors attacked the hegemony of strategic business unit (SBU) thinking, which tended to reduce the role of top management to a banking operation for a conglomerate, while competencies were 'imprisoned' in the SBUs.

It is, therefore, relevant to pause and consider the history of Eastman Kodak just before and just after 1989, if we wish to understand the anti-conglomerate strain in the competencies line of thought. Here was a typical, century-old giant US corporation. In 1986, it had entered the general consumer battery market; also, it established its Pharmaceuticals Division; and in the following year, the company entered what was then known as the electronic still-video market. Given this rapid diversification, it is understandable that top managers felt the need to rethink how the business was organised and how its boundaries were drawn. Furthermore, the 1980s saw rapid change in IT, in particular the shift from an emphasis on mainframe computing to PCs in networks. Incidentally, only seven years after entering the pharmaceutical sector, Kodak started pruning its activities and sold its non-imaging health-related businesses in 1994. Somehow, the rapid swing away from conglomerate diversification together with the shedding of internal support services cannot entirely be explained by changing technologies; there was a strong element of the effects of financial deregulation and the general increase in global trade volumes during that period.

Quinn and Hilmer (1994) have contributed to the development of the core competencies approach, but when Quinn (2000) later warmly recommended the outsourcing of innovation, the approach seemed to be unduly flexible. It can also be thought of as too concerned with competitors and too little with suppliers, as emerges from McIvor's (2003) study of a telecommunications equipment manufacturer, where the author concludes that analysis and use of the supply market can have a more significant influence on the outsourcing process than attempting to identify core competencies in that industry.

Instead, the complexity and pervasiveness of outsourcing requires a broader consideration of all major 'theories of the firm'. While the core competencies approach is associated with the resource-based view of the firm (Hoopes *et al.* 2003), there are two other fundamental views that have been used by many studies of outsourcing, sometimes in eclectic combinations of features from two or all three. The other two are institutional (or contract) economics, primarily transaction cost analysis (TCA) (Williamson 1985; Grover & Malhotra 2003) and network theories (Gadde & Håkansson 2001). A good example of how several bodies of theory can be integrated has been given by Kern and Willcocks (2001), who lean on transaction cost theory, relational contract theory and interorganisational (dyads and networks) theory for their 'relational advantage' conceptual structure. Moreover, Poppo and Zenger (1998) tested their questionnaire data for hypotheses derived from transaction cost theory as well as knowledge-based and measurement cost theories; their results can be interpreted as an argument for an eclectic approach.

The transaction cost approach is probably the most frequently used theoretical base for studies of outsourcing. As to the 'why' of outsourcing, the theory predicts that asset specificity, uncertainty and transaction frequency influence the choice between 'market' and 'hierarchy', where the decision to go from hierarchy (i.e. in-house) to market means outsourcing. However, operationalising the main concepts, which are actually one step removed from the core of the theory, that transaction costs and not (only) production costs matter, is acknowledged to be difficult. Nevertheless, reliance on transaction costs thinking offers a way of systematising a view of the outsourcing process, *ex ante* and *ex post* seen in relation to the signing of a contract.

It can be said, with some justification, that the assumptions underlying transaction cost analysis, in particular 'opportunistic behaviour' (Foss & Koch 1996), are more appropriate when explaining 'classical' contracts, and less efficient in sorting experiences from relational contracting within partnering arrangements. Nevertheless, we can add 'trust' to the transactions framework and conclude that, to take an example based on two oil industry case studies, issues such as organisational culture and the nature of various risks remain to be investigated (Meer-Kooistra & Vosselman 2000). Indeed, uncertainty is a major deterrent to outsourcing (Aubert *et al.* 2004). A useful analysis of risk mitigation in the context of IT systems outsourcing has been written by Willcocks *et al.* (1999).

Methods in research on outsourcing

A wide range of methods may be employed to investigate why and how firms outsource. Empirical studies of outsourcing are either large surveys or longitudinal case studies, although there are some intermediary approaches. Most surveys in this field suffer from low response rates and can be suspected of bias. One source of bias is that different levels in a management hierarchy might hold different views as to the desirability or success of outsourcing. It is a recurrent phenomenon that general managers tend to lean more towards outsourcing than their functional managers do at a lower level, although few investigations have acknowledged this.

Postal or e-mail surveys in this field tend to be weakened by low participation. In fact, the problem of low response rates in surveys of outsourcing has generated research of its own. Larson and Chow (2003) report their experimentation with follow-up mailings and monetary incentives for participants in a study of purchasing involvement in the management and outsourcing of logistics. They mailed 1800 members of the Purchasing Management Association of Canada subdivided into 18 groups, which were treated differently, and obtained response rates ranging from between 13 and 40 per cent.

There are only a few surveys that have used multivariate statistical techniques. A pioneering attempt to relate the degree of IT outsourcing to cost structures in firms relied on factor analyses and multiple regression, based on data from 55 US firms (Loh & Venkatraman 1992b). Trying to identify the effect of service quality and partnership attributes on IT outsourcing, Grover *et al.* (1996) analysed responses in 188 questionnaires to find whether service quality and partnership mediated between degrees of outsourcing of a range of information systems functions and outsourcing success. To take a more recent example, Quélin and Duhamel (2003) used factor analysis on 180 questionnaires to identify decision criteria for outsourcing.

Most of the literature is based on less sophisticated analyses of questionnaire data or on multiple case studies, sometimes based on longitudinal interview surveys of managers. As an instructive example, it is worth turning to Kern and Willcocks (2001) who explain in their methodology appendix how the Oxford database was created. At the other end of the range, we find investigations that focus on only one or two firms. This allows the introduction of elements of action research, as in a study of Aalborg Industries (Momme & Hvolby 2002), where one of the authors shifted from being an external case study researcher to also being a part-time project engineer, which then leads to a specific need for documentation and reflection on the effects of intervening in the process.

From traditional contracts to alliances

Many, if not all, writers on outsourcing assume or recommend a change away from contractual relationships based on strict and precise requirements. Instead, firms should strive for long-term relationships with an external supplier, and it is also common to suggest that it is advantageous to rely on as few providers as possible, ideally only one. Outsourcing to one supplier only, or to several, is an issue that is frequently discussed. Should there be multiple suppliers for the same service or for different services? There can also be a question of partial outsourcing of the same service because of variations in support demand over time.

When long-term relationships are considered, researchers often invoke the concept of trust. The argumentation is closely linked to a fundamental view of outsourcing as being either restricted to cost reduction for a given set of activities or a mechanism for transforming a business, bringing in the revenue side.

It is not unusual for a joint venture to be established, where the two parties can pool their skills while also allowing them to reduce certain types of risk, sharing costs and liabilities. There is thus almost a continuous scale ranging from in-house production to fully outsourced production, where staff, equipment and fixed assets have been transferred to the external provider. Neither is in-house production something that is uniquely determined; is a support service provided by a cost centre of its own, in the next phase by a profit centre, or has it been transformed into a subsidiary? Intermediary arrangements are not least important for public–private partnerships. Over time, there may be a succession of solutions where the in-house function is transferred to a subsidiary, which is then encouraged to compete for external customers, and ultimately sold to a group that operates on wider markets. But there is no historical necessity for such a process. Traditional contracts can be equipped with efficient regimes for dealing with change, as we shall see when we now turn to the process of outsourcing.

The process of outsourcing

Establishing an outsourcing relationship is a process that is typical of complex projects that are based on contracts. Only the basic outline can be given here. There are initial activities where a firm identifies its requirements and searches for potential suppliers. A call for tenders or request for proposals (RFP) is prepared, and in the case of service functions, it is usual to specify services as a basis for a service level agreement (SLA).

When formulating an SLA, there are several choices to be made. How precise should specifications be, and should they be directed towards output and performance, rather than prescribing inputs and activities in technical detail? The answers to such questions are obviously related to how far from, or close to, a long-term alliance the contractual relationship is intended to be. The type of service is another determinant, while the main source of ideas for writing an SLA remains the IT industry. A typical SLA, which forms part of the agreement between a service provider and the buyer, will include descriptions of particular services, their elements and levels. Key indicators of levels may cover characteristics such as availability periods and response times. Responsibilities of both parties and charging, if relevant, may also be included. However, and with few exceptions (see Buco *et al.* 2004), researchers have shown little interest in methods for service specifications and what they contain; instead, there are handbooks and recommendations published by consultants in the field (e.g. Hiles 2000).

Since outsourcing, as the concept is usually understood, might imply transfer of staff to an external supplier, an essential part of the process is dealing with people. This is also relevant when in-house employees are shifted from support functions to the core business, or when they leave for another new employer. However, there are only a few investigations that concentrate on the human aspect of outsourcing. A case study of a UK local authority raised post-transfer issues by asking questions such as: is work intensity higher ('yes'); are there higher career opportunities ('yes', but less than 'expected'); is there higher job satisfaction ('yes') (Kessler *et al.* 1999). Government regulation and union agreements can be crucial. Jeffers (1996) describes the effects in the UK of the Transfer of Undertakings (Protection of Employment) Regulations (TUPE), originally from 1981, an implementation of the Acquired Rights EC Directive. In fact, regulation intended to safeguard employee relationships can increase the volume of outsourcing. Changes in the US legal environment appear perhaps to lie behind 20 per cent of the rapid growth of temporary help employment between 1973 and 1995; Autor (2003) has shown that when and where state courts have limited employers' discretion to terminate employment relationships ('employment at will'), it has been reflected in a higher reliance on temporary help outsourcing. Autor (2003) also found that firms do not typically outsource jobs in which skill investments, in other words, costs for training, are large.

Once the contract has been signed, monitoring of supplier performance is considered to be a vital issue. Just as with specifications, there is a range of alternatives that reflect the essence of the relationship. Many authors emphasise the exchange of information during the period of a contract. Especially for the first contractual period, it is important to be able to adjust the distribution of risk between the parties, using price clauses and having defined opportunities for renegotiation (Kern *et al.* 2002). The issue of optimal contract duration cannot be resolved without an analysis of such mechanisms. Otherwise, we should expect contract duration to involve a trade-off between protecting the parties against hold-up of relationship-specific investment and reducing the flexibility that they have with which to respond to environmental changes (Brickley *et al.* 2003).

Relying on the theory of TCA, the economics of the process itself are worth studying. In his investigation into the costs of IT outsourcing, based on 50 cases of which the majority (76%) were still in their first contract, Barthélemy (2001) found that there were strong economies of scale in vendor search and contracting. Thus, he claims that, for contracts below $US10m, these costs are on average 6% of the total contract, whereas for contracts above $US100m, this proportion falls to 0.25%. The costs of managing the outsourcing effort (i.e. monitoring a contract) also display a strong effect of scale, the average being 8% of the yearly contract amount.

Industries and functions

Here, we shall look separately at a range of industries and fields of outsourcing. Although manufacturing claims historical priority, we begin with *IT systems*, by considering the pre-eminence of the 1989 Eastman Kodak decision to outsource. IT systems issues also dominate the outsourcing literature. Lee *et al.* (2003) have sketched how shifts in underlying technologies have been reflected both in outsourcing practices in the IT sector and in outsourcing research since the early experiences of time sharing in the 1960s. They offer a technology based explanation for the drift towards mutual exchange relationships, away from the initial emphasis on 'classical contracting'. Application services providers (ASPs) with monthly or per usage payments are in a sense a return to the access-to-remote-services four decades ago (Currie *et al.* 2004). In general, basing their conclusions on extensive multiple case studies, Lacity and Willcocks (2001) probably offer the best overview of IT outsourcing experiences. In their comparison of French and German IT outsourcing, Barthélemy and Geyer (2001) found similar patterns in both countries. Nevertheless, German firms were more likely to have outsourced networks and data centres than their French counterparts, which in turn were more ready to outsource applications development and 'microcomputers'. The authors ascribed a lower frequency of personnel transfers and lay-offs in Germany to the greater power of trade unions there and their opposition to outsourcing.

Business process outsourcing (BPO) can be identified as an extension of IT systems outsourcing. Finance, accounting (e.g. payroll, expense processing and tax issues) and human resources management are typical examples of support functions with activities that might be subject to BPO.

In a comparative perspective, *facilities management* (FM) is characterised by its immediate geographical link to the customer firm. In the particular context of IT functions, FM initially referred to operating and managing hardware and software assets, but without ownership of the facilities used, although there is a transfer of staff. Currently, however, FM usually refers to the co-ordination and operation of a range of building and office workplace support services. Outsourcing of FM shares many of the generic features of outsourcing arrangements. An early example from Rank Xerox (Houston & Youngs 1996) was among those that drew attention to large-scale FM outsourcing contracts. Bon and Luck (1999) identified development trends in Europe and North America during the 1990s. Katsanis (2002) stresses that the practice of outsourcing is likely to be more widely adopted in FM, since it is now possible for specialists to access an expanded market, unencumbered by geographical barriers. Davis (2004) recognises two variants in FM procurement strategy: output based competitive tendering and an open book partnership model with a declared profit margin and shared savings formulae, reducing the conflict of interest between parties. This is a theme that goes across many services.

Based on FM experiences from the Honeywell Group, Usher (2004) highlights the incompatibility of classical contracts and alliance thinking, as 'all semblance of "partnering" within a five-year contract, and the apparent commitment to togetherness, can be undone by the 90-day "termination at will" clause'. International comparisons of FM outsourcing show that firms in countries such as the UK and The Netherlands tend to rely on fewer suppliers for a broader range of FM services. One possible explanation is that markets where facilities are more dispersed will have less obvious economies of scale for external providers (Bröchner *et al.* 2002). FM services providers emanate from various industries. Furthermore, certain property owners have a history linking them with construction, others with an origin in the financial sector; this can be shown to have consequences for the bundles of FM services that are offered in heterogeneous markets (Bröchner *et al.* 2004).

In some ways *logistics* is akin to BPO, although there is, as in FM, a strong physical component in the services to be provided, but here in terms of goods to be moved and this in relation to fixed facilities. The concept of third-party logistics (3PL) is central to logistics outsourcing: 'a relationship between a shipper and third party, which, compared with basic services, has more customized offerings, encompasses a broader range of service functions, and is characterized by a longer term, more mutually beneficial relationship' (Knemeyer *et al.* 2003). According to their survey, the functions most often provided by third parties to US logistics customers, mostly in manufacturing, were outbound and inbound traffic control, carrier negotiating and contracting, freight consolidation and transportation planning and/or management. Lower frequencies were found for functions such as inventory management and pick-and-pack. However, although there were 388 respondents, the percentage response rate was low. Increasing the range of logistics services offered easily leads into a discussion of whether the third party should consider taking over assembly and other more manufacturing related activities from the customer. The boundary between logistics and manufacturing is being blurred or moved because of developments in IT and telecommunications, leading to supply chain management innovations. The geographical locus for the optimal supply chain, from raw materials to the end customer, is sensitive to how communication technologies develop. The literature review presented by Razzaque and Sheng (1998) ends in a eulogy of partnership between firms: it is all about offering faster deliveries, and information that is more accurate.

While there is a long tradition of outsourcing the production of components in the *manufacturing industry*, recent literature often emphasises risks associated with skills migrating abroad. One argument goes that since product and process technology are becoming indivisible in industries such as automobile, semiconductors and consumer electronics, design and manufacturing should be increasingly difficult to separate; the supplier learns the design skills of the outsourced business and 'industrial decline' continues (Bettis *et al.* 1992). A similar mechanism was identified in the relationship between market knowledge and manufacturing, in that it is more efficient to keep these close together. In other words, issues related to the transfer of specialised knowledge loom large. It has been noted that suppliers of parts to the car industry have weak bargaining power, meeting this by mergers and acquisitions; the underlying problem is excess capacity for many of them (Doig *et al*. 2001). Gilley *et al*. (2004) surveyed 86 small manufacturers and found that higher levels of perceived environmental dynamism and managerial risk aversion were associated with increased outsourcing activity. Mature firms would outsource more when their top managers were more risk averse. As already mentioned, McIvor (2003) in a study of a telecommunications equipment manufacturer found that analysis and use of the supply market could have a significant influence on the outsourcing process. Rapid development of technology led to a redefinition of the core business of organisations, whereas the analysis by Poppo and Zenger (1998), dealing with IT outsourcing, indicated that a high rate of change in a support technology was conducive to a higher rate of outsourcing. Thus the relative rate of technology change in the core and in support services appears to be important. Freytag and Kirk (2003) have studied as action research no less than fourteen firms in the metalworking and electronics industries, developing a framework for decisions on outsourcing, basically oriented towards a network approach.

However, the impression of industrial decline is also a consequence of domestic outsourcing of service functions from manufacturers. An input/output analysis of the UK manufacturing sector has shown a significant increase over the period 1992–1998 in purchases that manufacturing made from non-manufacturing, services and transport in particular (McCarthy & Anagnostou 2004).

Sharing characteristics of manufacturing and facilities management, there is the outsourcing of *industrial maintenance*. Here we have an investigation by Meer-Kooistra and Vosselman (2000), already mentioned because of their reliance on the concept of trust, where they compared two cases from the energy industry. Another study, in the food and chemical industry, indicated that firms that evaluated the option of outsourcing to full-service maintenance contractors, meaning those who offered comprehensive bundles of services were more concerned with the effect on general plant performance than with maintenance costs (Stremersch *et al*. 2001).

The public sector

Outsourcing from the public sector [for an overview, see Domberger (1998), Chapter 9] is often seen as belonging to the wider group of public–private partnerships. However, public sector use of private contractors may or may not involve the transfer of employees. In fact, local authorities in many countries often relied on private contractors and did not move services in-house until the second half of the nineteenth century; not to mention that tax collection in the late Roman republic was farmed out to private societies of publicans (Badian 1972). Water supply in France remains more or less privatised, however, and the wide range of contractual arrangements found there can be analysed using theories of contract economics (Ménard & Saussier 2002).

The Local Government Act 1988 introduced compulsory competitive tendering (CCT) in England, and it attracted followers in a number of other countries, while a longer US tradition of contracting with external providers appears to have had less international impact as a paradigm. A readable account of the long history of private involvement in the US prison systems has been presented by Schneider (2000). One sign that CCT has led to greater efficiency in services production is that English local authorities were less likely to copy administrative patterns from adjacent districts (Bivand & Szymanski 2000). Accountability, ethical issues and output qualities that are difficult to inspect complicate the practice of public sector outsourcing: health care services is a prime example (Preker *et al.* 2000), whether private or not; the concern with patients has been shown to affect attitudes to IT outsourcing in US health care (Lorence & Spink 2004).

Offshore outsourcing

On a small scale, outsourcing often involves moving staff, software or hardware away from the firm that outsources, perhaps because of a technology shift that reduces the effect of physical distance. Manufacturers moved production offshore much earlier than service firms due to lower production costs. However, it is when services are outsourced to providers outside the country that outsourcing acquires another political dimension. In 2004, the US Senate passed the Dodd amendment to the Jumpstart Our Business Strength (JOBS) Act, prohibiting companies on federal and state government contracts to outsource work abroad. However, Drezner (2004) has reviewed the offshore outsourcing issues and found that many authors have overstated the importance of the phenomenon.

Murray and Kotabe (1999), relying on a modified TCA, have studied how US service companies choose between internally and externally produced support services. Their statistical analysis established negative relationships between a service's market performance and both internal sourcing and foreign sourcing of supplementary services. The relationship between asset specificity and internal sourcing of supplementary services is mediated by inseparability and transaction frequency.

Clearly, the rapid development of information and telecommunications technologies has reduced the costs of remote provision of many services that do not need face-to-face contact with customers. A new international division of labour arises (Castells 1996). Time zone differences can even be turned into an advantage. The development of internationally recognised standards increases the mobility of the production of goods and now also of services. As the example of call centres located in India shows, the supply of educated people and affinities of language and legal systems matter. The slowly emerging consequences of what Macaulay stated in his 1835 Minute on Indian education, 'the English tongue is that which would be most useful to our native subjects', are evident today (Macaulay 1935). Leahy and Lau (2004) draw attention to the parallel in that although Chinese, Japanese and Korean are fundamentally unrelated languages, their writing systems, as represented by double-byte programming, facilitate outsourcing of software development to China. Differences in wage and salary levels remain important for determining the competitive advantage of any location.

Conclusions: studying global patterns of specialisation and integration

Having considered the origin of outsourcing, theories and methods of outsourcing research and reviewed a number of applications, the challenge is to reduce the complexity of outsourcing to a few basic principles. Practitioners tend to be unhappy when told that the reality is more complicated than they think. Columella, writing about agriculture under Nero, discusses why a large landowner should rely on tenant farmers rather than on his own slaves for the cultivation of certain parts (Heitland 1970, p. 253; Ste Croix 1981, p. 241). In short, he recommends owners to outsource to tenants those farms that are bad in quality, considering health and fertility, as well as farms that are out of easy reach. Columella appears to base his thinking on both the strength of individual incentives for farming work and the inspectability of work. In fact, Roumasset (1995) has provided evidence from Nepal, the Philippines and India that agricultural tenure choice is strongly related to land quality and distance. So here we have the rudiments of a simple theory for outsourcing: information costs and distance, which have implications for the cost of enforcing contractual obligations, should influence our decisions.

What this simplification lacks is tools that allow us to understand how markets for goods and services economise on specialised knowledge. We need a more straightforward understanding of the emerging global patterns of accumulated knowledge and how it is bundled for a heterogeneous set of customers. The growth of globally active, very large groups of service providers, often through acquisitions that spell horizontal integration, matches the vertical disintegration of traditional businesses. Today, this is a strong pattern that can be identified. Government regulation, not least in the fields of competition and the protection of intellectual property, is a strong formative influence for this international division of human skills.

In the background, the strongest driving force for outsourcing remains the development of information and communication technologies that shrink distances. However, these distances remain for most of the earth's population, who face barriers to movement. The new global patterns of production can be expected to mirror technology, markets and how both new and old welfare societies are transformed. We should not exclude the possibility that the trend to outsource will be weakened, not because new barriers are erected, but because old ones are torn down.

Retreating from the global perspective, there are other issues that should be on the agenda for research on outsourcing. Too much emphasis has been on understanding why companies choose to outsource, and too little on why external suppliers are able to be better producers; additionally, the possibility should be explored that actual patterns of outsourcing result from both features of the companies that outsource and conditions in the external supplier market. There should be a focus also on the transfer situation and on how to retain and develop an efficient procurement function, once the majority of those with the support function skills have left the company that outsources.

How the business cycle is related to willingness to outsource is poorly understood; relying on traditional thinking, the good years of the 1990s should have been accompanied by a rise in vertical integration when capacity use soared. Instead, perhaps because rapid technology change overshadowed the traditional integrating effect of the top of the cycle, companies chose to outsource.

Finally, it would be good to have better insights into fundamental differences between various types of services and types of industries, so that the best objects of comparison could be more readily identified. Moreover, this would be a stronger base for understanding how service level agreements and other contractual documents and practices should be designed.

References

Aubert, B.A., Rivard, S. & Patry, M. (2004) A transaction cost model of IT outsourcing. *Information and Management*, **41**(7), 921–932.

Autor, D.H. (2003) Outsourcing at will: the contribution of unjust dismissal doctrine to the growth of employment outsourcing. *Journal of Labor Economics*, **21**(1), 1–42.

Badian, E. (1972) *Publicans and Sinners: Private Enterprise in the Service of the Roman Republic*. Cornell University Press, Ithaca, NY.

Barthélemy, J. (2001) The hidden costs of IT outsourcing. *Sloan Management Review*, **42**(3), 60–69.

Barthélemy, J. & Geyer, D. (2001) IT outsourcing: evidence from an empirical survey in France and Germany. *European Management Journal*, **19**(2), 195–202.

Bettis, R.A., Bradley, S.P. & Hamel, G. (1992) Outsourcing and industrial decline. *Academy of Management Executive*, **6**(1), 7–22.

Bivand, R. & Szymanski, S. (2000) Modelling the spatial impact of the introduction of compulsory competitive tendering. *Regional Science and Urban Economics*, **30**(2), 203–219.

Bon, R. & Luck, L. (1999) Outsourcing of property-related management functions in Europe and North America, 1993–1998. *Construction Management and Economics*, **17**(4), 409–412.

Brickley, J.A., Misra, S. & Van Horn, R.L. (2003) Contract duration: evidence from franchise contracts. University of Rochester, William E. Simon Graduate School of Business Administration, Working Paper No. FR 03-08. February.

Bröchner, J., Adolfsson, P. & Johansson, M. (2002) Outsourcing facilities management in the process industry: a comparison of Swedish and UK patterns. *Journal of Facilities Management*, **1**(3), 265–271.

Bröchner, J., Olsson, H. & Sinik, D. (2004) Serviced offices: owner capabilities for FM coordination. *Facilities*, **22**(3/4), 74–78.

Buco, M.J., Chang, R.N., Luan, L.Z., Ward, C., Wolf, J.L. & Yu, P.S. (2004) Utility computing SLA management based upon business objectives. *IBM Systems Journal*, **43**(1), 159–178.

Castells, M. (1996) *The Rise of the Network Society*, Blackwell, Oxford.

Currie, W.L., Desai, B. & Khan, N. (2004) Customer evaluation of application services provisioning in five vertical sectors. *Journal of Information Technology*, **19**(1), 39–58.

Davis, R. (2004) Setting up and managing outsourcing contracts to deliver value and accommodate change. In: *Proceedings of Futures in Property and Facility Management II*, London, 25–26 March 2004. University College, London; pp. 148–153.

Doig, S.J., Ritter, R.C., Speckhals, K. & Woolson, D. (2001) Has outsourcing gone too far? *McKinsey Quarterly*, **4**, 25–37.

Domberger, S. (1998) *The Contracting Organization: A Strategic Guide to Outsourcing*. Oxford University Press, Oxford.

Drezner, D.W. (2004), The outsourcing bogeyman. *Foreign Affairs*, **83**(4), 22–34.

Foss, N.J. & Koch, C.A. (1996) Opportunism, organizational economics and the network approach. *Scandinavian Journal of Management*, **12**(2), 189–205.

Freytag, P.V. & Kirk, L. (2003) Continuous strategic sourcing. *Journal of Purchasing and Supply Management*, **9**(3), 135–150.

Gadde, L.E. & Håkansson, H. (2001) *Supply Network Strategies.* John Wiley & Sons, Chichester.

Gilley, K.M., McGee, J.E. & Rasheed, A.A. (2004) Perceived environmental dynamism and managerial risk aversion as antecedents of manufacturing outsourcing: the moderating effects of firm maturity. *Journal of Small Business Management*, **42**(2), 117–133.

Grover, V., Cheon, M.J. & Teng, J.T.C. (1996) The effect of service quality and partnership on the outsourcing of information systems functions. *Journal of Management Information Systems*, **12**(4), 89–116.

Grover, V. & Malhotra, M.K. (2003) Transaction cost framework in operations and supply chain management research: theory and measurement. *Journal of Operations Management*, **21**(4), 457–473.

Heitland, W.E. (1970) *Agricola: A Study of Agriculture and Rustic Life in the Greco-Roman World from the Point of View of Labour.* Greenwood Press, Westport, CT. [Original edn Cambridge University Press (1921)]

Hiles, A. (2000) *Service Level Agreements: Winning a Competitive Edge for Support and Supply Services.* Rothstein Associates, Brookfield, CT.

Hoopes, D.G., Madsen, T.L. & Walker, G. (2003) Why is there a resource-based view? Toward a theory of competitive heterogeneity. *Strategic Management Journal*, **24**(10), 889–902.

Houston, A. & Youngs, G. (1996) Proactive outsourcing – a strategic partnership: Rank Xerox Technical Centre. *Facilities*, **14**(7/8), 40–47.

Jeffers, R. (1996) Outsourcing and TUPE: problems and solutions. *Facilities*, **14**(7/8), 52–56.

Kakabadse, A. & Kakabadse, N. (2002) Trends in outsourcing: contrasting USA and Europe. *European Management Journal*, **20**(2), 189–198.

Katsanis, C.J. (2002) Outsourcing. In: *Building in Value: Workplace Strategies and Facilities Management* (eds R. Best, C. Langston & G. de Valence), Butterworth-Heinemann, Oxford; pp. 378–394.

Kern, T. & Willcocks, L.P. (2001) *The Relationship Advantage: Information Technologies, Sourcing, and Management.* Oxford University Press, Oxford.

Kern, T., Willcocks, L.P. & van Heck, E. (2002) The winner's curse in IT outsourcing: strategies for avoiding relational trauma. *California Management Review*, **44**(2), 47–69.

Kessler, I., Coyle-Shapiro, J. & Purcell, J. (1999) Outsourcing and the employee perspective. *Human Resource Management Journal*, **9**(2), 5–19.

Kleeman, W.B. (1994) Out-tasking: more widespread than outsourcing in the USA. *Facilities*, **12**(2), 24–26.

Knemeyer, A.M., Corsi, T.M. and Murphy, P.R. (2003) Logistics outsourcing relationships: customer perspectives. *Journal of Business Logistics*, **24**(1), 77–109.

Lacity, M.C. & Hirschheim, R. (1993) The information systems outsourcing bandwagon. *Sloan Management Review*, **35**(1), 73–86.

Lacity, M.C. & Willcocks, L.P. (2001) *Global Information Technology Outsourcing: In Search of Business Advantage.* John Wiley & Sons, Chichester.

Lacity, M.C., Willcocks, L.P. & Feeny, D.F. (1996) The value of selective IT sourcing. *Sloan Management Review*, **37**(3), 13–25.

Larson, P.D. & Chow, G. (2003) Total cost/response rate trade-offs in mail survey research: impact of follow-up mailings and monetary incentives. *Industrial Marketing Management*, **32**(7), 533–537.

Leahy, J. & Lau, J. (2004) China: frontier of a new global contest. *Financial Times*, June 1.

Lee, J.-N., Huynh, M.Q., Kwok, R.C.-W. & Pi, S.-M. (2003) IT outsourcing evolution: past, present, and future. *Communications of the ACM*, **46**(5), 84–89.

Linder, J.C. (2004) Transformational outsourcing. *Sloan Management Review*, **45**(2), 52–58.

Loh, L. & Venkatraman, N. (1992a) Diffusion of information technology outsourcing: influence sources and the Kodak effect. *Information Systems Research*, **3**(4), 334–358.

Loh, L. & Venkatraman, N. (1992b) Determinants of information technology outsourcing: a cross-sectional analysis. *Journal of Management Information Systems*, **9**(1), 7–24.

Lorence, D.P. & Spink, A. (2004) Healthcare information systems outsourcing. *International Journal of Information Management*, **24**(2), 131–145.

Macaulay, T.B. (1935) Minute on Indian education 1835. In: *Speeches by Lord Macaulay, with his Minute on Indian Education* (ed. G.M. Young). Oxford University Press, London; pp. 345–361.

McCarthy, I. & Anagnostou, A. (2004) The impact of outsourcing on the transaction costs and boundaries of manufacturing. *International Journal of Production Economics*, **88**(1), 61–71.

McIvor, R. (2003) Outsourcing: insights from the telecommunications industry. *Supply Chain Management: An International Journal*, **8**(4), 380–394.

Meer-Kooistra, J. van der & Vosselman, E.G.J. (2000) Management control of inter-firm transactional relationships: the case of industrial renovation and maintenance. *Accounting, Organizations and Society*, **25**(1), 51–77.

Ménard, C. & Saussier, S. (2002) Contractual choice and performance: the case of water supply in France. In: *The Economics of Contracts: Theory and Applications*, (eds E. Brousseau & J.M. Glachant). Cambridge University Press, Cambridge; pp. 440–462.

Momme, J. & Hvolby, H.-H. (2002) An outsourcing framework: action research in the heavy industry sector. *European Journal of Purchasing and Supply Management*, **8**(4), 185–196.

Murray, J.Y. & Kotabe, M. (1999) Sourcing strategies of US service companies: a modified transaction-cost analysis. *Strategic Management Journal*, **20**(9), 791–809.

Poppo, L. & Zenger, T. (1998) Testing alternative theories of the firm: transaction cost, knowledge-based, and measurement explanations for make-or-buy decisions in information services. *Strategic Management Journal*, **19**(9), 853–877.

Porter, M.E. (1985) *Competitive Advantage: Creating and Sustaining Superior Performance*. Free Press, New York.

Prahalad, C.K. & Hamel, G. (1990) The core competence of the corporation. *Harvard Business Review*, **78**(3), 79–91.

Preker, A.S., Harding, A. & Travis, P. (2000) 'Make or buy' decisions in the production of health care goods and services: new insights from institutional economics and organizational theory. *Bulletin of the World Health Organization*, **78**(6), 779–790.

Quélin, B. & Duhamel, F. (2003) Bringing together strategic outsourcing and corporate strategy: outsourcing motives and risks. *European Management Journal*, **21**(5), 646–661.

Quinn, J.B. (2000) Outsourcing innovation: the new engine of growth. *Sloan Management Review*, **41**(4), 13–28.

Quinn, J.B. & Hilmer, F.G. (1994) Strategic outsourcing. *Sloan Management Review*, **35**(4), 43–55.

Razzaque, M.A. & Sheng, C.C. (1998) Outsourcing of logistics functions: a literature survey. *International Journal of Physical Distribution and Logistics Management*, **28**(2), 89–107.

Roumasset, J. (1995) The nature of the agricultural firm. *Journal of Economic Behavior and Organization*, **26**(2), 161–177.

Ste Croix, G.E.M. de (1981) *The Class Struggle in the Ancient Greek World*. Duckworth, London.

Schneider, A.L. (2000) Public–private partnerships in the US prison system. In: *Public–Private Policy Partnerships* (ed P.V. Rosenau). MIT Press, Cambridge, MA; pp. 199–215.

Stremersch, S., Wuyts, S. & Frambach, R.T. (2001) The purchasing of full-service contracts: an exploratory study within the industrial maintenance market. *Industrial Marketing Management*, **30**(1), 1–12.

Usher, N. (2004). Outsource or in-house facilities management: the pros and cons. *Journal of Facilities Management*, **2**(4), 351–359.

Willcocks, L.P., Lacity, M.C. & Kern, T. (1999) Risk mitigation in IT outsourcing strategy revisited: longitudinal case research at LISA. *Journal of Strategic Information Systems*, **8**(3), 285–314.

Williamson, O.E. (1985) *The Economic Institutions of Capitalism: Firms, Markets, Relational Contracting*. Free Press, New York.

Part 3
Projects Milieu

Introduction

This part deals with the environment within which individual projects collectively reside (it therefore incorporates the notions of portfolio and programme management) and the interfaces between the corporate level of an organisation and the implementation of individual projects. Primarily, within this context the interfaces at the projects level centre on the implementation of corporate decisions, for example, with regard to governance and strategy, and encompass measuring performance, managing risk and uncertainty, achieving value, managing conflict and resolving disputes. Four specific topics have been singled out for further explanation: management of uncertainty, conflict management and dispute resolution, performance measurement and value management.

Management of uncertainty

Risk and uncertainty are inherent in all business activities and it is unwise to separate risk management from the management of a project and the management of the organisation. Both an organisation's corporate strategy on risk and its ability to manage uncertainties are paramount to how it will act on projects. In Chapter 10, Kalle Kähkönen discusses several shortcomings in contemporary risk management practice and presents compelling arguments for change. In particular, he highlights the need for a more holistic view facilitating a wider scope of risk and opportunity management, which in turn would be supported by localised risk and opportunity definitions pertinent to individual organisations.

Conflict management and dispute resolution

Two quotes are useful to describe the issue and the problem with conflict and dispute on commercial projects. The first is from an English Law Lord (Lord Donaldson) and the second from Abraham Lincoln, and they are separated temporally by more than 100 years:

> *I cannot imagine a... contract which does not give rise to some dispute. This is not a discredit of either party... what is to their discredit is if they fail to resolve those disputes as quickly, economically and as sensibly as possible.* (Lord Donaldson)

> *Discourage litigation. Persuade your neighbour to compromise... the nominal winner is often a real loser.* (Abraham Lincoln)

In Chapter 11, Peter Fenn constructs a compelling argument for the necessity of commercial managers to recognise both conflict management and dispute resolution. The chapter disentangles the strong connections between conflict and dispute, suggesting that while conflict is inevitable, disputes are not. It goes on to review the techniques available to resolve disputes and to propose an aetiological approach to the investigation of commercial disputes.

Performance measurement

Measuring organisational performance is important, as it could provide the necessary feedback required to control and improve actions, which are related to strategy. The measures themselves could, if designed properly, be indicative of an organisation's, or indeed a project's, effectiveness and efficiency. Moreover, they reflect the structure and management of the organisation. However, there is a clear tension between the wish and/or need for performance measurement and the problem of implementation. Malcolm Horner aptly illustrates this in Chapter 12 which provides an overview of what is meant by performance measurement, why it is important (but so often ignored) and the various means available to firms. It reviews past and current practice, with particular emphasis on 'what to measure'.

Value management

In essence, value management is concerned with establishing a common decision framework around which project participants can deliberate and communicate. This includes establishing a consensus on the merits of various courses of action with regard to achieving value for money or best value. Alternatively, value engineering is primarily concerned with achieving a given function at minimum cost. In Chapter 13, John Kelly describes and illustrates both value management and value engineering in the context of an integrated construction activity. In doing so, he considers several factors: a theory of integrated teams; understanding the client's view of a project; a theory of value management; the relationship between quality and value and its synthesis; the application of value management within the supply chain and value chain theory; and value engineering techniques and application. Five case studies are presented to illustrate these factors.

The contributor's profiles

Kalle Kähkönen

Dr Kalle Kähkönen is Chief Research Scientist at VTT Building and Transport, VTT Technical Research Centre of Finland, and was the chairman of Project Management Association Finland from 1998 to 2000. He holds degrees from two universities: MSc (Civil Engineering) from the Helsinki University of Technology and PhD from the University of Reading (UK). His main interests are advanced technology and solutions for modern project management and project business. Examples of recent projects and achievements include: Temper System – a computer tool for project risk analysis and management; DSM System – a solution and its computer application for complex design process management; VPC – software for the simulation and visualisation of a construction project in real time; and VTT Lumeportti – the development of PC Technology based VR-studio in VTT. Kalle is co-author of *Unknown Soldier Revisited – A Story of Risk Management* (Project Management Association Finland, 2000) and has published over 90 papers in both Finnish and English. The list of publications includes 23 books or monographs for which he has contributed as a principal author, co-author or editor.

Peter Fenn

Dr Peter Fenn is Senior Lecturer at the School of Mechanical, Aerospace and Civil Engineering at the University of Manchester. Peter is joint course director for the MSc in Commercial Management by distance learning; he is a Fellow of the Royal Institution of Chartered Surveyors, a Registered Mediator, an Adjudicator, a Conciliator and a Fellow of the Chartered Institute of Arbitrators. During the academic year 1997/98 he was a Visiting Professor at the University of Kentucky. Peter has edited and authored four books and written more that 30 technical and journal articles. He has carried out research into construction disputes; and has secured substantial research grants and formed an International Working Commission under the auspices of the International Council for Research and Innovation in Building and Construction. This International Working Commission recently produced a book on international dispute resolution and management co-edited by Peter: *Construction Conflict Management and Dispute Resolution: An International Review* (Routledge 1998). The research into construction disputes has been published in leading academic journals, including two publications in the prestigious journal *Nature*. The work has also been cited in two UK Government reviews: (1) Sir Michael Latham's Interim Review of the construction industry, *Trust and Money;* (2) Lord Justice Woolf's Interim Report to the Lord Chancellor on the civil justice system in England and Wales, *Access to Justice.*

Malcolm Horner

Professor Malcolm Horner has won research grants to a value approaching £3m, of which about one-half have been sponsored by UK government agencies and the remainder by industry. He has successfully managed more than 20 research projects, and published over 70 papers. His research encompasses lean thinking, labour productivity, project modelling, estimating, monitoring and controlling costs, whole life costing, whole life sustainability, maintenance management and risk management. He is Deputy Principal and Professor of Engineering Management at the University of Dundee where he leads the Construction Management Research Unit, and directs the Centre for Enterprise Management. He has been a member of the Foresight Construction Sector Panel, a member of the Editorial Board of *Construction Management and Economics*, Chairman of the Scottish International Resource Project and a member of the steering groups for the EPSRC and DETR LINK programmes *Meeting Clients' Needs through Standardisation* and *Integration in Design and Construction*. He is currently a member of the EPSRC Peer Review College and of the British Council Advisory Committee on Science, Engineering and the Environment. He has spent some 15 years in industry where he was a director of Atlantic Power, a North Sea oil project management company. He is currently a director of the Scottish Institute for Enterprise, Scottish Enterprise Tayside, and of Dundee Repertory Theatre, and Chairman of his own company, Whole Life Consultants Ltd.

John Kelly

Professor John Kelly holds the Morrison Chair in Construction Innovation at Glasgow Caledonian University. John trained with George Wimpey and Co. Ltd and practised as a quantity surveyor with a small architectural practice, before becoming an academic in 1976. He has researched, taught and practised in the field of value management for 25 years, focusing on the investigation of value strategies at the inception of projects. John has published widely, conducted considerable funded research and undertaken consultancy in strategic and project briefing, partnering and procurement strategies, with a wide spectrum of clients and projects varying from social housing to a football stadium.

10 Management of Uncertainty

Kalle Kähkönen

Introduction

Although its significance for modern business is widely acknowledged, risk and opportunity management is still an emerging discipline that is under continuous development and change. Enterprise-wide risk management solutions, or those covering the whole project life cycle, require wide-scope risk management solution, which is currently a key research and development target. Additionally, risk and opportunity management still suffers from conceptual complexity, while its models and tools are inadequate. These current and somewhat discouraging shortcomings need to be understood, but it should also be recognised that the evidence from many useful case studies, practice and tools provide a basis for successful risk and opportunity management. This chapter provides a discussion on the main body of risk and opportunity management, pinpointing several shortcomings and proposing improvements. In particular, it presents localised risk and opportunity definitions, a holistic paradigm for wide-scope risk and opportunity management, together with the core process where focus is on risk and opportunity identification, as new contributions to the subject. Risk and opportunity management is presented from the perspective of commercial management, where it is proposed that risk and opportunity management practice should be built around decision-making processes.

Increasingly, modern business and, in particular, projects are carried out under dynamic and turbulent conditions. More often than not, this phenomenon seems to be caused by top-level change in general business conditions; in particular, company structures and ownership are constantly changing. Also, competition is progressively more global and the emergence of new players can rapidly change the markets that are in our focus. For projects and their management this situation causes more pressure and threats, most of which can be characterised as project risks or potential problems that can dramatically affect project performance.

The situation described above appears in different ways. First, networked operations and the level of commitment in such operations are discussed within this chapter. During the past decade, outsourcing has perhaps been the most significant restructuring principle for companies wishing to focus on their core business. For projects, this means that operations are increasingly networked, since a single partner can only provide a very limited contribution. A contractor that provides such a contribution normally has many subcontractors, who in turn can have subcontractors of their own. This results in very complex networks, where each player's own contribution is less than before, since subcontractors increasingly provide supplements for each contribution. In practice, the development or delivery of one contribution is broken down into smaller contracts than used to be the case. The level of interest or commitment towards such smaller deals can be questionable and can change during the project life cycle, which can cause a serious threat to the project.

Second, the chapter discusses the influence of the business life cycle. Business viability of services or other products is increasingly difficult to anticipate. Thus, the time for completing projects is constantly shortening. In this situation, a project's start is based on assumptions rather than specifications, and all projects are constantly under threat of major changes or even of being halted.

The aspects outlined above may, at first glance, seem to be solely of a negative nature, but we should approach them as challenges. If they are managed successfully, we can meet the major requirements set by modern businesses. Importantly, it is obvious that these aspects have given rise to the current significance of project risk management. Presently, risk management is widely acknowledged as a main element of project management.

The title of this chapter 'Management of uncertainty' is a somewhat provocative one, and is intended to address the scope of risk management, its content and procedures. Risk management has increasingly been a popular research and development topic; it has had a broad presence in various training courses, and thus many business people and other experts have familiarised themselves with state-of-the-art risk management. At the same time, individuals within organisations are increasingly experimenting with risk management tasks and procedures. It is no longer a rare case to have risk management as a main heading in the standard meeting agenda or to organise special risk assessment and management workshops. Based on these experiments and real world research, we presently have evidence of risk management practices that work and have identified shortcomings and problems that originate from our current understanding of the content and procedures of risk management. This understanding is developing all the time and is providing a new and firmer basis for improved risk management practices. The objective of this chapter is to provide a step forward in the areas of risk management that seem to be stumbling blocks or problematic for successful practices. It is considered that there is still a need for fundamental improvements to be made in many aspects of risk management, in particular, those covering risk concepts, risk perception and related processes and procedures.

Conceptual foundation for risk management solutions

Traditionally, project risk is defined as an uncertain event that can have negative consequences. This demonstrates how people, in a general sense, associate the term risk with adverse outcomes or situations. Indeed, the above definition suffices if we limit our interpretation of risk management to cover potential threats to the success of a venture. If a broader risk management viewpoint is taken, it becomes apparent that such a narrow definition is no longer satisfactory. This may, for example, result in overpricing when the expected outcome of potential threats is analysed and summed up for inclusion in the project estimate.

As a consequence, an increasing number of researchers and other experts have proposed that risk management needs to cover, in a well-balanced manner, both potential downside problems and upside opportunities. The term 'uncertainty' has had a central role in this development. As the term risk has had a fundamentally negative association, the term 'uncertainty' is being explored as a more suitable concept to cover widely uncertain events and their management. However, 'Every word or concept, clear as it may seem to be, has only limited range of applicability' (Heisenberg 1959). This is not only the case with the term risk, but also true for the term uncertainty. Uncertainty is a universal term that can be understood to cover general variability and ambiguity around us, with or without our involvement.

It is generally considered that the biggest difference between the basic interpretations of risk and uncertainty, concerning business operations, lies in the level of individual or organisational involvement. Usually a risk is understood to be present in a situation where a certain level of involvement is present, for example, a pricing decision for a bid, while uncertainty means potential variability and ambiguity without context. Uncertainty can, however, be put into context by having, for example, a decision-making need or a project to be implemented as a starting point. Such a study of potential impacts arising from uncertainty, results in the identification of risks and opportunities. This underlying conceptual difference has caused a situation where, for practical reasons, we keep on using the terms risk and risk management, although an obvious conceptual dilemma exists when opportunities are included. A recent example of this is the Project Management Institute Body of Knowledge (PMBOK 2004).

For practical reasons it is very important to define the term risk in a way that provides practitioners with a common starting point for risk management tasks and as a basis for shared understanding. A multitude of definitions can be found from different textbooks. See Olsson (2001) for a review of the various existing definitions. It seems that researchers and other experts have tried to reach a single risk definition that would be generic enough to cover the scope of risk management in question, yet we have plenty of evidence from company practices where risk management has been built strongly on a situational context. This means that company management is focusing on particular risk management needs that they consider crucial. It is also a way to make risk management easier and more effective in the short time and available within the context of other work. For example, several companies have built their risk management practices around project work breakdown structure (WBS). Here the starting point and definition of risk and its management can be to 'identify WBS linked threats for meeting project objectives successfully'. Some examples of company practices for risk management can be found in Kerzner (2000).

It is considered to be the case that we need to acknowledge the value of localised risk and opportunity definitions and, furthermore, to include this approach in general risk and opportunity management models. Thus, it is proposed that companies need to be encouraged to use localised risk and opportunity definitions. Chapman and Ward (2002) are concerned that terms and concepts are being used that are restrictive and limit the potential for wide-scope risk and opportunity management. A generic conceptual foundation is needed, but on top of that we can build local definitions which can appear not to be a single definition but several situation-specific definitions. Together these localised definitions can cover wide-scope risk management, making the task of risk management easier and more effective (Fig. 10.1).

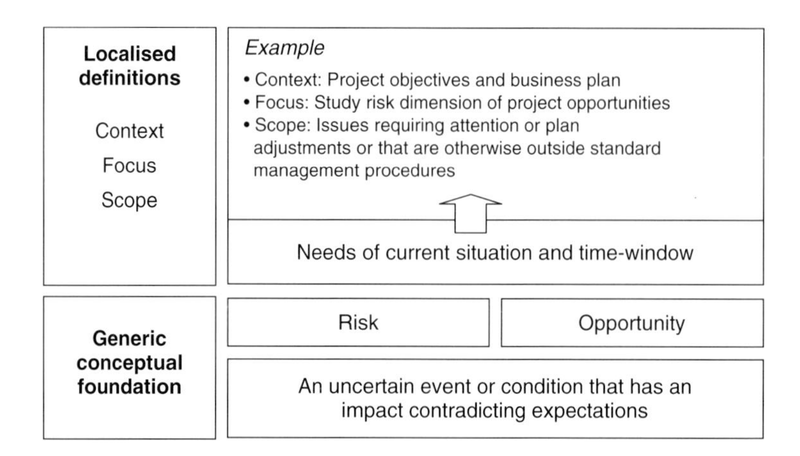

Figure 10.1 Localised definitions of risk and opportunity.

When pursuing localised risk and opportunity definitions, one needs to give attention to the three elements forming the content of each localised definition. These are context, focus and scope of risk and opportunity management operations. The context is an answer to the following questions:

(1) What is important in this situation?
(2) What is the information base to be used as a starting point?

The first question generates a statement of the project objectives for value-driven project risk management. The answer to the second question results in a definition of an information base that can, for example, be the WBS used as a common source for risk identification and for linking the identified risks to the project structure.

The focus of risk and opportunity management turns our attention towards the types of issues that need to be addressed in the ensuing exercise. Several studies have shown that negative effects (e.g. loss) and positive effects (e.g. gain) in general are unrelated in human risk perception (Rundmo 2002). This means that risks and opportunities require separate thinking modes and separate time. Thus, ideally, the participants in a meeting should focus either on risks or opportunities. If both are to be covered in a single meeting, then separate time slots and task definitions need to be organised. Hoffman (2002) provides useful examples of focus definition: opportunity–dimension-of-risk and risk–dimension-of-opportunity. For example, early project or product studies can be very opportunity-orientated with risk-orientated thinking too easily overlooked or avoided. During this phase it would be useful to study the risk–dimension-of-opportunity which can then be used as a localised focus definition.

The scope of risk and opportunity management is needed in order to create a statement concerning:

(1) The severity of risks or opportunities
(2) Range of risk or opportunity action.

The scope of risk and opportunity management explains what kinds of risks or opportunities are to be targeted. It can often be useful to define that we are primarily targeting risks or opportunities that are not, as such, present in our current plans or are outside standard managerial actions.

The discussion in this section of the chapter demonstrates how challenging and even problematic risk management is conceptually. The terminology used in this chapter also suffers from terminological difficulty. The need for wide-scope risk and opportunity management is acknowledged, but we are often bound to use the term risk management. Management of uncertainty is neither a well-established nor a well-known term, and it is not without its own conceptual weakness – also explained in this section. Risk and opportunity management is another alternative choice for a term with a wider scope, but the use of this term is also still rather limited.

Commercial management dimension of project uncertainty

Projects and their life cycles can be viewed using three main and different aspects (Forsberg *et al.* 2000):

- *Business aspect*: This is particularly linked to the early phase of project development where the business case is initiated and formed. In particular, it covers the strategic planning of the project and is closely linked to the management of a company and its strategies. During project execution, the business aspect incorporates control and decision flexibility. Decision flexibility is about reacting to the improvement options of project outcomes after the formal 'go/no go' decision has been made (Loch & Huchzermeier 2000). Business aspect also covers the contractual arrangement(s) for the realisation of the project.
- *Budget aspect*: Concerns understanding the total project cost, its elements and preparing the financing arrangements. It also provides grounds for making a decision based on commitments and targets.
- *Technical aspect*: Covers activities and events required for the execution of the project in an efficient way, where all objectives are met.

Primarily, the commercial management dimension of projects concerns business and budget aspects. The business aspect in projects is strongly dominated by decision-making challenges. After each positive 'go' decision the level of company commitment is higher (Fig. 10.2). Usually, several decisions are required before the final 'go/no go' decision is made. Research has shown that good decision makers think in terms of search and options (Baird 1989).

Opportunities for success and threats of failure are important general decision-making criteria to be studied under these decision points. In order to do this, localised risk and opportunity definitions for decision-making are required, applying the framework presented in the previous section.

When approaching the start of project operations, the commercial activities become heavily focused on negotiation and the preparation of contracts. Generally speaking, there are usually many more options available before signing the contract than there are after it has been signed. Experienced managers have expressed the opinion that the atmosphere in the project changes immediately after the signing of

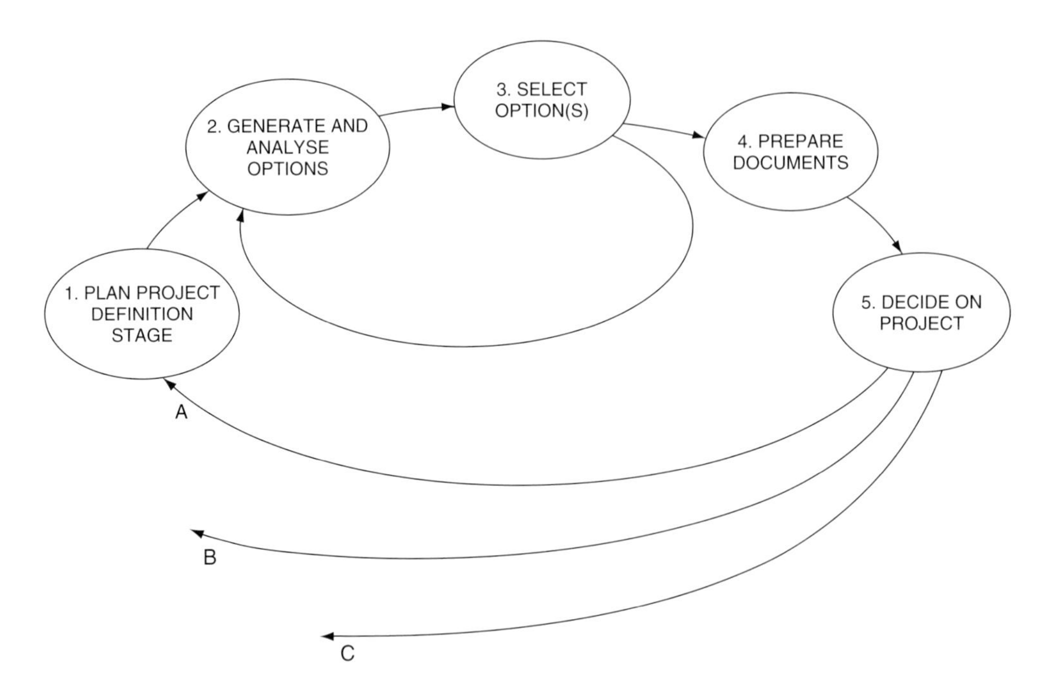

A = MAYBE (the same project definition stage will be carried out again but with a different option, assumption or working method)

B = GO or NO GO (project realisation starts or the project is cancelled)

C = MAYBE (move to the next project definition stage)

Figure 10.2 Internal repetitiveness of the project definition and decision process.

the contract, often leaving them with the feeling of only having obligations left. Furthermore, in a situation where contractual obligations are fully present, people tend to be more risk averse than during the early stages of a project, when the attention is more on project opportunities.

Proven principles and methods of risk management are generally applicable in different situations and for different purposes. It is obvious that one needs to give attention to the risk–dimension-of-opportunity assessment during the early phases of a project in order to have a well-balanced approach to commercial risk and opportunity management. Later on, i.e. in the phases where the company becomes fully committed and contracting starts, more attention should be given to opportunity–dimension-of-risk assessment and incorporating the results into the actual operations. Taking full advantage of the latter approach would require new improved decision flexibility, enabling the project to be adjusted according to the new emerging opportunities.

Need for new holistic paradigm

We are still in the pioneering stage in the area of project risk management. The discipline of project risk management is under continuous development and only gradually is it finding its role and position within other managerial work. During this pioneering effort we have learned a lot that can be used as a starting point for forming a 'new improved' project risk management discipline.

Generally, there is a need for risk management that has wide scope in terms of better management, concerning the positive and negative chances related to our business. In simple terms, this wide-scope risk management needs to cover the project life cycle from early studies to final business operations. Additionally, wide-scope risk management needs to provide solutions that meet the needs of the various organisational levels and stakeholders. The solutions termed enterprise-wide risks management (ERM) systems are moving in this direction. The ERM systems take a company value driven approach to managing risks at different organisational levels in order to maximise the total operational value of the company (Thieke 2000). Reports and recommendations by several financial institutions have strongly argued for company-wide solutions for improved internal control and risk management programmes (Hoffman 2002).

Company-wide risk management solutions are clearly needed, but they are very demanding targets. Such a system is supposed to cover many different risk types and their consistent measurement across the organisation. Without measurement, risks would only have very limited significance in the current or planned business context. Some risks can be rather straightforward and easy to model and measure, while others are very difficult or even impossible. Capital funding risk is an example of a risk that usually has direct links to quantitative data, providing a good starting point for risk measurement. However, many risks are associated with ambiguity, where data for scenarios, ranges and key variables for understanding variability are completely or partly missing. Total ambiguity is the most difficult situation, where the outcome is virtually impossible to predict. An example of total or near total ambiguity levels can be taken from country risks, such as entering the Iraq market in year 2005. Thus, we now face a dilemma, as the different risks require a variety of modelling and measuring approaches resulting in more complex solutions. The simple solutions tend to lose out on generality, while the generic solutions are complex and not acceptable in daily business. It seems obvious that information and communication technologies (ICT) can provide a solution for easily applying different risk modelling and measuring approaches, understanding the validity of results and, finally, integrating the results generated into the overall risk picture. However, existing risk modelling solutions seem to be uneven compared with each other, and more research and development are needed to create a solid foundation for the proposed solution.

Another aspect hampering the establishment of enterprise-wide risk management is human risk perception and the individual's inherent attitude to risk. Perception of risk is both complex and subjective, and involves factors such as an understanding of risk, a perception of loss and gain, cognitive biases and personality (Fenton-O'Creevy & Soane 2000). Risk perception research is an active community, but at present it is somewhat isolated from research and development addressing enterprise and project risk management. Thus, the findings from risk perception research are not widely known outside the research society itself. A number of industrial psychology studies show results indicating that employees' perceptions of risk are influenced by the context and culture of their work environment, leading to different 'worlds of risk' between professional groups and between levels of seniority within the work organisation (Mearns *et al.* 2001). Without a proper understanding and inclusion of this fundamental phenomenon it can be very difficult or impossible to reach shared attitudes, shared perceptions and unified managerial actions. In other words, the main implications of risk perception need to be carefully taken into account when enterprise-wide risk management programmes are prepared and implemented.

Understanding the organisational dimension, when coping with risks, is the extension of the human centred discussion above. Uncertainty and risks stemming from the lack of such an understanding appear to be a fundamental problem for complex organisations. The basic threat for any enterprise is the conditions where business success lies in interdependence with an environment which may be unco-operative (Zald & Scott 2003). A traditional strategy has been to smooth the input and output transactions together with buffers to provide tolerance for sudden, but rather limited, surprises. This is particularly the case in mass production manufacturing, but similar principles are often present in other lines of business and even in individual projects. Recent trends, particularly concerning product development projects in turbulent high technology areas, show several companies turning their attention towards the interpretation of weak signals arising from their clientele and markets generally. This is a quite different strategy, where the companies are targeting new business opportunities with rapid and flexible product development projects. These projects can be continuously started and killed. Many of them may fail, but the successful ones result in innovations and business volume improvements. In summary, this demonstrates that institutional risk management strategy can appear in different forms comprising the company's final risk management policy. The company's risk management policy is a rather delicate object, formed by managerial statements and affected by institutional issues and values. These institutional issues and values can be societal, in the business environment, or inside the company itself. The company and its management are affected by societal and business environment values, whereas these, together with company values, affect the company's employees. The logic here is the flow of values and resultant culture from higher organisational levels to lower levels. Company values must be seen as a true source for establishing business management procedures, operational standards and practices. Without this linkage, company values are negligible and cannot provide any working basis for the company risk management policy.

The paragraphs above have presented three important aspects for forming holistic enterprise-wide risk management: system, organisation and human. The system approach is the one we may call risk engineering, but without a careful inclusion of the two remaining aspects any enterprise-wide risk management is unlikely to succeed. A system centred approach will only allow for local solutions, thus meeting only the needs of a few groups of people and being capable of handling very limited risk types and situations. What is needed is a holistic paradigm that takes into consideration all three aspects, thereby creating a well-balanced approach for forming enterprise-wide risk management (Fig. 10.3).

SYSTEM	ORGANISATION	HUMAN
• Common R&O definitions • Rigorous R&O identification and measurement • R&O transparency and communication • Timeliness and quality of information • Clearly defined role for system and its use (capabilities, limits)	• The effectiveness of internal R&O policies and control • The degree of line management oversight • R&O management skills and knowhow • Diversification possibilities and avoidance of R&O concentrations	• R&O understanding: loss, gain • Cognitive biases: i) retrievability, ii) confirmation bias, iii) illusion of control • Personality: inborn set of dispositions affecting on preferences, habits and reactions

Legend: R&O = Risk & Opportunity

Figure 10.3 Holistic paradigm aspects for forming enterprise-wide risk and opportunity management.

Risk and opportunity management core and accessory processes

There are several state-of-the-art project management standards and other presentations in textbooks that provide a definition of the content of project risk management. The risk management content is generally defined as several stages or processes. Some of these merely describe the core risk management process, whereas others can be somewhat more descriptive in terms of explaining special characteristics of risk management with additional phases or stages.

Figure 10.4 introduces to the project process the core content of risk management. The basis of this model is in the recognition that some repetitive tasks seem to form the core effort required in various situations for completing the present needs for obtaining an improved understanding of the risks and/or opportunities contained within the project. Other processes around the core of risk management, called accessory processes, characterise in a more detailed manner the on-going required effort.

Core processes

Core processes include:

- *Risk identification*: determining the risks that are likely to affect the project, and documenting their characteristics
- *Risk estimation*: evaluating risks and risk interactions to assess the range of possible project outcomes
- *Risk response, planning and execution*: defining, developing, and executing enhancement steps for opportunities and responses to threats.

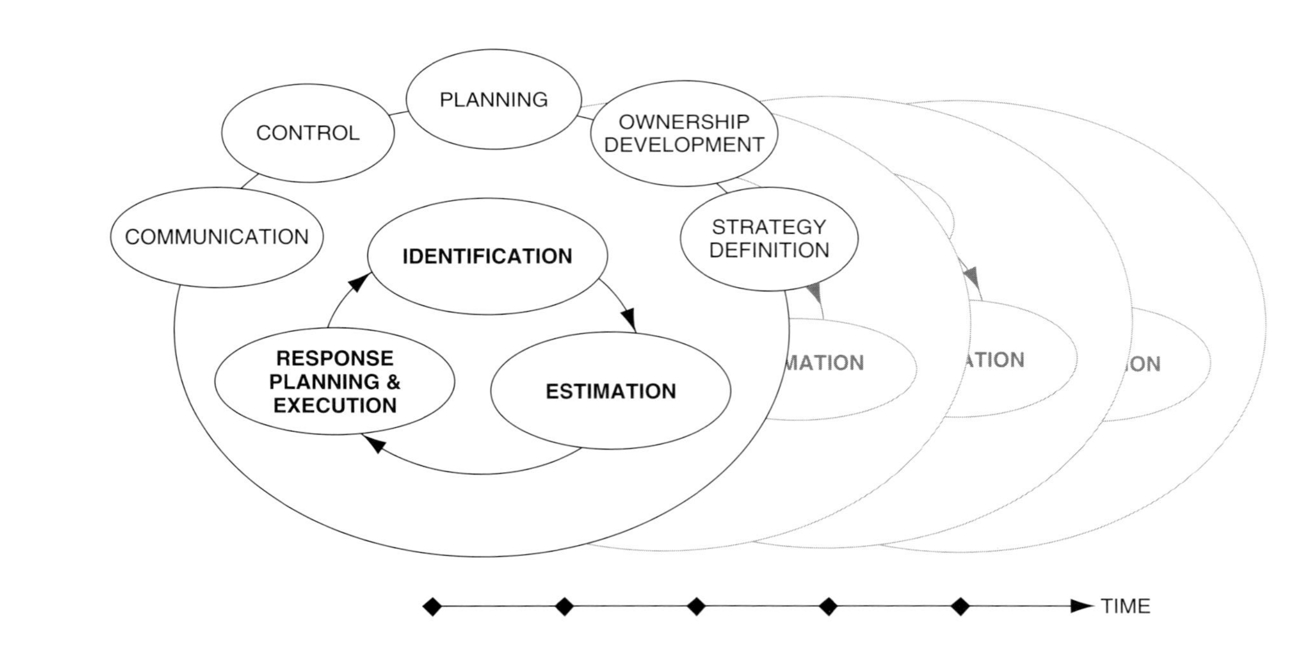

Figure 10.4 Risk and opportunity management core processes and accessory processes.

Accessory processes

Accessory processes include:

- *Risk management planning*: preparing and then deciding upon the appropriate risk and opportunity management approach, tasks and resources for the situation in question
- *Risk communication*: providing a shared starting point and understanding of risks and opportunities
- *Risk ownership development*: identifying the owners of the various risk and opportunity types or the identified risks and opportunities; examples of typical criteria are expertise and power
- *Risk management strategy*: defining the overall strategy and possibly specifying it in the terms of the most important risks, typical choices being to modify project objectives, risk avoidance, risk prevention, risk mitigation, develop contingency plans, keep options open, monitor situation, accept risk without any actions
- *Risk management control*: responding to changes in risk over the course of the project. As risk management is to be applied as a continuous process throughout the project life cycle, one important function for risk management control is to ensure that the process of identifying, estimating, and responding to risks is repeated during the project life in an iterative manner. Later in the chapter, while discussing the future of risk management, it is suggested that the scope of risk management control be extended to include activities that relate to the knowledge management aspects of gathering actual risk data from projects and transferring the gathered knowledge among projects in a multi-project environment.

Risk and opportunity identification

Risk (and opportunity) identification is the most crucial process of risk and opportunity management. During this process one can completely lose the game of having improved risk awareness for the project, or one can gain a substantially new understanding that, in turn, could easily result in improved performance. All succeeding tasks are based on the outcome from risk identification; of particular note are the achievements concerning human and group risk awareness that are usually gained during this phase. When human risk awareness improves, it usually influences one's thinking and work immediately. If key people are involved together in risk identification sessions on the projects, this phenomenon can result in straightforward project plan and management improvements. In simple terms, risks, identified and accepted as being important, stick with people and have influence over them.

The current knowledge and presentations of risk identification are rather tool-centred. Tools, such as risk checklists and inventories, expert interviewing, Delphi technique in brainstorming, and scenario thinking, are well known and widely documented in various textbooks. On the contrary, we do not have a well-established *entity of principles* for:

- setting up risk identification sessions
- running risk and opportunity identification
- applying various risk and opportunity identification tools.

Indirectly, one can see the presence of intuitively recognised risk and opportunity identification principles, particularly in company risk management cases (see, for example, Kerzner 2000; Hoffman 2002). Several company cases confirm the importance of localised risk and opportunity definition (see earlier definition and discussion on localised definitions in this chapter) for identification purposes. The main target is to transform the difficult and diffuse task of future thinking into a well-focused and well-grounded activity where all people feel comfortable from the start to the end of the identification exercise.

Start

Localised definitions can be built around the project plans and other material that is available for identification. After this, one can start applying the identification method that has been selected. It is worth noting that there is clear evidence from many company cases that the use of a carefully prepared question can facilitate a very productive risk and opportunity identification session. Usually it is good practice to present one question at a time instead of presenting several questions. Examples of questions are:

- *Assumptions*: Are there assumptions on plans or with the project set-up causing threats to success?
- *Objectives*: Are the objectives unclear or not established, resulting in a threat to success?
- *Missing knowledge*: For example, what aspects do we know too little about at the moment?
- *New to us*: Are there any critical actors or factors present that are new to us and that could threaten the success of the project?

A good question can open a broad and enlightening avenue of thinking without anchoring thoughts on a single topic or risk type. Such a question can provide the starting point for quick identification of eye opening issues, threats or opportunities, depending upon the task definition and question itself.

Momentum

Without continuous care and advanced planning, risk and opportunity identification can easily lose its momentum. In group work sessions, the facilitator's personal skills often play an important role. However, we should not consider project risk and opportunity identification as a sequential process. In practice, productive project risk identification is a very dynamic process, where the next step after a particular phase or piece of work must be based on the evaluation of the results gained. In certain situations it might be appropriate to think about and decide on risk responses immediately after risk identification. In other situations it may be more useful to undertake careful qualitative and quantitative risk analysis before any response planning is undertaken. This working practice is very dynamic and requires shifting between various risk management phases. It seems apparent that this working practice explains some important characteristics of proactive project risk management.

End

Clearly defining the scope for risk and opportunity identification at the start can help to complete the process in such a way that all participants feel comfortable. Failing to do so could lead to a position where the participants and interviewees in the identification session start to think about issues, hazards and problems of various types and levels of severity without knowing where to stop. In this situation people often lack confidence in the results so gained.

The conceptual difficulty of risk and opportunity management is particularly present in the process of task identification. Unclear definitions can quickly lower the level of motivation, as people do not have a proper common ground for communication and for guiding the process forward.

Risk and opportunity analysis

Risk analysis has too often proved to be a stumbling block for successful project risk management. In other words, when carrying out risk analyses people tend to get stuck with risk modelling and its results. *Risk and opportunity modelling inconsistency* is present in such situations. The following list of reasons explains how inconsistency can appear in different forms during risk and opportunity analyses:

(1) *Ambiguity and lacking information sources*: This is perhaps the most frequent problem in risk analyses. Usually a risk can be named, but understanding and modelling its characteristics and nature are other challenges. Here one should follow the modelling and analysis path that is possible for the risk in question. Risk identification can produce a list of risks that are very different from each other and that cannot be modelled using a single method. The possible modelling path can range from a discussion to quantitative analysis. In some cases a 'risk estimate' is a rather questionable result since it can be very difficult or impossible to reach.

(2) *Missing explanatory factors*: A quantitative risk estimate is not credible without a linkage to the main explanatory risk and opportunity factors behind the analysis. This linkage should be present as explicitly as possible during the life cycle of risk analysis and in any situation where the results are presented or reviewed.

(3) *Complex dependencies between different risks and opportunities*: Individual risks are often analysed separately from each other. This, however, is usually only a valid approach for the rough first step estimate. Any quantitative risk estimate is questionable without a proper consideration of the dependencies between risks and opportunities. However, detailed modelling of these dependencies can result in very complex models, if the level of detail is not limited in a reasonable way. A completely different approach is to prepare a risk estimate based on a 'chunk' of risks, that is, several risks are bundled together in order to prepare a joint estimate using, for example, scenario thinking.

(4) *Incomplete impact coverage*: In particular, this can be a problem with single point estimates (qualitative and quantitative) since the range of different outcomes is

not covered. The single point estimating method is not usually an approach to be selected when modelling large-scale risks, for example, risk of the client's bankruptcy.

Qualitative risk and opportunity modelling methods seem to work in many cases and, thus, qualitative modelling has a strong presence in many current risk management manuals and standards. Examples of well-known qualitative risk analysis methods are probability/impact grids and risk scoring, that is, a technique where one assigns a probability and impact score to each risk according to a predefined scoring list, for example, the scale 1–5. A simple calculation can be applied to define risk severity = probability × impact.

An insufficient link to the actual project plan is the main shortcoming of qualitative analysis. We should, therefore, keep on moving towards quantitative solutions for estimating the potential impacts of risks and opportunities. The main aspects that should be taken into account are the risk and opportunity modelling inconsistencies explained above and the different nature of risks and opportunities. Risks and opportunities need to be linked together, yet the discussion on opportunities needs to be partly separated from the risk discussion stream. This means that the human thinking mode is different when we move from risk-orientated thinking to opportunity-orientated thinking.

Basically, risk is the joint function of likelihood and impact. For preparing quantitative estimates of risk likelihood and impact we have learned from numerous live cases that range estimating is an approach with which people often feel comfortable. Classical risk range estimating is a well-known technique where one estimates the possible range of outcome without defining any probabilities for expressing the likelihood of the risk. Typically three possible outcomes are estimated:

(1) Optimistic/minimum impact
(2) Most likely impact
(3) Pessimistic/maximum impact.

The greatest benefit of this approach is that one does not need to estimate probabilities. However, based on assumptions of the probability distribution types beyond the estimated range, this analysis can also produce expected impact values covering the likelihood of various quantitative factors (Lichtenberg 1981).

An additional estimate of risk likelihood, together with an impact estimate, can produce a risk map where both the likelihood and the impact dimension are covered, thereby explaining the severity of a risk (Fig. 10.5). A risk map usually explains impacts and probabilities of individual risks, and has thus proved to be a good tool for communicating the overall risk picture and characteristics of each risk in question. Additionally, it is a useful tool for risk ranking and for supporting teamwork in risk analysis and management workshops.

Regarding the preparation of a risk map, we are back in a situation where one needs to estimate both risk impact and risk likelihood. A key challenge is to transform the preparation of the likelihood estimate into an easy task with which people feel comfortable. A promising approach is not to only use ranges for estimating impacts

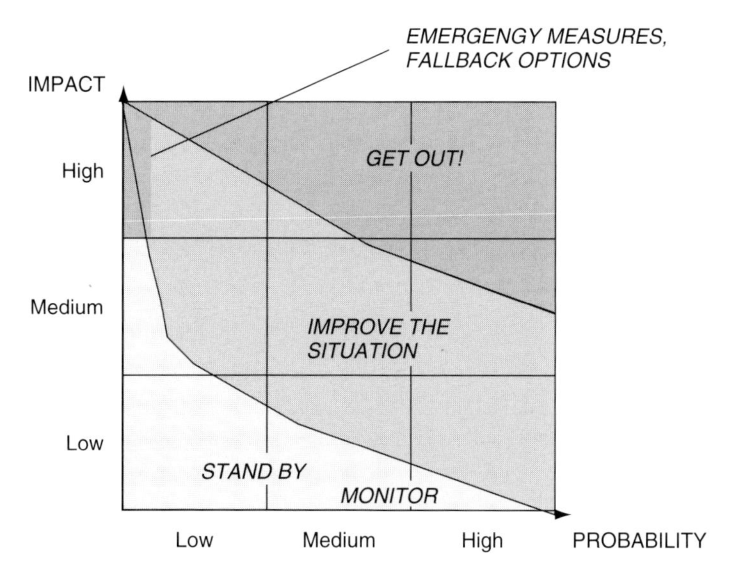

Figure 10.5 Risk map and various risk severity areas.

but also to estimate likelihood; Table 10.1 together with Figures 10.6 and 10.7 show a simple example of this approach. In this example, the range of potential outcomes in terms of the project margin are estimated. For example, risk number two 'Penalised milestones' could potentially lose in the range of 20%–60% from the project margin if the risk turns out to be a real problem. Here, the thinking over the range of possible outcomes is joined with the estimates on the probabilities.

Note how opportunity estimating is separated from risk estimating. Usually the measures needed to maximise the chances of meeting opportunities are different

Table 10.1 Estimating probabilities and impacts[a] of risks and opportunities with ranges.

	Impact (%)	Probability (%)
Risk		
Performance of the plant	20	30
	−150	10
Penalised milestones	−20	50
	−60	5
Major changes in the standards cause	−1	50
significant changes and re-design needs	−25	20
Opportunity		
Performance of the plant	20	30
	−150	10
Bonus from fast plant start-up	20	20

[a] Impact means percentage share in relation to project margin.

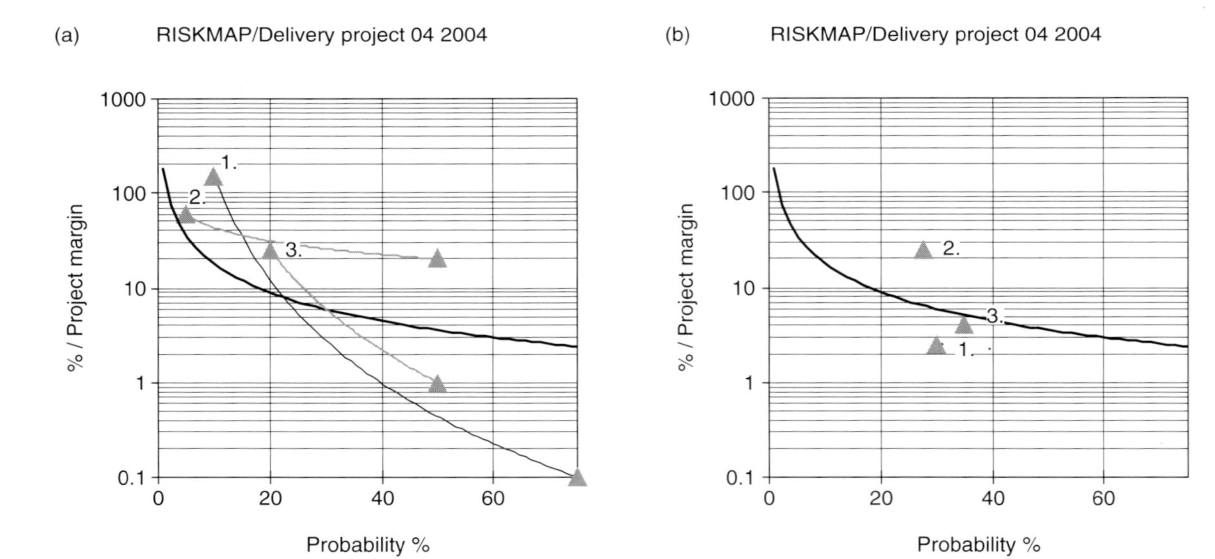

Figure 10.6 Risk map of estimates from Table 10.1. Risk map (b) provides a compressed presentation of risks as single points.

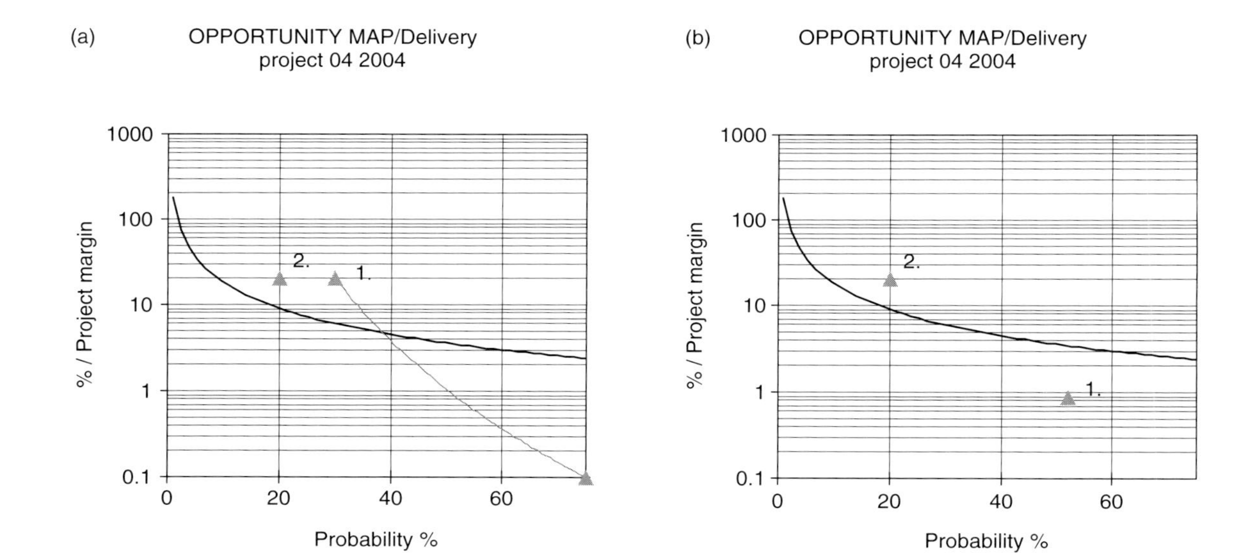

Figure 10.7 Opportunity map of estimates from Table 1. Opportunity map (b) provides a compressed presentation of opportunities as single points.

from those for minimising risks and, therefore, require separate treatment and presentation – like opportunity maps.

The example provided demonstrates how single risks and opportunities can be modelled and graphically presented for communication purposes. The estimations of the overall impact of various risks and opportunities require different approaches and presentation methods. Due to the complexity of risk dependencies, it was proposed earlier that an overall estimate should be prepared using a chunk of risks or opportunities; often this means the set of main risks (or opportunities). Having this as a starting point, a joint estimate is prepared with respect of main impact types, which, for example, can be time, cost, quality and commercial success.

Figure 10.8 presents a risk and opportunity management dashboard, facilitating a joint presentation of the main elements explaining different risk and opportunity

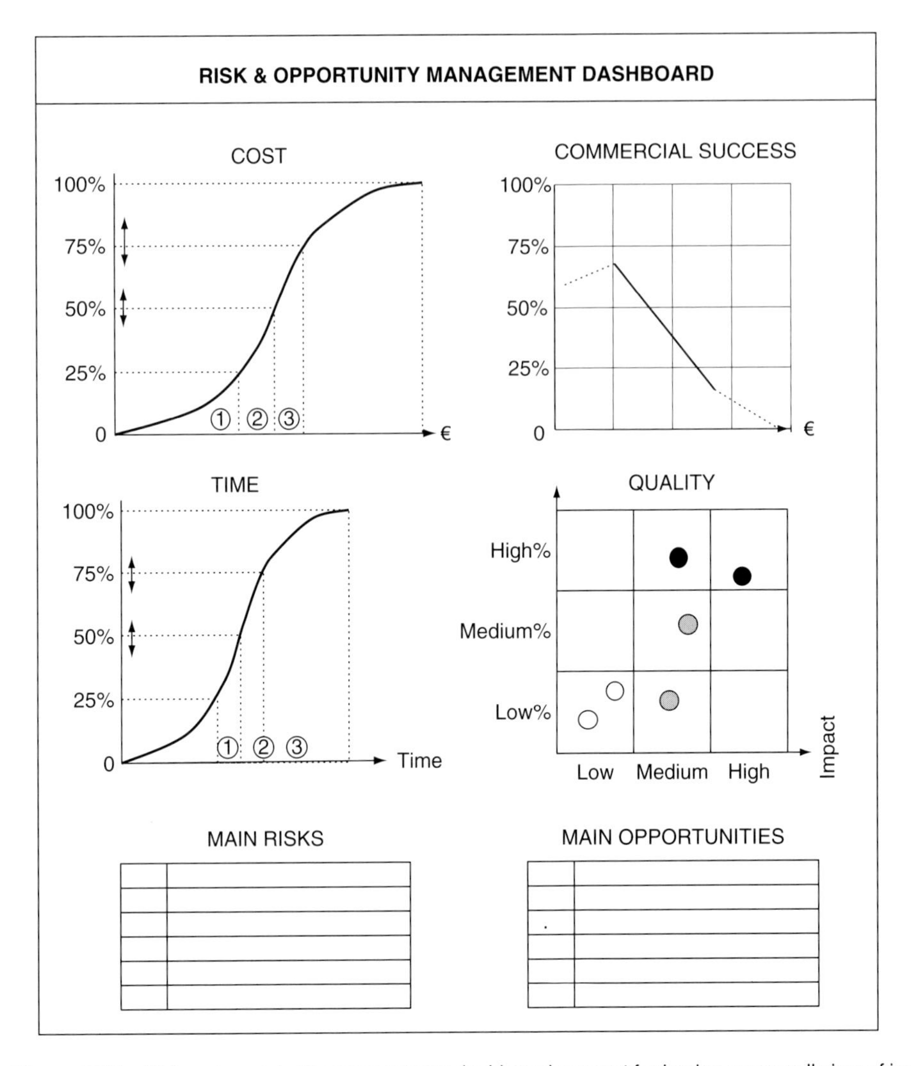

Figure 10.8 Risk and opportunity management dashboard concept for having an overall view of impacts of risks and opportunities.

dimensions for decision-making. Interrelationships between different elements (time, cost, etc.) can be rather complex for explicit modelling and should be covered on the side, usually by means of discussions. It is proposed that collective presentations such as standardised dashboards should be used as a decision-making tool concerning large scale business decision making, for example, pricing, setting cost, time and quality targets.

Risk knowledge models

True motivation for systematic risk management and particularly for risk identification can be difficult to reach. Access to experiences and lessons learned from past projects has proved to be an important motivational factor. In cases that are partly repetitive in nature, the importance of experiences from previous projects can prove to be the most important source for risk identification.

Intuitively, project managers and other project experts have started to use risk checklists as a technique for risk identification. These checklists are often frequently updated; thus, their contents reflect recent experiences to be shared with others. Frequent updating seems to matter most in terms of having checklists of interest. Project managers and other key people involved in risk management exercises are interested in looking at the most recent experiences, using them as a source for risk identification and ultimately for making additional improvements to their own project execution models.

In risk identification, a group of people first use their project model (designs, plans, requirements, objectives, assumptions, etc.) as a main context. The experiences can then be viewed in order to discover potential problems that are risks. This identification is a matter of processing cause–risk–effect chains and then discussing them. The main principle is to go beyond the risk and find the underlying cause. Basically the experiences can be presented in the form of cause–risk–effect chains. When going beyond this we may even encounter richer presentation of cause–risk–effect networks which can be expanded to cover potential risk responses (Fig. 10.9).

Project teams can be geographically dispersed, having few possibilities for intellectual interaction with each other. An obvious solution to this is the use of groupware technology, where intentional group processes are supported with software tools, predominantly utilising web-based intranet technology. This kind of tool can also be a solution for formalisation and dissemination of project risk management knowledge. In the following, some fundamental aspects are discussed that need to be taken into account when groupware applications are developed for this purpose.

Project risk management knowledge can emerge from different sources. The main ones are:

- Problems and best practice resulting in lessons learned
- Actual occurrences and their data from the current projects
- Plans and status report data from past projects.

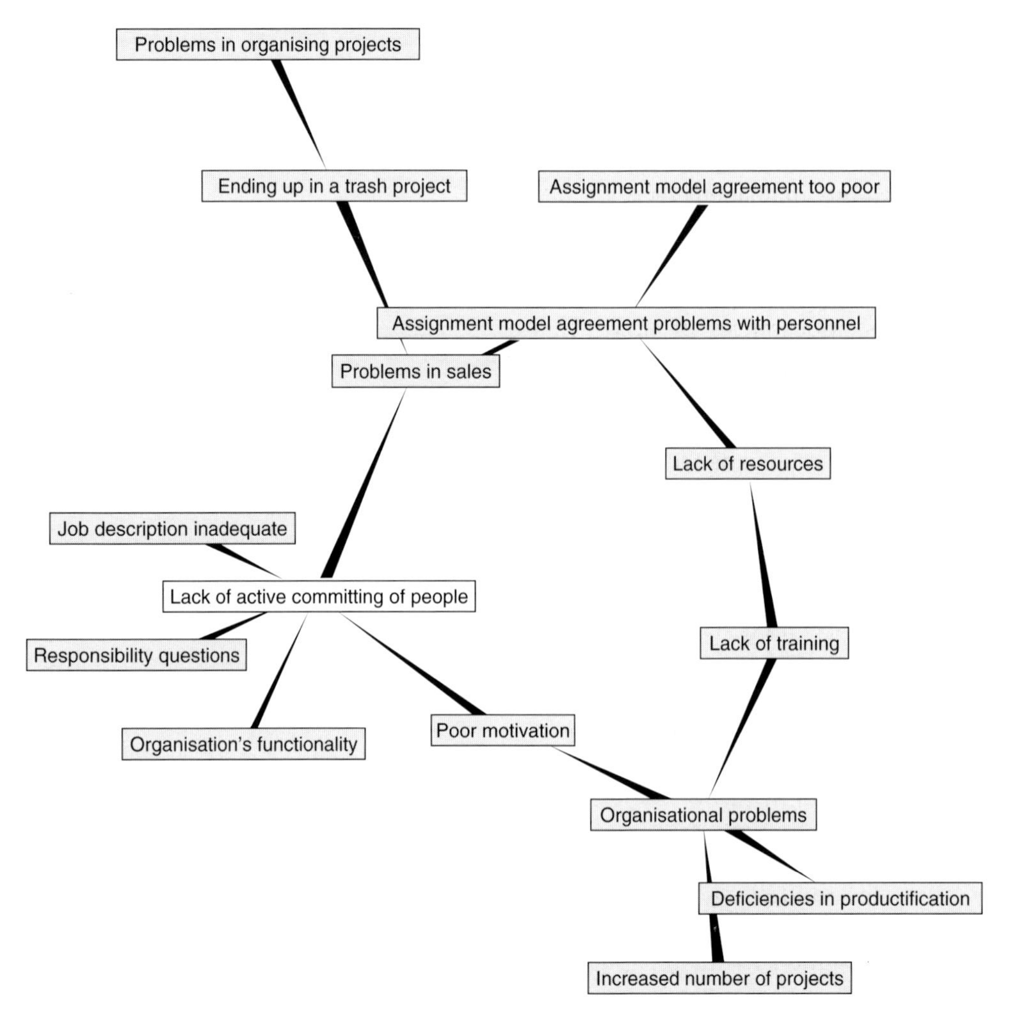

Figure 10.9 Interactive cause–risk–effect network for accessing risk knowledge.

This knowledge can be presented in many ways in order to be used by the individuals in charge of the current projects. The role of ICT in the area of knowledge management is still a somewhat questionable issue (Egbu & Botterill 2002; Wilson 2002; Wilson & Snyder 1999). The approach presented here proposes that a specific knowledge base can be developed for sharing project risk knowledge and for facilitating company or even industry-wide learning based on it. This application can be seen as partial application of a corporate memory system (Beckett 2000; Mentzas *et al.* 2001).

Discussion

The management of uncertainty, especially when encapsulating project risks and opportunities, has widely differing interpretations. One may even say that risk management and project management are fundamentally the same. Project management guides us through threats and performance variances towards the main commercial and other targets. However, it would be misleading to think that standard managerial procedures and practices are enough for successfully handling uncertain events and their potential impacts. It is very important to integrate risk and opportunity management with project management and company management, particularly concerning large-scale risks and opportunities.

This chapter has provided a discussion on the cornerstones of modern risk and opportunity management. It has not been the intention to cover or classify the various appearances of uncertainty management in project-orientated business; rather this chapter has presented recent insights for modern and continuously developing management of uncertainty. It seems obvious that, too often, we are still facing conceptual and terminological problems with the management of uncertainty; this can also be seen to be the case in this chapter. Despite the need for a holistic solution, the terms risk and risk management are still used widely and it seems that they are here to stay. Owing to various practical needs, it appears obvious that a simple single definition or term is not enough. It has been proposed in this chapter that companies applying systematic procedures for the management of uncertainty should move towards localised definitions of risks and opportunities. These situation-specific definitions can vary within a company.

Commercial risk management is concerned with the project budget and contracting. Within this context the qualitative risk analysis can easily seem insufficient, since it is not directly linked to the actual documents and their data. The simplest types of commercial risks are well-targeted threats which can be assessed separately from other issues. But we also face large-scale risks that can be combinations of many factors or actors. Several methods and approaches exist for modelling and estimating single well-targeted risks. These are well known and documented in many textbooks and standards. Modelling inconsistency was presented in this chapter to explain the problems people often face with quantitative risk analyses. For estimating the impacts of large scale and complex risks, one should be careful when using explicit, detailed models, especially concerning the dependencies between various risks when producing an overall impact estimate. Detailed modelling without clear guiding principles can be an avenue with a dead end, resulting in more questions than answers. Complexity should be increased only gradually with the first estimates based on a joint estimate of a 'chunk' of risks.

Conclusions

The management of project risks and opportunities is a continuously developing knowledge and skill area. Six critical areas of risk and opportunity management that demonstrate missing knowledge and are thus also fields for further research and development have been covered in this chapter. Some solutions have been presented here for filling these knowledge gaps. Project managers and other project experts should give more attention to these areas to create a better understanding of the nature of managing risks and opportunities. These six areas are:

(1) *Conceptual foundation based on localised definitions of risk and opportunity*: A clear shortcoming in the concepts and solutions presented both in textbooks and project management standards is the generality of concepts where localised needs are ignored.
(2) *Holistic paradigm for enterprise-wide risk and opportunity management*: Scaling risk and opportunity management vertically in organisations is very important and needs to be carefully taken into account according to the principles presented in this chapter.
(3) *Core and accessory processes for risk and opportunity management*: Core processes seem to appear as main drivers in risk and opportunity management. Accessory processes are integrating core processes internally (and externally if needed).
(4) *Need for the entity of principles for risk and opportunity identification*: This is the most important process that needs to be formed and structured.
(5) *Risk and opportunity modelling inconsistency*: Risk and opportunity modelling inconsistency can cause situations where the results of risk and opportunity analyses do not have credibility. For improved performance one needs to understand these causes and then carry out the analysis in an appropriate way.
(6) *Risk knowledge models and knowledge capturing*: Continuously updated risk knowledge storage is considered to be a valuable asset. Checklists as a source for risk identification are an embodiment of the risk knowledge model.

References

Baird, B.F. (1989) *Managerial Decisions Under Uncertainty*. John Wiley & Sons, ETM Wiley Series in Engineering & Technology Management, New York.

Beckett R.C. (2000) A characterisation of corporate memory as a knowledge system. *Journal of Knowledge Management*, **4**(4), 311–319.

Chapman, C. & Ward, S. (2002) *Managing Project Risk and Uncertainty*. John Wiley & Sons, Ltd., Chichester.

Egbu, C.O. & Botterill, K. (2002) Information technologies for knowledge management: their usage and effectiveness. *Journal of Information Technology in Construction*, **7**, 125–137.

Fenton-O'Creevy, M. & Soane, E. (2000) The subjective perception of risk. *Financial Times*, 25 April, pp. 14–15.

Forsberg, K., Mooz, H. & Cotterman, H. (2000) *Visualizing Project Management*. John Wiley & Sons, New York.

Hartman, F.T. (2000) *Don't Park Your Brain Outside*. Project Management Institute Inc., Newtown Square, PA.

Heisenberg, W. (1959). *Physics and Philosophy*. George Allen & Unwin Ltd, London.

Hoffman, D. (2002) *Managing Operational Risk*. John Wiley & Sons, New York.

Kerzner, H. (2000) *Applied Project Management*, John Wiley & Sons, New York.

Lichtenberg, S. (1981) Real world uncertainties in project budgets and schedules. *Proceedings of the Project Management Institute and International Project Management Association Symposium*, Boston; pp. 179–193.

Loch, C.H. & Huchzermeier, A. (2000) Hiding behind risk in fear of innovation. *Financial Times*, 23 May, 8–10.

Mearns K, Flin R. & O'Connor P. (2001) Sharing 'worlds of risk'; improving communication with crew resource management. *Journal of Risk Research*, **4**(4), 377–392.

Mentzas, G., Apostolou, D., Young, R. & Abecker, A. (2001) Knowledge networking: a holistic solution for leveraging corporate knowledge. *Journal of Knowledge Management*, **5**(1), 94–107.

Olsson, R. (2001) *On the handling of uncertainties in large complex product development projects: a study of benefits and implications*. Royal Institute of Technology, Department of Machine Design, Licentiate Thesis, Stockholm, Sweden.

PMBOK (2004) *A Guide to the Project Management Body of Knowledge* (3rd edn). Project Management Institute, Pennsylvania.

Rundmo, T. (2002) Associations between affect and risk perception. *Journal of Risk Research*, **5**(2), 119–135.

Thieke, S. (2000) Reflections of a risk manager. *Financial Times*, 27 June, 10–11.

Wilson, L.T. & Snyder, C.A. (1999) Knowledge management and IT: how are they related? *IT Professional*, **1**(2), 73–75.

Wilson, T.D. (2002) The nonsense of 'knowledge management'. *Information Research*, **8**(1), paper no. 144 (available at http://InformationR.net/ir/8-1/paper144.html).

Zald, M.N. & W.R. Scott (2003) *Organizations in Action*. Transaction, New Brunswick, NJ.

11 Conflict Management and Dispute Resolution

Peter Fenn

Introduction

There exists enormous interest in commercial disputes amongst the professions, industry and academia. The interest is mostly with the techniques used to resolve disputes; there exists little by way of research into conflict and dispute. This chapter considers four areas:

- A discussion of the difference between conflict and dispute
- A review of the techniques used to resolve disputes: this is the UK Government's current approach and that of other research based interests which seek to make savings by optimising efficiency in dispute resolution
- The lack of evidence that is apparent in any discussion of commercial disputes, and the chapter discusses the absence of an empirical base to the study of disputes; the UK construction industry is given as an example
- Understanding, explanation or prediction of commercial disputes. The lack of an empirical base means that there has been little consideration of the issues of understanding, explanation or prediction of commercial disputes. A research agenda is proposed where an aetiological approach to commercial disputes is employed; this, it is proposed, may help develop a mature and sophisticated research base, which may help industry performance.

Conflict and dispute

Disputes are time consuming, expensive and unpleasant. They can destroy client–supplier relationships which have been painstakingly built up over long periods of time. Disputes can add substantially to the cost of a project, even making a project unsuccessful, unfeasible or nullifying any benefits. Disputes need to be avoided; if disputes cannot be avoided then they should be resolved as efficiently as possible to manage the 'problem', negotiate a 'settlement', help 'preserve relationships' and maintain 'value for money'.

Is there a difference between conflict and disputes?

Many people would not recognise a distinct difference between the terms conflict and dispute. Certainly, most people would not concern themselves with any definition. Academics, and others, would usually make definition their starting point.

Definition provides structure, and structure may allow explanation and understanding.

Conflict and dispute studies do form academic disciplines. Any attempt here to summarise the various strands of academic disciplines would be doomed to failure and debate on definition. In an attempt to avoid this, it is suggested that the following are some of the areas of conflict and dispute studies:

- Peace and conflict studies
- Conflict management as an organisation management science
- Conflict management and dispute resolution.

Much useful material is available on the Internet: try www.socsci.uci.edu/gpacs/. However, many scholars do draw a distinction between the two terms.

Functional and dysfunctional conflict

Early conflict theory marked all conflict as a bad thing that should be avoided. Amongst the first to question this was Mary Parker Follett (1925); she said that effective conflict management ought not to conceive conflict as a wasteful outbreak of incompatibilities, but a normal process whereby socially valuable differences register themselves for the enrichment for all concerned. Three methods were advanced for dealing with conflict, of which only integration was strongly advocated. These included: domination, whereby there is a victory of one side over the other (a win–lose situation); compromise, whereby each side gives up something in the process (a lose–lose situation); and integration, whereby each side refocuses their efforts so that neither side loses anything and, in fact, each gains (a win–win situation).

Follett believed that domination should be avoided at all costs. Although application of this strategy requires little effort on the part of the parties and their agents, the long-term side effects can be devastating. Compromise carries with it the assumption that both parties will be happy because each will gain something, but each loses something as well, and this in turn creates the potential for further conflict. Integration was favoured simply because if both parties can become satisfied there will remain no issue or problem – obviously an ideal situation not easily attained.

Win–lose is often overused as a strategy for solving conflicts. It assumes the use of mental or physical power to bring about compliance; a lose–lose approach will also leave no one entirely happy. Compromise, side payments and submission of the issue to a neutral third party, as in the arbitration procedure, constitute examples of this latter approach. The win–win approach is now becoming more popular, although it is still misunderstood by many parties and their agents. This method yields solutions satisfactory to all, in that each party to the conflict wins something and the conflict is therefore resolved constructively. It could be suggested that important conflicts tend to be best managed with positive-sum (win–win) strategies, while more trivial issues merit no more than zero-sum (win–lose/lose–lose) strategies, with most situations calling for contingency or mixed modes (no win–no lose).

A further distinction between conflict and dispute that is particularly useful is the distinction made by Burton (1993) which distinguishes the two based on time and issues in contention. Disputes, Burton suggests, are short-term disagreements that are relatively easy to resolve. Long-term, deep-rooted problems that involve seemingly non-negotiable issues and are resistant to resolution are what Burton refers to as conflicts. Though both types of disagreement can occur independently of one another, they may also be connected. In fact, one way to think about the difference between them is that short-term disputes may exist within a larger, longer conflict. A similar concept would be the notion of battles, which occur within the broader context of a war. Other theorists talk of strategy and tactics; tactics win the battle but strategy wins the war.

From this analysis of conflict and dispute it can be argued that conflict is necessary and inevitable, but that disputes are to be avoided. The school of Western thought maintains that conflict (but not dispute) is inevitable (de Bono 1985). Conflict is part of Western societies and idioms; there is a Western dialectic argument idiom to use the academic jargon. Conflict is part of dynamic capitalism and an integral part of commercialism; conflict might be seen as the functional and necessary part. Dispute, on the other hand, only develops when conflict is not (or cannot be) managed; dispute is the unnecessary or dysfunctional element and logically there should be two areas for consideration:

- *Conflict management*: Here the emphasis is on the axiom that it must be in all parties' interests to avoid disputes by managing conflict in such a way that disputes do not arise; this is sometimes described as dispute avoidance
- *Dispute resolution*: Notwithstanding the emphasis on the desire to avoid dispute, there must be occasions where the parties have legitimate disputes and that the techniques of dispute resolution are employed to bring about the conclusion or resolution of the dispute.

The distinction between conflict and dispute is shown diagrammatically in Fig. 11.1.

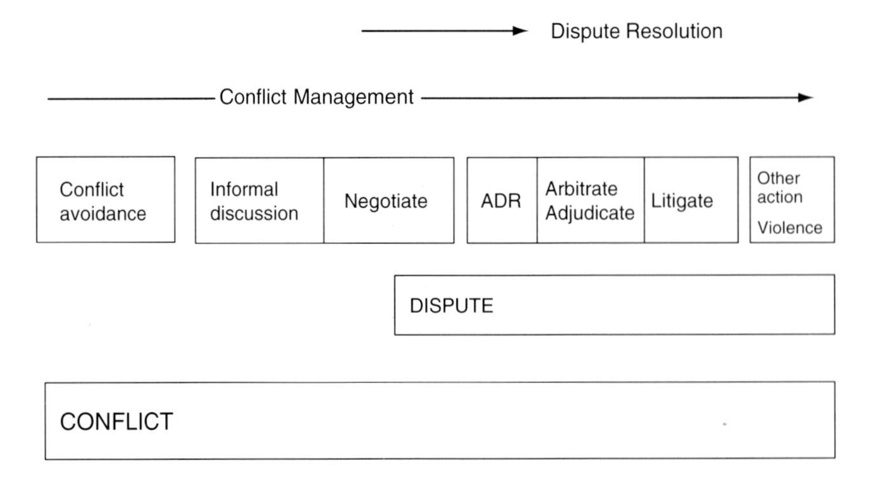

Figure 11.1 Conflict continuum.

Disputes on projects or contracts are more than unpleasant: they divert valuable resources from the overall aim, which must be completion on time, on budget and to the quality specified. In addition, they generally cost money and take time to resolve, and they can destroy relationships which may have taken years to develop.

The legal issue conflict or dispute: legal concerns about conflict or dispute

Although the esoteric discussion earlier on conflict and dispute is valuable, is there any pragmatic real-world issue in the distinction? The legal point is discussed by Brown and Marriott (1994) as the difference between behavioural conflict and justiciable dispute. The question as to whether or not a dispute exists is highly relevant where a contract provides that disputes are to be referred to arbitration or other dispute resolution.

The meaning of the word dispute would, at first sight, seem to be relatively straightforward; indeed cases such as *Hayter* v. *Nelson* (1990) and *Cruden* v. *Commission for New Towns* (1995) have stated that an ordinary English word such as dispute should be given its ordinary meaning. However, there has been a considerable amount of case law concerning the question of what constitutes a dispute. Much of that case law has been associated with arbitration and/or construction adjudication under the Housing Grants Construction and Regeneration Act 1996 (HMSO 1996), Section 108 of which provides that:

> *A party to a construction contract has the right to refer a dispute arising under the contract for adjudication under a procedure complying with this section. For this purpose 'dispute' includes any difference.*

It has been extremely common in adjudication cases for the responding party to allege that it has not previously been given the opportunity to review the case put forward by the referring party and, therefore, that there is no dispute capable of being referred to adjudication. On the basis of this, the responding party will contend that the adjudicator does not have jurisdiction to deal with the matter.

Recent cases

The recent case of *Cowlin Construction Ltd* v. *CFW Architects* (2003) considered the question of definition and, in doing so, the court provided a useful summary of the relevant cases.

It appears from recent cases that, whilst there is no special meaning to be given to the meaning of the word dispute, there are certain factors to take into consideration when deciding whether or not there is a dispute in an adjudication context. The approach adopted by the courts is one which attempts to prevent referring parties from being able to ambush the responding party, given that the timescales in adjudication are extremely tight. There seem to be two schools of thought as to what is required for the crystallisation of a dispute. In the wide approach advocated by *Halki Shipping Corporation* v. *Sopex Oils Ltd* (1998), a claim made and not admitted is sufficient [*Costain* v. *Wescol Steel Ltd* (2003) and *Orange EBS Ltd* v. *ABB*

Ltd (2003)]. The narrow approach advocated by *Beck Peppiatt* v. *Norwest Holst Construction* (2003) shows a reluctance to allow ambushes or for dispute resolution (at least adjudication) to be commenced prematurely.

Comparison of dispute resolution techniques

Introduction

There has been considerable recent interest in dispute resolution; the current approach seeks to make savings by optimising efficiency in dispute resolution. A central tenet of the UK Government's commitment is the pledge to alternative dispute resolution (ADR) which is given in full later in this section.

This section reviews the main techniques available and compares three key techniques.

Conflict management and dispute resolution techniques

The range of conflict management and dispute resolution techniques includes the following:

- *Conflict management and/or avoidance*: A variety of techniques, some used consciously and some subliminally, to avoid the escalation from normal conflict into dispute. Examples might include: risk management to ensure that risks are identified, analysed and managed; procurement strategies to ensure that risks are appropriately allocated, and contractual arrangements to allow sensible administration.
- *Negotiation*: This is easily the most common form of dispute resolution, carried out in many forms every day by just about everybody. In negotiation, the parties themselves attempt to settle their differences using a range of techniques from concession and compromise to coercion and confrontation.
- *Mediation*: A private and non-binding form of dispute resolution where an independent third party [neutral] facilitates the parties reaching their own agreement to settle a dispute. Mediation is often a structured process where the settlement becomes a legally binding contract.
- *Conciliation*: A process of mediation where the neutral proposes a solution. In the same way that we distinguished between a continuum of conflict and dispute, a continuum of mediation and conciliation shows mediation at one facilitative end and conciliation at the other evaluative end of the continuum (Fig. 11.2).
- *Med-arb*: This is a combination of mediation and arbitration where the parties agree to mediate, but if that fails to achieve a settlement the dispute is referred to arbitration. The same person may act as mediator and arbitrator in this type of arrangement.
- *Dispute Resolution Adviser (DRA)*: The concept of DRA is the use of an independent intervener. This independent intervener is paid for equally by the employer and

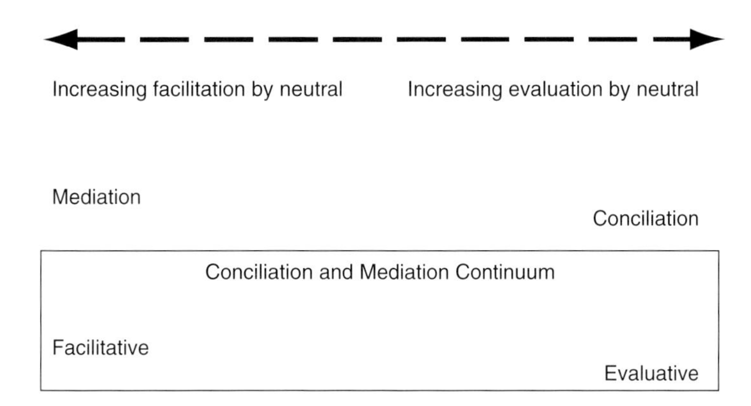

Figure 11.2 Conciliation and mediation continuum.

the contractor to settle disputes as they emerge, rather than wait until the end of the contract.

- *Dispute review board (and dispute review panel)*: A process where an independent board or panel evaluate disputes.
- *Neutral evaluation*: A private and non-binding technique where a third, neutral party (often legally qualified), gives an opinion on the likely outcome at trial as a basis for settlement discussions.
- *Expert determination (submission to expert, reference to an expert, expert adjudication)*: These are long-established procedures in English law and have been used across a number of industries. Examples include: accountants valuing shares in limited companies, valuers fixing the price of goods, actuaries carrying out valuations for pension schemes, certifiers of liability for on-demand performance bonds, and adjudicators who are said to be acting 'as expert and not as arbitrator'.
- *Mini-trial (or executive tribunal)*: This is a voluntary non-binding process. The parties involved present their respective cases to a panel comprising senior members of their organisation. The panel is assisted by a neutral facilitator and has decision-making authority. After hearing presentations from both sides, the panel asks clarifying questions and then the facilitator assists the senior party representatives in their attempt to negotiate a settlement.
- *Adjudication*: This refers to Statutory Adjudication in Construction Disputes as set out in the Housing Grants, Construction and Regeneration Act 1996 (HMSO 1996). Here decisions of an adjudicator are binding on the parties at least until a further process (arbitration or litigation) is invoked.
- *Arbitration*: A formal, private and binding process where disputes are resolved by an award of independent tribunal (third party or parties, the arbitrator or arbitrators). The tribunal is either agreed by the parties or nominated by a further independent body; for example, a court or a professional institution, The Chartered Institute of Arbitrators.
- *Litigation*: The formal process whereby claims are taken through court and conducted in public; judgements are binding on the parties subject to rights of appeal.

This is by no means an exhaustive or exclusive list; there are undoubtedly others. Indeed, one definition of ADR is 'appropriate dispute resolution' and there may be a 'killer application' yet to be devised.

The stages of conflict management and dispute resolution

The stages of conflict management and dispute resolution are usefully described in a document produced by the Office of Government Commerce: Dispute Resolution Guidance (http://www.ogc.gov.uk). The stages are:

- *Stage 1*: Negotiation
- *Stage 2*: Non-binding techniques and processes
- *Stage 3*: Binding techniques and processes.

This epitomises the current approach which seeks to make savings by optimising efficiency in dispute resolution; there is strong support for this in the UK by the Government and internationally by research teams such as those at Harvard (Sander 1976) and Cornell (Lipsky & Seeber 1998). The principal stages and the dispute resolution options are shown at Figs. 11.3 and 11.4.

Outline of the three dispute resolution techniques

This section provides an outline of the three dispute resolution techniques predominantly practised:

- arbitration
- construction adjudication
- mediation.

In many commercial contracts, arbitration is included as the default dispute resolution technique; there are many reasons for this. Commerce and commercial managers soon became disillusioned with the courts and took their disputes to arbitrators; they sought confidentiality, commercial decisions and, often, continuing relationships.

Arbitration

Arbitration is a process, subject to statutory controls, whereby formal disputes are determined by a private tribunal of the parties' choosing. According to Stephenson (1998), Lord Justice Sir Robert Raymond provided a definition some 250 years ago which is still considered valid today:

> *An arbitrator is a private extraordinary judge between party and party, chosen by their mutual consent to determine controversies between them, and arbitrators are so called because they have an arbitrary power; for if they observe the submission and keep within due bounds, their sentences are definite from which there lies no appeal.* (Stephenson 1998)

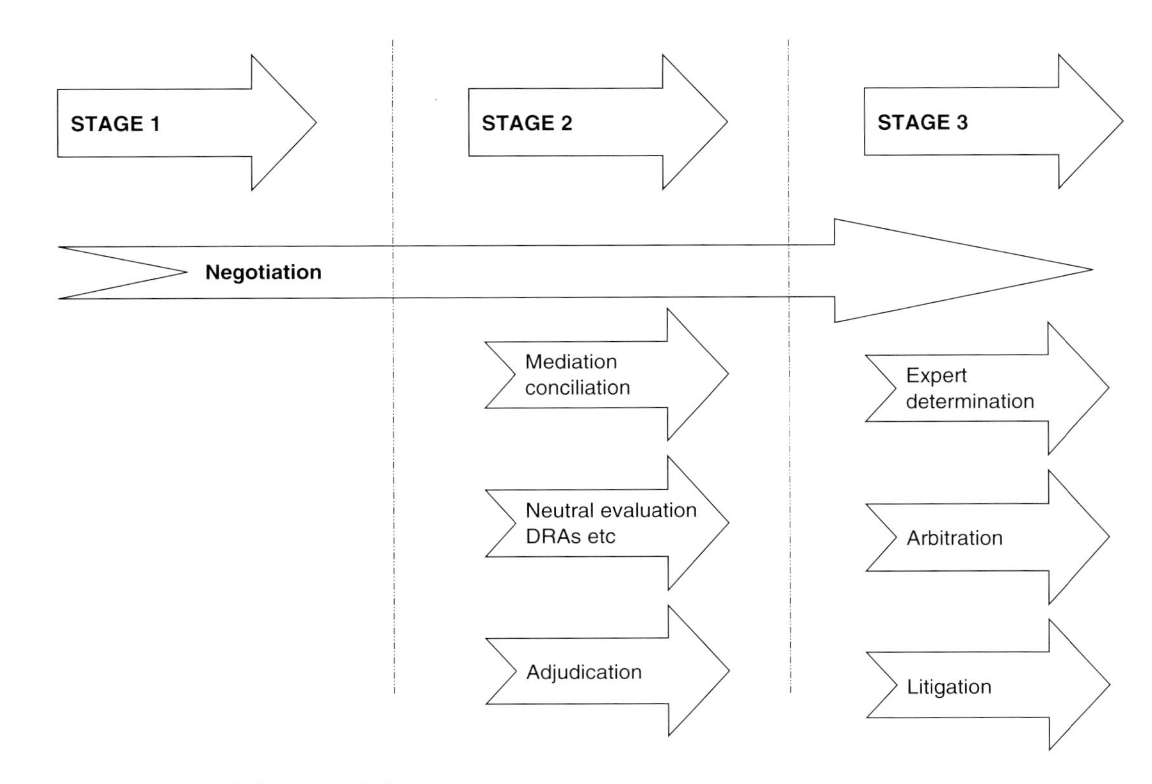

Figure 11.3 The principal stages of dispute resolution.

METHOD	COMMON LAW/ STATUTE BASIS	FREQUENCY OF USE	SPEED	COST	CONFIDENTIALITY	BINDING	ADVERSARIAL	SPECIAL FEATURES
					Stage 1			
Negotiation	No	Very common ubiquitous	Varies	Low	Yes	No	No	Can continue throughout the dispute
					Stage 2			
Mediation	No	Common	Fast	Low	Yes	No (unless agreed)	No	
Conciliation	No	Fairly common	Fast	Low	Yes	No (unless agreed)	No	Often included with mediation
Neutral evaluation	No	Infrequent	Fast	Low	Yes	No	No	
Adjudication	Yes	Common	Fast	Low	Yes	Yes (until completion or Arb/Lit)	Yes	Statutory adjudication is construction specific
					Stage 3			
Arbitration	Yes	Common	Contingent	Contingent	Yes	Yes	Yes	
Expert determination	No	Fairly common	Fast	Moderate	Yes	Yes	Yes	
Litigation	Yes	Common	Slow	High	No	Yes	Yes	

Figure 11.4 The dispute resolution options.

Some argue that arbitration fell into the very traps that had first brought commercial disputes from the courts: arbitration became slow and expensive and mimicked the courts. In the UK, the Arbitration Act 1996 (HMSO 1996) tried to undo the damage that had been done.

The five main aims that underlie the Arbitration Act are:

(1) To ensure that arbitration is fair, cost-effective and rapid
(2) To promote party autonomy, in other words to respect the parties' choice
(3) To ensure that the courts' supportive powers are available at the appropriate times
(4) To ensure that the language used is user friendly and clearly accessible
(5) To follow the model law wherever possible.

The first of these objectives is included in Section 1 of the Act:

The provisions of this part are founded on the following principles, and shall be construed accordingly –

(a) The objective of arbitration is to obtain the fair resolution of disputes by an impartial tribunal without unnecessary delay or expense;
(b) The Parties should be free to agree how their disputes are resolved, subject only to such safeguards as are necessary in the public interest;
(c) In matters governed by this part of the Act the court should not intervene except as provided by this part.

Construction adjudication (under the Housing Grants Construction and Regeneration Act 1996)

At the same time that the Arbitration Act sought to return arbitration to its roots, the Housing Grants Construction and Regeneration Act received Royal Assent during July 1996. Those parts relating to construction (Part II of the Act) commenced on 1 May 1998. The Act sets out a framework for a system of adjudication; all construction contracts must meet minimum criteria and if they fail, the *Scheme for Construction Contracts* will apply.

Under Part II of the Housing Grants, Construction and Regeneration Act 1996 (HMSO 1996) a party to a construction contract is unilaterally given the right to refer a dispute arising under the contract to adjudication. The Act only applies to 'construction contracts' which fall within the detailed definition of Section 104. For example, 'architectural design, surveying work or to provide advice on building, engineering, interior or exterior decoration or the laying out of landscape in relation to construction operations' are included within the scope of the Act, whilst contracts of employment are expressly excluded. In addition, a construction contract is defined to include an agreement to carry out 'construction operations'. Construction operations are further defined in Section 105 to include a wide variety of general construction related work together with a list of notable exceptions. A further notable exception is a construction contract with a residential occupier. The provisions only apply where the construction contract is in writing.

Mediation

Mediation is a way of settling disputes in which a third party, known as a mediator, helps both sides to come to an agreement which each considers acceptable. Mediation can be 'evaluative', where the mediator gives an assessment of the legal strength of a case, or 'facilitative', where the mediator concentrates on assisting the parties to define the issues. When a mediation is successful and an agreement is reached, it is written down and forms a legally binding contract, unless the parties state otherwise.

Many trainers teach a facilitative model in training courses on the basis that this is considered the most successful of the techniques. Any settlement which occurs is the parties' own, and the analogy of chemical catalysts is often made. A catalyst makes a reaction take place between two or more chemicals; the catalyst is not affected or changed by the reaction. Sometimes the reaction will take place without the catalyst and the effect is to speed reaction and sometimes the reaction will not take place without the catalyst. The analogies with mediation are obvious.

Mediation is the most widely used and accepted ADR technique. Whilst there is no prescriptive mediation process, the typical stages in a mediation might be:

(1) A brief written summary of the matter in dispute is presented in advance to the mediator
(2) The parties meet with a mediator for an initial joint meeting including perhaps a brief oral presentation by the parties
(3) Caucus sessions take place, where the mediator has private meetings with the parties in turn. During the caucuses the mediator often shuttles backwards and forwards to clarify issues and search for settlement possibilities. This process is often termed shuttle diplomacy
(4) Plenary sessions are called to either continue negotiations directly, to conclude agreement, or where the process is unsuccessful, to conclude a mediation.

Most mediators agree to a contingency approach to mediation; that is, there is no set procedure but the procedure is tailored to suit the parties and the dispute in question. This often means that mediation is conducted without joint meetings, and the mediators play a variety of roles. The mediator may act as a mere facilitator, there purely to assist communications. Alternatively the mediator acts as a deal maker, to assist the parties in finding overlap in their bargaining positions or encouraging concession and compromise. Perhaps the mediator may act more as a problem solver assisting the parties in designing and searching for creative solutions. The mediator may act as a transformer, transforming the dispute by allowing the parties a fresh insight into the issues and their positions. The final role of the mediator may be as an adjudicator or assessor to provide the parties with an appraisal of the merits of their cases on a legal, technical or even common sense standpoint.

Comparison of litigation, with construction adjudication, arbitration and mediation

It is useful to compare and contrast the major dispute resolution techniques in areas where the characteristics of each technique are highlighted. Litigation, construction adjudication, arbitration and mediation are compared under the following headings:

- Formality
- Speed
- Flexibility
- Cost
- Confidentiality
- Relationships
- Control and choice
- Solutions.

Formality

Mediation is an informal process; the parties may agree to certain mediation rules but they are at liberty to amend any rules. There is no requirement to produce specified information before the mediation can commence, neither is there a requirement to spend resources filing and serving documents. Mediation is informal and uncomplicated. Adjudication is an informal process and the procedure is, within the bounds of natural justice, at the discretion of the adjudicator. Arbitration has been criticised for mimicking litigation; many steps have been taken to redress this and arbitrations are less formal; nevertheless arbitration may be considered formal and complicated when compared with mediation. Litigation is, properly, a highly formalised process with specialised rules; non-compliance may prevent litigation proceeding. Resources have to be committed in filing and serving documents. Litigation is a highly formal and complicated process.

Speed

In mediation, the timing is within the control of the parties; subject to the availability of suitable and acceptable mediators, mediation may take place as quickly as the parties desire. The length of the mediation is similarly in the control of the parties; they can agree to stay as long, or as briefly, as required. The great majority of mediations are restricted to one working day or less. Adjudication operates under very tight timescales laid down by the Act; the maximum time from notice to decision is 35 days which may be extended by agreement to 49 days. Speed is often claimed as a feature of arbitration; however, the reality is that the availability of all the parties involved, not least the arbitrators, dictates that the process is often protracted. Litigation is often an infuriatingly slow process; in many jurisdictions advisors talk in terms of years rather than months as the timescale for trial dates. Although many great strides have been taken in many countries to address this (in the UK the Civil Procedure Rules (Department for Constitutional Affairs) following the review of civil justice (Woolf 1996) is a particular example), time continues to be an issue.

Flexibility

Mediation is a flexible process; all arrangements can be changed if it becomes apparent that this is necessary. Arbitration can share much of this flexibility and the

Arbitration Act 1996 has given arbitrators wide ranging powers to achieve flexibility. Adjudicators, too, have much scope for flexibility. Litigation is an inflexible process; specific steps must be taken to initiate and progress matters.

Cost

Mediation is inexpensive; this is achieved and facilitated by the informality and speed of the process. The amount of lawyer involvement can be reduced if the parties agree, and in many cases the cost of preparing for mediation is marginal to the other preparation. The parties can share the mediator's costs and the cost of the venue in an agreed fashion. Adjudication can be an inexpensive process as a result of the tight timescales. Arbitration can certainly help in reducing costs and dealing with a dispute in a proportionate manner. In comparison to litigation, it must be remembered that while the state pays for the judge and rooms in many cases, in arbitration the parties must pay the arbitrator's costs. Litigation is expensive; this is dictated by the formality and slowness of the process. There are many, many examples of the disproportionate costs of litigation: amongst the most famous is the example of Jarndyce v. Jarndyce in *Bleak House* (Dickens 1852–1853) where the parties disputing a will expended the entire legacy in legal costs!

Confidentiality

Here things are clear; in adjudication, arbitration and mediation all matters are confidential. This is an important issue for commercial disputes where the parties often wish to avoid publicity and to keep commercial confidentialities. There is an issue often where arbitration awards are the subject of appeal or referral to the courts; then all matters will become public. Litigation is a public matter and though civil commercial litigation seldom attracts tabloid press interest, it is clear that litigation can expose confidential issues.

Relationships

Again there are clear differences. Mediation is a non-adversarial process, while litigation and arbitration are both adversarial. Adjudication may avoid the dysfunctional aspects of adversarialism. In mediation, the parties do not seek to convince the neutral that they are in the right, or that others are in the wrong. The emphasis of facilitative mediation is on the parties' interests as opposed to parties' rights. As a result, mediation need not affect working relationships in an adverse manner; often, mediation can improve relationships as parties achieve an improved understanding of underlying interests and concerns. Litigation and arbitration, on the other hand, are not conducive to even maintaining relationships, let alone improving them. Opposing parties aim to convince the tribunal that the law and the facts support their argument to the detriment of the other side; this seldom helps relationships and often destroys them. Adjudication allows the power imbalance in relationships to be dealt with in that weaker subcontractors have a clear route to deal with more powerful contractors.

Control and choice

In mediation, the control of the dispute always remains with the parties and the choice is theirs. Who will be the mediator; where will the mediation take place; when will it take place; and who will attend? Mediation is a voluntary process and the parties remain in control. This control means that the parties have to 'buy in' to the settlement, and any resolution becomes their own settlement. Litigation, adjudication and arbitration hand over the dispute to the lawyers and the judge, or adjudicator, or arbitrator. The process passes control and choice in a similar fashion.

Solutions

The essential difference is that mediation allows for creative solutions to disputes; during a mediation a wide range of issues can be addressed or uncovered. These issues can include past unresolved matters and even future intentions. The solutions to the issues can take many forms, and are not restricted to payment of money; they can be as creative as the parties to the dispute. Mediated agreements have included:

- apologies
- future business arrangements
- revamped commercial arrangements.

Litigation and arbitration cannot allow for creative solutions, but must be limited to the legal remedies available. Adjudication is similarly restricted by legal remedies, but does allow prompt solutions which permit the project to be completed.

The comparison is shown in outline in Table 11.1.

The UK Government's pledge to ADR

In the UK, the result of this comparison of techniques, and a desire to make savings by optimising efficiency in dispute resolution, was that on 23 March 2001 the Government made a pledge that government departments will only go to court as a last resort. Instead, they will settle their legal disputes by mediation (or arbitration) whenever possible. Government departments and agencies will settle legal cases by

Table 11.1 Litigation compared with construction adjudication, arbitration and mediation.

Dispute technique area	Mediation	Adjudication	Litigation	Arbitration
Formality	Informal	Informal	Formal	Formal
Speed	Fast	Fast	Slow	Contingent
Flexibility	Good	Good	Poor	Contingent
Cost	Inexpensive	Inexpensive	Expensive	Contingent
Confidential	Yes	Yes	No	Yes
Adversarial	No	Contingent	Yes	Yes
Party control	Yes	No	No	No
Party choice	Yes	No	No	Contingent
Creative solutions	Yes	No	No	No

ADR techniques in all suitable cases whenever the other side agrees. The pledge is worth repeating in full.

Government Departments and agencies make these commitments on the resolution of disputes involving them:

- *Alternative Dispute Resolution will be considered and used in all suitable cases wherever the other party accepts it.*
- *In future, Departments will provide appropriate clauses in their standard procurement contracts on the use of ADR techniques to settle their disputes. The precise method of settlement will be tailored to the details of individual cases.*
- *Central Government will produce procurement guidance on the different options available for ADR in Government disputes and how they might be best deployed in different circumstances. This will spread best practice and ensure consistency across Government.*
- *Departments will improve flexibility in reaching agreement on financial compensation, including using an independent assessment of a possible settlement figure.*
- *There may be cases that are not suitable for settlement through ADR, for example cases involving intentional wrongdoing, abuse of power, public law, human rights and vexatious litigants. There will also be disputes where, for example, a legal precedent is needed to clarify the law, or where it would be contrary to the public interest to settle.*
- *Government departments will put in place performance measures to monitor the effectiveness of this undertaking.*

The principle that litigation should be a last resort has the approval of the courts; in *Frank Cowl and other* v. *Plymouth City Council* (2001), Lord Woolf said:

insufficient attention is paid to the paramount importance of avoiding litigation whenever this is possible.

Why commercial contracts go wrong (or an aetiological approach to commercial disputes): the construction example

It certainly seems that commercial contracts do go wrong; everyone knows that it is one of the problems of commerce. As an example we might consider the construction industry in the UK, where the problems are writ large. These problems have intrigued, one might say obsessed, the industry and government for 50 years. Reports on construction are nothing new and a list (incomplete) of reports since World War II makes depressing reading:

1944 *Report of the Committee on the Placing and Management of Building Contracts* (The Simon Report) (Ministry of Works 1944)
1949 British and American Productivity Council
1950 *Report of the Working Party on the Building Industry* (The Phillips Report) (Ministry of Works 1950)

1962 *Survey of the Problems before the Construction Industries* (The Emmerson Report) (Ministry of Works 1962)
1964 *The Placing and Management of Contracts for Building and Civil Engineering Work* (The Banwell Report) (Ministry of Public Building and Works 1964)
1965 *Communications in the Building Industry* (Higgin & Jessop 1965)
1967 Action on the Banwell Report (Economic Development Committee for Building 1967)
1975 *Public Client and the Construction Industries* (The Wood Report) (Building and Civil Engineering Economic Development Councils 1975)
1980s *Faster Building for Industry; Faster Building for Commerce* (Building Economic Development Committee 1983, 1988)
1992 *Building Towards 2001* (Building Employers Confederation 1992)
1993 *Trust and Money* (Latham 1993)
1994 *Constructing The Team* (Latham 1994)
1998 *Rethinking Construction* (Egan 1998)
2001 *Modernising Construction* (National Audit Office 2001).

What is the problem with construction? That much is easy. A simplistic analysis of all the above reports finds three key areas: the industry has a poor image and is renowned for products which are:

- of poor quality
- over budget (expensive)
- over programme (late).

Stella Rimington tells this with stunning clarity in her book about her time in MI5, *An Open Secret*:

> *Like all huge building projects, particularly in the public sector, the Thames House Refurbishment had been fraught with difficulties... It was clear that dealing with the building industry was just as tricky as dealing with the KGB.* (Rimmington 2002, p. 229)

Besides construction projects being of poor quality, late and expensive, another feature is the preponderance of disputes, often described as 'adversarial attitudes'. Many would question the evidence for any of these problems, but here let us restrict the questioning to the issue of disputes.

The adversarial attitudes problem is often expressed as received wisdom about construction. Simply put, there exist received wisdoms that:

(1) Construction suffers more contractual disputes than other industries
(2) The occurrence of disputes has risen recently and continues to rise
(3) The performance of the industry is adversely affected by the disputes.

The received wisdoms are repeated throughout the construction literature. For a contemporary confirmation in the UK see Latham (1994) and National Audit Office (2001); for an international synopsis see Kumaraswamy (1997a). However, there is little empirical work to test the received wisdoms, and random theorising is allowed to pass unchallenged. For a detailed discussion on this see Fenn (1997) and Alkass *et al.* (1998). It is intriguing that in an area accustomed to the rigours of evidence, both legally and scientifically (the law and academia), this has been allowed to pass with barely a murmur of protest. This evidential sloppiness reached its zenith, or nadir, during the debate which accompanied the introduction of the Housing, Grants Construction and Regeneration Act 1996 in the UK; regarding the introduction of adjudication, the Department of the Environment (1996) claimed *inter alia*:

> *there is compelling anecdotal evidence that adjudication would reduce overall project costs*

It is suggested that the phrase 'compelling anecdote' is an oxymoron. Where is the evidence that adjudication would reduce overall costs (future); and now that some years' experience exists, where is the evidence that it has reduced overall costs (past)? As to the received wisdoms, where is the evidence that construction suffers more contractual disputes than other industries; that the occurrence of disputes has risen recently and continues to rise; and crucially that the performance of the industry is adversely affected by the disputes? The problem with anecdote is that one person's anecdote is different from his neighbour's anecdote.

The literature on why construction contracts go wrong

A first step in any academic research is to review the literature in the field. A literature review of the contemporary work into construction disputes reveals some researchers attempting to carry out empirical work to set up a classification of sources of disputes. Bearing in mind the size of the industry and its importance in the nation's economy, the literature might be thought to be a bit thin. The researchers, by country, are shown in summary in Table 11.2. The literature review found 38 studies from 10 countries, either empirical or theoretical.

The literature and the findings regarding the types, the factors and the impact areas of construction disputes, the context and their 'reasons' or 'causes' are chronologically listed in Table 11.3. It might be argued that rather than identifying

Table 11.2 Synopsis of literature on why contracts go wrong.

Country	Studies
Australia	Watts & Scrivener (1992, 1994)
Canada	Revay (1992), Semple *et al.* (1994), Bristow & Vasilopoulous (1995)
China	Liang (1999)
Hong Kong	Kumaraswamy (1997a, 1997b)
Jordan	Al-Momani (1999)
Nigeria	Kehinde & Aiyetan (2002)
Saudi Arabia	Assaf *et al.* (1995)
Taiwan	Lee (1994), Hu (1998), Tsai (2001), Wang (2001), Yao (2001), Wu (2002), Yan (2002), Public Construction Commission (2003)
UK	Lewis *et al.* (1992), Rhys Jones (1994), Construction Industry Council (1994), Vidogah & Ndekugri (1998), Gould *et al.* (1999), Fenn (1999)
USA	Diekmann & Nelson (1985), Mathews & Ashley (1985), Diekmann *et al.* (1994), Jergeas & Hartman (1994), Molenar *et al.* (2000), Mitropoulous & Howell (2001)
Global views	Murdoch & Hughes (1996), Sykes (1996), Sykes (1999), Kululanga *et al.* (2001), Ren *et al.* (2001), Lo (2002), Chang & Ive (2003)

Table 11.3 Chronological listing of literature on why contracts go wrong.

Researchers	Context	Findings (sources of disputes)
Diekmann & Nelson (1985)	Empirical: 427 claims on 22 projects in USA	The two most common causes of contract claims were (1) design errors (46%) and (2) discretionary or mandatory changes (26%). Other claim headings include: differing site conditions; weather; strikes and value engineering
Mathews & Ashley (1985)	Empirical: in USA	96 items were concluded as construction disputes within contract clauses
Watts & Scrivener (1992)	Empirical: 72 judgments from 56 construction litigation cases in Australia	59 categories of disputes and 117 sources of disputes within which the subgroups are: (1) determination of the agreement, (2) payment, (3) the site and execution of work, (4) time, (5) negligence and nuisance, (6) final certificate/payment
Lewis *et al.* (1992)	Empirical: cases study in UK	Five causes of conflict: (1) one of the potential risk events occurs, (2) one or more of the parties suffers some loss as a result of it, (3) the damaged party had not identified the risk as relevant to the project, (4) the risk was identified but insufficient steps were taken to mitigate its effects, (5) the allocation of risks between the various parties to the contract was not clearly established in the first place
Revay (1992)	Empirical: in Canada	Seven most frequent causes for claims: (1) inadequate site and/or soil investigation prior to starting the design, (2) starting design efforts and/or soil investigation too late prior to starting the design, (3) calling for bids with an incomplete set of drawings, (4) endeavouring to complete the design through shop drawing review, (5) introducing untimely design revisions without allowing commensurate time extension for the completion of the project or without recognising the contractor's right to impact costs, (6) interfering both with the sequence and the timing of construction (e.g. to compensate for the delay in the delivery of owner-supplied equipment/material), (7) continuing to introduce changes under the disguise of correcting deficiencies
Construction Industry Council (1994)	Empirical: general survey for dispute resolution within construction industry in the UK	Six categories of main reasons for disputes: (1) general, (2) consultants, (3) client, (4) contractor, (5) subcontractors, (6) manufacturers and suppliers

Table 11.3 (Continued).

Researchers	Context	Findings (sources of disputes)
Diekmann *et al.* (1994)	Empirical: survey from the participants of 159 projects in the construction industry and lawyers in the USA	Three areas: (1) people, (2) process, (3) project
Jergeas & Hartman (1994)	Empirical: authors' experiences in North America	Well known reasons by which claims arise: (1) increase in scope of work, (changes, extras and errors), (2) inadequate bid information, (3) faulty and/or late owner-supplied equipment and material, (4) inferior quality of drawings and/or specifications giving rise to ambiguities in contract requirements, (5) insufficient time for bid preparation, (6) stop-and-go operations because of lack of coordination, design information, equipment, or material, (7) work in congested areas and overcrowding, (8) acceleration to regain schedule, (9) inadequate investigation before bidding, (10) unbalanced bidding and underestimation
Lee (1994)	Empirical: in Taiwan	Disputes from contract problems: (1) unfair contract clauses, (2) vague definition of contract documents (in terms of performance period, payment, quality and variations), (3) not comprehensive stipulation
Rhys Jones (1994)	Empirical: survey of construction industry and lawyers in UK	Ten factors in the development of disputes: (1) poor management, (2) adversarial culture, (3) poor communications, (4) inadequate design, (5) economic environment, (6) unrealistic tendering, (7) influence of lawyers, (8) unrealistic client expectations, (9) inadequate contract drafting, (10) poor workmanship
Semple *et al.* (1994)	Empirical: examined 24 projects in Western Canada	The most common contributing factors in claims are: (1) increase in scope of the work, (2) weather, (3) restricted access, (4) acceleration
Watts & Scrivener (1994)	Empirical: comparative study on judgments between the courts of Australia and the UK	290 sources of disputes are identified from 60 cases in each country and the 21 categories are grouped into five sub-groups. The most frequent source of disputes in the UK is negligence, while in Australia failure and determination have the highest occurrence.
Assaf *et al.* (1995)	Empirical: large public building projects in the eastern province of Saudi Arabia	56 factors, which were grouped into nine major areas: (1) materials, (2) manpower, (3) equipment, (4) financing, (5) environment, (6) charges, (7) government relations, (8) contractual relationships, (9) scheduling and controlling techniques

Bristow & Vasilopoulous (1995)	Empirical: in Ontario, Canada	Five primary causes of claims: (1) unrealistic expectations by the parties, (2) ambiguous contract documents, (3) poor communications between project participants, (4) lack of team spirit among participants, (5) a failure of participants to deal promptly with changes and unexpected conditions
Murdoch & Hughes (1996)	Theoretical	Background to disputes: (1) motivation factors of individuals, (2) pre-conceptions about roles, (3) project success or failure, (4) the roots of contractual disputes, (5) business relations
Sykes (1996)	Theoretical	Two major sources: (1) misunderstanding due to lack of clarity, (2) unpredictability of unforeseen circumstance
Kumaraswamy (1997a)	Empirical: in Hong Kong	Heads of claim categories: (1) cost, (2) time extension
Kumaraswamy (1997b)	Empirical: in Hong Kong	Two categories: (1) root causes, (2) proximate causes
Hu (1998)	Empirical: in Taiwan	Factors of construction disputes: (1) defective performance in quality, (2) unfair contract clauses, (3) conservative attitude of public employer and supervising engineers, (4) negligence of design and supervision
Vidogah & Ndekugri (1998)	Empirical: in UK	Eight heads of claims likely to be disputed (in rank order): (1) cost of disruption, (2) head office overheads, (3) interest and finance charges, (4) cost of preparing claims, (5) loss of profit, (6) inflation of costs, (7) on-site overheads, (8) others
Al-Momani (1999)	Empirical: 130 public projects in Jordan	Seven categories: (1) poor design, (2) change orders, (3) weather, (4) site condition, (5) late delivery, (6) economic condition, (7) increase in quantity
Fenn (1999)	Empirical: litigation cases, questionnaires survey and statistics analysis in UK	Three independent variables predicted construction disputes well: (1) the level of variations, (2) the success of the employer's advisors on a past project, (3) the tender period
Gould *et al.* (1999)	Empirical: evaluation of UK practice by the Partners in Technology Research project of the Department of Environment, Transport and the Regions	Nine areas of issues in dispute: (1) project delays, (2) change in the scope of the work, (3) payment issues, (4) differing site conditions, (5) design issues, (6) defective work or product, (7) site administration problems, (8) property damage, (9) personal injury

Table 11.3 (Continued).

Researchers	Context	Findings (sources of disputes)
Liang (1999)	Empirical: in China	Four types of most frequently seen claims: (1) scope of work claim, (2) delay claim, (3) change of site condition claim/adverse physical conditions or obstructions, (4) acceleration claim
Sykes (1999)	Theoretical	Origins of disputes: (1) omission and unforeseen events, (2) lack of capacity to settle claims, (3) different expectations
Molenar *et al.* (2000)	Empirical: in USA	Three factors found to have influence on or be closely related to dispute potential: (1) people issue, (2) project complexity, (3) owner management ability
Mitropoulous & Howell (2001)	Empirical: in USA	Basic factors driving the development of disputes are: (1) project uncertainty, (2) contractual problems, (3) opportunistic behaviour
Kululanga *et al.* (2001)	Theoretical	Four basic sources: (1) contract documents due to errors, defects and omissions, (2) failure to appreciate the real cost of a project at the beginning, (3) changed conditions, (4) stakeholders involved in a project
Ren *et al.* (2001)	Theoretical	Three factors: (1) social, (2) industrial, (3) project
Tsai (2001)	Theoretical: from literature	Five areas of disputes: (1) performance period, (2) payment, (3) quality, (4) quantity, (5) contract
Wang (2001)	Empirical: in Taiwan	Grounds for claims: (1) unfairness of contract/unfairness of risk allocation, (2) variations, (3) defective contract documents, (4) delay claim, (5) circumstance changes, (6) breach of obligations of employers, (7) termination of contract
Yao (2001)	Empirical: in Taiwan	Disputes can be categorised into two stages: (1) before contract awarded, (2) after contract awarded
Kehinde & Aiyetan (2002)	Empirical: investigate 52 projects and conduct questionnaire survey in Nigeria	Highest source of contractual claims in most building contracts is constituted by (1) variations, (2) additional works

Lo (2002)	Theoretical: from literature	Causes of construction conflict: (1) differences in goals and objectives of parties in the project, (2) differences in contract interpretation between the construction manager and contractor that have to be compromised by serious negotiation which may take a long period of time, (3) lack of understanding about the needs of others also involved in the planning, design and construction process, (4) uncertainty about role, responsibility, authority and procedure ambiguity, (5) unclear reward structure or opportunity for the project participants, (6) specific allocation of limited resources such as materials, capital, labour, etc., (7) excessive demands on resources normally depended on to assist to the resolution of conflict, (8) incorrect assumptions made from biased perceptions, (9) demands for higher quality than specified, (10) failure to provide products in conformance with user requirements, (11) insufficient time to make required decisions, (12) inability to do the job, (13) subcontractor performance problems, (14) frustration over a lack of control of events affecting performance, (15) desire to take advantages of those in a weaker position, (16) averse relationship between client and contractors, (17) work slowdowns and strikes, (18) interpersonal conflicts, (19) regulatory problems, (20) lack of communication
Wu (2002)	Empirical: in Taiwan	Factors of disputes: (1) unreasonable contract clauses, (2) defectiveness of contract
Yan (2002)	Empirical: interview survey in Taiwan	Construction disputes in Taiwan fall into four categories: (1) problems occur in planning and designing stage, (2) problems occur in contract performance stage, (3) problems occur in completion and acceptance stage, (4) problems occur in maintenance stage; four sources of disputes: (1) contractual factors, (2) technological factors, (3) external factors, (4) artificial factors
Chang & Ive (2003)	Theoretical	Natures of two types of dispute of construction disputes are: (1) pure cognitive dissonance, (2) opportunism intention to take advantage of one party's vulnerability
Public Construction Commission (2003)	Empirical: from 865 mediation cases of PCC in Taiwan	Dispute types in mediation cases fall into four categories: (1) contract awarding stage, (2) performing of construction contract stage, (3) acceptance stage, (4) maintenance stage. Performance period issue was greatest, with 226 cases (26.13%) among all issues

the 'causes' of disputes, attention could be better directed at the identification of the 'causes' of successful projects. There is a rich literature in this area too. A further literature review found 25 studies regarding project success criteria and/or project success factors. The methodologies of the studies can generally be classified into three sources: theoretical, empirical (from experience) and empirical (from surveys), see Table 11.4. The findings of these studies are listed chronologically in Table 11.5.

Table 11.4 List of studies on project success.

Methodologies	Studies
Theoretical	Sayles & Chandler (1971), Cleland & King (1983), de Wit (1988), Parfitt & Sanvido (1993), Munns & Bjeirimi (1996), Straight (1999), Chan (2001)
Empirical: from experience	Martin (1976), Hayfield (1979), Baker *et al.* (1983), Locke (1984), Kerzner (1987), Belassi & Tukel (1996), Westerveld (2003)
Empirical: from surveys	Morris & Hugh (1986), Ashley *et al.* (1987), Pinto & Slevin (1988), Russell & Jaselskis (1992), Sanvido *et al.* (1992), Songer & Molenaar (1997), Molenaar & Songer (1998), Chua *et al.* (1999), Lim & Mohamed (1999), *Cooke-Davies* (2002) Chan *et al.* (2004)

Table 11.5 Chronological listing of literature on project success.

Researchers	Context	Findings (success criteria/success factors)
Sayles & Chandler (1971)	Theoretical	Critical success factors: (1) Project manager's competence (2) Scheduling (3) Control systems and responsibilities (4) Monitoring and feedback (5) Continuing involvement in the project
Martin (1976)	Empirical: from experience	Critical success factors: (1) Define goals (2) Select project organisational philosophy (3) General management support (4) Organise and delegate authority (5) Select project team (6) Allocate sufficient resources (7) Provide for control and information mechanisms (8) Require planning and review
Hayfield (1979)	Empirical: from experience	*Definition of a successful project*: the one that is completed on time, within budget and to performance specification

		(1) Macro factors: (a) Realistic and thorough definition of project – what? (b) Efficient manner of project execution – how? (c) Comprehension of project 'environment' – context (d) Selection of organisation realizing project – by whom? (2) Micro factors: (a) Formulation of sound project policies – policies (b) Clear and simple project organisation – framework (c) Selection of key personnel – human resources (d) Efficient and dynamic management controls – controls (e) Reliable management information systems – information
Baker *et al.* (1983)	Empirical: from experience	Critical success factors: (1) Clear goals (2) Goal commitment of project team (3) On-site project manager (4) Adequate funding to completion (5) Adequate project team capability (6) Accurate initial cost estimates (7) Minimum start-up difficulties (8) Planning and control techniques (9) Task (vs social orientation) (10) Absence of bureaucracy
Cleland & King (1983)	Theoretical	Critical success factors: (1) Project summary (2) Operational concept (3) Top management support (4) Financial support (5) Logistic requirement (6) Facility support (7) Market intelligence (who is the employer?) (8) Project schedule (9) Executive development and training (10) Manpower and organisation (11) Acquisition (12) Information and communication channels (13) Project review

Table 11.5 (Continued).

Researchers	Context	Findings (success criteria/success factors)
Locke (1984)	Empirical: from experience	Critical success factors: (1) Make project commitments known (2) Project authority from the top (3) Appoint competent project manager (4) Set up communications and procedures (5) Set up control mechanisms (schedules, etc.) (6) Progress meetings
Morris & Hugh (1986)	Empirical: from surveys	Factors grouped under ten headings: (1) Project definition (2) Planning and design (3) Politics (4) Schedule duration (5) Schedule urgency (6) Finance (7) Legal agreements (8) Contracting (9) Project management (10) Human factors
Ashley *et al.* (1987)	Empirical: from surveys	*Definition*: Results much better than expected or normally observed in terms of cost, schedule, quality, safety, and participant satisfaction. Six criteria most frequently used to measure construction project success were: (1) Schedule performance (2) Budget performance (3) Employer satisfaction (4) Contractor satisfaction (5) Functionality (6) Project manager/team satisfaction Results of the study indicated that statistically significant differences existed between average and outstanding construction projects in such important areas as planning effort (construction and design), project team motivation, project manager goal commitment, project manager technical capabilities, control systems, and scope and work definition
Kerzner (1987)	Empirical: from experience	*Definition of project success*: (1) Within time (2) Within cost or budget (3) At the desired performance or quality level (4) Within the original scope or mutually agreed upon scope change

		(5) Without disturbing the corporate culture or corporate values (6) With well-documented, post-audit analysis Six critical success factors for successful projects: (1) Corporate understanding of project management (a) Executive level (b) Middle level (c) Employee level (2) Executive commitment to project management (a) Project sponsor (b) Life-cycle management (3) Organisational adaptability (a) Simple, lean (b) Informal, formal (4) Project manager selection criteria (a) Results-orientated (b) Interpersonal skills (c) Organisational understanding (d) Corporate values (5) Project manager's leadership style (a) Authority (b) Responsibility (c) Adaptability (d) Interface management (6) Commitment to planning and control (a) Planning (b) Controlling (c) Review process
de Wit (1988)	Theoretical	*Definition*: The project is considered an overall success if the project meets the technical performance specifications and/or mission to be performed, and if there is a high level of satisfaction concerning the project outcome among: key people in the parent organisation, key people in the project team, and key users or clientele of the project effort.
Pinto & Slevin (1988)	Empirical: from surveys	Critical success factors to project success: (1) Project mission (2) Top management support (3) Project schedule and plans (4) Employer consultation (5) Personnel (6) Technical expertise

Table 11.5 (Continued).

Researchers	Context	Findings (success criteria/success factors)
		(7) Employer acceptance (8) Monitoring and feedback (9) Communication (10) Troubleshooting An additional four external factors to project success: (1) Characteristics of the project team leader (2) Power and politics within the organisation (3) Environmental events (4) The urgency of the project
Russell & Jaselskis (1992)	Empirical: from surveys	(1) The model predicts the probability of contractor failure at the project level (2) *Failure is defined* as a significant breach of contract (3) A predictive contractor failure model: four variables: (a) The amount of owner–contractor evaluation (b) Whether cost monitoring was performed by the owner (c) The level of support received by the project manager from the contractor's senior management throughout the course of the project (d) The early involvement of the contractor's project manager
Sanvido *et al.* (1992)	Empirical: from surveys	Four critical success factors: (1) A well-organised, cohesive facility team to manage, plan, design, construct and operate the facility. Team chemistry was typically developed by common goals and activities (2) A series of contracts that allows and encourages the various specialists to behave as a team without conflicts of interest and differing goals. These contracts must allocate risk and reward in the correct proportions (3) Experience in the management, planning, design, construction and operations of similar facilities (4) Timely, valuable optimisation information from the owner, user, designer, contractor and operator in the planning and design phases of the facility

		If project participants can predict probability of success better, they can take steps to: (1) Avoid unsuccessful projects (2) Identify good projects worth pursuing (3) Identify problems on current projects and take correct action
Parfitt & Sanvido (1993)	Theoretical	Major factors: (1) The facility team (2) Contracts, obligations and changes (3) Facility experience (4) Optimisation information
Belassi & Tukel (1996)	Empirical: from experience	Ranking of critical success factors for construction industry: (1) Technique background (2) Employer (3) Co-ordinate (4) Communication (5) Technology (6) Economic
Munns & Bjeirimi (1996)	Theoretical	Critical success factors for projects: (1) Human parties (2) Relations with employer (3) Politics (4) Legal agreements (5) Contracting (6) Project administration (7) Efficiency (8) Profit (9) Objectives
Songer & Molenaar (1997)	Empirical: from surveys	(1) Five primary project characteristics: (a) Well-defined scope (b) Shared understanding of scope (c) Owner construction sophistication (d) Adequate owner staffing (e) Established budget (2) Public-sector criteria of success: (a) Staying on budget (b) Conforming to user's expectation (c) Staying on schedule
Molenaar & Songer (1998)	Empirical: from surveys	(1) Performance criteria: (a) Budget variance (b) Schedule variance (c) Conformance to expectations (d) Administrative burden (e) Overall user satisfaction

Table 11.5 (Continued).

Researchers	Context	Findings (success criteria/success factors)
		(2) Project characteristics in which are significant correlations with success include: (a) Scope definition (b) Schedule definition (c) Budget definition (d) Project complexity (e) Agency experience (f) Agency staffing (g) Owner design input (h) Design-build market (i) Design-builder pre-qualification (j) Method of selection
Chua *et al.* (1999)	Empirical: from surveys	Project success is determined by: (1) Inherent characteristics of the project can be thoroughly understood (2) Appropriate contractual arrangements are adopted (3) Competent management team is assigned (4) Sound monitoring and control system is established
Lim & Mohamed (1999)	Empirical: from surveys	Criteria for macro viewpoint of project success: completion and satisfaction Criteria for micro viewpoint of project success: completion
Straight (1999)	Theoretical	Contractor's past performance is the best predictor of future performance. It can be observed from the following aspects: (1) Meeting specification and standards (2) Maintaining cost control (3) Maintaining quality (4) Meeting schedule or timeliness goals (5) Exhibiting cooperative behaviour (6) Emphasising customer satisfaction (7) Maintaining a satisfactory business relationship (8) Providing service to the end-user
Chan (2001)	Theoretical	Six project success factors: (1) Project team commitment (2) Contractor's competencies (3) Risk and liability assessment (4) Employer's competencies

		(5) End-users' needs (6) Constraints imposed by end-users
Cooke-Davies (2002)	Empirical: from survey	Twelve factors: Project management success On-time performance: (1) Adequacy of company-wide education on the concepts of risk management (2) Maturity of an organisation's processes for assigning ownership of risks (3) Adequacy with which a visible risk register is maintained (4) Adequacy of an up-to-date risk management plan (5) Adequacy of documentation of organisational responsibilities on the project (6) Keep project (or project stage duration) as far below three years as possible (one year is better) On-cost performance: (7) Allow changes to scope only through a mature scope change control process (8) Maintain the integrity of the performance measurement baseline Project success: (9) The existence of an effective benefits delivery and management process that involves the mutual co-operation of project management and line management functions Critical to consistent corporate success: (10) Portfolio- and programme management practices that allow the enterprise to resource fully a suite of projects that are thoughtfully and dynamically matched to the corporate strategy and business objectives (11) A suite of project, programme and portfolio metrics that provides direct 'line of sight' feedback on current project performance, and anticipated future success (12) An effective means of 'learning from experience' on projects, that combines explicit knowledge with tacit knowledge in a way that encourages people to learn and to embed that learning into continuous improvement of project management processes and practices
Westerveld (2003)	Empirical: from survey	Project success criteria: (1) Project results: time, costs, quality/scope (2) Appreciation employer (3) Appreciation project personnel

Table 11.5 (Continued).

Researchers	Context	Findings (success criteria/success factors)
		(4) Appreciation users (5) Appreciation contracting partners (6) Appreciation stakeholders Critical success factors for projects: (1) Leadership and team (2) Policy and strategy (3) Stakeholder management (4) Resources (5) Contracting (6) Project management (a) Scheduling (b) Budget (c) Organisation (d) Quality (e) Information (f) Risks (7) Success criteria (8) External factors
Chan *et al.* (2004)	Empirical: from survey	Critical success factors (CSFs) are identified: (1) Project-related factors (2) Procurement-related factors (3) Project management actions (4) Project participants-related factors (5) External environment

Conclusion

Disputes and conflicts on projects or contracts are unpleasant: they divert valuable resources from the overall aim, which must be completion on time, on budget and to the quality specified. In addition, they generally cost money and take time to resolve, and they can destroy relationships which may have taken years to develop. There is a school of Western thought which maintains that conflict (but not dispute) is inevitable. Conflict is part of Western societies and idioms; to use the academic jargon, there is a Western dialectic argument idiom. Conflict is part of dynamic capitalism and an integral part of commercialism; conflict might be seen as the functional and necessary part. Dispute on the other hand only develops when conflict is not (or cannot be) managed; dispute is the unnecessary or dysfunctional element and logically there should be two areas for consideration:

- *Conflict management*: Here the emphasis is on the axiom that it must be in the interests of all parties to avoid disputes by managing conflict in such a way that disputes do not arise; this is sometimes described as dispute avoidance

- *Dispute resolution*: Notwithstanding the emphasis on the desire to avoid dispute, there must be occasions where the parties have legitimate disputes and that the techniques of dispute resolution are employed to bring about the conclusion or resolution of the dispute.

Commercial management needs to recognise both conflict management and dispute resolution.

Research agenda: explanation or prediction of commercial disputes

The axiom that dispute avoidance is to be preferred is evident, if not explicitly stated, in most of the literature. One clear example is in Latham (1994):

The best solution is to avoid disputes.

If we seek to avoid disputes it is axiomatic that we seek to predict, because by prediction we can take the necessary action to avoid. Prediction is at the very heart of the scientific method of research and we might look to other disciplines for guidance on predictive research. Medicine is an example of a mature scientific discipline where predictive techniques have developed a concept of preventative medicine. Medical science has a well-established branch of aetiology: the study of the causes, for example, of a disease. The word comes from the Greek 'aitia', a cause, plus 'logos', a discourse.

A literature search found only one use each of the word aetiology in construction and legal literature. Indeed, there was little evidence of any aetiological studies outside medical literature. The construction and legal references are Knocke (1995) and Ormrod (1990). Knocke (1995) uses the term aetiology in the discussion of construction defects and failures. Ormrod (1990) describes an aetiological approach to the law as the question, 'What has caused the law on a given topic to be what it is?' He states that lawyers rarely concern themselves with this kind of question and that it receives little research interest. Ormrod's interest in this line of legal logic may be influenced by the fact that before becoming a member of the English Court of Appeal he practised medicine.

An aetiological approach might throw new light on construction disputes. Continuing the analogy with medicine, construction disputes represent the dysfunctionality of conflict and, therefore, the disease on the body of construction. Aetiology has thrown new light on many problems of medical disease, providing important clues to the understanding of the nature of the disorder and promoting advances in diagnosis, treatment and prevention.

Diagnosis is the act of identifying a disease from its symptoms or signs; in construction disputes, the symptoms of dispute currently are not considered: only when a dispute has manifested itself are the parties concerned with the dispute and then only with the resolution of that dispute. Treatment of the dispute, following most dictionary definitions and not the medical concept, is the application of the techniques or actions in specified situations. Treatment of disputes is, therefore, the application of the techniques of conflict management and dispute resolution. Prevention is the act that is used to avoid disease and, in the argument presented here, prevention is made possible by the prediction of the occurrence of disputes.

References

Alkass, S., Mazerolle, M. & Harris, F. (1998) Rigour in research and peer review, a reply. *Construction Management and Economics*, **16**, 401–423.

Al-Momani, A.H. (1999) Construction delay: a quantitative analysis. *International Journal of Project Management*, **18**, 51–59.

Ashley, D., Lurie, C. & Jaselskis, E. (1987) Determination of construction project success. *Project Management Journal*, **18**(2), 69–79.

Assaf, S.S., Al-Khalil, M. & Al-Hazmi, M. (1995) Causes of delay in large building construction projects. *Journal of Management in Engineering*, **11**(2), 45–50.

Baker, B., Murphy, D. & Fisher, D. (1983) Factors affecting project success. In: *Project Management Handbook* (eds D.I. Cleland & W. King), Van Nostrand Reinhold, New York; pp. 669–685.

Belassi, W. & Tukel, O.I. (1996) A new framework for determining critical success/failure factors in projects. *International Journal of Project Management*, **14**(3), 141–151.

Bristow, D. & Vasilopoulos, R. (1995) The new CCDC2: facilitating dispute resolution of construction projects. *Construction Law Journal*, **11**, 245–258.

Brown, H. & Marriot, A. (1994) *ADR Principles and Practice*. Sweet and Maxwell, London.

Building and Civil Engineering Economic Development Councils (1975) *Public Client and the Construction Industries (The Wood Report)*. HMSO, London.

Building Economic Development Committee (1983) *Faster Building for Industry*. NEDO, London.

Building Economic Development Committee (1988) *Faster Building for Commerce*. NEDO, London.

Building Employers Confederation (1992) *Building Towards 2001*. BEC, London.

Burton, J.W. (1993) Conflict resolution as a political philosophy. In: *Conflict Resolution Theory and Practice: Integration and Application* (eds H. van der Merwe & D.J.D. Sandole), Manchester University Press, Manchester and New York; pp. 55–64.

Chan, A. (2001) Design and build project success factors: multivariate analysis. *Journal of Construction Engineering and Management*, **127**(2), 93–99.

Chan, A.P.C., Scott, D. & Chan, A.P.L. (2004) Factors affecting the success of a construction project. *Journal of Construction Engineering and Management*, **130**(1), 153–155.

Chang, C.Y. & Ive, G. (2003) Discussion of 'Model for understanding, preventing, and resolving project disputes' by Panagiotis Mitropoulos and Gregory Howell. *Journal of Construction Engineering and Management*, **129**, 223–231.

Chua, D.K.H., Kog, Y.C. & Loh, P.K. (1999) Critical success factors for different project objectives. *Journal of Construction Engineering and Management*, **125**(3), 142–150.

Civil Procedure Rules. Department for Constitutional Affairs: Justice, Rights and Democracy. www.dca.gov.uk/civil/procrules_fin/menus/rules.htm (accessed 1 August 2005).

Cleland, D.I. & King, W.R. (1983) *Project Management Handbook*. Van Nostrand Reinhold, New York.

Committee on Industrial Productivity (1949) *First Report of the Committee on Industrial Productivity*. HMSO, London.

Construction Industry Council (1994) *Dispute Resolution – A report by the Dispute Resolution Task Force of the Construction Industry Council which identifies the disputes that arise in the construction industry and existing methods of their resolution*. Construction Industry Council, London.

Cooke-Davies, T. (2002) The 'real' success factors on projects. *International Journal of Project Management*, **20**(3), 185–190.
de Bono, E. (1985) *Conflicts*. Penguin, London.
de Wit, A. (1988) Measurement of project success. *International Journal of Project Management*, **6**(3), 164–170.
Department of the Environment (1996) *Making the Scheme for Construction Contracts: a consultation paper*. Department of the Environment, London, p. 47.
Dickens, C. (1852–1853) *Bleak House* [2003 edition (ed. N. Bradbury), Penguin, London].
Diekman, J. & Nelson, M. (1985) Construction claims: frequency and severity. *Journal of Construction Engineering and Management*, **111**(1), 74–81.
Diekmann, J., Girard, M. & Abdul-Hadi, N. (1994) *DPI – Disputes Potential Index: A Study into the Predictability of Contract Disputes*. Construction Industry Institute, University of Texas at Austin, Austin, TX.
Economic Development Committee for Building (1967) *Action on the Banwell Report (A survey of the implementation of the recommendations of the committee under the chairmanship of Sir Harold Banwell on the placing and management of contracts)*. HMSO, London.
Egan, J. (1988) *Rethinking Construction*. HMSO, London.
Fenn, P. (1997) Rigour in research and peer review. *Construction Management and Economics*, **15**(4), 383–385.
Fenn, P. (1999) *An aetiology of construction disputes*. PhD thesis (unpublished), University of Manchester Institute of Science and Technology, Manchester.
Follett, M.P. (1925) Constructive conflict. In: *Scientific Foundations of Business Administration* (ed. H.C. Metcalf). Williams and Wilkins, Baltimore, MD.
Gould, N., Capper, P., Dixon, G. & Cohen, M. (1999) *Dispute Resolution in the Construction Industry*. Thomas Telford, London.
Hayfield, F. (1979) Basic factors for a successful project. In: *Internet Congress Garmisch-Parten-Kirchen, FRG*.
Higgin, G. & Jessop, N. (1965) *Communications in the Building Industry*. Tavistock, London.
HMSO (1996) *Housing grants and Construction Regeneration Act*. HMSO, London.
Hu, W.L. (1998) *Dealing with Public Construction Contract Disputes*. Construction World, Taiwan (in Chinese).
Jergeas, G. & Hartman, F. (1994) Contractors' construction-claims avoidance. *Journal of Construction Engineering and Management*, **120**(3) 553–560.
Kehinde, J. & Aiyetan, O. (2002) A study of the nature of contractual claims in building contracts in Nigeria. *Journal of Financial Management of Property and Construction*, **7**(3), 197–202.
Kerzner, H. (1987) In search of excellence in project management. *Journal of Systems Management*, **38**(2), 30–39.
Knocke, J. (1995) *Post Construction Insurance and Liability*. E. & F.N. Spon, London.
Kululanga, G., Kuotcha, W., McKaffer, R. & Edum-Fotwe, F. (2001) Construction contractors' claims process framework. *Journal of Construction Engineering and Management*, **127**(4), 309–314.
Kumaraswamy, M. (1997a) Conflicts, claims and disputes in construction. Engineering Construction and Architectural Management, **4**(2), 95–111.
Kumaraswamy, M. (1997b) Common categories and causes of construction claims. *Construction Law Journal*, **13**(1), 21–34.
Latham, M. (1993) *Trust and Money*. HMSO, London.
Latham, M. (1994) *Constructing the Team*. HMSO, London.
Lee, J.C. (1994) Construction disputes and arbitration practice. *Commercial Arbitration*, **39**, 24–29 (in Chinese).

Lewis, J., Cheetham, D.W. & Carter, D.J. (1992) Avoiding conflict by risk management – the role of the client's project management. In: *Construction Conflict: Management and Resolution* (eds P. Fenn & R. Gameson), Chapman Hall, London; pp. 72–95.

Liang, J. (ed) (1999) Construction claims of international projects. In: *Management of International Projects*. Shu-Shin Publishing, Bejing and Taipei (in Chinese).

Lim, C.S. & Mohamed, M.Z. (1999) Criteria of project success: an exploratory re-examination. *International Journal of Project Management*, **17**(4), 243–248.

Lipsky, D.B. & Seeber, R.L. (1998) *The appropriate resolution of corporate disputes: a report on the growing use of ADR by US corporations*. Institute on Conflict Resolution, Cornell, Ithaca, NY.

Lo, C.H. (2002) *Comparing Western and Eastern conflict management and dispute resolution*. MSc thesis, University of Manchester Institute of Science and Technology, Manchester.

Locke, D. (1984) *Project Management*. St. Martin Press, New York.

Lord Chancellor (2001) *The Pledge: Settlement Of Government Disputes Through Alternative Dispute Resolution*. Department for Constitutional Affairs: Justice, Rights and Democracy. www.dca.gov.uk/civil/adr/adrmon03.htm#part1 (accessed 1 August 2005).

Lord Woolf (1996) *Access to Justice* (Final Report). Department for Constitutional Affairs: Justice, Rights and Democracy. www.dca.gov.uk/civil/final/contents.htm (accessed 1 August 2005).

Martin, C.C. (1976) *Project Management: How to Make It Work*. Amacom Books, New York.

Mathews, J.J. & Ashley, D.B. (1985) *Contract Clause Study Data*. University of Texas at Austin, Department of Civil Engineering, Construction Engineering and Project Management, Report No. UTCEPM-85-1. Austin, TX.

Ministry of Public Building and Works (1964) *The Placing and Management of Contracts for Building and Civil Engineering Work (The Banwell Report)*. HMSO, London.

Ministry of Works (1944) *Report of the Committee on the Placing and Management of Building Contracts (The Simon Report)*. HMSO, London.

Ministry of Works (1950) *Report of the Working Party on the Building Industry (The Phillips Report)*. HMSO, London.

Ministry of Works (1962) *Survey of Problems Before the Construction Industries (The Emerson Report)*. HMSO, London.

Mitropolous, P. & Howell, G. (2001) Model for understanding, preventing and resolving project disputes. *Journal of Construction Engineering and Management*, **127**(3), 223–231.

Molenaar, K. & Songer, A. (1998) Model for public sector design–build project selection. *Journal of Construction Engineering and Management*, **124**(6), 467–479.

Molenaar, K., Washington, S. & Diekmann, J. (2000) Structural equation model of construction contract dispute potential. *Journal of Construction Engineering and Management*, **126**(4), 268–277.

Morris, P. & Hugh, G. (1986) *The Pre-conditions of Success and Failure in Major Projects*. Major Projects Association, Templeton College, Oxford.

Munns, A. & Bjeirimi, B. (1996) The role of project management in achieving project success. *International Journal of Project Management*, **14**(2), 81–87.

Murdoch, J. & Hughes, W. (1996) *Construction Contracts – Law and Management* (2nd edn). E. & F.N. Spon, London.

National Audit Office (2001) *Modernising Construction*. The Stationery Office, London.

Ormrod, R. (1990) An aetiological approach to the law. *Arbitration*, **56**, 227–241.

Parfitt, M.K. & Sanvido, V.E. (1993) Checklist of critical success factors for building projects. *Journal of Management in Engineering*, **9**(3), 243–249.

Pinto, J. & Slevin, D. (1988) Critical success factors across the project life cycle. *Project Management Journal*, **19**(3), 67–74.

Public Construction Commission (2003) *Disputes factors and types analyses for mediation procedure of government procurement*. Research report submitted to the Public Construction Commission of the Executive Yuan of Taiwan, Taipei, Taiwan (in Chinese).

Ren, Z., Anumba, J. & Ugwu, O.O. (2001) Construction claims management: towards an agent-based approach. *Engineering, Construction and Architectural Management*, **8**(3), 185–197.
Revay, S. (1992) Can construction claims be avoided? *Building Research and Information*, **12**, 56–58.
Rhys Jones, S. (1994) How constructive is construction law? *Construction Law Journal*, **10**(1), 28–38.
Rimmington, S. (2002) *Open Secret*. Arrow Press, London.
Russel, J. & Jeselskis, E. (1992) Predicting construction contractor failure prior to contract award. *Journal of Construction Engineering and Management*, **118**(4), 791–811.
Sander, F. (1976) *The Muti-door Court House*. 70 F.R.D. 111, Harvard.
Sanvido, V., Grobler, F., Parfitt, K. & Guvenis, M. (1992) Critical success factors for construction projects. *Journal of Construction Engineering and Management*, **118**(1), 94–111.
Sayles, L. & Chandler, M.K. (1971) *Managing Large Systems*. Free Press, New York.
Semple, C., Hartman, F.T. & Jergeas, G. (1994) Construction claims and disputes: causes and cost/time overruns. *Journal of Construction Engineering and Management*, **120**(4), 785–795.
Songer, A. & Molenaar, K. (1997) Project characteristics for successful public-sector design–build. *Journal of Construction Engineering and Management*, **123**(1), 34–40.
Stephenson, D.A. (1998) *Arbitration Practice in Construction Contracts* (4th edn). Blackwell Science, Oxford.
Straight, R. (1999) Measuring contractors' performance. *Journal of Supply Chain Management*, **35**(2), 18–28.
Sykes, J. (1996) Claims and disputes in construction: suggestion for their timely resolution. *Construction Law Journal*, **12**(1), 3–13.
Sykes, J. (1999) *Construction Claims*. Sweet and Maxwell, London.
Tsai, Y.S. (2001) *An auxiliary information system coping with contract executive disputes for public contract: dispute modulation as an example*. MSc thesis, National Taiwan University, Taipei, Taiwan (in Chinese).
Vidogah, W. & Ndekugri, I. (1998) Improving the management of claims on construction contracts: consultant's perspective. *Construction Management and Economics*, **16**(3), 363–372.
Wang, P.C. (2001) Guidance for employers in coping with construction claims. *Formosan Brothers Law Journal*, **9**, 2–14 (in Chinese).
Watts, V. & Scrivener, J. (1992) Review of Australian building disputes settled by litigation. In: *Construction Conflict: Management and Resolution* (eds P. Fenn & R. Gameson), Chapman Hall, London; pp. 209–219.
Watts, V. & Scrivener, J. (1994) Building Disputes Settled by Litigation – Comparison of Australian and United Kingdom Statistics. Australian Institute of Building Papers, Canberra.
Westerveld, E. (2003) The project excellent model: linking success criteria and critical success factors. *International Journal of Project Management*, **21**, 411–418.
Wu, S.K. (2002) *Risks and Disputes of Construction Contract* (in Chinese).
Yan, A.Y.M. (2002) Dispute management in construction: work towards dispute prediction and avoidance. In: *Proceedings of the Chinese Decision Science Conference*, Taipei, Taiwan.
Yao, N.J. (2001) The often seen public construction disputes and their prevention. *Taipei Bar Journal*, **265**, 2–7.

12 Performance Measurement

Malcolm Horner

Introduction

If performance isn't measured, how do we know whether we're meeting our objectives? Performance measurement is central to control: 'what ain't measured ain't managed'. These obvious and simple statements lead to surprisingly complex issues about what should be measured, how it should be measured, how often it should be measured, at what level of detail, by whom, for whom, and for what purpose. For example, is the purpose to impress shareholders, for statutory accounts, to effect control or to satisfy a political agenda with or without a small 'p'? Is it for government, for the client, for the board of directors, or for the project manager? Should performance be measured only at the end of a project, annually, monthly, weekly or even daily? Do we need to measure everything, or can we use surrogates or indicators? Like any other project management system, performance measurement must be 'fit for purpose'.

There is clearly a tension between the need for performance measurement and the difficulty in doing it. By and large, difficulty wins: performance is under-measured in most simple, never mind complex projects (Tucker 1986). However, contemporary forces are driving organisations to investigate more closely their attitudes towards performance measurement. Global competition, the focus on share price and shareholder value, governments' preoccupation with productivity, new procurement methods such as the private finance initiative (PFI) and prime contracting, the spread of lean thinking from its birth in Toyota are all creating new pressures for improved performance measurement.

This chapter explores what is meant by performance measurement, why it is important, and why it is so often ignored. It reviews past and current practice, identifies the shortcomings, and sets out the research agenda. Although it focuses on the project, it is set in the wider context of client and user needs, not least because one of the research challenges is to develop a hierarchical approach which allows performance measures at the project level to be 'rolled up' into national statistics. The challenges are similar for all process-based industries, of which one of the most common is construction. For this reason, construction is used as a typical example to illustrate the issues and myths surrounding performance measurement.

Performance measurement in context

Horses for courses

Before a performance measurement system is designed, its purpose must be defined. At government level, industrial performance must be measured to inform economic, environmental and social policy, and to provide a means of benchmarking against international competition. At the organisation level, companies must convince shareholders to invest in their activities, they must measure the level of satisfaction of their clients, they must demonstrate the profitability and sustainability of their operations, they must benchmark their performance against that of their competitors, and increasingly, they must measure the performance (and satisfaction) of their key resource, their employees. At the project level, management will need to focus on two sets of measures, one outward-facing, designed to ensure that the customer's goals are achieved and that other people affected by the project at least suffer no adverse effects; the other inward-facing, designed to ensure that the organisation's goals are met. Clearly, some, if not all of these measures are interdependent, but as yet, no system has been designed which sensibly links them all together.

Effecting control

In truth, the fundamental purpose of performance measurement is to effect control. Control means planning what should be achieved, monitoring what is achieved, and crucially, taking action to minimise any difference or variance between the two. Planning commonly consists of two elements: a model, and the data that are applied to it. For example, the duration of a project can be predicted by first drawing a network showing all the activities necessary for completion and their logical interdependencies (the model), and applying to this the estimated duration of each activity. A typical control loop, using project costs as an example is shown in Fig. 12.1.

Overview of performance measures

National scale

At a national scale, economic performance is most usually measured in terms of gross domestic product (GDP), the total value of the goods and services produced by a country. It is often used as a measure of a nation's wealth, particularly when it is expressed per head of population. GDP divided by the population of a country approximates to the average wage of a nation, and allows rapid comparisons of relative wealth and poverty. For example, the GDP/head in 2003 in Japan was $US27 989, whilst in India it was $US2909. An alternative is total factor productivity (TFP). This is defined as output divided by input. In order to measure the outputs of different industries and processes in a common unit, the value of goods produced or services rendered is usually taken as the yardstick. To calculate TFP, it is necessary to measure all the inputs, which may include the costs of labour, materials, plant and

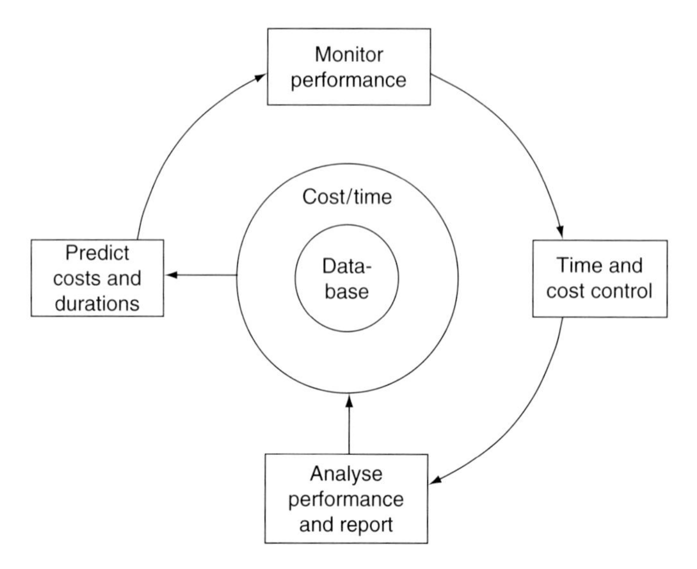

Figure 12.1 Control loop.

equipment, energy, and investment. Thus TFP is the value of industrial output divided by the costs of all the inputs. Partial factor productivity (PFP) is the ratio of total output to any one or more of the inputs. This is not such a useful measure but can be used for international comparisons or to compare the labour efficiency of one industry with another. A measure of growing popularity is 'value added', which is the total value of goods produced minus the cost of materials, equipment and other services bought in.

Increasingly, governments are concerned with environmental performance. This is most frequently measured in terms of green house gas or CO_2 emissions.

In terms of social performance, a measure gaining some credence is that of social capital, but this is so new that thus far a widely accepted definition has not been agreed (Portes 1998). In the absence of referenda, social or public satisfaction, it may be argued, is only measured rigorously at election times, though opinion polls are frequently used to sample the electorate's view of government's performance.

Industry scale

At the industrial scale, total volume of output, such as tons of steel or barrels of oil may be used. Some of the national level performance measures may also be relevant. Almost uniquely, the construction industry has developed a set of national key performance indicators (KPIs) (Constructing Excellence 2004); they are shown in Table 12.1. As we shall see, the higher the level of performance indicator, the more difficult it is to carry out comparisons on a like-for-like basis, since more variables are

Table 12.1 Key performance indicators for the construction industry.

Client satisfaction – product
Client satisfaction – service
Defects
Predictability – cost
Predictability – time
Profitability
Productivity
Safety
Construction cost
Construction time

hidden. For example, cost/m^2 may vary because of differences in quality, complexity, or functionality.

Company scale

At the company scale, the most common measure of performance is profitability, though there are a plethora of financial ratios which are beloved by accountants (Haskins *et al.* 1996). There is growing pressure for companies to report their environmental and social performance, though as yet, these are impossible to measure unequivocally. Forward looking companies are taking increasing pains to measure client satisfaction, on which their competitive edge depends, and employee satisfaction, on which their ability to produce the necessary goods and services relies.

Project scale

At the project scale, the most frequently quoted measures of performance are cost, time and quality. However, as we shall see shortly, these are crude indicators. Emerging measures include client satisfaction, functional performance (does the facility and/or service do what it was designed to do?), productivity and waste, whilst health and safety remain the top priority for reputable organisations. More sophisticated measures, fuelled in part by the concept of the private finance initiative/ public–private partnership (PFI/PPP), consider whole life as well as initial costs, leading to considerations of operability, reliability, maintainability and availability. The remainder of this chapter will focus on these and other issues in more detail, but before that, it is necessary to provide a theoretical framework within which to couch the discussion. The most recent of these is lean thinking, which we will use as our starting point.

Lean thinking

Origins

In the early 1980s, the American car manufacturing industry realised that the Japanese were producing cars at lower cost and better quality than any other country in the world. The Massachusetts Institute of Technology (MIT), one of the foremost American research institutes, was commissioned to undertake a five year, $US5m research programme to find out why. Their findings are chronicled in the seminal book by Womack, Jones and Roos (1990) *The Machine that Changed the World*. They discovered that, after making allowance for differences in different models, the Japanese could build cars in half the time, using half the space and with fewer than half the defects than their American counterparts. Of course, the $64 000 question was why? In a nutshell, the answer lay in the elimination of waste, or 'muda'. Krafcik (1988), one of the members of the MIT team, is credited with coining the term 'lean' as the antithesis of the 'fat' or waste that lean thinking strives to eliminate.

Fundamental principles

Lean thinking is about removing waste in all its forms, in all parts of the process: in design, in production, in operation and maintenance, and even in eventual disposal. There are three types of waste: first there is the waste which arises by doing activities which are not needed at all; second, there is the waste that arises from activities which are apparently necessary for the process to be successful, but which add no customer value; and third, there is the waste which arises through doing necessary activities at less than optimum efficiency.

The research by MIT identified a number of characteristics and attributes of the Toyota production system which were central to the elimination of waste. These included:

- customer focus
- value pull and flow
- integrated supply chains
- powerful project management
- empowering and multi-skilling the workforce
- the concept of 'price minus'.

The relationship between these characteristics is illustrated by the conceptual model in Fig. 12.2.

The underlying principle is to do no work that does not add value for the customer, and to make the customer 'pull' value through the supply chain. In other words, nothing is done until absolutely necessary. In project management terms, every activity starts at its latest possible start date.

The terms in Fig. 12.2 are self explanatory, with the exception of 'price minus'. The typical Western approach to determining the price of a product or project is

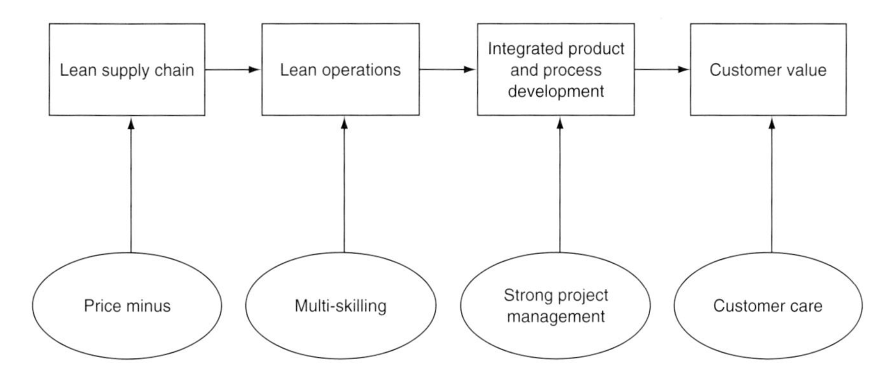

Figure 12.2 A model of lean production and value pull.

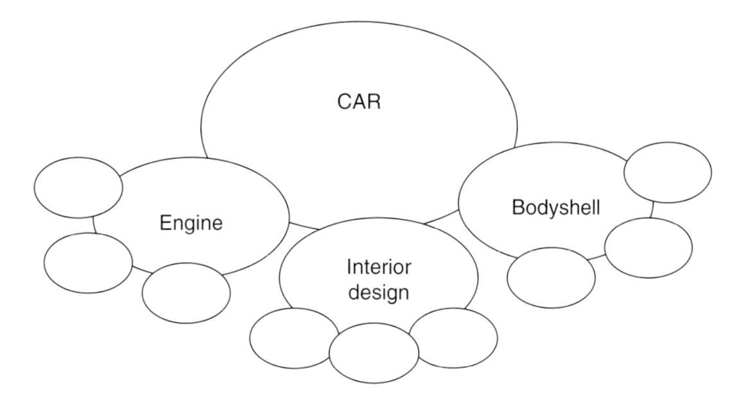

Figure 12.3 Design clustering.

to calculate the cost and add on a sum for profit: the so-called 'cost plus' approach. The lean approach is quite the opposite. First, the market price is determined. The design and manufacture or construction teams are then set the task of producing the product or project at a cost which allows everyone in the supply chain the opportunity to make a decent profit. The profit of each supplier is proportional to their contribution towards driving down costs. This leads to the concept of design clustering illustrated in Fig. 12.3.

In the Toyota production system, the principal design team consists of leaders of the engine, bodyshell, interior design, suspension, traction systems and so on. Thus the interaction and interfaces between the various systems can be explored in depth and optimised. Each design cluster involves the supplier of the parts that make up each system. The interior design cluster might include the seat designer, interior trim designer and the instrument layout designer. The seat designer might himself lead a cluster comprising the mechanism designer, safety belt designer, and ergonomic expert. Thus everyone who has an influence on the final outcome is involved in the design process.

Performance measurement

Generic issues

This framework gives rise to a number of unresolved issues. First, how do we measure lean? What are the characteristics of a lean organisation? How does a 'lean' project or process differ from one which is simply well-managed? How do we measure value to the customer, which may be multi-dimensional, involving not only cash, but certainty of delivery, reliability, after-sales service, reputation or status and many more?

Nevertheless, lean thinking does point modern performance measurement into some new directions. The measurement of waste, the degree of integration of the supply chain, customer satisfaction, are all gaining currency in the modern world of project management.

Application to projects

Clearly, the car manufacturing industry, which spawned lean thinking, is very different in character from project-based industries. Car manufacture is high volume: most projects, except for housing, are 'one off'. Car manufacture takes place in a stable environment with a more or less permanent workforce: projects are generally executed by temporary teams pulled together solely for the duration of the project. In car manufacture, the lead partner is the designer: in projects, it may be a contractor from outside the commissioning organisation. The market for cars is relatively stable, and the market price can easily be defined. The demand for projects fluctuates wildly, and there is, in general, no such thing as a market price.

At the process level, there is one fundamental difference between manufacturing and project-based industries. In manufacturing, the product flows over the resources, which are, to a first degree of approximation, constant, since in a typical production line, the number of workstations and their output capacity is fixed. This is illustrated in Fig. 12.4(a). In project-based industries, on the other hand, the resources, which are variable, flow over the product, which is fixed in position as illustrated in Fig. 12.4(b).

It follows therefore that, whilst in manufacturing it is theoretically possible to eliminate waste by designing the process so that the resources are in perfect balance, and increasing the flow of product until the resources are perfectly saturated, the same is not possible in projects. It is not possible to produce a design which allows one trade to finish its work without a return visit, or even to ensure that the various activities undertaken by one trade all consume exactly the same amount of resources. In any case, if we were to maximise the flow of product, that is, if we were to execute the project as quickly as possible, we would end up with too many resources all clamouring to be in the same place at the same time. In projects, then, the problem becomes one of minimising wastage by constantly re-optimising the use of resources. The difference is illustrated in Fig. 12.5.

The special case of construction

Despite the lack of a theoretical framework, *Rethinking Construction* (Egan 1998) and its successor, *Accelerating Change* (Strategic Forum for Construction 2002), adopted many of the principles of lean thinking. This is hardly surprising, since Professor

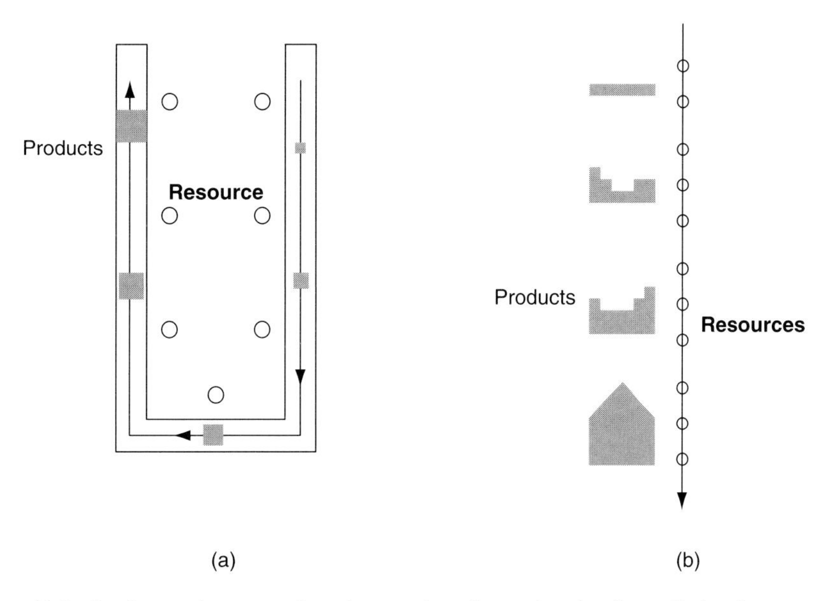

Figure 12.4 Product and resource flows in manufacturing and project-based industries.

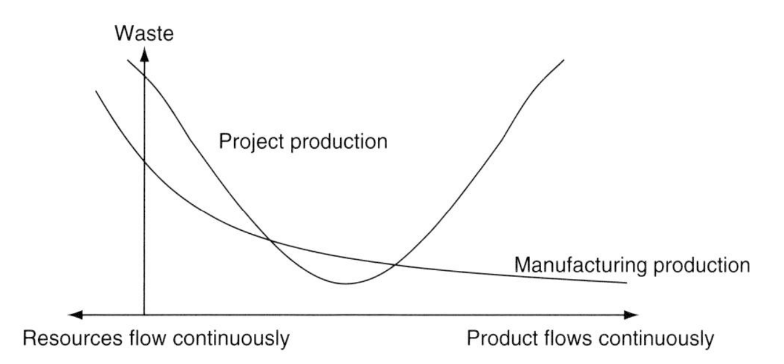

Figure 12.5 The fundamental difference between manufacturing and project-based problems.

Dan Jones, the Jones in Womack, Jones and Roos, sat on the Egan panel, and had considerable influence on its deliberations. Unfortunately, Dan Jones had no experience of the construction industry, and other members of the panel had no expertise in lean thinking, so some of the proposals fail to recognise the difference between manufacturing and construction, and there is no attempt to tailor lean thinking for application in the construction industry. Nevertheless, *Rethinking Construction* gave rise to a number of useful initiatives including the Construction Best Practice Programme and the Movement for Innovation, now united under the 'Constructing Excellence' banner (www.constructingexcellence.org.uk) leading to the important set of KPIs set out in Table 12.1.

Implications for performance measurement

The philosophy of lean thinking has numerous implications for performance measurement at the project level. It can be argued that since there is no widely accepted, all-embracing definition of lean, it is impossible to measure how lean an organisation or a project really is. Nevertheless, the philosophy of lean thinking relies heavily for its success on measurement. In Japan, there is a saying, 'no data, no discussion'. Certainly it seems that for whatever reason, the emphasis on measurement in Japanese lean thinking is in stark contrast to the relative dearth of measurement in the management of complex projects.

Considering first the outward-facing measures, customer satisfaction is a prime concern. In manufacturing industry, of principal concern is customer loyalty. This can be relatively easily measured, albeit crudely, by market share and the amount of repeat business. At the project level, this is theoretically possible, but except in the relatively infrequent cases of serial clients such as retail chains, the time between successive orders from the same client is so long that there are too many intervening variables for any conclusions to be drawn about customer satisfaction and loyalty. Nevertheless, reputation and customer feedback remain important, if slightly remote indicators of the quality of performance at the project level. But in the lean approach, customers include any third party who may be affected by the activity. Third party issues may be considered at both an organisational and project level. They are most frequently concerned with the environment, both in the immediate vicinity of the project (noise, dust fumes, aesthetics), and remote (pollution, embodied energy, CO_2 emissions, greenhouse gases and global warming). Increasingly there are concerns about social effects, dynamics of local communities, travel-to-work distances and methods, crime, vandalism, terrorism and public safety, all of which may be adversely or advantageously affected by project level activities.

Inward-facing measures are potentially far more numerous, and to an extent, somewhat simpler to handle. Many measures focus on continuous improvement and total quality management which are inherent in lean thinking. Exposing performance for all to see is a powerful way of motivating a workforce. Thus displays of outputs per shift, defects, safety incidents, number of suggestions for improvement are commonplace in a manufacturing environment. If lean thinking is adopted by project managers, they will be obliged to measure at an intensity and level of detail to which they are largely unaccustomed.

Practical challenges

What to measure

The section 'Horses for courses' highlighted the need to specify the purpose of measurement. Potential purposes include control, continuous improvement, benchmarking and the fulfilment of statutory requirements. We will deal with each of these in turn.

Cost, time and quality

The traditional modelling approach

Referring to Fig. 12.1, it is obvious, but rarely recognised, that what needs to be measured is a function of the model that we use for control. If, for instance, the project model consists of no more than the ubiquitous 'S' curve of anticipated project expenditure, then all that needs to be measured is project expenditure. This, of course, leads nowhere, principally because it tells us 'what', but not 'why'. If control means taking action to minimise any variance between planned and actual performance, then the control system must provide some diagnostic information. It is impossible to decide rationally what action to take unless the cause of the variance is known. This raises a number of questions such as, what do we want to control, and what can we control? Controlling profit may be an admirable goal: it's also an impossible one. Profit is the difference between price and cost. At a project level, there is very little control over price, except when the contract is varied. The only way to control profit therefore is to minimise costs. To minimise costs, we need to minimise waste in all its forms. This means first optimising the design of the project, second optimising the process, third optimising the balance between resources (normally material, labour and plant costs) and finally making sure that each resource is used to maximum effect. So just as measuring cost serves no useful purpose from a control point of view, neither does measuring profit.

Cost, time and quality are the three dimensions that control systems most frequently seek to address. The cost model is often a list of items to which costs are assigned; the time model is the programme; and the quality model is the specification. There are three fundamental problems.

(1) The cost and time models are heavily interdependent, but normally quite different. Cost packages may be based on the materials content of a project, whilst the programme may be based on operations or activities. Yet they are both heavily dependent on the same input data: the amount of work to be done, the assumed productivity (often the principal source of risk), and the number of resources.

The duration of an activity is:

$$\frac{\text{quantity of work}}{\text{productivity} \times \text{number of resources}}$$

Its cost is:

$$\frac{\text{quantity of work} \times \text{cost of resource per unit time}}{\text{productivity} \times \text{number of resources}}$$

In other words,

$$\text{cost} = \text{duration} \times \text{cost of resources per unit of time}$$

Thus, productivity is the common denominator of both cost and time, where productivity is defined as output per unit time.

(2) The cost and time models contain so many elements that to try to measure them all is a daunting if not impossible task. Cost models may contain thousands of elements, for no other reason than to facilitate the pricing of variations. This is not a lean approach. Programmes may contain thousands of activities, simply because computers can handle large amounts of data in very short periods of

time. In neither case is the person who develops the model the same as the one who has regularly to collect the data needed to effect control. This raises serious issues about the level of detail at which the project should be modelled and at which performance measurements should be taken.

(3) The quality model, the specification, has until recently usually been at pains to describe in detail the quality of materials and standards of workmanship rather than what the function of the project should be. This is a matter of deciding with whom the risk should lie. To over-simplify the choice, if the client wishes the contractor to take the risk of how well the completed facility performs, then he needs to provide a functional specification, and to measure the performance of the facility on completion. If on the other hand, the client intends to take the performance risk, he needs to measure the quality of materials and standards of workmanship at every stage of the project's realisation.

The cost/time/resource approach

In the petrochemical industry, attempts to control cost and time are often made using the cost/time/resource or CTR approach. This is simple in principle, and harks back to manufacturing industry, where each operation is given a job card identifying the resources needed to complete the activity (materials, man hours of each trade, any specialised equipment needed and the time allowed). This has the benefit of integrating the time and cost models, and can be very useful for estimating purposes.

It is difficult to use for the purposes of control, however, for two reasons. First, it is difficult to measure how much of the work on the job card has been completed at any given point in time. If, for instance, the job card involves cutting, preparing and welding seven 3 m long 50 mm diameter steel pipes with four bends to one side of a steel manifold, and three 0.5 m long 100 mm diameter pipes with no bends to the other side, how much of the job is complete when one of the 50 mm pipes has been welded in position, the other two have been cut and prepared, and one of the 100 mm pipes has been cut to length? This example illustrates the difficulty of measuring not only outputs, but also their corresponding inputs. Unless the time spent by each resource on each activity (cutting, preparing and welding) for each pipe diameter is measured, it is impossible to diagnose the cause of any variance between planned and actual progress. The second difficulty revolves around the frequency of measurement. Progress is frequently only measured when an activity or job card is complete, because it is much easier to measure the total input to a completed activity than to monitor the input to every constituent sub-activity. However, from a control point of view, this is completely useless. It is impossible to control what has already happened. Only the future can be controlled. So the question arises, how frequently should performance be measured?

Root cause analysis

Lean thinking requires project managers not simply to determine the first or obvious cause of variance, but to uncover its root cause. Eliminating the first cause may provide a temporary fix, but only by eliminating the root cause can a reoccurrence of the problem be avoided, and the pursuit of perfection brought a step nearer. If a team's productivity is lower than anticipated, it may be because the target is wrong, the team lacks skill or motivation, or is unbalanced, too much overtime is being worked, the necessary materials, plant or information is not available or a whole host of other reasons. Until more detailed measurements identify the root cause of the problem, the chances of permanently rectifying the situation are remote. This, too, raises questions about the level of detail at which performance needs to be measured if the control system is to be sufficiently diagnostic to allow the appropriate remedial action to be determined.

Summary

Performance measurement for the purposes of control raises the following questions.

- How can we create a culture of value-adding performance measurement?
- What performance should be measured?
- At what level of detail should performance be measured?
- How can performance measurement be made sufficiently simple for the cost of data collection to be acceptable, yet sufficiently comprehensive to allow the root causes of variance between actual and planned performance to be diagnosed?
- How frequently should performance be measured?
- Can cost and time models be integrated?

Continuous improvement and benchmarking

Continuous improvement and benchmarking go hand-in-hand. When Thomas Camp (1989) of Xerox first articulated the notion of benchmarking, it was for the purpose of improving organisational performance. Increasingly, it is being used in the context of projects. In its purest form, benchmarking is concerned with mapping the processes in one company and comparing them with those in another. Whilst outputs may also be compared, the emphasis is on process. In recent times, and partly stimulated by lean thinking, people have been encouraged to view project realisation as a process (Koskela 2000; Koskela & Vrijhoef 2001). Process mapping, and its lean derivative, value stream mapping, is an important tool in the modern performance measurer's armoury. It exposes fat and provides clues as to how it may be eliminated. A comprehensive description of the approach may be found in 'The seven value stream mapping tools' (Hines & Rich 1997). Its widespread uptake at the project level remains a challenge.

Continuous improvement requires a longitudinal, long-term approach. It implies a knowledge-centric, learning organisation. At the project level, this translates into passing the lessons learned from one project on to the next. Post-project evaluation is a rarity, which probably explains why 80% of all problems occurring on projects are not new. Its rarity no doubt stems from doubts about its value, about how to measure the performance of one site in a way which is valuable for the next, how to capture the lessons learned, and most importantly, from the preference of project managers to fight today's fires rather than prevent tomorrow's.

Statutory requirements

Financial performance

Statutory requirements relate principally to financial performance, health, safety and the environment. The presentation of statutory accounts is beyond the scope of this chapter. However, from a performance measurement point of view, one of the challenges is to integrate the financial and cost management systems. The purpose of financial accounts is both to satisfy statutory requirements, especially with reference to information to be provided to shareholders, and to control the flow of cash. Cost management systems, on the other hand, need to help management, and especially project managers, to effect project control. They must therefore satisfy the conditions set out earlier in the section 'Cost, time and quality'. Yet although both rely on the same basic input data, there is a notable absence of any attempt to eliminate the fat occasioned by running two systems rather than one. The blame for this does not lie solely at the feet of industry, but at the feet of a political, fiscal and professional system that too often seeks to 'hide the wood in the trees'.

Health, safety and the environment

The control of safety has been left until the end of this section, not because it is least important: on the contrary, the opposite is true. However, strictly speaking, it is not the responsibility of the commercial manager, though a poor safety record will surely have serious commercial consequences.

It is not so easy to think of safety in terms of a model. Yet to do so is not impossible. Setting targets for fatalities, serious accidents, minor accidents and near-misses is, in principle, no more than modelling desired outcomes. Two special problems arise: if the safety system is working well, serious accidents are too infrequent to serve as a useful performance indicator for the purposes of control; and minor accidents, which are significantly more frequent, are prone to under-reporting (Health and Safety Executive 2001).

The statutory requirements for environmental performance include pollution of land, water and air, and if we include planning requirements, aesthetics too. Measurement of environmental performance presents its own special challenges. A plethora of methods is available: BREEAM, BEES, Eco-footprint, LEED, ATHENA, EcoHomes, Ecoprofile, and many more. However, most use fairly arbitrary relative measures, and none of them offer an absolute measure. A further problem is that most improvements in the environment come at an extra cost, but since environmental impact cannot be measured in terms of cost, there is no way of combining economic and environmental value.

Although there is currently no statutory requirement to measure the social performance of projects, there is increasing pressure to do so. Not only are social impacts difficult to measure (How can a reduction in crime or vandalism be valued?), it is impossible to measure, in a common unit of currency, the so-called triple bottom line: economic, environmental and social issues.

In addition to the difficulties in measuring the immediate impacts of a project on sustainability, there is the problem of internalising external effects. No method is currently available for including in a project's assessment its effect on the world at large, on global warming, climate change or rising sea levels. Sustainability accounting is an area of performance measurement in its infancy at the organisational level (Atkinson 2000; Bebbington *et al.* 2001). It is no more than a 'twinkle in the eye' at the project level.

In forward-thinking companies, especially in the petrochemical industry, a single risk management system deals with health, safety, environmental and commercial risks. So although the statutory requirements for measuring performance in safety go no further than accident reporting, commercial managers may yet find themselves involved in the measurement of many more than mere costs.

The state-of-the-art

Measuring cost performance

Most companies measure cost performance in one form or another. The client wants certainty that the cost of their project is not going to exceed budget. The contractor wants to make sure that their costs do not exceed the price they get paid. Strangely, designers, who arguably have the greatest influence on cost performance, can be least interested in costs. In the construction industry, designers often place satisfaction of their aesthetic aspirations ahead of value for money. They have a limited understanding of buildability, not least because a widely accepted measure is lacking.

Traditional approaches

At feasibility stage, clients want to know how much their project is going to cost, and how they can optimise their return on investment. This requires some 'what-if' analyses. Most of the time, costs are rather crudely expressed in terms of usable area of the project, or in terms of some key parameter such as barrels of oil per day, number of trains per minute, or numbers of hospital beds. Such parametric estimates rely on historical data from previous, similar projects. In crude terms, only the total project cost and the project output need to be measured. Unfortunately, the range of costs per unit of output can be as much as 100% for projects which are notionally similar. Measuring costs at this level of detail is therefore necessary, but insufficient to provide the cost certainty that clients require. A further approach is to break the design into elements such as steam generators, boilers, cooling water system, and price each element separately. This helps, but accuracy may still be no better than ±20% (Barnes 1989; Flanagan & Norman 1983). The reason is because elements as large as 'cooling water systems' hide a host of other variables such as quality of water available, intake and outlet conditions, and plant layout. The problem then is how to break down a project into sufficient elements that each can be costed to an acceptable level of accuracy without making the process so complex that collecting and analysing the necessary data becomes unmanageable. We return to this vexed question in the next section.

Once work starts on site, the contractor will be required to report his costs against a cost model, which will take some form of bill of quantity, job cards or 'line items'.

Traditionally, performance is measured only to ensure that costs do not exceed budget. The contractor's costs are compared with the tender prices. This is fine from a financial management point of view, but provides no information about where the project is going astray, or if it isn't, whether the opportunity exists to make it even more profitable. It provides no diagnostic capability.

A very few companies take a slightly more rational approach by breaking the project down into work packages, and tracking the cost performance of each package. Although an improvement, the level of detail is still too great to eliminate confounding variables and to provide the information needed to diagnose the project's health.

The petrochemical industry has, for the past 50 years or so, applied an approach called 'earned value' whose use is only just beginning to penetrate into other industries. Again, this is dealt with in the next section.

As far as post contract-evaluation is concerned, the client will record the extent to which the project fulfils its function, though even then, only gross measures are reported, and the user is rarely consulted. This limits the scope for continuous improvement. Some members of the design team, notably cost consultants, will record and archive costs for future use, but only at the level of detail at which the contractor is obliged to report. Most British project managers carry out practically no post-contract evaluation, though there is anecdotal evidence that European and Asian project managers are more enlightened. Under these circumstances, it is hardly surprising that successive governments and the reports prepared on their behalf paint a gloomy picture of the construction industry's performance.

Recent developments

Measuring performance at the design stage

As we have previously noted, 80% of the cost of a project is committed once the design is complete. This is a critical stage in a project's life. The first requirement is to find some way of measuring value to the client. This means eliciting their needs and assessing their relative priorities. Barrett and Stanley (1999) have suggested how the briefing process can be improved, and the emergence of multi-criteria decision-making techniques (Bardos *et al.* 2001) is a useful development, especially for complex projects. Value management (see Chapter 13) is also gaining in popularity as a way of ensuring that the design maximises the return on the client's investment. In any case, it is necessary to develop measures which allow the relative merits of alternative design solutions to be assessed with confidence. Increasingly, too, it is necessary to consider not simply initial costs, but the whole life costs of a project. Since the cost of operations and maintenance can be as much as seven times the capital cost (Spedding 1994), and since the value of the output from a project can be 500 times the capital cost (Evans *et al.* 1998), it is foolish to focus on a solution which simply minimises the cost of the initial investment.

There are numerous emergent approaches to addressing these problems. First is the development of simpler cost models of complex projects. It has been shown that for any set of data, most information is obtained for least effort by considering only that data whose value is greater than the average. This provides a simple way of identifying the cost-drivers for any project. What is more, for any class of project such as retail store, offshore oil platform, water purification plant or road, these cost drivers are the same for all projects in the class, and more interestingly, represent a surprisingly consistent proportion (typically 80% ± 5%) of the total cost (Horner & Asif 1988). The cost drivers are typically fewer than 10% of the items in a conventional cost model. The model for reinforced concrete bridge projects shown in Table 12.2 demonstrates that 13 items consistently account for 82% of the total cost.

The total cost of a project can be determined simply by calculating the cost of the cost drivers and multiplying by a factor to take account of the large number of small value items which have been omitted from the model. The cost-drivers can be expressed either as elements or as work packages. Elements are important for measuring whole life performance, as we shall explain shortly, whilst work packages are important for measuring performance particularly during the construction phase. There are two great advantages of this approach. First, cost performance needs to be measured only in those elements or work packages which have been identified as cost drivers. Second, at the early design stage, it is necessary to design and optimise only the elements which are cost drivers together with any performance-critical elements in order to measure the return on the client's investment.

Measuring whole life performance is a new departure for project managers. It is hampered by two constraints: first, the need to collect large amounts of data relating

Table 12.2 Simplified cost model[a] for preliminary design stage of reinforced concrete bridges.

CSWP No.	Cost-significant work package	Unit
1.1	Establishment of piling equipment for piles in the main piling	ITEM
1.2	Moving piling plant for piles in the main piling	No
1.3	Supply and driving of piles for main piling	lin m
1.4	Load testing of piles in main piling	No
2	Vertical (85°–90°) formwork ≥300 mm wide for end supports and intermediate supports	sq m
3	Horizontal (0°–5°) formwork ≥300 mm wide for deck	sq m
4	Curved formwork at any inclination ≥300 mm wide for intermediate supports	sq m
5	In situ concrete	cu m
6	Precast concrete	cu m
7	Bar reinforcement	tonne
8	Paving in paved areas to surfaces ≥10° to horizontal	sq m
9	Waterproofing on surfaces ≥300 mm sloping up to 45° to horizontal	sq m
10	Supply of parapets	lin m
11	Drainage of end supports	ITEM
12	Imported fill deposited adjacent to structures including around structural foundations	cu m
13	Void former	sq m

[a]Cost model factor = 0.82.

to every component of an asset or project, and second, the discontinuity between the capital cost model, the life-cycle replacement cost model, the facilities management (operation and maintenance) cost model and the demolition cost model. Maintenance management operating systems are widely used in petrochemical, utilities, and increasingly infrastructure projects, but it can cost £250 000 simply to tag every component of a complex asset. The cost of collecting data on every component can inhibit the extent to which the systems realise their full potential, either in terms of planning or in analysing performance with a view to effecting improvement.

The new ISO Standard 15686-5 *Building and Constructed Assets – Service Life Planning, Part 5; Whole Life Costing* incorporates a unique, elemental cost and data breakdown structure which, for the first time, is consistent throughout every phase of a project's life. When this is coupled with knowledge of the cost drivers, a performance measurement system can be designed which is sufficiently accurate for rational decisions to be made, yet sufficiently simple to ensure that data collection is not too onerous.

Measuring performance during construction, erection or installation

The petrochemical industry has been using the concept of earned value for at least 50 years, but only recently and infrequently has it been used in the construction industry. In its simplest form, earned value is a way of combining cost performance across a range of activities. For example, if a gang of pipefitters installs 20 m of 100 mm diameter pipework, 60 m of 50 mm pipework, two boilers and a condenser in a week, and if the tender price of labour for these activities is £15/m, £13/m, £500 each and £300 each, respectively, then the value 'earned' by the gang is:

$$20 \times £15 + 60 \times £13 + 2 \times £500 + 1 \times £300 = £2380$$

If the cost of the gang for the week is £2000, then the contractor knows that his labour costs, at least in this activity, are under control. The concept of earned value has wide application in performance measurement. It can be used as a measure of productivity by comparing earned and actual hours, or it can be used for measuring time and cost performance simultaneously. These applications together with their shortcomings are described in the following section 'Combined cost and time models'.

An alternative approach is to set up a system which operates at the task level, and which addresses the performance in terms of labour, plant and material. This certainly simplifies the process of diagnosing the causes of variances, but can demand the collection of very large volumes of data. The problem can be ameliorated by measuring performance only in the project cost drivers and by applying the principle of earned value to combine outputs in different activities. A typical cost summary system is illustrated in Table 12.3, and the breakdown to labour costs in Table 12.4. Table 12.3 illustrates that the cost of wiring is £420 higher than it should be. The breakdown of labour costs in Table 12.4 shows that higher than anticipated labour costs account for £320 of the overspend, implying that there is an additional overspend of £100 in either plant or materials. Table 12.4 also shows that the hourly cost of labour is £12/h (720/60), which is 20% higher than the budgeted cost of £10/h, and that productivity is 3.3 m/h (200/60), which is 33% lower than the budgeted productivity of 5 m/h. Similar tables can be designed to track plant and materials costs and wastage.

Table 12.3 Overall cost record.

Element	Quantity	Budget cost/unit (£)	Budget total cost (£)	Quantity complete	Earned value (£)	Actual cost (£)	Variance (£)	Remarks
Wiring	1000 m	4.00	4000	200 m	800	1220	+420	*Breakdown required*
Pipe racks etc.	100 m	38.00	3800	10 m	380	280	–100	

Measuring time performance

Traditional approaches

Time performance is almost always measured by reference to a programme, most usually a bar chart, but in complex projects a network or precedence diagram is more likely to be employed. These approaches are widely used and well documented (Lester 2003). They suffer from a number of shortcomings.

- Neither offers any element of a diagnostic system, nor do they provide data in a form which can be used to improve future performance.
- In both, an estimate has to be made of the percentage complete of any activity which is incomplete at the time of measurement. Without some further breakdown this can lead to considerable inaccuracies unless the concept of earned value is introduced.
- Bar charts are easy to use, and relatively easy to update, especially on smaller projects. However, because they show no logic links, the consequential effects of variances from planned progress on subsequent activities are difficult to adduce.
- Because networks and precedence diagrams are designed for computer applications, it is easy to be fooled into creating a work breakdown structure containing a thousand or more activities. This militates against sensible data collection and performance measurement, and helps to explain why such techniques are less widely used now than they were 30 years ago.
- Progress in complex projects expressed in many thousands of activities may only be measured at monthly intervals. This is certainly too infrequent to successfully control activities which last no longer than about eight weeks, since by the time the cause of a variance has been diagnosed and remedial action designed and implemented, the activity is already complete.

Recent developments

'Last Planner' is a system which has grown directly out of lean thinking. In essence, it demands the production of detailed two week rolling programmes that contain only those activities for which all the necessary resources and information are guaranteed to be available. Performance is measured in terms of the percentage of planned

Table 12.4 Labour cost control.

				Budget						Actual					
Element	Total Quantity	Unit	Prod'ty	Hours	Hourly cost (£)	Total cost (£)	Quantity complete	Earned hours	Earned value (£)	Earned hours	Var.	Cost (£)	Var. (£)	Prod'ty	Var.
Wiring	1000	m	5 m/h	200	10	2000	200 m	40	400	60	+20	720	320	3.3 m/h	–1.7 m/h

activities that are completed each week (Ballard & Howell 2003). Whilst this has much to commend it, there is a danger of 'picking the low hanging fruit', i.e. of only programming the activities that are easily achieved. These are not necessarily the ones which are critical to progress.

However, there is a growing body of evidence to suggest that short term planning holds the key to optimising resources, the holy grail of lean construction, and simple methods have been developed to involve in the planning process all those who have a role to play, and to provide feedback on actual performance, particularly in diagnosing why things did not proceed according to plan (Horner & Duff 2001).

Combined measurement of cost and time

Productivity

Because productivity is the common link between cost and time, because improving productivity is central to improving project performance, and because there is growing political interest in measuring productivity, it deserves a special mention. In the UK, productivity is most frequently defined as the ratio of output to input, but in America and American based industries, it is defined as input divided by output. Whichever definition is used, it is necessary to measure both inputs and outputs.

Input may be measured in three ways as shown in Table 12.5. Available time is the time during which the labour force is actually able to work. Productivity in terms of available time will improve if:

- interruptions or disruptions are reduced
- the skill and/or motivation of the team improve
- a better way of doing the work is implemented
- project complexity is reduced.

This raises the question of how the skill and motivation of the team can be measured, and how the complexity of a project can be assessed.

Outputs in some activities, such as pipelaying or brickwork, are easy to measure. In other activities, measurement may be more difficult. Guidance on overcoming some of these difficulties has been offered by Horner and Duff (2001).

Table 12.5 Labour productivity inputs.

Measure	Definition	Purpose
Total time	Total paid time	Estimate costs
Available time	Total time minus unavoidable delays	Measure management performance
Productive time	Available time minus avoidable delays	Measure intrinsic capability of labour force and project complexity

There are two occasions when it may be desirable to combine the outputs in two or more activities:

- when we want to calculate the 'global productivity' on a site
- when we want to aggregate outputs in two 'sub-activities' so that we don't have to allocate the manhours input into each sub-activity separately: one such example would be a gang of electricians running conduit and cabling at the same time.

In these circumstances, earned value can be used to combine outputs in different activities. Productivity is then defined as the ratio of earned to actual hours. The problem is that the method depends on the existence and accuracy of 'norms', the standard time for each activity. Although some norms do exist, especially in the petrochemical industry (e.g. Page 1999), they were first produced over 50 years ago and take no account of changes in work practices. One option is to use in-house or published estimating data, but until a sufficiently large and consistent bank of historical data is built up, the approach is unlikely to command the confidence it merits outside the petrochemical industry.

There are several ways of measuring and calculating productivity, without engaging in expensive and time-consuming observation and record keeping. One method, which lies at the heart of the Building Research Establishment's 'Calibre' system, is called 'activity sampling'. In brief, it involves an observer touring the site at regular intervals, recording the number of men at work, and what they are doing, and the number of men who are idle. From these measurements, it is possible to calculate the amount of productive and non-productive time. This, of course, is not a measure of productivity as it is defined in this chapter, because there has been no measure of output. Nor does it take any account of the fact that longer interruptions cause disproportionately greater productivity loss than shorter ones, or that interruptions shorter than 15 minutes have no effect on output (Noor 1999). Nevertheless, it does offer a quick and reliable way of identifying the reasons for delays which have been shown to be the major cause of low productivity.

Another alternative is to use the simplified models described in the earlier section 'Recent developments' to identify the few cost driving activities in which it is worthwhile measuring productivity. The process is defined in more detail by Horner and Duff (2001).

A coarser, yet more sophisticated measure introduced as a construction industry KPI is value added. This is simply defined as the value of work completed minus the costs of goods and services supplied divided by the labour input. It is relatively simple to calculate, especially at project completion, but shares with many other performance measures the lack of a diagnostic capability.

The way we measure productivity will depend on why we want to measure it. Table 12.6 shows how the objectives may affect the method and frequency of measurement.

Combined cost and time models

Many people recognise the advantages of using a single model for controlling cost and time. Cost and time are frequently interdependent, and are linked by the prime

Table 12.6 Matching objectives and methods of measurement.

Purpose	Method	Frequency
Compare global site performance	Earned value or added value	End of project
Progress towards completion	Measured outputs	Monthly
Improve performance and provide estimating data	Measure productivity in significant activities	Weekly
Calculate bonus	Measure output of each gang in every activity	Weekly
Efficiency of labour utilisation	Record idle time	Daily

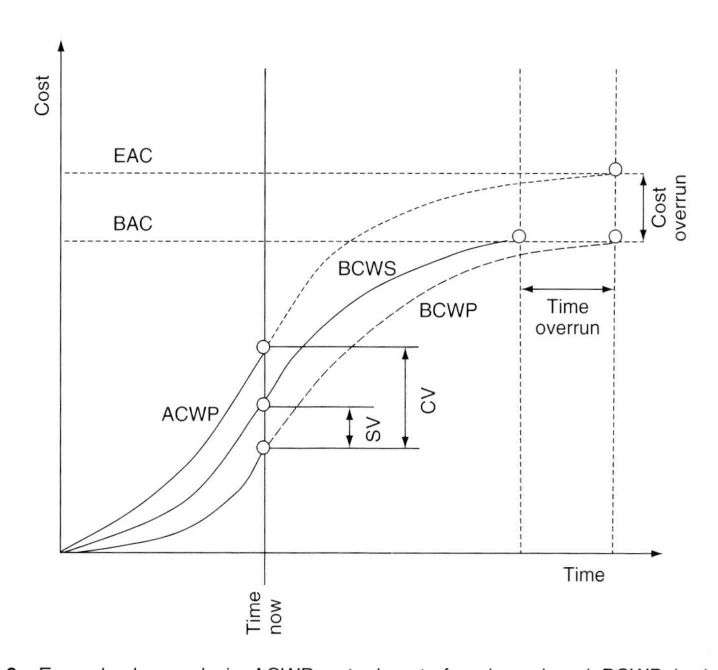

Figure 12.6 Earned value analysis. ACWP, actual cost of work produced; BCWP, budgeted cost of work produced; BCWS, budgeted cost of work scheduled; BAC, budget at completion; EAC, estimate at completion; SV, schedule variance; CV, cost variance.

variables of quantity and productivity. A single model reduces the amount of data collection and the opportunity for error. However, attempts to develop such approaches have, by and large, failed. Horner and Zakieh (1996) demonstrated how quantity significance might hold the key, but there has been little uptake across the industry. Most recently, Dauber (2003) has proposed a system for electronically linking bonus measures to productivity analysis, estimating, planning and wage systems, but the work is still in its infancy.

Figure 12.6 illustrates how earned value analysis can be used to measure cost and time performance simultaneously, and whilst, as we have already remarked, the technique is well known in the petrochemical industry, it does not enjoy the same popularity in other industries, mainly because of the absence of reliable norms.

Measuring quality performance

Quality is perhaps the most difficult of all the dimensions of performance to measure, largely because it has no units of measurement, and because it is affected by people's expectations. In lean thinking, the quality of the production process can be measured in terms of the number of defects. In projects, records of defects are rarely maintained except immediately before hand-over when the inevitable snagging list is produced. There is a move now towards a zero defects regime, but there is little real effort yet to record defects and their causes when they arise, so that they can be eliminated from the process. (Toyota achieves this by empowering their employees to stop the production line as soon as a defect is detected.) Surveys of client satisfaction are becoming more commonplace, but they rely on expectations rather than absolute values.

ISO 9000, Quality Management Systems, was introduced as a way of improving quality, but it requires only compliance with a pre-defined process. It measures the quality neither of the process itself, nor of its output. Nevertheless, the number of 'non-conformances' is often used as a surrogate measure for quality.

Lean thinking and the growing importance of PFI/PPP procurement methods have encouraged the use of functional rather than prescriptive specifications. These do offer a real basis for measuring the quality of the output of the project, but this is a function not only of the quality of the design but also of the way the plant is operated and maintained. Unless the specification covers operation and maintenance, or unless the constructor is also responsible for maintenance and operation, the cause of failure to comply with a functional specification will be open to debate.

Most recently, again springing from lean thinking, the cost of quality approach has been tried. Here, attempts are made to measure the cost of every error or variance from plan, and to display the results in a prominent position such as the site canteen (Hall & Tomkins 2001). This is a useful way of heightening awareness of the importance of error-free working, but the results are often overshadowed by the consequential costs of increased liquidated damages rather than the immediate cost of rectification.

Measuring safety performance

In the section 'Health, safety and the environment', the inadequacy of measures linked to statutory requirements was noted. The petrochemical industry has developed a series of approaches based on constructive challenge. They are, by and large, variants of Dupont's original Safety Training Observation Programme (STOP), which encouraged employees to enquire and record every time they observed a hazardous practice. As with ISO 9000, performance is not measured directly. The number of observations, and the number of potentially hazardous situations identified are the typical performance outputs from this system. Duff *et al.* (1997), working in the construction industry, have developed a different approach. First, they developed a set of surrogate measures, such as number of ladders incorrectly fixed, number of incidences of upturned nails in waste timber; and second, they used the technique known as reinforcing positive behaviour to motivate the workforce to report on potential hazards and near misses. Rowlinson (2004) provides a useful overview of recent thinking on safety, and notes that the balanced scorecard approach (see section 'Balanced scorecard') can also be applied to the measurement of safety performance.

The CDM Regulations (1994) were designed to ensure that designers assumed as much responsibility for the safety of projects as constructors. The regulations place considerable emphasis on risk assessment, and it is increasingly recognised that the management of safety is not dissimilar to the management of risk which is dealt with in Chapter 13.

Measuring sustainability performance

Sustainability performance is a political hot potato. Sustainability requires that our demands on the natural world do not outstrip what it can supply (Wackernagel & Rees 1996). The Rio Earth Summit (1992), Kyoto Protocol (1997) and Johannesburg Summit (2002) committed many governments to improving the environmental performance of their industries. But, as noted earlier, the measurement of sustainability remains elusive. However, a research team working within the Engineering and Physical Sciences Research Council (EPSRC) Sustainable Urban Environment initiative is starting to address this problem (www.sue-mot.org.uk). The recently introduced ISO 14000 Environmental Management is designed to do for the environment what ISO 9000 has done for quality. It suffers, however, from all the same shortcomings.

Whole life performance

As if all this were not difficult enough at the pre-commissioning stage, the need to measure whole life performance adds yet another layer of complexity. Thus far, there is no generally accepted approach to measuring whole life performance. Although in housing, the Housing Association Property Manual (HAPM) has been collecting data on the life expectancy of common components for many years, there remains a lack of understanding of how these components are affected by operating conditions and maintenance regimes. These confounding variables also have serious effects on the ability to measure in a meaningful way the reliability of most standard components.

The Whole Life Cost Forum has recently been established to provide a portal for whole life cost information, but this initiative is in its infancy, and its success remains to be seen.

Although there are sporadic attempts to collect whole life data relating to environmental and social performance, no systematic approach has been developed, not least because no commonly accepted measurement framework exists.

Balanced scorecard

Kaplan and Norton (1996) developed the balanced scorecard approach which seeks to integrate four dimensions of performance measurement: finance, customers, internal business processes, and learning and growth. In each dimension, there is flexibility for an organisation to determine its own KPIs against which to measure its progress. Designed originally as a strategic planning tool to focus on future rather than past performance, it can be adapted for a variety of purposes, as we have seen in the section 'Measuring safety performance'. Although not widely used at the project level, it can drive the development of aggressive targets for continuous improvement in, for example, customer relations, reducing waste at the design–construction interface, productivity improvement, and re-skilling the workforce. It articulates well with the notion of lean thinking, but considerable care is needed to ensure that the chosen KPIs truly reflect the goals of the organisation.

Conclusions

Performance measurement of complex projects leaves much to be desired. Historically it has focused on cost, time and quantity, without considering the important dimensions of employee and client satisfaction, health and safety, environmental impact and sustainability. Most performance measurement systems have not been designed with 'fitness for purpose' in mind. They are either so detailed that no one wants to implement them, or so coarse that their accuracy is unacceptably low. None finds the right balance between simplicity and answers which are accurate enough.

Lean thinking places a greater emphasis on detailed performance measurements than has hitherto been common, but there is still much to do to adapt successfully the way lean thinking is applied in manufacturing to the management of complex projects.

There remains a sad lack of integration in modern performance measurement systems. There are discontinuities between cost and financial systems, cost and time systems, between the systems used in different industries and between systems used at project and national levels.

The private finance initiative and public–private partnerships have forced project managers to think beyond initial conditions and to consider performance, including sustainability, throughout the whole life of their projects. This presents a series of challenges which so far have hardly been addressed.

And finally, the concept of a 'learning organisation' is still in its infancy. As a result, the same mistakes are made time and time again, and post-project evaluation remains the exception rather than the rule.

Research challenges

This final section of the chapter draws together the challenges for the research community from the foregoing discussion. The length of the list demonstrates the enormous opportunity for improving the performance of the industry. It is set as a series of questions whose purpose is to stimulate the design of suitable research programmes. Many of them are interdependent.

(1) Performance is under-measured. How can a measurement-averse industry be encouraged to develop, use, and preferably share a consistent set of performance indicators?

(2) Is it possible to develop project models which are sufficiently simple to encourage data collection, yet sufficiently accurate to produce data which are of real value?
(3) As an incentive to measurement at a project level, can a hierarchical system of KPIs be developed which allow measurements at project level to be rolled up into national statistics?
(4) Is it possible to define what we mean by 'lean' and to design appropriately lean performance measures?
(5) How can stakeholder and user values be measured before the design phase? How can their fulfilment be measured post-commissioning?
(6) Can value-stream mapping be adapted for use in project-based industries? Can it yield valuable performance measures?
(7) How can the performance of members of integrated supply chains be measured in a way which allows appropriate rewards to be distributed?
(8) How can the effect of the design on the performance of the users of a building be measured?
(9) Can a single system measure finance, cost, time and productivity simultaneously with the same set of input data? Can it be linked to bonus, wages, planning, estimating and risk management systems to provide a single organisational business measurement system? Can it also address issues of health and safety, the environment and sustainability in general?
(10) How can the three dimensions of sustainability, economics, and environmental and social performance be measured in a common unit of currency?
(11) Is it possible to develop a generic framework for the measurement of whole life performance measurement addressing reliability, availability, maintainability and sustainability?
(12) How can the context of a project's operating and maintenance environment and its effect on functionality be measured?
(13) How can the performance, especially motivation and skill of operatives and managers, be measured?
(14) Should buildability be one dimension of the performance measurement for designers? If so, how can it be measured consistently?

If the list seems somewhat daunting, we can take comfort from the words of James Thurber.

It is better to know some of the questions than all of the answers.

References

Atkinson, G. (2000) Measuring corporate sustainability. *Journal of Environmental Planning and Management*, **43**, 235–252.

Ballard, G. & Howell, G.A. (2003) Lean project management. *Building Research and Information*, **31**, 119–133.

Bardos, P., Lewis A., Nortcliff, S., Mariotti, C., Marot, F. & Sullivan, T. (2001) *Review of decision support tools and their use in Europe: report of CLARINET working group.* Final Report, DEFRA. r^3 Environmental Technology Ltd, UK.

Barnes, N.M.L. (1989) Financial control of construction. In: *Control of Engineering Projects* (ed. S.H. Wearne). Thomas Telford, London; pp. 125–145.

Barrett, P. & Stanley, C. (1999) *Better Construction Briefing*. Blackwell Science, Oxford.

Bebbington, J., Gray, R., Hibbitt, C. & Lirk, E. (2001) *Full cost accounting: an agenda for action*. ACCA Research Report No. 73. Association of Chartered Certified Accountants, London.

Camp, T.C. (1989) *Benchmarking: The Search for Industry Best Practices that Lead to Superior Performance*. American Society for Quality, Quality Press, Milwaukee, WI.

Constructing Excellence (2004) *Construction Industry Key Performance Indicators*. Constructing Excellence, London.

Construction (Design and Management) Regulations (1994) (CDM Regulations).

Dauber, V.F. (2003) *Measuring and minimising waste in construction by applying lean thinking*. PhD thesis. University of Dundee.

Duff, A.R., Marsh, T.W. Phillips, R.A. & Robertson, I.T. (1997) Safety measurement and goal-setting: theory and practice. In: *Safety and Health on Construction Sites*. CIB Publication 209, Working Commission 99; pp. 76–83.

Egan, J. (1998) *Rethinking Construction*. Report of the Construction Task Force. HMSO, London.

Evans, R., Haryott R., Haste, N. & Jones, A. (1998) *The Long Term Costs of Owning and Using Buildings*. The Royal Academy of Engineering, London.

Flanagan, R. & Norman, G. (1983) The accuracy and monitoring of quantity surveyors' price forecasting for building work. *Construction Management and Economics*, **1**, 157–180.

Hall, M. & Tomkins, C. (2001) A cost of quality analysis of a building project: towards a complete methodology for design and build. *Construction Management and Economics*, **19**, 727–740.

Haskins, M.E., Ferris, K.R. & Selling, T.I. (1996) *International Financial Reporting and Analysis: A Contextual Emphasis*. Irwin, London.

Health and Safety Executive (2001) *Levels and Trends in Workplace Injury: Reported Injuries and the Labour Force Survey*. Health and Safety Statistics Unit, London.

Hines, P. & Rich, N., (1997) The seven value stream mapping tools. *International Journal of Operations and Production Management*, **17**, 46–64.

Horner, R.M.W. & Asif, M. (1988) Simple models for estimating and controlling project costs. In: *From Conception to Completion, Proceedings of the 9th World Congress on Project Management*, Glasgow; pp. 95–105.

Horner, R.M.W. & Duff, A.R. (2001) *More for less – A Contractor's Guide to Improving Productivity in Construction*. Construction Industry Research and Information Association (CIRIA), London.

Horner, R.M.W. & Zakieh, R. (1996) Characteristic items – a new approach to estimating and controlling construction projects. *Construction Management and Economics*, **14**, 241–252.

International Organization for Standardization (ISO)/DIS 15686-5 Buildings and Constructed Assets – Service Life Planning, Part 5: Whole Life Costing. TC 59 – Building Construction, SC 14 – Design Life. International Organization for Standardization, Geneva.

International Organization for Standardization (ISO) 9000:2000 Quality Management Systems. International Organization for Standardization, Geneva.

International Organization for Standardization (ISO) 14000: Environmental Management. The ISO 14000 Family of International Standards. International Organization for Standardization, Geneva.

Kaplan, R.S. & Norton, D.P. (1996) *The Balanced Scorecard: Translating Strategy into Action*. Harvard Business School Press, Boston, MA.

Koskela, L. (2000) *An exploration towards a production theory and its application to construction*. PhD thesis, Helsinki University of Technology, Helsinki.

Koskela, L. & Vrijhoef, R. (2001) Is the current theory of construction a hindrance to innovation? *Building Research and Information*, **29**, 197–207.

Krafcik, J. (1988) Triumph of the lean production system. *Sloan Management Review*, **30**(1), 41–52.

Lester, A. (2003) *Project Planning and Control*. Butterworth-Heinemann, London.

Noor, I. (1999) *A study of the variability of labour productivity in building trades*. PhD thesis, University of Dundee.

Page, J.S. (1999) *Estimator's Installation Man-Hour Manual*. Gulf, Houston.

Portes, A. (1998) Social capital: its origins and applications in modern sociology. *Annual Review of Sociology*, **24**(1), 1–14.

Rowlinson, S. (2004) *Construction Safety Management Systems*, E. & F.N. Spon, London.

Spedding, A. (ed.) (1994) *CIOB Handbook of Facilities Management*. Chartered Institute of Building, Longman Scientific & Technical, London.

Strategic Forum for Construction (2002) *Accelerating Change*. Rethinking Construction c/o the Construction Industry Council, London.

Tucker, R.L. (1986) Management of construction productivity. *Journal of Management in Engineering*, **2**, 148–156.

Wackernagel, M. & Rees, W.E. (1996) *Our Ecological Footprint: Reducing Human Impact on the Earth*. New Society, Gabriola Island, BC.

Womack, J.P., Jones, D.T. & Roos, D. (1990) *The Machine that Changed the World*. Rawson Associates, New York.

13 Value Management of Complex Projects

John Kelly

Introduction

Value management is commonly viewed from the position of the client whose value system dominates the project from inception to completion. Value for money is obtained when the client's value criteria are satisfied. Commercial management is seen as a facilitation vehicle to realise this value for money objective, and in the context of this chapter is considered between the point of engagement of the commercial manager to the signing off of the project by the client. From the perspective of the construction project, value engineering is commonly considered to be a key commercial management activity. However, there is considerable merit in understanding the parts to be played by both value management and value engineering as essential ingredients of an integrated process. The application of an integrated construction process has occupied the industry for some years, and is typified by the following quotation from Sir John Egan's report to the Deputy Prime Minister.

> *If we are to extend throughout the construction industry the improvements in performance that are already being achieved by the best, we must begin by defining the integrated project process. It is a process that utilises the full construction team, bringing the skills of all the participants to bear on delivering value to the client. It is a process that is explicit and transparent, and therefore easily understood by the participants and their clients. The rationale behind the development of an integrated process is that the efficiency of project delivery is presently constrained by the largely separated processes through which they are generally planned, designed and constructed.* (Egan 1998)

This put the point of view that for construction to give a value for money service to the client, then that service has to be integrated. Whether a forward thinking construction company, a project manager or the client leads the integrated team is a debate beyond the scope of this chapter, although as described below, construction companies are currently leading development in the UK.

If a primary aim is to deliver value to the client through an explicit and transparent process, then that process has to be formalised into accepted procedures, tools and techniques. In this chapter it is argued that in all cases the construction activity is a project running parallel to the client's core business. For the project to be successful it must achieve perfect strategic fit with that core business.

This chapter will therefore describe and illustrate both value management and value engineering in the context of an integrated construction activity. Several factors are considered, namely:

- a theory of integrated teams in modern construction
- understanding the client's view of a construction project
- a theory of value management in the context of construction projects
- the relationship between quality and value and its synthesis
- the application of value management within the supply chain and value chain theory
- value engineering techniques and application.

The above are illustrated with three value management and two value engineering case studies.

Integrated teams in modern construction

In 1985, Professor John Bennett stated that problem-solving organisations that see the construction process as an integrated whole rather than as a collection of separate parts, were the organisation structures of the future. Recognising that there is always considerable straightforward construction for which simpler procedures apply, Bennett (1985) concluded that the increasing amount of complex construction required larger or increased numbers of knowledge-based organisations. During the intervening 15 years, the construction industry became more integrated with changes in roles and responsibilities within the team. Langford and Male (2001) observed that the traditional role of the consultant is to act as an intermediary in the market exchange relationship between the client and contractor, a system which has become formalised and institutionalised over a number of years. However, the increasing demand by clients for design and build, with or without the finance and operate elements, at a fixed or guaranteed maximum price has, in many instances, reversed these institutionalised distribution channels, weakening the power of consultant advisers who are increasingly part of the contractor's supply chain. The move of consultants to the contractor's supply chain necessarily alters the nature of technical advice for the client. Winch (2002) stated that the trend towards concession contracting, i.e. design, build, finance, operate (DBFO), in both the UK and elsewhere, places enormous demands on client decision-making. Clients have to think beyond the physical building to define precisely the services to be provided by the facility. This requires a briefing process that includes the preparation of an output specification; a statement of what the client wants in terms of its business needs rather than a method of its achievement. However, as Winch (2002) stated, it is always easier for the construction team to design and build a school rather than facilitate community education.

The following, taken from the website of a leading construction management contractor gives a flavour of what to expect from a knowledge-based integrated construction organisation.

> [Our organisation] *is a professional management company, which works predominantly in the property and construction sector. Our business is the management and delivery of the development, design and construction process from project inception to post completion aftercare.*
>
> *We are an independent company, working in collaboration with clients and the industry. Our vision is to become the best of the best. We offer a broad range of services and products and these can be tailored to meet individual clients' needs.* [Our organisation's] *operations are structured into sectors, both to offer clients a knowledge-based, value-added service, and also to facilitate a total solutions offer if appropriate.*

The above supports the contention that to move forward in providing value for the client, the industry must operate as an integrated organisation focused on the client's value criteria.

The client's view of a construction project

The *Oxford English Dictionary* defines a project as being a plan, a scheme, or a course of action. Börjeson (1976) defined a project as 'a temporary activity with defined goals and resources of its own, delimited from but highly dependent upon the regular activity'. Morris and Hough (1987) defined a project as 'an undertaking to achieve a specified objective, defined usually in terms of technical performance, budget and schedule'. Therefore, a project is the 'investment of resource by the client for a quantifiable return'. In this context, investment is defined as being financial, manpower and/or material; the return is either commercial or social. This is a useful definition, as it does not restrict the project to any particular industrial, commercial or social sector.

The construction client's view of a construction project therefore is as a separate, temporary activity, often unrelated in character to that of the client's core business, illustrated in Fig. 13.1. For example, consider an international hotel chain experienced in the procurement of new buildings. The client requires a new hotel in order to expand the core business of selling bedrooms, meals, conventions, etc. The construction of the hotel is a non-core project, a temporary activity in order to achieve a strategic aim and to add value to the core business of the client. The project

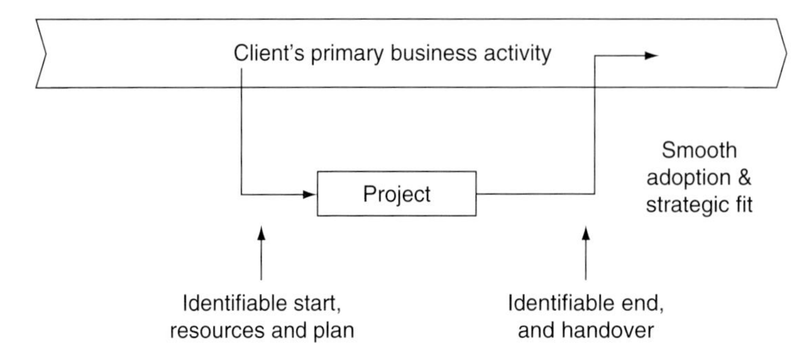

Figure 13.1 The project in relation to client's core business (Source: Kelly *et al.* 2004).

has by definition a start date, a completion date, resources for its undertaking, a method for its smooth integration into the core business, and, ideally, performance indicators which allow its impact on the core business to be measured. The client is core business focused, and the client's team is likely to be a permanent team striving for long term business success. This in some respects sits uneasily with the construction project team that is by definition a project focused temporary team striving for short term success.

A further distinction between projects and core business is the smoothness or unevenness of workflow. Core business usually involves a process for which an employee has been specifically trained and tends to adapt smoothly to demand, although subjected to peaks and troughs. Projects start, progress and end, often involving skills not present in the core business organisation. This presents a logistical problem for those involved with project work who often become fixated with work smoothing by having a number of projects at various stages of development to exactly match the resources available. Unfortunately, for those involved with project work, it is often a case of feast or famine.

The monitoring of core business performance will rely on a quality scheme such as balanced scorecard, benchmarking, European Foundation for Quality Management (EFQM), Six Sigma, Total Quality Management (TQM), etc. However, once a quality scheme has revealed an opportunity to improve performance, there is then a high likelihood that a project will need to be launched to investigate. This distinction between projects and core business is important, because value management can only be applied to projects where projects are defined as temporary activities as above.

In summary, a construction project is defined as an investment by a client organisation on a temporary activity to achieve a core business objective within a programmed time that, by reference to the client's value system, returns added value to the business activity of the organisation. Commercial management has to operate within this definition. The tools that may be used by the commercial manager include value management, reviewed in the next section.

A theory of value management in the context of construction projects

The concept of managing value as well as cost has a history of over half a century, although its application as a formal technique in construction is relatively recent. In the late 1940s the manufacturing industry of the USA evolved a technique for value engineering that focused on the elimination of unnecessary cost in manufacturing through the analysis of product function. For 30 years the technique remained embedded in the manufacturing industry. In the USA in the early 1970s, the value engineering technique mutated for use in construction; over the past 30 years the technique of what is now called 'value management' has become more sophisticated, structured and international in its application. In 1996, President Clinton signed Public Law 104–106 (United States Public Laws 1996), which requires *inter alia*, all Federal projects to be the subject of value management studies with the aim of improving performance, reliability, quality, safety and life cycle costs. Other countries followed with similar directives; for example, mandatory provisions were instituted in Australia in 2001. With strong support from French and German authorities, the European Value Management standard EN 12973 was issued by CEN in 1999 and the British Standard BS EN 12973: 2000 *Value Management* was published in 2000 (European Committee for Standards (CEN) 2000). Value management was defined by Kelly *et al.* (2004) as:

> *The name given to a process in which the functional benefits of a project are made explicit and appraised consistent with a value system determined by the client.*

This brief definition leads to a number of questions.

What is the value management process?

Value management studies normally comprise short intensive workshops at which the project stakeholders are guided through a structured analytical process by a skilled value management facilitator. The stages of the process are often termed 'the job plan' and comprise:

- A pre-workshop orientation phase where the client representative, the key stakeholders and the facilitator agree the focus of the study and the membership of the stakeholder team
- An information phase, the first activity of the workshop, where all information relating to the focus of the study is clearly described and each component part analysed to determine its function
- A value exposition phase, where the client's value system is made explicit in a form which can be used to appraise options
- An innovation phase, during which the stakeholder team explore options and evaluate them by reference to the client's value system
- Action planning, during which options that at first sight seem advantageous are listed and assigned to the person best able to develop them in detail
- A post workshop development phase, during which options are examined in more detail
- A post workshop implementation meeting at which the key stakeholders meet to agree the way forward to the next stage.

What are functional benefits?

In a value management exercise the project is disassembled into its component parts and the function of each part determined to ensure maximum value. In this context, maximum value is achieved when the component part delivers maximum function at the required quality for the least cost. If a part of the project is found to be delivering no functional benefit to the business as a whole or the project in particular, then it may be concluded that unnecessary cost has been incorporated into the project. In a hospital, for example, high circulation space leads to excessive patient journeys by trolley, unnecessary capital cost and the unnecessary recurrent costs of cleaning, heating, lighting, porterage, etc., none of which gives any functional value to the business of healing the patient.

What is the client's value system?

Simplistically, the client's value system is an algorithm or expression that incorporates all of the factors of project success, each expressed with reference to their degree of importance. For example, a boiler house refurbishment project might be judged a success if it was to be completed on time, within the capital budget, with minimum disruption to service, and on completion had the characteristics of low operating costs, comfortable environment for boiler house staff, low carbon emissions, flexible heating/hot water regimes, etc.

The client's value system should be defined for each project and circulated to all project participants. Each strategic decision, design development, construction or facilities management plan should be audited against the client value system.

Relationship between quality and value

Burt (1975) stated that maximum value is achieved when the required level of quality is obtained at the least cost, the highest level of quality is achieved for a given cost, or an optimum compromise is achieved between the two. A number of value engineering texts concur, stating that value denotes a relationship between function, cost and worth (e.g. Parker 1977) or function, cost and quality (e.g. Dell'Isola 1997). Seeking options to meet users' needs, desires and expectations, with varied cost, quality and worth attributes is undertaken usually for the purpose of improving value. This concept can be summarised as '*the lowest cost to reliably provide the required functions or service at the desired time and place and with the essential quality*' (Mudge 1989). Value is expressed in the US value engineering literature as:

- *Exchange value*: Relates to worth or the sale price of a product under the voluntary conditions of a willing buyer and a willing seller
- *Esteem value*: Relates to the monetary amount to be paid for functions of prestige, appearance and/or other non-quantifiable benefits. It may also refer to the monetary measure of the functions of a product that contribute to its desirability or sale
- *Use value*: Relates to need or the life cycle cost, considering user function only, or those properties that accomplish a use by some work or service.

Fallon (1980) quoted from Adam Smith's *An Enquiry into the Nature and Causes of the Wealth of Nations* (Smith 1776) on the paradox that extremely useful goods, such as water, have little or no exchange value, whereas certain other goods of comparatively little use, such as diamonds, have great exchange value. Miles (1989) stated that the degree of value in any product depends on the effectiveness with which every useable idea, process, material and approach to the problem has been identified, studied and utilised. Value is therefore considered to be good if a product of lower or equivalent cost to its competitor contains a somewhat better combination of ideas, processes, materials and functions.

In summary, value is defined here as a relationship between cost, time and quality, where quality comprises a number of variables determined by the client corporate and including such aspects as esteem, exchange and use. The value system made explicit is therefore the representation, at a fixed point in time, of a discrete range of variables against which all decisions affecting core business or a project can be audited. These variables are encapsulated by time, cost and quality.

Quality in the context of the construction project

Definitions of quality, even in textbooks and articles on the subject, are few in number. Bicheno (2002) quoted Deming (1986) who stated that quality can only be defined in terms of customer satisfaction: there is no absolute measure – two customers may perceive a product or service differently. Juran and Gryna (1988) defined quality as 'the totality of features and characteristics of a product or service that bear on its ability to satisfy stated needs or implied needs. Quality consists of freedom from defects'. This definition of quality is repeated in standards. Vorley (1998) quoted Juran and Gryna's definition of quality and commented that quality needs to cover more than function, citing aesthetics, the method of distribution, initial and running costs, user awareness or knowledge.

In his book *Zen and the Art of Motorcycle Maintenance*, Pirsig (1974, p. 252) stated:

> *...why does everybody see Quality differently? ...Quality is shapeless, formless, indescribable. To see shapes and forms is to intellectualise. Quality is independent of any such shapes and forms. The names, the shapes and forms we give Quality depend only partly on the Quality. They also depend partly on the a priori images we have accumulated in our memory. We constantly seek to find, in the Quality event, analogies to our previous experiences... The reason people see Quality differently is because they come to it with different sets of analogies.*

If this is true, then it is logical to assume that individuals setting targets for attributes of quality must have sufficient experience or knowledge of an analogous event and have the tools and measures to assess that event. For example, someone working with polyester resins might be in a good position to be able to define adequate ventilation in a workshop where polyester resins were being used. However, without appropriate tools and measures, they may be unable to specify a quality ventilation system.

Bicheno (2000) described the Kano model developed by the Japanese quality guru Dr Noriaki Kano who states that maximum quality is attained when targeted characteristics are achieved and the customer is delighted. There are three variables within the model. These are 'basic factors', 'performance factors' and 'delighters', which have a relationship to the presence of quality characteristics and customer satisfaction. These variables are included in the Kano model, illustrated in Fig. 13.2.

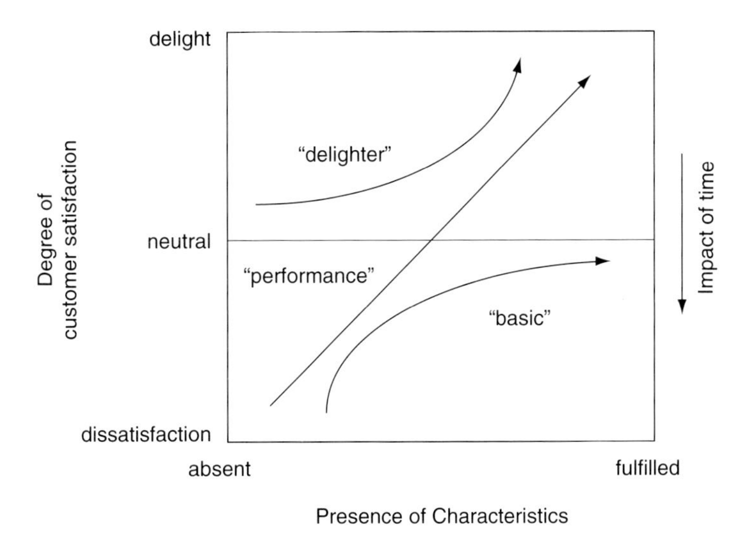

Figure 13.2 The Kano model (adapted from Bicheno 2000).

In the Kano model, a basic characteristic is expected to be present; the customer will be dissatisfied if it is absent and only neutral if the characteristic is completely fulfilled. The performance characteristic relates to the essential function and assumes the more of this facet the better, i.e. the customer will be more satisfied if higher levels of performance are achieved. The delighter is that extra, often personal, characteristic experienced by the customer. There is, however, a time dimension to the model such that the three variables will tend to sink over time, i.e. what once delighted is now expected, and higher levels of performance are always sought. For example, power steering on small cars as a standard feature once delighted customers, but now power steering is expected as a basic characteristic and its absence would lead to dissatisfaction.

In summary, 'quality' is defined here as the degree to which stated objectives, characteristics and/or attributes have been met. This is often associated with a degree of excellence, and is the provision of all basic functions at the required level and all performance functions at the highest level. Delight functions provide the added value. Quality is thus a construct of comparability.

Synthesis of value and quality in the context of a construction project

There is sufficient evidence to accept the premise that value denotes a relationship between time, cost and quality, and that quality relates to the extent to which objectives, characteristics and/or attributes have been met. The objectives, characteristics and/or attributes can be classified as being basic, performance or delight, and are defined by individuals or groups by reference to a comparator.

A framework for the representation of a value system representing time, cost and quality in relation to a construction project is as follows:

- *Time*: From now until the completion of the project, the point at which the project is absorbed into the core business of the client
- *Capital cost*: The amount to be paid for the project
- *Operating cost*: The amount to be paid to maintain the change brought about by the project for as long as it remains a part of core business
- *Esteem*: The admiration and respect brought by the project to the core business
- *Exchange*: The amount for which the project may be sold
- *Use/utility/comfort/reliability*: The user satisfaction measure
- *Environment*: The client's environmental principles regarding the impact of the project on the environment beyond the legislated minimum
- *Politics/community*: The impact of the project on the local community and the political popularity of the project
- *Flexibility*: The extent to which it is necessary for the project to adapt to continuous changes in core business.

In the development of the framework it has been necessary to avoid variables that are highly correlated. Following the logic of comparison and marginal utility, the tool developed uses paired comparison to determine a ranking in client values. Fieldwork trials determined that this approach is robust (Kelly *et al.* 2004). Figure 13.3 illustrates a value system for a particular hospital and demonstrates that in this instance the client was not seeking a building with high esteem, but was certainly expecting the design and construction team to meet the challenge of the provision of flexible space.

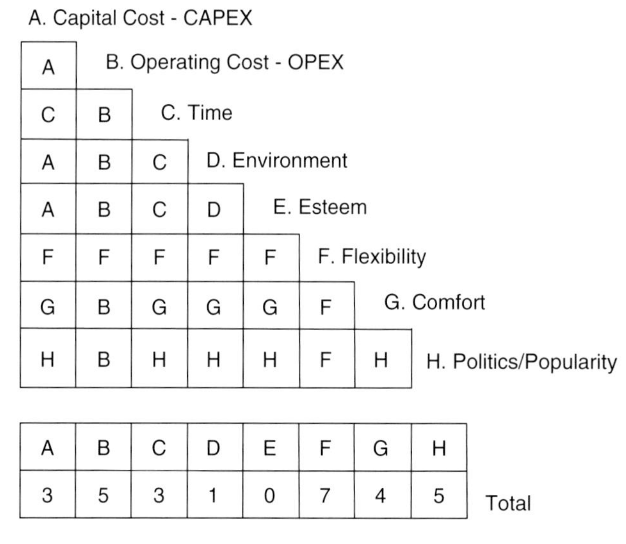

Figure 13.3 A paired comparison approach to determining a client's value system, showing the ranked importance of each variable in the total line.

Value management within the supply chain and value chain theory

A supply chain is a representation of the processing and movement of material or the provision of services, from the raw source to the hands of the client. All of the organisations within the supply chain are independent firms but are commercially dependent upon each other. Supply chain management is usually viewed from the position of an organisation within the supply chain in terms of its management of upstream and downstream organisations for mutual corporate success. Supply chain management should not be confused with vertical integration, which is seen as being the ownership of processes upstream and downstream of the core business activity.

A common mistake made by organisations in supply chain management is to view the supply chain from their organisation downward rather than understanding their position in the ultimate client or customer's total supply chain. The ultimate client or customer may be several positions above the organisation.

Many construction projects are undertaken by a number of organisations operating in silos as illustrated in Fig. 13.4. The value system is strong between client and design team leader but becomes weaker as it is passed between silos. Supply is illustrated in Fig. 13.4 by the black arrows, denoting 'over the wall' thinking with no value chain in place.

In contrast, if the silos were destroyed and the various members of the supply chain positioned so that they could all see the client, then everyone would be equally aware of the client's value system. By understanding the client value system, each member of the supply chain ensures that the goods and services they supply contribute added value to the next supplier in line, and so on until incorporated as a part of the project. This principle is illustrated in Fig. 13.5.

In this context, the core project team have an important responsibility in the structuring of the supply chain to ensure compatible relationships that create and maximise value. Each supply chain member should feel sufficiently comfortable and have the confidence to identify non-value-adding operations which exist anywhere within the value chain. It will require confidence for a minor supplier to suggest that their

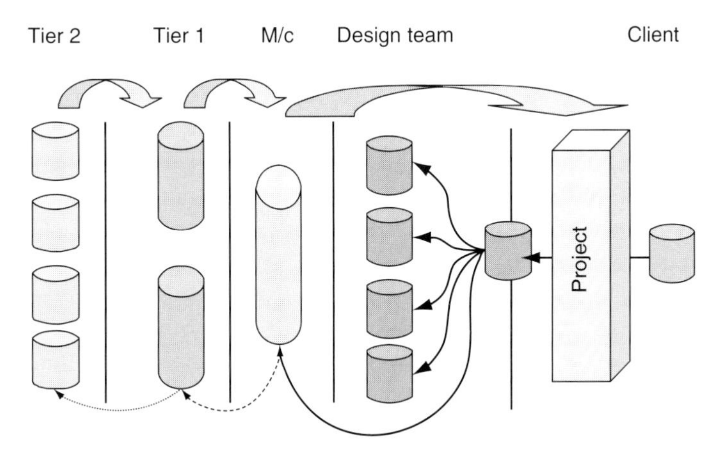

Figure 13.4 Supply chain divided into operational silos. M/c, main contractor.

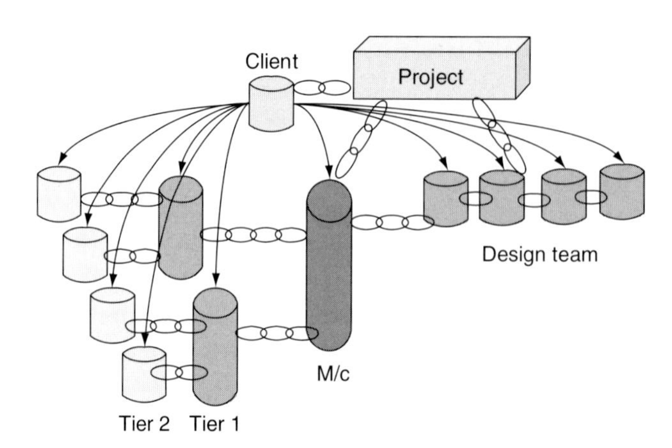

Figure 13.5 Client facing structure: each member of the supply chain receives the client's value system and contributes to the value chain. M/c, main contractor.

paymaster has a non-value-adding activity that is impacting on the project. Value chains work best within a partnering environment. Each organisation in the value chain accepts, as a fundamental requirement of the partnering agreement, the necessity to undertake activities to secure the elimination of waste by co-operation and also to participate in exercises aimed at preventing things from going wrong.

Value engineering

Value engineering is a specific activity that takes place as soon as there is some project detail or design to analyse. A definition of value engineering is:

> *an organised approach to providing the necessary functions at the lowest cost without compromising quality* (Miles 1972)

or alternatively:

> *an organised approach to the identification and elimination of unnecessary cost without compromising quality* (Miles 1972)

In this context, unnecessary cost is that which does not provide use, life, appearance or customer benefits. The reference to quality is important as value engineering can only be recognised as such when the product or process under review can be modified but still delivers the same quality, i.e. specification reduction is not value engineering unless the specification itself comes under review.

Value engineering activity follows the same 'job plan' process as value management carried out within a facilitated workshop. However, because the study is normally narrowly focused, workshops tend to involve few people, are short in duration and are self contained in terms of decision making such that action planning is limited and implementation workshops are rare.

Value engineering revolves around five questions:

(1) What is the process or product under review?
(2) What is its function?

(3) How much does it cost?
(4) What else will perform the function to the same quality?
(5) How much does that cost?

Value engineering is extensively used in manufacturing where the opportunities for savings in repetitive operations are high, but it also has a place in construction in the analysis of process, or product in terms of space, elements and components.

Value management case studies

The following three case studies illustrate the beneficial application of value management to project strategy, answering the question 'what precisely are the client's project objectives?' This approach necessarily requires the explicit definition of the client's value system and the functional strategic specification of the project.

Refurbishment of a cement works

A contractor was asked by the client to negotiate a design and build solution for the refurbishment of a cement works with a 30-year-old kiln. The major part of the project was the decommissioning, demolition and replacement of the kiln. A project brief had been prepared which detailed the work to be undertaken. A major risk item for the contractor was derived from one of the client's requirements: the need to maintain production whilst refurbishing the works. The contractor requested a value management study to investigate the strategic mission of the project and specific technical elements of the project brief. A two-day workshop was held with the client's technical team and the contractor's design and construction team. The contractor's project manager introduced the workshop by explaining that production could only be maintained if a new kiln and associated infrastructure were to be built in series with the existing one. The workshop re-examined the project mission and decided that the original mission was incorrect. The desire of the cement company was to maintain supplies to customers; whether the kiln was up and running was of no interest to the customer provided that they received cement. A solution was found to the new mission that allowed the construction of the new kiln on the site of the existing one.

Bottling plant extension

A manufacturer of alcoholic drink was failing to meet demand targets and instituted a project to increase the number of bottling lines, which required a large extension to the bottling plant. An architect was engaged to take a brief for the new building. At a facilitated workshop the client stakeholders defined the project mission as 'the efficient bottling, labelling, packing, storage and distribution of alcoholic drink to meet peak demands'. In a review of the mission, a number of stakeholders questioned whether the problem was truly in the bottling area since the mission also included other activities, particularly storage and distribution. Further analysis revealed that a warehousing policy of minimising stock resulted in a large number of product changes on the bottling lines, resulting in changes to carriers on the machines and labelling, sometimes three or four 1-hour changes per day. The problem was therefore redefined as inadequate warehousing and the technical project redefined as a warehouse extension.

New manufacturing plant

A bespoke furniture manufacturer had a workshop and showroom in a poor inner city location. The manufacturer had sufficient workshop capacity and a loyal workforce and was acknowledged by existing customers to be a reliable supplier of high quality products. However, it was apparent that tenders from new customers were not invited where that tender was preceded by a visit by the customer to the workshop and showroom. 'Lack of prestigious premises' was the problem identified by the marketing department. Reluctantly, the managing director sourced a plot for a replacement workshop on a prestigious out of town commercial park. An architect was engaged to take the brief. The mission of the project was confirmed as being 'to enhance the image of the company through relocation to a prestigious building'. Further analysis of the mission highlighted that the problem lay in the prestigious location of the showroom alone and not the prestigious location of the showroom and workshop. The project was cancelled and the manufacturer rented showroom space in the central business district of the city. Subsequently, the existing showroom was demolished to make way for a workshop extension to meet increased demand.

Value engineering case study – process

The following case study illustrates the use of the value engineering technique in the analysis of process. The contractor had recognised that, as a general rule, savings in process add directly to the bottom line and, if repeatable, could significantly enhance the profitability of the company. An example study was undertaken on the issue of a particular architect's instruction requiring a variation. The project was a six storey office building for an owner occupier. Joinery work had begun when the client, on a visit, indicated a dislike to the ironmongery chosen by the architect. The architect issued the instruction illustrated in Fig. 13.6.

The first stage of the value engineering exercise required the listing of all activities which followed the issuing of the architect's instruction. These are shown in Fig. 13.7.

Architect's Instruction

(1) For all hardwood veneered doors in all office and reception areas change the specified lever handle sets, Type 1 satin chrome, to Type 2 satin stainless steel.

(2) Omit all latch sets where indicated on the schedule and add lock sets.

(3) Enclosed: revised Ironmongery Schedule SW/004/P21/106c.

Figure 13.6 Example of architect's instruction.

(1) Acknowledge receipt of instruction
(2) Copy instruction 4 times
(3) File original instruction in the master file
(4) Send copy of instruction to site QS
(5) Copy ironmongery schedule 4 times
(6) File original schedule in the master file
(7) Update drawing register
(8) Site QS completes purchase request form
(9) Site QS sends purchase request form with copy of instruction and ironmongery schedule to purchasing
(10) Site QS sends copy of instruction to site stores requesting return of Type1 ironmongery
(11) Site project manager sends copy of instruction and schedule to finishes foreman instructing careful removal of fixed Type 1 and the fixing of Type 2
(12) Finishes foreman issues instruction to joiner
(13) Purchasing send purchase order to Supplier 2 for Type 2 ironmongery
(14) Purchasing send copy of purchase order to site stores
(15) Purchasing send copy of purchase order to accounts
(16) Purchasing send copy of purchase order to site QS
(17) Joiner removes Type 1 and returns to site stores
(18) Site stores issue returned goods receipt to joiner with copy to file
(19) Site stores send unfixed and previously fixed Type 1 ironmongery to supplier
(20) Site stores raise returned goods voucher
(21) Site stores send returned goods voucher to supplier including request for refund
(22) Site stores send returned goods voucher to accounts
(23) Supplier sends credit note to accounts
(24) Accounts modify payment file for supplier
(25) Copy of modified payment file sent to site QS
(26) Supplier 2 send Type 2 ironmongery to site
(27) Joiner collects Type 2 ironmongery from stores and fits
(28) Supplier 2 send invoice to accounts
(29) Site stores receive Type 2 ironmongery and complete materials received summary
(30) Materials received summary sent to site QS
(31) Materials received summary sent to accounts
(32) Accounts check materials received summary against purchase order and invoice
(33) Accounts raise cheque to pay Supplier 2
(34) Accounts send cheque
(35) Joiner completes timesheet cafor removal and refitting of ironmongery
(36) Joiner gives timesheet to finishes foreman for checking and signature
(37) Site cost clerk deducts joiner's time removing and refitting ironmongery from bonusable work
(38) Site cost clerk credits time removing and refitting ironmongery to joiner's pay record
(39) Site cost clerk completes daywork record of time spent
(40) Site cost clerk passes daywork record of time spent to site QS
(41) Site QS extracts data from purchase order, daywork record of time spent, suppliers modified payment file, breakdown bill of quantities priced by the estimator
(42) Site QS calculates additional time and materials involved in ironmongery exchange and completes daywork sheet.
(43) Daywork sheet passed to site project manager for signature
(44) Site project manager signs daywork sheet
(45) Finishes foreman indicates time spent is daywork
(46) Finishes foreman signs timesheet and passes to site cost clerk
(47) Copy of daywork sheet returned to site QS for filing
(48) Daywork sheet sent to clerk of works/architect
(49) Daywork sheet signed by clerk of works/architect as an accurate record of time spent and materials expended
(50) Signed daywork sheet sent to site QS
(51) Site QS prices daywork sheet
(52) Site QS includes cost of daywork in interim valuation submission
(53) Site QS sends interim valuation submission to client's QS
(54) Site QS and client's QS meet to agree interim valuation including architect's instruction requiring a variation
(55) Client's QS completes interim valuation
(56) Client's QS sends interim valuation to architect for completion of interim payment certificate
(57) Client's QS completes cost report, noting additional cost due to ironmongery exchange at amount agreed in interim certificate
(58) Architect completes interim payment certificate and sends to contractor's accounts department
(59) Accounts department raise VAT invoice and sends it with interim payment certificate to client for payment
(60) Accounts send copy of invoice and payment certificate to site QS
(61) Client sends cheque to accounts department in payment of invoice
(62) Accounts department send payment received summary to site QS
(63) Site QS notes architect's instruction now paid
(64) Site QS includes amount of agreed payment in final account

Figure 13.7 Activities following the issue of an architect's instruction.

(1) ****Acknowledge receipt of instruction
(2) Copy instruction 2 times
(3) File original instruction in the central master file
(4) Copy ironmongery schedule 2 times
(5) File original schedule in the central master file
(6) Update drawing register
(7) ****Site project manager sends copy of instruction and schedule to finishes foreman instructing careful removal of fixed Type 1 and the fixing of Type 2 3515
(8) Finishes foreman telephones supplier of Type 1 to explain circumstances of return
(9) Finishes foreman telephones supplier 2 to order Type 2 ironmongery
(10) ****Finishes foreman issues instruction to joiner
(11) ****Joiner removes Type 1 and returns to site stores
(12) ****Site stores send unfixed and previously fixed Type 1 ironmongery to supplier
(13) Supplier sends credit note to accounts
(14) Accounts modify payment file for supplier
(15) ****Supplier 2 sends Type 2 ironmongery to site
(16) ****Joiner collects Type 2 ironmongery from stores and fits
(17) Supplier 2 sends invoice to accounts
(18) Accounts raise cheque to pay supplier 2
(19) Accounts send cheque
(20) Site QS prepares application for interim payment based on costs to date plus a fee
(21) Site QS sends details to accounts department
(22) Accounts department raise VAT invoice and send to client
(23) ****Client sends cheque to accounts department in payment of invoice

Figure 13.8 Revised activities following the issue of an architect's instruction.

In the value engineering study, the 64 operations were analysed for function; it was discovered that 41 of the operations related to three functions:

- To comply with company quality policy
- To guard against fraud
- To facilitate measured work bonus scheme.

During the innovation phase and in the context of the study it was decided to adopt a position of trust without compromising the essential elements of the company's quality policy. It was further suggested that the bonus scheme should be based upon the profitability and safety record of the project and paid to all operatives and staff. The procedure was re-written as shown in Fig. 13.8. It could be argued that only activities 1, 7, 10, 11, 12, 15, 16 and 23 (annotated **** in Fig. 13.8) add value from the perspective of the client, i.e. only eight out of 23 activities or 35% efficient. However, in the previous situation this was eight out of 64 or 12.5% efficient. Admittedly, this approach is fairly simplistic as it ignores the time spent on each activity and therefore its cost, but it illustrates how a structured value engineering (process study) can add significantly to the bottom line of the company.

Value engineering case study – product

In the context of building, most value engineering work is undertaken at the level of space, elements and components. The case study involved a major refurbishment of an office building, constructed in the 1960s, comprising reinforced *in situ* concrete frame and floors, precast concrete dado panels with single glazed, horizontal sliding aluminium windows. The south elevation of the building faced a busy High Street and the north elevation backed onto a service yard with houses and gardens beyond. The east and west elevations were precast panels requiring new mastic pointing but

(1) Transmit light	(2) Separate climates	(3) Enhance esteem
(4) Control ventilation	(5) Control solar gain	(6) Enhance aesthetic
(7) Attenuate noise	(8) Restrict heat loss	(9) Secure space
(10) See outside	(11) Control glare	(12) Minimise maintenance
(13) Enclose space	(14) Control cold radiation	(15) Facilitate cleaning

Figure 13.9 Element functions for the element 'window'.

otherwise, with cleaning, were in good condition. The project was design and build with a novated design team. The original design involved the stripping back of the building to the original frame and floors with new glass curtain walling fixed to the front and rear elevations. Full air conditioning was to be integrated but because of the restricted floor to floor heights was to involve an innovative up and down distribution through a raised access floor, necessitating considerable diamond drilling of the floors and fire damping. Extensive finishing treatment was to be required to a now exposed concrete soffit, there being no room for a suspended ceiling. The lowest contractor's bid was well in excess of the cost plan and the project as designed was not viable. The contractor convinced the client that there was a solution and suggested a value engineering study. The following is a description of the action on the curtain walling element.

Stage 1 of the study was a holistic, generic analysis of the element function – window. The functions generated at the workshop are as listed in Fig. 13.9.

Stage 2 of the study was to ascribe the functions to the office building. A debate highlighted the fact that control ventilation and attenuate noise were mutually exclusive and could not both be delivered by the element window to the south elevation. At this point it became clear that the functions were different for the north and south elevations.

Stage 3, the innovation stage, involved brainstorming, evaluation of options and recommendations for development. Following the logic that all functions except four applied to the south elevation, whereas only 1, 2, 4, 8, 9, 10, 12, 13, 14 and 15 applied to the north elevation, the recommendations for development were to replace the rear elevation windows with modern opening double glazed units incorporating ventilation. For the front elevation a higher quality solar controlled glass curtain walling was selected, set off 300 mm from the face of the building, which allowed space for ventilation ducting fixed to the edge of the floor slab. Comfort cooling cassettes were suggested for the office ceilings in place of full air conditioning. The raised access floor was omitted.

Stage 4 was development and implementation. The ideas from the workshop were tested for economic viability, technical feasibility, functional suitability and client acceptability. The lower cost, but in many ways higher quality, solution was accepted by the client.

Discussion of case studies

There is a view that value management and an understanding of the client's value system are the preserve of those consultants appointed by the client. Newer procurement systems such as design and build, public–private partnership/private finance initiative, management contracting, prime contracting, Procure21, and the growing adoption of partnering give an opportunity for the forward thinking commercial manager to offer the benefits of an integrated service direct to the client. This chapter has highlighted the importance of a full understanding of value management illustrated by three case studies, all of which expose some of the shortcomings of existing briefing. These studies were all triggered by an overspend after the appointment of the contractor. Obviously this is not ideal, but it does highlight the need for an early integrated approach and opportunities were the contractor to offer an integrated service.

Value engineering is a much misused term, often being seen by many quantity surveyors and commercial managers as a synonym for specification reduction. The two case studies above illustrate the use of value engineering in two ways: process and product, the latter being more common. For reasons which defy logic many contracting organisations are unwilling or unable to spend the time and resources to examine their processes, even though this is the quickest and most effective way of adding to their bottom line. The final case study is typical of many element orientated studies. These are generally quick (if done correctly), economic of resources and very effective, with pay backs many times the cost of undertaking the study. As in the final case study, the action of undertaking the study can often highlight the obvious and perhaps this is its downfall. After all, most of us have perfect 20/20 hindsight.

Conclusion

This chapter commenced by referring to the fact that commercial management is an ideal vehicle to facilitate the introduction of value management and value engineering into current construction processes. The reference to construction is deliberate, although it is accepted that commercial management is applicable to any endeavour where the objective is to pursue a project. Although manufacturing was the genesis of value engineering and has subsequently taken value engineering to high levels of sophistication, the pursuit of value management has largely been ignored. The reasons for this are not well understood, but are suggested here to be a result of the lack of integration between marketing and production. Currently, it is the production departments which promote and execute value engineering, but examples of value management within marketing and customer orientated product innovation are few in number. In this respect the 'lack of integration' criticism may well be directed at manufacturing.

The greatest opportunity for value management, however, does not lie with manufacturing but with the public sector. Early research work with the public sector has demonstrated a perfect strategic fit between value management and the development of public sector projects. A recent experiment, considering the local authority responsibilities with respect to early childhood services in a cross-cutting study involving education, social services and health, demonstrated the power of the value management approach in focusing on the mission of the project and innovative ways of satisfying the mission whilst maximising the benefit to all involved stakeholders.

Research work in value management and value engineering is required into the manufacturing areas of customer orientated product innovation, as mentioned above, and also into the performance specification of production space. Currently, space for production is considered to be a fixed overhead. Early research in this area has demonstrated that this treatment ignores the full potential and liability of production infrastructure. In construction, attention is required in the areas of strategic and project briefing. Although some work has been done in this area, the application of primarily value management is limited; this needs investigation. Finally, as stated above, the prime opportunity lies in the public sector in the theories and applications which lie behind Best Value legislation.

References

Bennett, J. (1985) *Construction Project Management*. Butterworth, Cambridge.

Bicheno, J. (2000) *The Lean Toolbox* (2nd edn). PICSIE Books, Buckingham.

Bicheno, J. (2002) *The Quality 75: Towards Six Sigma Performance in Service and Manufacturing*. PICSIE Books, Buckingham.

Börjeson, L. (1976) *Management of Project Work*. Swedish Agency for Administrative Development, Statskontoret, Gotab, Stockholm.

Burt, M.E. (1975) *A Survey of Quality and Value in Building*. Building Research Establishment, Watford.

Dell'Isola, A. (1997) *Value Engineering: Practical Applications for Design, Construction, Maintenance and Operations*. R S Means, Kingston, MA.

Deming, W.E. (1986) *Out of the Crisis*. Massachusetts Institute of Technology, Cambridge, MA.

Egan, J. (1998) *Rethinking Construction*. HMSO, London.

European Committee for Standards (CEN) (2000) *BS EN 12973:2000 Value Management*. British Standards Institute, London.

Fallon, C. (1980) *Value Analysis* (2nd edn). Miles Value Foundation, Washington DC.

Juran, J.M. & Gryna, F.M. (1988) *Juran's Quality Control Handbook* (4th edn). McGraw-Hill, New York.

Kelly, J., Male, S. & Graham, D. (2004) *Value Management of Construction Projects*. Blackwell Publishing, Oxford.

Langford, D. & Male, S. (2001) *Strategic Management in Construction* (2nd edn). Blackwell Science, Oxford.

Miles, L.D. (1972) *Techniques of Value Analysis and Engineering* (2nd edn). McGraw-Hill, New York.

Miles, L.D. (1989) *Techniques of Value Analysis and Engineering* (3rd edn). Miles Value Foundation, Washington DC.

Morris, P.W.G. & Hough, G.H. (1987) *The Anatomy of Major Projects: A Study of the Reality of Project Management*. John Wiley & Sons, Chichester.

Mudge, A.E. (1989) *Value Engineering: A Systematic Approach*. J Pohl Associates, Pittsburgh, PA.

Parker, D.E. (1977) *Value Engineering Theory*. Miles Value Foundation, Washington DC.

Pirsig, R.M. (1974) *Zen and the Art of Motorcycle Maintenance*. Vintage, London.

Smith, A. (1776) *An Enquiry into the Nature and Causes of the Wealth of Nations*. W. Strathan and T. Cadell, London.

United States Public Laws (1996) *National Defense Authorization Act for Fiscal Year 1996: Public Law (PL) 104–106*. Public Laws enacted by the 104th Congress of the United States of America.

Vorley, G. (1998) *Quality Management (Principles and Techniques)* (3rd edn). Quality Management and Training, Guildford.

Winch, G. (2002) *Managing Construction Projects*. Blackwell Science, Oxford.

Part 4
Project Milieu

Introduction

This last part focuses on the interfaces at the specific project level. These comprise interfaces with the external environment (purchasers, competitors and suppliers) and the internal environment, regarding the implementation and monitoring of organisational governance and strategy, and interaction with the divisions of the organisation responsible for project execution. The following five specific topics have been selected for further explanation: project governance; contract management; bidding; purchasing, logistics and supply; cost–value reconciliation.

Project governance

Projects, regardless of size and scope, inherently involve relationships of contract and subcontracts. These could be between one or several suppliers and their clients in a network of relationships. The governance of the project is affected by how the organisation manages transactions within these relationships, be they external or internal. In Chapter 14, Graham Winch addresses how the different transactions that take place over the project life cycle can be co-ordinated and controlled. Drawing upon transaction cost economics, he provides a theoretical framework that allows for all transactions to be taken into consideration.

Contract management

Contract management has been defined as 'the process which ensures that all parties to a contract fully understand their respective obligations enabling these to be fulfilled as efficiently and effectively as possible to provide even better value for money' (Central Unit on Procurement 1997). This process commences with the identification of the purchaser's needs and concludes with the completion of the contract. Further, the process has two dominant characteristics:

- *Risk identification, apportionment and management*: related to contract performance
- *Relationship management*: between the purchaser and supplier.

Underlying the law of contract are a few basic assumptions: a contract is based upon a mutual understanding between the parties, the contract terms express the intentions of the parties, parties are free to choose their terms of business, and privity of contract restricts the effect of a contract to the two parties who sign it. However, as shown by Will Hughes in Chapter 15, practice is a long way from being straightforward and frequently diverges from these ideals. Drawing on experience from the construction sector, this chapter provides an absorbing examination of the commercial context of business-to-business contracts as it seeks to explain the relationship between law and procurement.

Bidding

Notwithstanding the increased use of partnering and open book arrangements, competitive bidding is still the most commonly used approach to supplier selection. Moreover, existing and anticipated EU legislation advocates the use of competition. It is widely held that, for the supplier, the bidding process involves two crucial decisions: first, whether or not to bid (tender) for a project, and second, the determination of the bid price. In Chapter 16, David Lowe and Martin Skitmore assert that this statement is overly simplistic, as it does not consider the fact that suppliers continually review the decision to bid (d2b) from prospect identification through to bid submission, and it implies that price is the only criterion used to evaluate bids. Taking a generic approach, they draw upon research from a variety of industry sectors to set bidding decisions within the context of the supplier's business plan; review the factors that influence the d2b process and evaluate several models developed to inform the process; examine the process of supplier selection, in particular the bid evaluation criteria adopted by purchasers and the transition from lowest price to multi-criteria selection; analyse the factors that influence the determination of the bid price; and appraise the various models developed to derive the optimum bid price.

Purchasing, logistics and supply

In Chapter 17, Andrew Cox and Paul Ireland provide an overview of the key schools of thought contained within the procurement and supply chain management literature. They contend that many of the recent 'best practice' tools and techniques that have been developed within project-based industries to allow practitioners to manage project relationships effectively, may not deliver the anticipated benefits in all cases. Using evidence from the UK construction industry, the chapter provides practitioners with a theoretical framework for understanding an industry and supply chains that have primarily a project-based demand and supply structure. This framework, which may be applied across other project-based industries, provides practitioners with a way of thinking about the attributes of buyer and supplier power and leverage in construction, the appropriateness of different relationship management approaches, and the circumstances under which supply chain management approaches, recommended by recent government and industry reports, can be implemented successfully.

Cost–value reconciliation

Accounting regulations for organisational governance require organisations to identify the income and expenditure that arises from their projects. Project budgets are used as a basis for reporting value allowances and associated cost items, and cost–value reconciliation reports are prepared by the production teams for communication to senior management for decision-making. Andrew Ross and David Hugill suggest in Chapter 18 that a research opportunity exists to consider these communications from site and argue that information that arises from this process may be structured in a manner which may assist in the predictive process. They suggest that the barriers to the use of accounting data as a basis for process modelling relate to the power differential that exists between site and head office, and that this is reflected in the design of the information systems used by construction organisations. Using an analysis of a number of cost–value reconciliation forms, they suggest that accounting data could form a useful resource for modelling organisational and interorganisational costs.

The contributor's profiles

Graham Winch

Graham M. Winch, BA, MA, MSc, PhD, MAPM, PMI, is Professor of Project Management in Manchester Business School, the University of Manchester, and a director of the Centre for Research in the Management of Projects. A social scientist by training, he has run construction projects and researched various aspects of innovation and project management across a wide variety of engineering sectors. He is the author of *Managing Production: Engineering Change and Stability* (Oxford University Press 1992), a study of the implementation of CAD/CAM systems, *Innovation and Management Control* (Cambridge University Press 1985), a study of new product development in the car industry, and most recently, *Managing Construction Projects, an Information Processing Approach* (Blackwell 2002). These are complemented by over 30 refereed journal articles, numerous book chapters, conference papers, and research reports. He has held numerous ESRC and EPSRC awards, and is currently

Principal Investigator on *ProRIde: Project Risk Identification* and a co-investigator on *Rethinking Project Management: Developing a New Research Agenda.*

Will Hughes

Dr Will Hughes, BSc, PhD, MCIOB, is a Reader in Construction Management and Economics in the School of Construction Management and Engineering, University of Reading, having started his career as a joiner in the West Midlands. He has an international reputation in the field of construction contract law and project organisation. His PhD is on *Organisational analysis of building projects*, and he has researched and published extensively on this topic. His current research interests are in the costs of the commercial processes in construction and contracting based upon paying for the performance of a building, rather than for what it is made of. Previous research into contractual and procurement issues has been undertaken for: the Engineering and Physical Science Research Council, the Féderation Internationale des Ingénieurs Conseils, European International Contractors, Joint Contracts Tribunal, the Reading Construction Forum and the Design Build Foundation (now both part of Be) and the Office of Government Commerce. He has also carried our consultancy work for the Office of Government Commerce. He has co-authored seven books on construction management and contractual issues, including a best-selling textbook, *Construction Contracts: Law and Management* (Spon 1992, 1996, 2000). He is editor of the international journal *Construction Management and Economics*, a member of the editorial boards of the *Journal of Construction Research* and the *Journal of Asian Architectural and Building Research*. He is an active member of the Association of Researchers in Construction Management, having been chairman from 1998 to 2000, and remains responsible for developing and maintaining their information resources for researchers in this field. Dr Hughes is joint director of the Reading Innovative Construction Research Centre and director of the PhD Programme in the School of Construction Management and Engineering. In his spare time, Will is Chairman and Trustee of Woodley Concert Band, a local community wind band, in which he plays trumpet and cornet.

David Lowe

Dr David Lowe, BSc, MSc, PhD, is Senior Lecturer in Commercial Project Management and leads the Management of Projects Research Group in the School of Mechanical, Aerospace and Civil Engineering at the University of Manchester. He is Programme Director for the highly successful MSc in the Management of Projects, which recruits around 65 graduate students each year. He is also joint programme director for a distance-learning MSc in Commercial Management, a bespoke programme for a blue-chip telecommunications company. He is a Fellow of the Royal Institute of Chartered Surveyors. Consultancy work includes benchmarking the engineering and project management provision of an international pharmaceutical company. His PhD, completed at UMIST, investigated the development of professional expertise through experiential learning. Further, recent research projects include: an investigation of the cost of different procurement systems and the development of a predictive model; a project to assist medium sized construction companies develop strategic partnerships and diversify into new business opportunities offered by public and private sector clients; and an investigation into the function of commercial management in the telecommunications and construction industries. He is a member of the editorial board of the *Journal of Financial Management of Property and Construction*, and a member of the RICS Foundation *Research Paper Series* Review Panel. David has published over 40 refereed papers and articles. His book contributions include: Contract management, in *The Wiley Guide to Managing Projects* (Wiley 2004), and Human factors in estimating, in *Estimating Capital Cost of Projects* (Institution of Civil Engineers/Thomas Telford 1995).

Martin Skitmore

Professor Martin Skitmore completed his MSc and PhD at Salford University in the UK where he was a Professor and Director of Research and Postgraduate Studies. In 1995 he moved to Queensland University of Technology, taking over the position of Head of the School of Construction Management and Property until 1997. He is currently Research Professor and Post Graduate Coordinator in the School of Urban Studies. Martin has been party to more than 20 successful research grant applications and has supervised 18 PhD and Research Masters students to completion. He has authored or co-authored several books and over 100 academic journal papers, mainly concerned with price and cost modelling, contractor and consultant selection, and various aspects of project management. Martin is also Co-Editor-in-Chief of the *Journal of Construction Innovation*.

Andrew Cox

Andrew Cox, BA, MA, PhD, FCIPS, FRSA, is Professor in Business Strategy and Director of the Centre for Business Strategy and Procurement (CBSP) within Birmingham Business School. Professor Cox has been awarded a large number of major grants from the Engineering and Physical Science Research Council and other funding agencies to undertake dedicated research in all aspects of business strategy, effective supply chain management and procurement competence. This has involved working closely with a large number of public and private sector organisations in the development of audit tools for the strategic and operational alignment of supply chain management. He has also undertaken extensive surveys of the effective use of business tools and techniques by major multinational companies and the impact of e-business on the strategy and internal processes and procurement, logistics and supply chain management functions of European companies.

Andrew Cox is also Chairman of Robertson Cox, a competence development, consultancy and publishing company. Robertson Cox are currently working closely with a number of leading American and European multinationals within a wide variety of industries on the effective alignment of overall corporate strategy with procurement and supply chain management strategies. Professor Cox is a Fellow of the Chartered Institute of Purchasing and Supply in the UK, holder of the Swinbank Medal for Outstanding Services to Purchasing and Supply, and Fellow of the Royal Society of Arts, Commerce and Manufacture. He has also acted as adviser and consultant to the British Government and the European Commission. He has published numerous books and articles on business strategy and procurement and supply management including *Supply Chain Management: A Guide to Best Practice* (2003) and *Business Relationships for Competitive Advantage* (2004).

Paul Ireland

Dr Paul Ireland, BEng, MSc(Eng), MBA, PhD, is a Research Fellow in the Centre for Business Strategy and Procurement (CBSP) within Birmingham Business School, University of Birmingham. He has undertaken detailed research focused on analysing the structures of power and market dynamics within a wide range of supply chains, including construction, information technology, pharmaceuticals, aerospace, oil and gas, automotive and medical equipment. This research, aimed at the development of appropriate procurement and supply strategies within these sectors, has been funded by leading private sector firms, public sector organisations and industry bodies.

In addition to the study of effective business strategy and supply chain management within these industries, Paul's recent research activities have also focused on effective supply management and business leverage in project-based industries, the impact of the Internet and IT applications on the structure of organisations and supply chains, effective demand and supply management in the UK National Health Service, and the effective sourcing of indirect expenditure. This work has been presented widely at academic and practitioner conferences and disseminated in numerous books and articles including *The E-Business Report* (2001), *Supply Chains, Markets and Power* (2002), *Supply Chain Management: A Guide to Best Practice* (2003) and *Managing Construction Supply Chains: Understanding Appropriate Sourcing Choices* (2005).

Andrew Ross

Dr Andrew Ross, BSc, MSc, PhD, MCIOB, MRICS, is the head of postgraduate programmes at the School of the Built Environment at Liverpool John Moores University. He has a wide range of industrial experience, working as a commercial manager with an international contractor before undertaking an MSc in Information Technology. His PhD comprised an investigation of contractors' project economic organisation. He has many years' experience teaching in the area of construction economics to undergraduate and postgraduate students. His research interests are in construction information management and procurement. He co-authored *Building Design Cost Management* (2002) and has published over thirty national and international refereed research papers. He is a member of the editorial board of the *Journal of Construction Procurement*.

David Hugill

Dr David Hugill, BSc, MBA, PhD, is an Associate Partner of EC Harris, consultants in the real estate, infrastructure, industrial and construction sectors. He is a Chartered Quantity Surveyor and Chartered Builder. He was previously a lecturer at the University of Manchester Institute of Science and Technology. During his period in academe, David's research interests focused on commercial management, spanning management accounting, project management and transaction cost economics. He has published in the areas of quality management and communications, in particular commercial negotiation, and has attempted to adopt an ethnomethodological orientation in his work. Prior to his academic career, David was the Business Manager for a large consultancy operating in the construction and property sectors. He has a wealth of experience on building and civil engineering projects ranging in value from £0.1m to £2.0bn.

Reference

Central Unit on Procurement (CUP) (1997) *Contract Management*. CUP Guidance No. 61. HM Treasury, London.

14 The Governance of Project Coalitions – Towards a Research Agenda

Graham M. Winch

Introduction

How are project coalitions governed? In other words, how are all the different transactions that take place over the project life cycle co-ordinated and controlled so that the client's requirements can be met? The aim of this chapter is to develop further and generalise a framework first presented in Winch (2001) that attempts to provide a consistent set of answers to these questions. It does this by drawing on the body of literature known as transaction cost economics, and, thereby, addresses the problem of the costs of doing business in a project mode. The chapter explores the concept of the project coalition as a *nexus of treaties* governed through the choice of appropriate organisational and legal arrangements. The chapter will first address some issues of definition, before moving on to present the underlying theoretical perspective deployed. Adaptation of this perspective to the particular issues of project management follows, and some indicative research propositions are derived. A brief case study indicates how the governance perspective provides additional insight into commercial relations in a project context. Some suggestions for future research avenues form the conclusions in an attempt to define an aspect of the discipline of commercial management at the interface of law, economics and management in the context of complex projects.

What do we mean by 'governance'?

The term 'governance' has become widely used in many areas of corporate and public life in recent years, and so it is appropriate to clarify how the term will be used in this chapter. Generally, 'governance' refers to the processes of control and direction of an organisational entity, frequently with an additional connotation of accountability to stakeholders. The concept of corporate governance is well established and developed by Goergen and Renneboog in Chapter 5. However, 'governance' has been applied to project organisations in two distinct ways, rendering the notion of 'project governance' ambiguous. First, it refers to accountability for a project to the corporate entity which is sponsoring it. This is the meaning which derives from the literature on corporate governance and accountability, and specifically, the recommendations for the reform of internal control and audit promulgated by the (English and Welsh) Institute of Chartered Accountants in the *Turnbull Report* of 1999 (Smith 2003). The Association for Project Management has recently published a guidance note on what it calls the 'governance of project management' which applies the concept in this manner and clearly distinguishes it from the second meaning (APM 2004).

This second meaning relates to the management of the relationships between the various parties engaged in the temporary multi-organisation (Cherns & Bryant 1984) that is a project. This meaning is derived from the literature on institutional economics. The early work of Coase (1937) identified the market and the firm as alternative modes of allocating resources in a market economy, and this insight has been extensively developed by Williamson and his followers in what has become known as transaction cost economics as one branch of institutional economics (Williamson 1985, Ch. 1). Williamson argues that, where markets are used for resource allocation, spot-contracts in which the parties simultaneously agree terms and exchange goods or services are rare; contracts are typically the result of a continuous process of negotiation both before and after formal contracts are exchanged and therefore require active management rather than passive response to market forces. From this perspective, *commercial management* can be defined as the function of selecting and maintaining appropriate governance arrangements between independent firms; its internal, or hierarchical, counterpart is *human resource management*.

There are, however, links between these two concepts of project governance. Turner and Müller (2003) have argued that the role of the project manager (or project management team) can be conceptualised as the agent for the client of the project, using another conceptual framework from institutional economics – agency theory (Eisenhardt 1989). From this perspective, the assurance of corporate accountability for projects is a principal/agent problem, where the governance challenge is that the agent (the project manager) has more information about the true state of progress on the project than the principal (the corporate entity sponsoring the project).

Understanding the governance of project transactions

In the classic definition, a transaction occurs whenever 'a good or service is transferred across a technologically separable interface' (Williamson 1981, p. 552). A technologically separable interface is one that is not constrained by the nature of the production technology – in other words, the production technology chosen does not foreclose the possibility of allocating two different parts of the production process to different parts of the organisation, or to separate organisations. The fundamental insight of the transaction cost approach is that, in order to economise on the total cost of a good or service, both production costs and transaction costs must be taken into account. Thus total costs are the sum of the costs of production and the costs of governing the transactions inherent in that choice of production organisation. A choice of production organisation that has the lowest production costs might not be the economising choice if transaction costs are also taken into account.

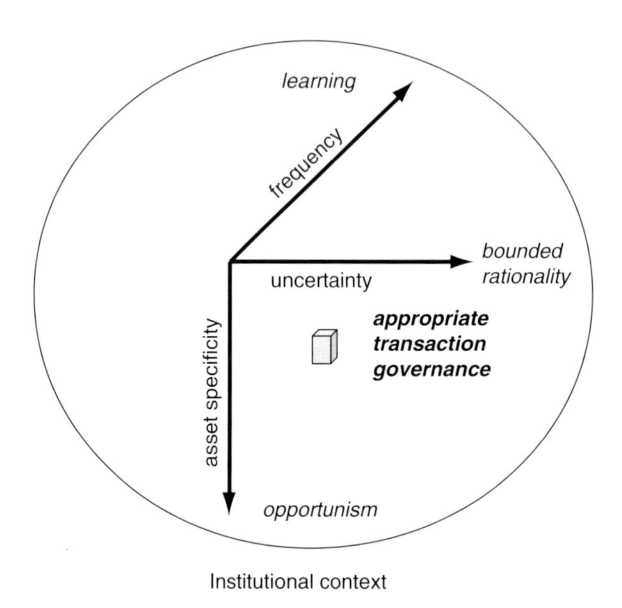

Institutional context

Figure 14.1 The generic transaction governance model (source: Winch 2001).

The basic framework was first presented by Williamson in 1975, and has been elaborated since without losing its initial insight. The elaborated framework consists of three elements:

- *Contingency factors*: The features of the transaction under consideration, which are uncertainty, frequency and asset specificity[i]
- *Behavioural factors*: The ways in which managers typically respond to these factors, which are, respectively, bounded rationality, learning,[ii] and opportunism
- *Context*: Originally called 'atmosphere' by Williamson, this is the sectoral and national business system within which the transaction is embedded, which in turn is situated within the broader national socio-cultural context.

This framework can be summarised in the model presented in Fig. 14.1 which shows how the appropriate choice of transaction governance mode occupies a three-dimensional space as a function of the three contingency factors. These contingency factors are only troublesome in interaction with each other: in the absence of uncertainty, complete contracts can be written in advance to negate opportunistic behaviour arising from asset specificity; in the absence of asset specificity, negotiations to handle unforeseen events can take place when they occur. Frequency is the critical factor in determining the return on investing in transaction-specific governance modes. Differences between sectoral and national business systems contexts will shift the whole space, as will the evolution of business systems through time.

Through extensive research and debate, the market/hierarchy dichotomy that Williamson proposed in 1975 has been elaborated in many ways. There is still debate around the range of governance modes that is possible, but the following can be considered to be the main options (Williamson 1985, Ch. 3):

- *Market*: Where the relationship between the parties is entirely externally governed through the national laws of private property, contract, and dispute resolution
- *Hierarchy*: Where the relationship between the parties is entirely internally governed through administrative means within a unified organisational structure where dispute resolution is resolved by fiat through the exercise of an authority relation
- *Trilateral*: Where third parties are mobilised to arbitrate in the case of dispute between the two principal parties, or more generally to facilitate the relationship. This is a distinctive feature of transaction governance in many project-orientated industries (Maher 1997), and Williamson himself notes the role in dispute resolution role of the architect in construction contracts (Williamson 1985, p. 75)
- *Bilateral*: Where the principal parties negotiate their own governance arrangements.

The process of governance refers to practice on specific transaction-sets (the micro-analytic level), or 'projects milieu' as defined by the editors of this book. Williamson (1996) distinguishes this from the broader set of issues at the macro-analytic (or institutional) level, which set the 'rules of the game' in the national business system and the broader socio-economic context, thereby reducing uncertainty in organisational and individual decision making. This is the 'external milieu' as defined by the editors. As North (1990, p. 3) argues, 'institutions are the rules of the game in a society – in consequence they structure incentives in human exchange, whether political, social or economic'. The range of governance options open to any firm is therefore limited by the institutional context within which it trades. Major features of the institutional context include the security of property rights, the availability of arrangements for the enforcement of contracts, the nature of regulatory initiatives by the state, and arrangements for the protection of workers. There is an extensive literature on the distinctive features of national business systems – see, for instance, Hall and Soskice (2001) and Whitley (1999) for relevant analyses. However, any cross-national comparative analysis of governance arrangements would need to take these into account. Similarly, the sectoral business system constrains the available governance options. For instance, governance arrangements on construction projects are constrained *inter alia* by the range of standard forms of contract available, regulatory arrangements related to land-use planning, and choices by the state as principal client for construction projects in how it procures (Winch 2000a).

As is implicit in Fig. 14.1, the governance level also interacts with the behaviour of individuals and groups within the governance arrangements (Williamson 1996, Ch. 9) or the 'project milieu' in the terms used by the editors. The argument is that the choice of appropriate governance mode is both a function of contingency factors – notably bounded rationality – and also generates particular behaviours such as trust and opportunism. The relationship is not deterministic, and any given governance arrangement can work relatively well or badly depending on the values and behaviours that individuals bring to its implementation from outside the particular governance mode selected. These factors are also influenced by broader issues of organisational and professional culture associated with the transacting organisations. While it is widely asserted that particular contractual arrangements generate particular

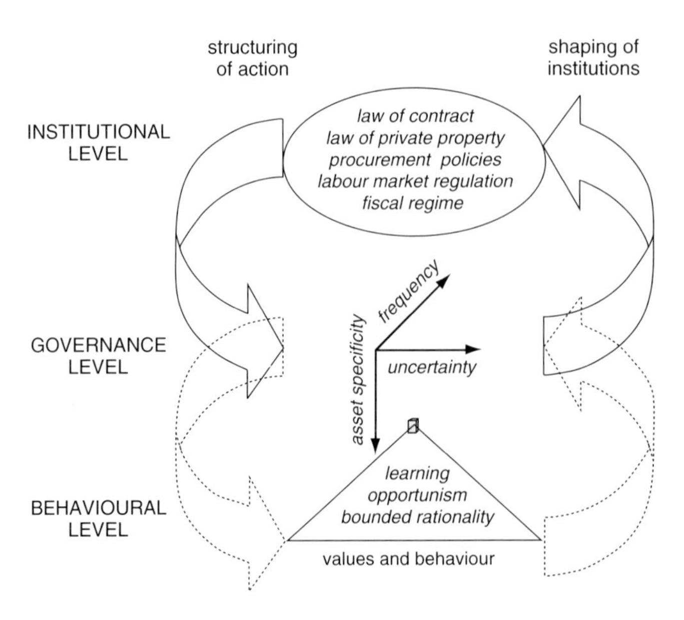

Figure 14.2 The context of transaction governance.

behaviours – positive or negative – there is remarkably little research that actually tests these assertions. Figure 14.2 locates the generic governance model from Fig. 14.1 in this broader context of institutions and behaviour by defining it as the *governance level* mediating between the *institutional level* and the *behavioural level*. Through this mediation, governance choices by firms are both structured by the institutional context and shape that context over time. Similarly, individual behaviours and values are both influenced by the choice of governance mode and influence that choice. This simultaneous process of structuring and being structured can be conceptualised as a process of *structuration* (Giddens 1984), and more specifically as the *tectonic* approach to organisation (Winch 1994). There are, of course, also direct interactions between the institutional and behavioural levels, but it can be suggested that these are beyond the scope of the discipline of commercial management.

The focus of this chapter is on the micro-analytic governance level, for which it presents a conceptual framework to facilitate the more trenchant analysis of the governance of project coalitions. We now turn to the distinctive features of projects which require the adaptation of the generic governance model from Fig. 14.1 so that deeper insight into the governance of projects can be obtained.

The nature of project organisation

Project organisation began to emerge as a distinctive form of business organisation during the nineteenth century as society began to create complex systems at either private or public initiative. The railways were catalytic here, followed by electrical generation and distribution systems. By the 1950s, project organisation was emerging as a fully fledged organisational form attracting significant research interest (Morris 1994). Success with the Polaris, Atlas and Apollo projects meant that project organisation and the associated tools and techniques such as critical path analysis began to diffuse widely. Development faltered as the institutional context of project organisation became more difficult, particularly with the development of the environmental lobby and other stakeholder activism (Horwitch 1987). By the 1990s, however, project organisation was again diffusing rapidly as its value for new product development became acknowledged (Bowen *et al.* 1994) and products became ever more complex (Hobday 2000; Winch 2000b). Most recently, government agencies such as the UK's Office of Government Commerce have begun to appreciate the importance of effective project management for the achievement of national policy objectives.

Traditional forms of production organisation have tended to be functional. That is to say, the different resources required for production are held within discrete organisational units and deployed in sequence. This tends to be favourable so far as the efficiency of resource use is concerned, but can comprise production effectiveness as flows between units are poorly co-ordinated. Attempts to improve the customer focus of the functional approach have led to organisation by product rather than by function. While this can improve production flows and effectiveness, it tends to compromise the efficiency of resource use. Product-orientated organisations also tend to regress to functional ones over time. Matrix organisation was developed in order to have the best of both these worlds, but is seen as unstable and conflict-ridden, and is not widely used (Winch 1994). However, a version of the matrix form – project organisation – is distinguished from matrix organisation in that it is temporary in character (Bryman *et al.* 1987; Lundin & Söderholm 1995). While the life of a matrix organisation is normally indeterminate, the life of a project organisation is clearly determinate with the termination date identified as part of its declared objectives. Thus project organisation has evolved in order to overcome the limitations of the traditional organisation of production around functionally organised resources where information and materials flow sequentially from one unit to the other and finally to the customer, and product organisation where the emphasis on flows tends to compromise resource use efficiency. The aim of a project organisation is to mobilise the capital and human resources required to meet a client's specific demands from the *resource bases* that hold them. These resource bases are usually either functional departments within the client firm, or independent specialist firms and form the skill containers (Kristensen 1996) upon which project organisations draw in order to meet project objectives. Thus resource bases participate in portfolios of projects, and projects mobilise networks of resource bases.

The time component of project organisation lends a particular characteristic – a determinate life cycle. At the inception stages of a project, uncertainty is very high – the asset of the future is little more than an idea and possibly a few sketches and notes. How high depends upon a number of factors, such as the extent to which the asset is a copy of ones already existing, the extent to which standardised components and solutions can be used, and the extent of the requirement for new technologies to solve the particular problems posed by the project. This may be thought of as the level of *mission uncertainty* inherent in the project. As the project moves through the life cycle, uncertainty is reduced as more information becomes available – ambiguities in

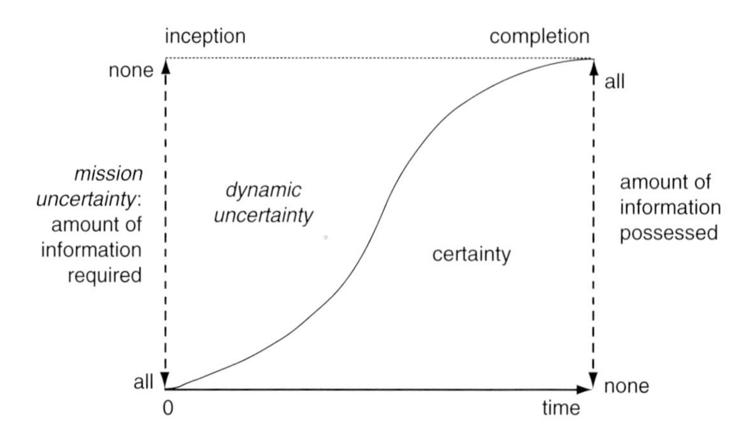

Figure 14.3 The project process (source: Winch 2001).

design are resolved, technology trials are completed, regulatory approval is obtained, component suppliers provide their shop drawings, and contractors successfully complete their tasks. The level of uncertainty at a particular point in the project life cycle relative to earlier and later points in the project life cycle may be thought of as the level of *dynamic uncertainty* on the project. This framework is illustrated in Fig. 14.3 which shows how uncertainty is progressively reduced through time until all the information required for the project is available at completion and embodied in the asset created. The area to the left of the s-curve represents information still to be acquired, i.e. uncertainty; the area to the right represents what is known, i.e. certainty.

These considerations help us to define the distinctive characteristics of project organisations:

- Project organisations operate as inter-firm networks. Construction is often seen as the classic network industry, but volume industries such as car manufacture also work in networks for their new product development projects. For instance, 85% of the investment value for the Twingo development project was spent externally to Renault (Midler 1993, p. 151). Thus it is appropriate to talk of the network of independent firms and internal resource-bases participating in the project organisation as the *project coalition* which comes together on a temporary basis to achieve a specific end (Winch 1989).
- Project organisations all display the temporary matrix structure described above, where the project organisation mobilises the separate resource bases – be they internal or external to the client firm – for a predefined period to deliver value for the client.
- Project organisations move through the distinctive life cycle shown in Fig. 14.3 and change in their organisational characteristics through that life cycle.
- Project organisations are not economic actors in their own right – they are capitalised by their clients, whose resources they use to recompense the resource bases.
- All project organisations are determinate, although some may last a lot longer than intended due to schedule overruns, and others may be abandoned before they have run their intended course. Obviously, the risks of this are greatest in the earlier stages, when less capital has been invested, but in some cases, particularly

in areas such as pharmaceuticals, abandonment can come very late as a new drug fails to perform as expected in trials (Giard & Midler 1993, p. 51).

Having presented the generic governance framework and specified the distinctive characteristics of project organisation, the argument now brings them together to develop a governance framework for the commercial management of complex project organisations.

The project coalition as a nexus of treaties

There have been a number of attempts to apply concepts from transaction cost economics to project organisations over the years, such as the work of Reve and Levitt (1984), Stinchcombe and Heimer (1985), Winch (1989; 1995), Stinchcombe (1990), Clark and Ball (1991), Masten *et al.* (1991), Pietroforte (1997), Walker and Lim (1999), Lai (2000), Chang and Ive (2001), Turner and Keegan (2001), Turner and Simister (2001), and Lindqvist (2004). However, these attempts have all been partial in the sense that they have tended to focus upon a particular transaction, typically that between the principal contractor and its client or between the client and its project manager, rather than examining the full range of transaction governance modes over the project life cycle. The governance of project coalitions perspective builds upon and synthesises this work, and draws upon some recent developments in institutional economics and economic sociology, to propose a coherent conceptual framework which encompasses the entire project life cycle. This is important, because for too long, our understanding of the project process through the life cycle has focused on particular transactions – typically principal contractor procurement – without placing that transaction in the broader project process context. Thus the argument here is akin to that developed in the work of Aoki *et al.* (1990) which sees the firm as a 'nexus of treaties', where the term 'treaty' is preferred to 'contract' due to the rather narrow legalistic perception of the latter. Within this perspective, the *project coalition* can be conceived as a *nexus of internal and external treaties*.

Project organisations are therefore complex, temporary, dynamic coalitions of firms which move through a distinctive life cycle as they define and deliver the project mission (Winch 2005). How can the framework summarised in Fig. 14.1 be applied to such a 'messy' situation? In order to make the argument clearer, it is useful to distinguish between vertical and horizontal transaction governance. Vertical governance focuses on the shifting set of transactions between the client and its first tier suppliers. Depending on the level of integration of the coalition structure chosen, the number of vertical transactions can vary from unity for a completely integrated project coalition where the client writes a 'turnkey' contract with a supplier, to more than 50 for a mediated project coalition where the client hires an independent project manager to co-ordinate a number of resource packages while retaining bilateral contractual relationships with each package supplier.[iii] These organisations form what can be called a *project chain*, in analogy to Porter's (1985) value chain.

Behind each one of these vertical transactions there may be one or more horizontal transactions as the first tier suppliers assure the resource inputs required to meet

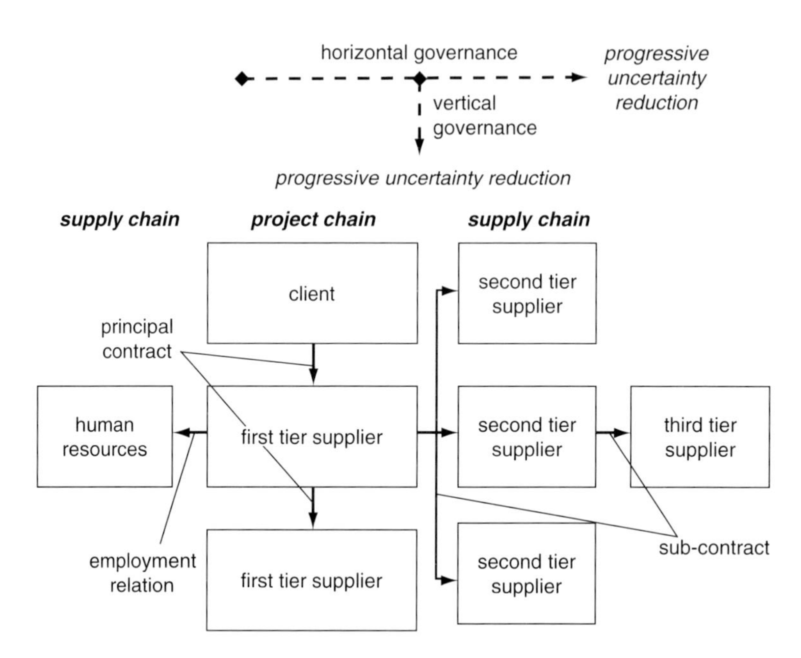

Figure 14.4 The project coalition as a nexus of treaties (developed from Winch 2002).

their treaty obligations to the client. Where all resources mobilised are in-house to the first tier supplier, then no further external transactions need to be governed and supply contracts switch to employment contracts as human resources are mobilised directly in what Williamson (1975) calls hierarchical governance. Where an extensive *supply chain* develops, there may be a large number of horizontal transactions which need to be governed through sub-contracts – this dimension is discussed in more detail by Cox and Ireland in Chapter 17. This nexus of potential treaties is illustrated schematically in Fig. 14.4. As can be seen, the economic concept of the project coalition as a nexus of vertical and horizontal treaties maps onto the organisational concept of a project organisation as a matrix of projects and resource bases (see Winch 2005) which, in essence, derives from Galbraith (1977). We now turn to considering the principal features of this nexus through the governance lens in terms of uncertainty, frequency and asset specificity.

The implications of uncertainty for choice of governance mode

The propositions on *uncertainty* derive from organisation theory. In essence, the mission of a project organisation is the progressive reduction of uncertainty through time as information is acquired which transforms uncertainty through risk into certainty (Winch 2005). It follows that transactions later in the project chain as it moves through the project life cycle will be under lower levels of uncertainty than those earlier. We would therefore expect to see different governance modes deployed because the contingencies under which economising occurs are different.

A review of the construction procurement literature (Winch 2002, Part III) confirms that this is indeed the case. The same proposition also applies to horizontal governance. Consider the case of task uncertainty (Perrow 1967). If an individual or team faces task uncertainty, then it follows that – in the absence of impacted information – those responsible for managing that task cannot face any less uncertainty. Indeed as managers are responsible (in essence) for co-ordinating between tasks, they are likely to face greater uncertainty, due to multiple task uncertainties, than the individuals or teams responsible for executing those tasks separately. It then follows that the client letting the contract probably faces greater uncertainty than the supplier taking the contract, and so on. The corollary is that main contracts are designed to handle greater uncertainty than sub-contracts.

How are project coalitions governed in the vertical dimension of the project chain? A number of propositions can be suggested:

U1 Project coalitions facing high mission uncertainty will be governed differently from coalitions facing lower mission uncertainty
U2 Under high mission uncertainty, the emphasis will be on flexibility and effectiveness; under low mission uncertainty, the emphasis will be on efficiency
U3 Inter-firm relationships later in the project life cycle, when dynamic uncertainty has been reduced, will be more tightly defined than earlier in the project life cycle
U4 Trilateral governance will be a feature of high asset specifity/high uncertainty transactions.

How are project coalitions governed in the horizontal dimension of the supply chain? We can suggest the following propositions:

U5 Contracts for subcontract will be more tightly drawn than principal contracts
U6 Trilateral governance will be more a feature of principal contracts than subcontracts.

The implications of asset specificity for choice of governance mode

The propositions on *asset specificity* derive from economic theory. Asset specificities occur when, due to the small numbers of players, a fully competitive market does not operate. These may occur *ex ante*, where supply is limited because market structure factors and competitive tenders cannot be obtained. This is a particular problem in low-volume production processes (Cox *et al.* 2002), many of which are associated with project forms of organisation. Even where the market for supply is competitive, problems of asset specificity in a project context often arise *ex post* through the 'fundamental transformation' (Williamson 1985, Ch. 3) after the contract has been agreed, as multilateral negotiations distil into bilateral treaties. For instance, a particular problem on projects is what Masten *et al.* (1991) call *temporal specificity* – the ability of suppliers to hold up the project programme and hence disrupt the production of other suppliers. Clients are therefore exposed to the switching costs of opportunistic behaviour up to the full replacement cost of the supplier; inversely, suppliers make transaction-specific investments which would have to be written off if they abandoned the project (Winch 2002). In distinction to the uncertainty contingency, asset specificities can, in principle, lie anywhere within the matrix depending on the specialist resource inputs required and their position on the critical path. However, it can be suggested that there is a systemic tendency for asset specificities to migrate from the supply chain to the project chain because of the risks for first tier suppliers caught between opportunism by subcontractors and treaty obligations to clients. Second tier suppliers owning significant asset specificities are therefore often invited to form consortia with first tier suppliers (Clark & Ball 1991) rather than act as subcontractors.

Asset specificities are the source of the *switching costs* that clients face when dissatisfied with supplier performance, and are a source of power in the supply chain for the holder of the specific asset (Cox *et al.* 2002), allowing higher than average profits to be extracted. In project organisations, switching costs are particularly troublesome *ex post*. For the client, the costs of replacing a supplier can be quite large, and this gives the supplier a significant margin for opportunistic behaviour. The costs of switching include:

- transaction costs of re-tendering
- inability to recover the additional costs generated by the original supplier's failure
- associated litigation
- the premium likely to be charged by the new supplier to complete the works because of the uncertainties around what work has actually been successfully completed.

Thus the margin for opportunistic behaviour by suppliers is, in formal terms, equal to the client's switching costs. These problems exist even where there is no change in the contracted scope of works, but where it is necessary to renegotiate the contract. In such circumstances, this opportunism can be given full play up to the cap of the client's switching costs, and suppliers are well practised in exploiting this margin. This is illustrated in Fig. 14.5, which shows the overall build-up of costs. Some switching costs may be absorbed in contingency; the balance appears as budget overruns. Arguably, one of the main aims of alliancing and partnering with suppliers for clients is to control switching costs.

The existence of these switching costs has two very important implications for the governance of project coalitions. First, in the situation where performance is perfunctory, it tends to be indulged so long as the additional costs generated by poor

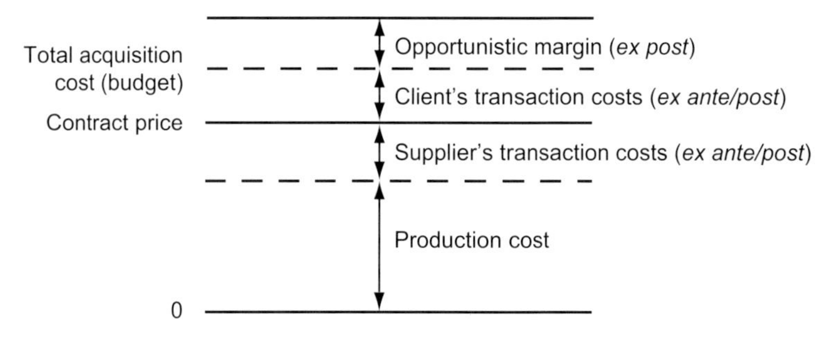

Figure 14.5 Switching costs in *ex post* transaction governance.

performance lie within the opportunistic margin. Second, where uncertainties need to be resolved in the contract as more information is acquired through the life cycle, or actual changes in the contract are required due to new information arriving, then the negotiations around the adjustment of the contract allow ample scope for opportunism up to the switching cost limit. While clients are not entirely helpless in this situation – and indeed may withhold information from suppliers – the information asymmetries in this situation typically favour the supplier.

Drawing on these considerations, the following propositions may be made:

A1 In the face of post-contract asset specificities, principals will prefer trilateral governance arrangements in the project chain
A2 Principals will tolerate opportunistic behaviour up to the threshold of switching costs; above that threshold, principals will seek to replace suppliers
A3 Where asset specificities in the supply chain are high, suppliers will seek joint venture relationships with key suppliers
A4 Prices charged for additional and changed works tend to carry a higher profit margin than those for the works as originally specified.

The implications of frequency for choice of governance mode

The propositions on frequency are derived from organisational psychology. Project organisations are temporary organisations. Transaction *frequencies* tend, therefore, to be low – often effectively unity for most principal/supplier dyads. It is these inherently low levels of transaction frequency that most clearly distinguish the governance of project coalitions from the governance of most other business transactions which are repetitive. In turn, this lack of repetition reduces opportunities for learning and hence the possibilities for the generation of trust as an economising mode of governance. From a governance perspective, the development of partnering is an attempt to raise transaction frequencies in project coalition.

In the context of transaction governance, *trust* is the confidence that the parties to the transaction will not take advantage of asset specificities to behave opportunistically (Lyons & Mehta 1997), either by withholding information or by seeking monopoly rents. Trust is discussed in more detail in Chapter 8. Where transactions are made under high uncertainty, trust is essential for their effective governance (Nooteboom *et al.* 1997). Two types of trust can be distinguished (Lyons & Mehta 1997):

- *Self-interested trust*: This is essentially future orientated in terms of the expectation that one's transaction partner is trustworthy and will not behave opportunistically in future transactions. It lies at the governance level in terms of Fig. 14.2. Axelrod (1990) provides an extended discussion.
- *Socially orientated trust*: This is past orientated in that it is generated through obligations generated through social and family networks. In the terms of Fig. 14.2, it lies at both the institutional and behavioural levels. Fukuyama (1995) provides an extended discussion.

Self-interested trust predominates in business transactions, but can be supported at crucial points by socially orientated trust.[iv] The generation of self-interested trust is largely a function of frequency, because only through learning in repeat transactions can parties come to know each other, and only when there is the prospect of further transactions does enlightened self-interest preclude opportunism.

From these considerations, the following propositions may be derived:

F1 Self-interested trust can only emerge where transaction frequency is high
F2 Disputes in the context of high frequency will tend to be resolved bilaterally
F3 Where frequency, asset specificity and uncertainty are high, then hierarchical governance will be preferred.

The testing of these propositions associated with the effects of uncertainty, asset specificity and frequency, on the choice of appropriate transaction governance mode requires extensive empirical work using a variety of methodologies ranging from the analysis of standard forms of contract through to econometric modelling, and cannot be attempted here. However, in order to suggest some of the ways in which this perspective can illuminate governance problems, the case of maintenance contracts in the UK rail sector is helpful.

An integrating case: rail maintenance contracts

The privatisation of British Rail under the Railways Act (HMSO 1993) is widely seen as a failure of either principle, or execution, or both, depending on political viewpoint. Most commentators agree that the fragmented organisation structure of the industry fits poorly with the inherently integrated nature of the rail network as a complex system. A governance analysis would doubtless yield many insights into the apparent failings of this organisation structure. However, this case will focus on one area where the failings have been recognised, and a marked shift from market to hierarchy has taken place – contracts for the maintenance of the permanent way.

British Rail had carried out its own rail maintenance as it presided over a slowly declining system. The policy of introducing 'market disciplines' into the operation of the network when Railtrack[v] took over as network operator in 1994 led to the outsourcing of maintenance to a number of construction companies on the basis of 20 term contracts dispersed geographically around the network. Contractors were contracted, in essence, to maintain the permanent way to the standards existing at the time of privatisation by providing track fit for purpose against predetermined standards. However, those standards were already in decline by the time the new companies took over during the mid 1990s: for instance, rail breaks per million train-miles had been rising since 1989 and reached crisis levels in late 1998 (Anon. 2000). By 1995, the track replacement rate was the lowest that it had been since the 1940s, and had not surpassed the steady state rate of renewal of 800 km per year since 1983 (NAO 2004).

Although performance did improve once Railtrack entered the private sector and additional investment funds became available, concern was expressed that it was not spending enough, and was focusing its efforts on performance criteria that led to bonuses for reducing delays to trains, rather than addressing the underlying quality of its assets which continued to deteriorate (NAO 2000). Moreover, Railtrack – like British Rail before it – had a poor idea of the overall condition of the network, which made both its own setting of priorities difficult and the job of the Office of the Rail Regulator (ORR), charged with overseeing its performance, nearly impossible in this area. The most detailed knowledge lay with the contractors rather than Railtrack itself, yet Railtrack also moved to an asset renewal policy based on actual condition rather than age (NAO 2000). At the same time, utilisation of the network increased dramatically, with passenger traffic increasing by 27% and freight by 35% between 1995 and 1999. This poor asset management interacted with rising utilisation of the network to create the crisis with came to a head as a result of the fatal Hatfield derailment in October 2000.

Railtrack was effectively forced into administration by the Government in October 2001 because it proved incapable of meeting the challenge of the crisis in maintenance. Its successor, a not-for-profit company called Network Rail, started to address the problem by reviewing the governance of rail maintenance contracts almost as soon as it took over in October 2002. In a little over a year it had entirely reversed the outsourcing policy in a shift that can be usefully interpreted through the governance framework. In order to understand this, we first need to identify the principal contingencies of the transaction to be governed. Rail maintenance – as compared to renewal and upgrade – operations combine high frequency and asset specificity and medium levels of uncertainty because they are:

- repetitive
- small scale
- require well-located depots with few alternative uses
- geographically dispersed around the network
- typically conducted at night against tight deadlines to ensure network availability the following day.

Although task uncertainty itself is relatively low, the physical dispersion and constrained 'possession' periods combine with the more familiar contingencies to make transaction governance in the absence of high trust very costly. Direct supervision of contractors by the client – a classic transaction cost – would be prohibitively expensive, so consummate performance has to be left to the contractor. The need for trust is reinforced by the performance nature of the contract: it is up to the contractor to determine fitness for purpose. However, the repetitive nature of transactions also presents the possibility for learning.

The initial rationale offered by Network Rail for in-sourcing maintenance contracts was a learning one. It argued that it did not know enough about the cost drivers of rail maintenance to act as an informed client, and so decided to take in house one contract that was a 'microcosm of the network' when it expired in March 2003 (Anon. 2003a). This was followed by the announcement some months later that two further contracts would be taken in house. A senior director of Network Rail stated: 'Our objective is clear – to drive down maintenance costs and become a more informed and intelligent buyer. The maintenance contracts we have inherited from Railtrack do not give us a clear understanding of cost and efficiency issues. That is why we are changing them' (Anon. 2003b). By October 2003, the argument had moved on – Network Rail had undergone its learning and concluded that in-house maintenance was cheaper. It was announced that all rail maintenance contracts would be taken in house as part of the New Maintenance Programme, with implementation over the following 12 months. It was claimed that this could save £0.3bn on the £1.3bn annual maintenance budget (Anon. 2003c). It is, perhaps, also worth noting the argument which Network Rail did not rely upon – that the changes to contractual relationships offered advantages in safety in the febrile atmosphere of the aftermath of the Hatfield accident and a later fatal one at Potter's Bar in May 2002, which was also related to the rail maintenance regime.

What are the sources of this potential saving? Clearly, Network Rail had learned what the contractors already knew – the production cost structure of maintenance contracts. That these were lucrative is indicated by the large falls in the share prices of the contracting companies, which typically derived between 10% and 30% of their turnover from this source (Anon. 2003c), when the decision was announced. However, it is unlikely that profits were at the level of over 20% on turnover under the watchful eye of the ORR, and the potential for efficiency savings is limited due to the fragmented nature of the work. It can, therefore, be suggested that many of the savings were in transaction costs:

- Multiple layers of inspection in a context where there was very low trust of the contractor's ability to perform work to the expected standards and where Network Rail was already committed to raising levels of inspection and planning (Anon. 2003d)
- Simplified administration of contracts in a context of high transaction frequency
- The efficiency benefits of administrative fiat in a context of many low-level decisions needing to be made
- The opportunity to generate trust through an employment contract, rather than a commercial contract.

It is also worth noting that Network Rail were very clear that they had no plans to take in house renewal and upgrade work for two reasons:

(1) Frequency levels are much lower in renewal and upgrade transactions, and projects are significantly larger
(2) Transaction costs as a proportion of total costs are lower, because 'you can measure performance very clearly in renewals. You can structure renewals contracts so that your contractor has a real incentive to show he's improving efficiency' (Anon. 2003c).

The implementation of the New Maintenance Programme was completed in July 2004 with a projection of savings of a more modest £700m over the following five years. This announcement was accompanied by reports that train delays had fallen by 21% in those areas that had been taken in house earlier (Anon. 2004).

Towards a research agenda

The approach to commercial management on complex projects presented here represents a break from the traditional analysis of standard forms of contract and procurement routes which currently predominates: Masterman (2001) provides a typical analysis. This work has tended to be narrowly focused, relying on the specification of a set of project attributes to predict the most appropriate procurement route (Tookey *et al.* 2001), and disconnected from broader perspectives in management and economics. This has started to change: Stinchcombe and Heimer (1985) provide the first analysis of standard forms of contract from a transaction cost perspective, while Hughes and Murdoch (2001) link the analysis of procurement routes to organisation theory. It is hoped that the framework put forward in this chapter will help crystallise this change, and that the propositions suggested will provide an initial stimulus for further research in commercial management.

This will not be straightforward. Obtaining good quality data on the 'costs of doing business' is difficult, in that firms do not routinely collect these data, resulting in transaction costs being the 'hidden factory' (Miller & Vollmann 1985). As a result, the influence of transaction costs on managerial decision-making is typically based on *perceptions* of the relative transactions costs of alternative governance modes, not hard data (Buckley & Chapman 1997). However, the body of empirical work on the economics of transaction governance is growing (Masten 1996; Shelanski & Klein 1995), and methods are being developed of operationalising the propositions deriving from the theory. Chang and Ive (2001) have also done important development work in analysing the options for the empirical analysis of alternative governance modes. What all these successful implementations have operationalised is the basic perspective of transaction cost economics as an exercise in comparative statics – poor data mean that the absolute levels of transaction costs are unlikely to be available, but the analysis requires only data on the comparative level of transaction costs between discrete alternatives (Williamson 1996, Ch. 4).

As usually articulated, the transaction cost approach takes the economics of production as read: it regards them as important, but already well understood. It focuses, instead, on understanding the drivers of transaction costs. However, Turner and Simister (2001) have argued that, in a project context, these are not necessarily independent, and that appropriate governance arrangements can motivate the search for reduced production costs. It follows that relatively costly governance arrangements could be economising due to their favourable impact on production costs[vi]. It can be suggested that alliance contracts (Scott 2001) provide one example of this effect. Thus a full analysis of alternative governance modes in project-based industries cannot confine itself to transaction costs, but must also bring in comparative production costs. A broader analysis is also required to deepen understanding of the processes of structuration summarised in Fig. 14.2.

This chapter has attempted to develop further the project governance framework, first presented in Winch (2001), and to indicate its potential to provide new insight into some classic contracting problems in commercial management. The brief empirical case can, at best, provide an indication of fruitful paths for empirical enquiry. Much more rigorous data collection and formalisation of the propositions are required before we can be confident of the benefits of the proposed framework for understanding the governance of project transactions.

As the growing list of conceptual contributions indicates, many believe that there is potential in this approach, but this potential remains to be empirically realised. If it can be realised, commercial management will have been placed on a sound empirical and theoretical foundation.

References

Anon. (2000) *Railway Gazette*, November.http://www.railwaygazette.com (accessed 11 November 2004).

Anon. (2003a) *Financial Times*, 14 January. http://www.news.ft.com/home.uk (accessed 11 November 2004).

Anon. (2003b) *Financial Times*, 12 August. http://www.news.ft.com/home.uk (accessed 11 November 2004).

Anon. (2003c) *Financial Times*, 25 October. http://www.news.ft.com/home.uk (accessed 11 November 2004).

Anon. (2003d) *Financial Times*, 22 August. http://www.news.ft.com/home.uk (accessed 11 November 2004).

Anon. (2004) *Financial Times*, 29 July. http://www.news.ft.com/home.uk (accessed 11 November 2004).

Aoki, M., Gustafsson, B. & Williamson, O.E. (1990) *The Firm as a Nexus of Treaties*. Sage, London.

APM (2004) *Directing Change*. Association for Project Management, High Wycombe.

Axelrod, R. (1990) *The Evolution of Co-operation*. Penguin, Harmondsworth.

Bowen, H.K., Clark, K.B., Holloway, C.A. & Wheelwright, S.C. (1994) *The Perpetual Enterprise Machine*. Oxford University Press, New York.

Bryman, A., Bresnen, M., Beardsworth, A.D., Ford, J. & Keil, E.T. (1987) The concept of the temporary system: the case of the construction project. *Research in the Sociology of Organizations*, **5**, 253–283.

Buckley, P.J. & Chapman, M. (1997) The perception and measurement of transaction costs. *Cambridge Journal of Economics*, **21**, 127–145.

Chang, C.-Y. & Ive, G.J. (2001) A comparison of two ways of applying a transaction costs approach. *Bartlett Research Paper 13*, University College, London.

Cherns, A.B. & Bryant, D.T. (1984) Studying the client's role in construction management. *Construction Management and Economics*, **2**(2), 177–184.

Clark, I. & Ball, D. (1991) Transaction cost economics applied? Consortia within process plant contracting. *International Review of Applied Economics*, **5**, 341–357.

Coase, R.H. (1937) The nature of the firm. *Economica*, **4**, 386–405.

Cox, A., Ireland, P., Lonsdale, C., Sanderson, J. & Watson, G. (2002) *Supply Chains, Markets and Power: Mapping Buyer and Supplier Power Regimes*. Routledge, London.

Eisenhardt, K.M. (1989) Agency theory: an assessment and review. *Academy of Management Review*, **15**, 57–74.

Fukuyama, F. (1995) *Trust: The Social Virtues and the Creation of Prosperity*. Hamish Hamilton, London.

Galbraith, J.R. (1977) *Organization Design*. Addison-Wesley, Reading, MA.

Giard, V. & Midler, C. (1993) *Pilotages de Projet et Entreprises*. Economica, Paris.

Giddens, A. (1984) *The Constitution of Society: Outline of the Theory of Structuration*. University of California Press, Berkley, CA.

Hall, P.A. & Soskice, D. (2001) *Varieties of Capitalism: The Institutional Foundations of Comparative Advantage*. Oxford University Press, Oxford.

HMSO (1993) *Railways Act*. HMSO, London.
Hobday, M. (1998) Product complexity, innovation and industrial organisation. *Research Policy*, **26**, 689–710.
Hobday, M. (2000) The project-based organisation: an ideal form for managing complex products and systems? *Research Policy*, **29**, 871–893.
Horwitch, M. (1987) Grands programmes: l'expérience Américaine. *Revue Française de Gestion*, Mars-Avril-Mai, 54–69.
Hughes, W. & Murdoch, J. (2001) *Roles in Construction Projects: Analysis and Terminology*. Construction Industry Publications, Birmingham.
Knott, T. (1996) *No Business as Usual: An Extraordinary North Sea Result*. British Petroleum, London.
Kristensen, P.H. (1996) On the constitution of economic actors in Denmark: interacting skill containers and project coordinators. In: *The Changing European Firm* (eds R. Whitley & P.H. Kristensen). Routledge, London; pp. 118–158.
Lai, L.W.C. (2000) The Coasian market–firm dichotomy and sub-contracting in the construction industry. *Construction Management and Economics*, **18**, 355–362.
Lindqvkist, L. (2004) Governing project-based firms: promoting market-like processes within hierarchies. *Journal of Management and Governance*, **8**, 3–25.
Lundin, R.A. & Söderholm, A. (1995) A theory of the temporary organization. *Scandinavian Journal of Management*, **11**, 437–455.
Lyons, B. & Mehta, J. (1997) Contracts, opportunism, and trust: self-interest and social orientation. *Cambridge Journal of Economics*, **21**, 239–257.
Maher, M.E. (1997) Transaction cost economics and contractual relations. *Cambridge Journal of Economics*, **21**, 147–170.
Masten, S.E. (1996) *Case Studies in Contracting and Organization*. Oxford University Press, New York.
Masten, S.E., Meehan, J.W. & Snyder, E.A. (1991) The costs of organization. *Journal of Law, Economics and Organization*, **7**, 1–25.
Masterman, J. (2001) *An Introduction to Building Procurement Systems* (2nd edn). E. & F.N. Spon, London.
Midler, C. (1993) *L'Auto qui n'existait pas*. InterEditions, Paris.
Miller, J.G. & Vollmann T.E. (1985) The hidden factory. *Harvard Business Review*, **63**(5), 142–150.
Morris, P.W.G. (1994) *The Management of Projects*. Thomas Telford, London.
NAO (2000) *Office of the Rail Regulator: Ensuring that Railtrack Maintain and Renew the Rail Network*. National Audit Office, London.
NAO (2004) *Network Rail: Making a Fresh Start*. National Audit Office, London.
Nooteboom, B., Berger, H. & Noorderhaven, N.G. (1997) Effects of trust and governance on relational risk. *Academy of Management Journal*, **40**(2), 308–338.
North, D.C. (1990) *Institutions, Institutional Change and Economic Performance*. Cambridge University Press, Cambridge.
Perrow, C. (1967) A framework for the comparative analysis of organizations. *American Sociological Review*, **32**, 194–208.
Pietroforte, R. (1997) Communication and governance in the building process. *Construction Management and Economics*, **15**, 71–82.
Porter, M. (1985) *Competitive Advantage*. Free Press, New York.
Reve, T. & Levitt, R.E. (1984) Organization and governance in construction. *Project Management*, **2**, 17–25.
Scott, B. (2001) *Partnering in Europe: Incentive-based Alliancing for Projects*. Thomas Telford, London.
Shelanski, H.A. & Klein, P.G. (1995) Empirical research in transaction cost economics: a review and assessment. *Journal of Law, Economics and Organization*, **11**, 335–361.
Smith, R. (2003) *Audit Committees: Combined Code Guidance*. Financial Reporting Council, London.

Stinchcombe, A.L. (1990) *Information and Organizations*. University of California Press, Berkeley, CA.
Stinchcombe, A.L. & Heimer, C.A. (1985) *Organization Theory and Project Management: Administering Uncertainty in Norwegian Offshore Oil*. Norwegian University Press, Oslo.
Tookey, J.E., Murray, M., Hardcastle, C. & Langford, D. (2001) Construction procurement routes: re-defining the contours of construction procurement. *Engineering Construction and Architectural Management*, **8**, 20–30.
Turner, R.J. & Keegan, A. (2001) Mechanisms of governance in the project-based organization: roles of broker and steward. *European Journal of Management*, **19**, 254–267.
Turner, J.R. & Müller, R. (2003) On the nature of the project as a temporary organization. *International Journal of Project Management*, **21**, 1–8.
Turner, J.R. & Simister, S.J. (2001) Project contract management and a theory of organization. *International Journal of Project Management*, **19**, 457–464.
Walker, A. & Lim, C.K. (1999) The relationship between construction project management theory and transaction cost economics. *Engineering, Construction and Architectural Management*, **6**, 166–176.
Whitley, R. (1999) *Divergent Capitalisms: The Social Structuring and Change of Business Systems*. Oxford University Press, Oxford.
Williamson, O.E. (1975) *Markets and Hierarchies: Analysis and Anti-Trust Implications*. Free Press, New York.
Williamson, O.E. (1981) The economics of organization: the transaction cost approach. *American Journal of Sociology*, **87**, 548–577.
Williamson, O.E. (1985) *The Economic Institutions of Capitalism*. Free Press, New York.
Williamson, O.E. (1996) *The Mechanisms of Governance*. Oxford University Press, New York.
Winch, G.M. (1989) The construction firm and the construction project: a transaction cost approach. *Construction Management and Economics*, **7**, 331–345.
Winch, G.M. (1994) *Managing Production; Engineering Change and Stability*. Oxford University Press, Oxford.
Winch, G.M. (1995) Project management in construction: towards a transaction cost approach. *Le Groupe Bagnolet Working Paper 1*, University College, London.
Winch, G.M. (2000a) Construction business systems in the European Union: editorial introduction. *Building Research and Information*, **28**, 88–97.
Winch, G.M. (2000b) The management of projects as a generic business process. In: *Projects as Business Constituents and Guiding Motives* (eds R.A. Lundin & F. Hartman). Kluwer, Dordrecht; pp. 117–130.
Winch, G.M. (2001) Governing the project process: a conceptual framework. *Construction Management and Economics*, **19**, 799–808.
Winch, G.M. (2002) *Managing Construction Projects: an Information Processing Approach*. Blackwell Science, Oxford.
Winch, G.M. (2005) Rethinking project management: project organizations as information processing systems? In: *Innovations: Project Management Research 2004* (eds D.P. Slevin, D.I. Cleland & J.K. Pinto). Project Management Institute, Newtown Square, PA; pp. 41–55.

Endnotes

[i] Asset specificity refers to the situation where buyers have limited or no choice of supplier and vice versa. Pre-contract, this is the situation of near or actual monopoly/monopsony; the post-contract situation is discussed later.

[ii] Williamson and his followers tend to ignore transaction frequency in their analyses, and Williamson himself never associated a behavioural factor with it. I propose learning as that missing behavioural factor.

[iii] An example of this from the construction industry is the *construction management* procurement route.

[iv] The classic problem is how to govern the last transaction in a series. If self-interested trust relies upon the expectation of future relationships, yet it is known that this is the last time the parties will do business, then the whole development of self-interested trust can unravel.

[v] Railtrack was formed as a government-owned company in 1994 and successfully floated on the London Stock Exchange in 1996, completing the transfer of rail network assets in Great Britain to the private sector.

[vi] The importance of this point should not be underestimated. Although, in formal terms, the objective of transaction cost analysis is to understand how firms economise on the sum of production and transaction costs, even Williamson himself (e.g. Aoki *et al.* 1990, p. 13) tends to simplify the analysis by focusing on transaction costs alone.

15 Contract Management

Will Hughes

Introduction

The aim of this chapter is to explain the relationships between law and procurement practice in the context of the construction sector. Business-to-business contracts are, of course, widespread and common. The interesting things about the construction sector concern the volume of contractual agreements that have to be put in place for a construction project, and the way that complex standard forms are used and adapted so extensively. While drawing on experience from the construction sector, this chapter provides an interesting exploration into the commercial context of contracts between businesses. The underlying theme is the development of a strategy that will enable conflict to become a positive agent for innovation. The law relating to construction contracts has grown into a specialist branch of contract law, which can be bewildering to construction teams. These teams need to act as integrated wholes, and the law should facilitate this. Underlying the law of contract are a few straightforward assumptions: a contract is based upon a mutual understanding between the parties, the contract terms express the intentions of the parties, parties are free to choose their terms of business, and privity of contract restricts the effect of a contract to the two parties who sign it. Practice is a long way from these ideals. Frequently contracts are signed in an atmosphere of uncertainty; the standard-form contracts have often been the subject of such lengthy development that they no longer reflect standard practice, and procurement is based upon the premise that a team of people will be planning, designing and building the project, each of which needs to be tied to the client, the project and to each other. In order to make practice more effective, new contracts are needed which will more accurately reflect intended project team structures, while mitigating the sometimes-adverse effects of power imbalance between suppliers and purchasers. Whenever enough people perceive that there are problems surrounding the use of contracts, the traditional response of the construction sector has been to produce a new form of contract. But in nearly every case, the new form of contract has been predicated on the assumption that the client will pay for builders' work and materials, not for a finished building. There is a gulf between these two sets of expectations, and this causes problems. Generally, in most other sectors, most of the time, purchasers pay for what things do, rather than what they are made of, and this is the main distinguishing characteristic of building and construction contracts, although few in the construction sector realise or acknowledge it.

Construction contracts and the law

Construction contract law has become a specialist branch of contract law. This is clear from the number of lawyers who set up construction departments and develop a special expertise. It is needed because of the unique technical nature, size and complexity of projects. Many small construction subcontracts run to hundreds of pages, even without drawings.

Disputes are common in the construction sector, judging by the growth in publications and reports dealing with construction cases. However, since the emergence of the Housing Grants Construction and Regeneration Act 1996 (HMSO 1996) the signs are that arbitration and litigation are dwindling and being replaced by adjudication, which is now the primary dispute resolution procedure for payment disputes. Although arbitration and litigation occur far less frequently, there are probably just as many disputes about payment as there were before the Act, if not more.

The principles of contract in English law, and the way that they apply to construction contracts, are covered extensively elsewhere (see, for example, Murdoch & Hughes 2000). It is worthwhile reflecting on the basic principles of contract: offer, acceptance, consideration, legality, frustration. A contract is a legally enforceable agreement. An agreement is usually defined in terms of an 'offer' made by one party and an 'acceptance' of that offer by the other. This need to find an offer and an acceptance will cause few, if any, problems where it is clear that the parties' contractual relationship is to begin only after a formal document is signed. However, life is not always this simple and, even in those cases where such a document is signed by both parties, it is quite likely that this document will be seen merely as the formal record of a contract which already exists. Indeed, it is by no means uncommon to find a project running for months or even years before 'the contract' is finally executed. An offer is merely the final stage in a negotiation to which the answer is an unconditional 'yes'. Other statements are merely stages in the negotiation. It is also important to remember that a contract for an illegal act cannot be enforced, so, for example, pay-when-paid clauses may still arise in contracts, even though they have no legal effect. These clauses have long been a bone of contention in the construction sector, where subcontracting and subsubcontracting are widespread. Contractors often imposed payment conditions on their subcontractors that meant the subcontractor would only be paid once the main contractor was paid, resulting in the transfer of risk of non-payment down the contracting chain to those least able to bear it.

Assumptions underlying law of contract

The law of contract is based upon a few straightforward assumptions: a contract is based upon a mutual understanding between the parties (known by lawyers as a 'meeting of minds' or *consensus ad idem*), the contract terms express the intentions of the parties, the parties are free to contract on any terms they wish, and privity of contract restricts the effect of a contract to the two parties who sign it (Murdoch & Hughes 2000). A few simple examples will highlight how far practice departs from these assumptions.

Meeting of minds

In interpreting and applying contract terms, the courts assume that the parties who have signed the contract will have discussed and agreed the terms. This seems a reasonable expectation. However, in practice, many people have an aversion to contracts, particularly as most of them are written in an impenetrable style. Moreover, the use of standard-form contracts overcomes the need for individuals to read the terms, as they often feel that they already have an understanding of them, based on previous experience and, perhaps, the courses they studied when they qualified.

Many contracts eventually come into being as a result of a 'battle of the forms' (the situation where offers are met with counter-offers until work is started, or materials are accepted). An example of this is the case of *Sauter Automation Ltd* v. *Goodman (Mechanical Services) Ltd* (1986), in which a quotation submitted by a subcontractor for supplying and installing boiler equipment was subject to their standard conditions, which included a retention of title clause. The main contractors accepted this quotation with an order stating, 'terms and conditions in accordance with the main contract', which did not contain a retention of title clause. When the subcontractors delivered the equipment to the site, it was held that this amounted to an acceptance by them of the main contractor's counter-offer. In contrast to this, in *Chichester Joinery Ltd* v. *John Mowlem & Co plc* (1987), a quotation for joinery was submitted by subcontractors, with their standard conditions. The main contractors sent a purchase order containing their own standard terms, which stated, 'any delivery made will constitute an acceptance of this order'. The subcontractors delivered their joinery to the site, after they had sent the main contractors a printed form headed 'acknowledgment of order', which stated that the order was accepted 'subject to the conditions overleaf'. On this occasion it was the subcontractors' conditions which prevailed, as it was held that, by accepting the joinery, the main contractors had accepted the terms on which it was delivered.

Although each case needs to be viewed in the context of the whole of the correspondence, it is clear that contract terms can be accepted by conduct, and that very often the last piece of paper proffered will prevail. This is as far from a meeting of minds as could be imagined, especially when one considers that the people who are 'thrusting' these pieces of paper at each other are often delivery drivers and storekeepers on site. Good contract practice would demand that agreements truly reflect the specific negotiations that took place between the parties, and that the parties really have understood the contracts they execute.

Uncertainty of contract terms

Contracts are frequently signed in an atmosphere of uncertainty, where neither party wishes to direct attention to the terms of the contract for fear of appearing litigious; indeed, work often starts without the formal signing of a contract. The case of *Rosehaugh Stanhope* v. *Redpath Dorman Long* (1990) was an interesting example of this: although the dispute concerned some £8 m worth of work, the contract was not formally signed until after the formal contract completion date! This was not an issue, as it happened, but it highlights the extent to which the precise contractual terms are not formalised until very late in the process.

Freedom of contract

The parties to a contract are free to contract on any terms they choose. Businesses must be allowed to choose their contract terms, within the limits of legality, otherwise they will be unable to compete in a commercial manner. This means that the courts assume that when a contract has been drafted and executed, it constitutes an exhaustive statement about the intentions of the parties. Frequently, many construction cases illustrate that too little attention is given to contractual clauses, but the courts are reluctant to add extra clauses except to make effective what the parties clearly intended. An example of this judicial reluctance to intervene is *Bruno Zornow (Builders) Ltd* v. *Beechcroft Developments Ltd* (1989), a case in which obscure drafting made it very difficult even to discover the express terms of the contract. In the end, the judge was prepared to imply a completion date (on the basis that no commercial party would contemplate a contract without one); however, he absolutely refused to imply a term for sectional completion, even though it appeared that this was what the parties had intended.

Privity of contract

A serious problem confronting the courts is the knowledge that, although there is no contractual link between the client and the subcontractor, it is clear why the work is being done and where the money is coming from. The lack of a contractual link is a problem because the principle of privity of contract means that a contract can only impose obligations and duties upon those who sign it. However, construction procurement is based upon the premise that a team of people will be planning, designing and building the project, each of whom needs to be tied to the client, the project and to each other. Although contracts tie pairs of people together, the relationships are subject to complications because of the law of tort (a civil wrong independent of contract), which can be used to establish liabilities without contractual links. That this is a problem is highlighted by a series of cases, probably the most notable (or notorious) of which was *Junior Books Ltd* v. *Veitchi Co Ltd* (1983), since it established a subcontractor's liability in tort to an employer for financial losses suffered due to negligent performance of the sub-contract; despite the long established legal principle that tort actions could not extend to purely financial losses. Although many cases since have avoided following Junior Books, no court has overruled it or declared it to be bad law. Therefore, there is ambiguity surrounding this particular area of the law. Currently, the courts are strongly of the view that tort should not be used to improve poor contractual situations, and this view has reduced the utility of tort, and increased the dependence upon contracts.

This increasing dependency on contracts was given a further push in the Contracts (Rights of Third Parties) Act 1999 (HMSO 1999) which gives rights to those who are not party to a contract. However, in order for this Act to apply, the contract must expressly provide such rights, either to individuals or to specific classes of person. In fact, nearly all of the current standard-form building contracts specifically exclude such rights. Therefore, this Act has had little, if any, impact on the construction sector.

Drafting conditions of engagement

Agreements are usually drafted defensively, and this can cause problems. The practice in the past has been that each interest group, apart from contractors, negotiates its own agreement with the client. Professional terms of engagement have, in the past, been standardised by the institutions for all their members' appointments, regardless of scale, complexity and level of involvement. Fortunately, this position is changing. The RICS Building Surveyors' conditions of engagement allow for a range of alternative types of appointment. The *Architect's Appointment* (RIBA 1992) also allows for some flexibility in the services being paid for by the client. However, each institution assumes a certain pattern of leadership, and sticks to it. To a large extent, they do not consider that project management and co-ordination are anything other than discrete responsibilities, and that they are either in or out of their agreement. Clearly, effective project teams require co-ordination, communication and integration at all levels. It is all too easy for people to avoid such responsibility by only engaging to produce certain documents. Architects seem to have little design responsibility, and their co-ordination and inspection duties are greatly reduced in their newer form of appointment. Worse, the proliferation of versions of the standard forms of appointment continues to grow, particularly with the publication of an architect's plan of work that was not officially sanctioned by the RIBA council (Hughes 2003).

The progressive reduction of each individual's liability often leaves the client dangerously exposed and unable to recoup losses which become unattributable to others, due to their careful avoidance of liability. This has led to cases in which the plaintiffs tried to sue everyone connected with a project, until liability could be made to stick. Such practice has been severely curtailed since *Murphy* v. *Brentwood District Council* (1990), but even though the problem may have been clarified in legal terms, it still remains a difficult management issue.

That contracts define the terms of potential disputes, and enable obligations and duties to be codified is a basic function of contracts. That disputes need resolving is not a contractual problem, but a management problem (see Chapter 11). The principle of 'set-off' serves to highlight this. Set-off, in the construction context, is the right which may be written into a contract for an employer to withhold money owed to a contractor because of delay or defective work. Only very limited rights to do this are available, and it is a difficult legal mechanism (Murdoch & Hughes 2000). The use of a properly drafted set-off provision makes sense to a management contractor who has a valid reason for withholding money from a subcontractor. If the reason is entirely arbitrary, however, this can be a very onerous provision from the point of view of the subcontractor. Therefore, it is not such clauses which in themselves are onerous, but the way in which they are exercised by the parties. This highlights the way in which many so-called contractual problems are, in fact, management problems.

These days, there are many steps being taken towards collaborative team working on construction projects. One aim of these developments is to avoid altogether the need for expensive litigation or arbitration. If good business relationships can be maintained, particularly from one project to another, there are many advantages. Communications between people are easier if they are familiar with each other, common working practices can help to avoid waste and the need to develop everything from first principles, the use of design and construction details repeated from one project to another can provide huge productivity gains, and a developing ethos of trust (see Chapter 8) between the parties can lead to the financial benefits of these gains being shared.

Project management

While the primary purpose of a contract is to enable contracting parties to have recourse to the law in the event that either of them fails to perform, the purpose of a project is somewhat more productive. Contracts can be seen as the glue that binds participants from different organisations into the project process. In order to understand how contracts can best be managed, it is important to place them into the wider context of the management of projects generally.

The established view of project management is that a project must be managed in the context of the client's requirements concerning time, cost and quality (TCQ). Research has shown that this view is too limited to take account of the complexities of modern construction projects (Hughes 1989a). The established view is indeed a reflection of the establishment, typified by the involvement of architects, quantity surveyors and contractors. Each of these professions influences the project in a different way. Put crudely, architecture can be linked with quality, quantity surveying with money and contracting with time. This may be too crude a characterisation but, increasingly, contractors concentrate upon time management. Certainly, the approach adopted to construction management in the USA emphasises that the construction manager is chiefly responsible for scheduling and time-related management (Bennett 1991).

Beyond 'time, cost and quality'

It may be that the established view of project management is one of the problems, rather than one of the solutions, to the recurring problems in construction projects. As a model of what architects, builders and quantity surveyors offer collectively, TCQ is useful because it demonstrates the needs and processes of project management split into some elemental parts. But, because it is rooted in the professions, it serves to preserve traditional roles, even if it enables a shift of the power balance between them. Although contingency theories of management clearly state that organisational structure should be established as a response to the environment in which it operates, the TCQ view is based on an institutionalised perspective of the industry and, as such, cannot be responsive to individual project environments (Hughes 1989a).

The approach to managing any particular project can only be developed by examining the basis upon which construction projects come into existence, and by the application of analytical principles to the context of construction project management. Based on organisational theory, and using case studies of UK building projects, research has shown that project management can be seen as the control of the project in relation to its environment, or context (Hughes 1989a, b). The relationships between the concepts are illustrated in Fig. 15.1. An analytical approach to project management requires that project objectives are established in terms of the major contextual influences, which are characterised as having political, economic, physical, cultural and social facets. Each of these has a more specific counterpart in the task environment surrounding the project, and these are policy, financial, technological, aesthetic and legal, respectively. If the project's control systems are to be effective then they must relate the management of the project to the environment through a series of control systems, each of which matches one of the facets of the task environment. These lead to the identification of five control systems – time, budget, function, quality and conflict – necessary for effective project management. In this context, 'function' defines the objective characteristics of what is to be built, and 'quality' defines the degree of excellence in a purely subjective sense: the purpose of this is to preserve the distinction between the physical and the aesthetic aspects of a building (see Fig. 15.1).

These control systems are at the heart of project documentation and control, and thus they form the basis for agreements, expressing intentions in such a way that they can be planned and managed. There is a great deal of experience of managing time and budget in the construction sector and there are many tools and techniques available to the practitioner, such as bar charts and network planning for modelling time, and cost plans, standard methods of measurement and bills of quantity for modelling and controlling cost. However, there is a fundamental problem concerning function and quality control because there is a widespread approach to quality control which involves defining quality as 'conformance to requirements' (which assumes that requirements can be reduced to a set of objective and measurable criteria). But the basic thrust of architecture is to create a built environment

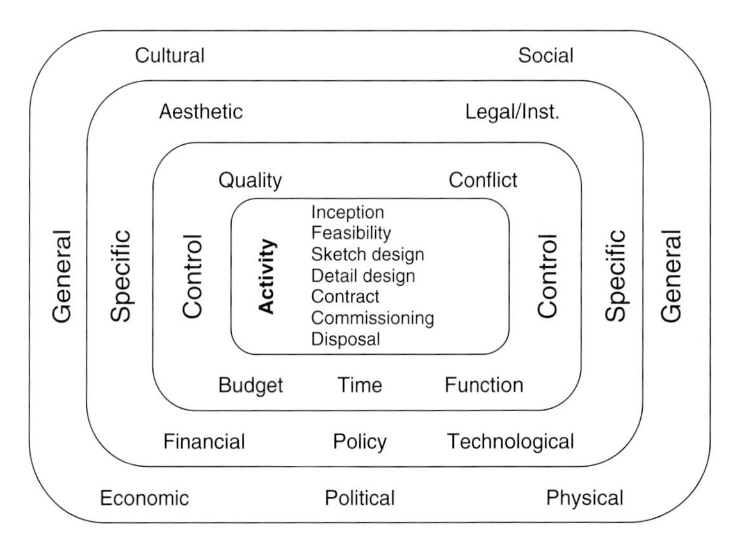

Figure 15.1 Context of project management and control systems.

based upon wider considerations than the functional requirements of a client, not least of which should be the aesthetic. There is a long-running debate on the relative merits of subjective and emotional criteria versus objective and measurable criteria (Seymour & Pheng 1990) which is beyond the scope of this chapter, but there is clearly a substantive difference between achieving objective (measurable) targets and creating an aesthetic output. Although it is a simplification, it is useful to think of the functional as objective and the aesthetic as subjective. The fifth strand in this view of project management, and a useful context for thinking about contracts and contract management, is conflict control, and for this there are few traditional tools to help the project manager, particularly in the construction literature.

The evidence for these five kinds of control is the documentation already used on most projects. The programme, bills and cost statements, brief, drawings and specifications, all interact to achieve the first four control systems. Conflict control has traditionally led to a dependence on contracts and the law. In this sense, it has been reactive and negative because it was only used to help contracting parties out of a problem by apportioning blame and seeking redress. Good conflict control will reduce reliance on litigation, and increase the likelihood that contracts and agreements accurately reflect the intentions of those who participate in construction. In many ways, recent developments to construction management practice in the UK, growing from the reports of Latham (1994) and Egan (1998) provide increasing opportunities to deal with conflicts before they become destructive disputes.

Negotiating contracts

The most important stage in developing a no-dispute approach to construction contracting is in the initial stages of setting up the agreements on scope and price. Traditionally, tendering processes in construction have been predicated upon the professional roles alluded to above. One major feature of general contracts such as JCT98, ICE7 and so on is that the design work is carried out before the contractor is appointed. Another is that the selection of the contractor has been based on the lowest bid. Both of these customs have been at the root of major contractual problems for decades.

The relationship between design and construction

Commentators on the construction sector have often noted that it is unusual to find whole industrial sectors where design is separated from construction (fabrication/manufacture/production). The development of the professions in construction is connected with the complexity of the design task, not least the need for aesthetic considerations in many projects. Architectural skills seem somewhat vague and impenetrable to non-architects. Moreover, planners have a powerful role to play in dictating what kind aesthetic appearance should prevail in many developments. All of this means that it is typical for most of the design work to have taken place before the contractor is appointed. Even in design and build projects, where it might be reasonable to assume a higher level of integration, design is often carried out before the contractor is appointed, and then the contracts between the client and designers are 'novated' to the builder when he or she is appointed.

Managing a contractor's responsibilities

A key feature of building projects is the involvement of the architect in a design and co-ordination role, the quantity surveyor in a cost planning and control role, and the contractor in a production role. Theoretically, the contractor is not liable for design, but only for workmanship. However, in practice, the division between design and workmanship is not always clear-cut. Although the design must be documented in order for the contractor to be able to tender for the work, the documentation will not cover everything. Many detailed aspects are within the skill and knowledge of a competent contractor. Therefore, the exact position of individual nails in a floor, or the location of joints between different pours of concrete and their formation, are matters for the contractor. There are also many factors which are not documented simply because no one thought of them before the site work started. In these cases, it is essential for the contractor to seek clarification from the architect. A contractor who makes assumptions (even when based upon common sense and experience) will incur design liability for those choices. This leaves the employer exposed because the contractor may not carry indemnity insurance for such design decisions. The consequence of this is that contract management is a complicated and difficult issue in the construction sector. In those cases where the completed building does not perform according to the client's expectation, it is frequently the case that this is not the fault of the contractor, for the reasons given above.

Another important point about the differences between design decisions and workmanship decisions is that there is no substantive difference between them. Clearly, the contractor focuses on *how* to build, and chooses from among alternative courses of action in making these decisions. Similarly, a designer makes choices among possible solutions about *what* to build. Although designers may go through different processes from contractors in accomplishing their work, the practice of management relates to outputs rather than to thought processes. The application of the law also relates to what has been produced, rather than what was in the minds of those who produced it. The importance of focusing on output is underlined by views represented in legal literature, where the focus is on the consequences of design (Cornes 1988). Thus, design is generally recognised as the exercise of informed choice. As this definition is simply a means of attributing liability, it makes quite clear that the only difference between design and workmanship is that they are carried out by different organisations.

Early involvement of the contractor

In seeking to develop better contract management practices, the construction sector has for many years been experimenting with different ways of procuring work. The greatest successes in this direction have been in developments to collaborative working practices. Collaborative approaches to contracting nearly all focus on the early involvement of the contractor. Indeed, many of the current initiatives in collaborative working involve the contractor in a design and build capacity (with real design responsibility). This is interesting because, by definition, as the scope of the work is not known when the contractor is selected, the selection of the contractor cannot be based on the lowest price for the finished building. This may not accord with widely held views about European procurement regulations, but those regulations make clear that lowest price is not the only acceptable means for choosing suppliers and/or contractors. In fact, any means is acceptable, provided that the basis of the competition is made clear from the outset, and that the decision is made on the grounds that were initially declared. There are useful documents explaining how contractors can be selected for such collaborative processes; for example, see Jackson-Robbins (1999).

One effect of early contractor involvement is that tendering becomes a very different and protracted matter. Traditionally, contractors were given about four weeks in which to price a fully designed and specified building, based on forms of documentation that were familiar to all involved. In collaborative contracting, contractors may spend months, up to a year, in designing and negotiating the scope of the project. If this design is subject to competition at the end of the process, clients will cause contractors to incur huge costs, which the contractors can only recoup from winning bids. Thus, there is scope for a lot of waste, unless the contractor is chosen before the design process is commenced in earnest.

On the positive side, a contractor who has been appointed early can put a team together that develops the design and manages the process. This overcomes the historical divide between design and construction. Clearly, it also challenges the traditional dominance of the architectural role, and may also eliminate the need for some of the professional roles in construction (Foxell 2003).

Replacing contract management with a performance obligation

With the development of private finance initiative (PFI) procurement for public sector projects, the focus of contract management has shifted. By getting the private sector to procure the finance, invest in a building, and operate the facility in return for a unitary service charge, the public sector client has moved to a position where purchasing decisions are about buying a service, rather than a building. This approach is becoming more widespread and is echoed across a number of different public sector procurement practices. If a similar process were applied to private sector procurement, then instead of paying for buildings and managing contracts for builders' work and materials, clients would pay for services without necessarily taking on the ownership of buildings. Already it is possible to procure lighting and floor coverings through a leasing arrangement where ownership of the materials and components remains with the supplier and the client only pays for them when they are working and/or needed.

Performance-based contracting offers clients the possibility of procuring their accommodation in the same way as they purchase anything else, that is, by paying for what things do, rather than by paying for what they are made of. Moreover, if the facility fails to perform, then the supplier has an obligation to put it right or refund the money. This is a radical departure from traditional construction practice. Further, it can only be offered by suppliers with very different financial structures from those which currently exist in the construction sector. Historically, construction has been a cash flow business. Performance-based contracting, however, places much greater financial liability on suppliers, who would have to be paid considerably more in order to be able to carry this risk. Indeed, paying for a whole building on the basis of its function is more of a property sector proposition than a construction sector proposition. But within the existing construction sector, there are great possibilities for dealing with subsystems of the building on a performance basis. This would change the basic and construction business proposition; it will be interesting to see how the construction sector responds to this demand, which is an inevitable consequence of current movements towards a more efficient construction process.

Conclusions

The practice of contract management in the construction sector has evolved into a complicated, supplier-orientated process that demands a large array of professional consultants. All of this has happened as a consequence of industrialisation. In a post-industrial society like the UK, it may be that radical changes to the process could result in a challenge to the traditional roles of the professions. If the people who buy buildings deal directly with the people who make buildings, and if they are only paid for when they fulfil the clients' expectations, then the process will be much more lean and perhaps more efficient. On the other hand, the professions in construction add value at a number of points in the process. Clearly, they all need to become more adept at explaining to their clients exactly what kind of value they are adding to the process.

A future research agenda should deal with the need for professional roles in construction, new financial structures for companies in the supply chain that can develop efficient and responsive subsystems, and new contractual mechanisms to give effect to these radical new business models that are emerging at the leading edge of practice. The professional institutions seem to have an in-built tendency towards protectionism and self-preservation. While there is no doubt that these institutions have a valuable and positive role to play in developing the construction sector, research is needed into the way that such institutional forces might be balanced with the wider interests of buyers and suppliers in the construction sector. In particular, the role of architects seems to have been significantly eroded over the past century or so. We need to understand why. Liability for building design need not lie with architects, and there are plenty of examples from other countries where different patterns of liability can reduce or even eliminate the need for professional indemnity insurance, leaving contractors with complete responsibility and authority over the design and building process; the biggest problems arise when responsibility and authority lie with separate organisations. In order to develop the kind of business models that are needed to underpin this sort of performance liability, research is needed into the financial structures of construction firms. Contractors may need to take on full design responsibility with strict product liability and robust guarantees of performance. But contractors are cash businesses that rely heavily on subcontracting and have little capital investment. Such firms will find that there is no future for them in the new construction sector. More interaction is needed between academics and practitioners if we are to develop the kind of business models that will enable the construction sector to become a thriving and modern industry suited to the business needs of its clients. While liability is a dangerous thing, satisfied clients do not sue their suppliers. The problem of client satisfaction is more closely related to marketing than it is to quality management. This industry urgently needs research into the relationship between marketing and quality management if we are to move away from outdated notions of factory management that have been widely misapplied and misunderstood in the construction sector. Although our past approach has been to develop documents that enable suppliers to demonstrate that they have discharged their contractual obligations with little residual liability, we should investigate the extent to which we can move to more effective models of client satisfaction and performance guarantees. Commercial management brings a much needed reality check to the process of developing effective business models for a future construction sector.

References

Bennett, J. (1991) *International Construction Project Management*. Butterworth Heinemann, Oxford.

Cornes, D. (1989) The concept of design. In: *Construction Contract Policy: Improved Procedures and Practice*, (eds J. Uff & P. Capper), Centre of Construction Law and Management, King's College, London; pp. 67–73.

Egan, J. (1998) *Rethinking Construction*. HMSO, London.

Foxell, S. (ed.) (2003) *The Professionals' Choice: The Future of the Built Environment Professions*. Building Futures (a joint initiative of RIBA and CABE), London.

HMSO (1996) *Housing Grants Construction and Regeneration Act*. HMSO, London.

HMSO (1999) *Contracts (Rights of Third Parties) Act*. HMSO, London.

Hughes, W.P. (1989a) *Organizational analysis of building projects*. PhD thesis (unpublished), Liverpool Polytechnic, Liverpool.

Hughes, W.P. (1989b) Identifying the environments of construction projects. *Construction Management and Economics*, **7**, 29–40.

Hughes, W. (2003) A comparison of two editions of the RIBA Plan of Work. *Engineering, Construction and Architectural Management*, **10**(5), 302–311.

Jackson-Robbins, A. (1999) *Selecting Contractors by Value*. CIRIA Special Publication 150. Construction Industry Research and Information Association, London.

Latham, M. (1994) *Constructing the Team*. HMSO, London.

Murdoch, J.R. & Hughes, W.P. (2000) *Construction Contracts: Law and Management* (3rd edn). E. & F.N. Spon, London.

RIBA (1992) *Architect's Appointment*. RIBA Publications, London.

Seymour, D.E. & Pheng, L.S. (1990) The quality debate. *Construction Management and Economics*, **8**(1), 13–29.

16 Bidding

David Lowe and Martin Skitmore

Introduction

For the supplier (contractor), according to Skitmore (1989) and Shash (1993), the bidding[i] (tendering) process involves two crucial decisions: first, whether or not to bid (tender) for a project, and second, the determination of the bid price. While the latter has been subject to intense research interest [Stark and Rothkopf (1979), for example, were able to catalogue approaching 500 papers associated with modelling the bid price decision and currently there are well in excess of 1000 papers (Seydel 2003)], there has been comparatively little in the way of objective research into the former. Moreover, the opening statement is overly simplistic; it does not consider the fact that suppliers continually review the decision to bid (d2b) from prospect identification through to bid submission, while it implies that price is the only criterion used to evaluate bids. Both decisions require the assessment of a variety of factors which potentially influence the decision outcome, particularly those concerning the purchaser (client) and the competition.

This chapter is divided into four sections. Section 1 seeks to set bidding decisions within the context of the supplier's business plan and introduces the concept of the purchaser's 'buying centre'; Section 2 reviews the factors that influence the d2b process and evaluates several models developed to inform the process; Section 3 examines the process of supplier selection, in particular the bid evaluation criteria adopted by purchasers, and the transition from lowest price to multi-criteria selection. It considers the various pricing policies, systems and strategies available to suppliers, analyses the factors that influence the determination of the bid price, and appraises the various models developed to derive the optimum bid price; Section 4 proposes a research agenda for this area of commercial management. The chapter seeks to take a generic approach, drawing upon research from a variety of industry sectors. However, having said this, a significant proportion of the research into bidding decisions has been from the perspective of the construction industry.

The supplier's bid strategy

According to Tweedley (1995), no successful supplier enters a bidding opportunity without initially developing a plan for 'winning' the work. The bid process is dynamic and proactive, requiring the supplier to continually test and evaluate alternative solutions to the problems and opportunities presented in each bid. Further, he contends that a good bid strategy reflects a good business plan, which comprises both quantifiable and intangible elements. Central to any strategy, therefore, must be the objective of winning new and profitable business for the organisation, while each bid should support the adopted business strategy, which is generally either purchaser or product and/or service focused.

A good bid strategy integrates both business and competitive strategies into a cohesive 'win' plan. The competitive forces influencing the bid process include:

- *Purchaser factors*: award criteria, preferences and buying-power
- *Supplier factors*: price, supply chain and sources of alternative work
- *Competitor factors*: economic environment – newcomers, old rivals and emulators.

It is widely held that a supplier should apply competitor profiling, so as to develop a thorough understanding of the competition and to inform the establishment of a strategy to counter any threat they may pose, with the purpose of building a profile of each competing bidder that includes information on the nature of their bid, its chance of being successful, their assessment of the threat posed by your bid, and the possible tactics they may adopt. In several surveys, e.g. Mochtar and Arditi (2001), the most common form of assessment undertaken as part of the mark-up[ii] decision is that of the competition.

Likewise, suppliers should assess the purchaser's needs, highlighting key values that they will expect or require; for example, does the proposed solution provide the purchaser with a competitive advantage in their own field of operations, or will it make their operations easier? The term 'buying centre' is used to describe the group of individuals directly involved in the purchase of products or services. The identification of these key individuals and the influence they surreptitiously exert is important, as it will affect not only the supplier's negotiating style but also the pricing and marketing strategies adopted prior to these negotiations.

According to Mattson (1988), the purchaser's 'buying centre' comprises three dimensions:

- *Height*: The hierarchy within the purchaser's organisation that exert an influence on the 'buying centre'
- *Width*: The number of functional areas or departments that are involved in the decision to purchase
- *Depth*: The total number of individuals engaged in the purchasing decision.

In order to develop an understanding of their purchaser's buying centre, in addition to establishing the identity of the key decision makers, Kennedy and O'Connor (1997) suggest that the supplier has to consider the purchaser's business environment (mission, objectives and markets; communication network; decision-making processes; award criteria both declared and hidden) and establish the roles played by the individuals involved in purchasing the product and/or service. However, they discovered sharp divergences between the least and most successful bidding organisations across a range of activities involved in understanding the purchaser's buying centre. For example, only one in ten of the least successful firms made any attempt to understand the purchaser's mission, objectives and markets, while only a sixth of the least successful organisations rated themselves as very effective in getting a feel for the roles played by the individuals and interest groups that influence the final choice of supplier, compared with over two-thirds of the most successful companies (Kennedy & O'Connor 1997).

Based on their interpretation of these data, each supplier will respond to a project opportunity in one of the following ways:

- *Passively*: by adopting previously applied strategies and tactics
- *Aggressively*: by attacking the market with the intension of obtaining the work by adopting an aggressive position and/or aggressive pricing
- *Defensively*: by adopting a protective approach to the bid, predominantly when perceived to be threatened by the competition and unable to adopt an aggressive approach.

The final stage in the process is to translate this information into an attack plan for the supplier's own bid and incorporate it into the bid strategy, the aim being to challenge the competition on issues that are critical to the purchaser and where the supplier stands the best chance of being awarded the project.

The decision to bid

The d2b process (alternatively referred to as project selection, pre-bid analysis, project screening or the bid/no-bid decision) is both complex and dynamic, involving many factors (Shash 1993), and is interactive and iterative (Odusote & Fellows 1992). Moreover, the selection of the most appropriate projects for which to bid is fundamental to a successful commercial strategy. It also forms the basis for generating the supplier's marketing approach to the project in terms of bidding and/or solution development (Cova *et al.* 2000). Moreover, the decision, as with that of determining the project mark-up, is very important, as success or failure of a supplier's business lies in the outcome derived from these decisions. What evidence there is, however, suggests that this decision is usually determined by subjective rather than objective information (Fellows & Langford 1980; Ahmad & Minkarah 1988; Shash 1998).

Odusote and Fellows' (1992) respondents claimed that when invited to bid they submitted bids for an average of 73.1% of projects and that they had an average success rate of 22.41% (based on an annual mean of 284 project opportunities from range of 50 to 2850). Likewise, Tucker *et al.* (1996) reported that most major construction contractors were successful in one out of six tenders submitted. Any improvement in the supplier's selection of projects would, therefore, give significant benefit to both the supplier and consequently to their clients. Moreover, a suitable decision support model would be a strategic tool in determining the most appropriate projects to seek and for which to submit a bid, resulting in fewer but more successful bids.

The d2b process

Commercial organisations are required to be selective, choosing which work they will seek out and bid for from a continually changing array of potential projects, due primarily to the availability of sufficient resources (Smith 1995). While few suppliers will actually decline an invitation to tender (ITT), there is still the need to make a strategy decision on whether to submit a *bona fide* bid or not.[iii] This decision is extremely important to the supplier, as, beside the issue of resource allocation, the preparation of a *bona fide* bid requires the supplier to commit to some outlay, which is only recovered if the bid is successful. For example, for a typical bid within the UK construction industry Fellows and Langford (1980)[iv] estimate this to be 0.25% of annual turnover or alternatively as 1% of the projected contract sum for each bid submitted. This could be significantly higher, especially where considerable design or development work is required prior to the submission of a bid. In addition to achieving strategic objectives, these statistics demonstrate the potential financial benefit that can be realised by organisations through adopting an effective and systematic approach to decision-making when deciding whether or not to bid for a project.

The d2b decision is predominantly a group assessment, especially on large, complex projects (Odusote & Fellows 1992), involving several managers, each of whom will have a different organisational perspective. When considering the decision, these decision makers make judgements to balance market opportunities and risks (Thorpe & McCaffer 1991), the primary objective being the continued existence and further development of the company (Skitmore 1989). Several sources suggest that this evaluation involves assessing a number of readily discernible features, for example, company objectives and policies (the potential contribution of the contract to the company's turnover in a particular market segment, the overhead recovery and anticipated profit); contract conditions and/or details; workload; type of work; resources needed (the financial resources required to support the project and the availability of resources); bid documentation; the cost of preparing a bid; contract size; location of contract; and the contract buyer and/or purchaser (Chartered Institute of Building 1997; Thorpe & McCaffer 1991; Ward & Chapman 1988).

Additionally, Skitmore (1989) highlights special objectives, which include market, supply, production, financial, personnel and organisational aims; he collates these under the headings of monetary, non-monetary and market-related objectives. Likewise, Ward and Chapman (1988) believe that recent information concerning competitors' workload, business strategy, strengths and weaknesses, is required to account for non-price factors. Kwakye (1994) considers the concept of competitive advantage, while Park and Chapin (1992) suggest that the number of competitors has a direct effect on the probability of winning the bid. However, one of the most important issues in the d2b decision is the supplier's need for work (Smith 1995).

The d2b process is, therefore, multi-bid in scope and seeks to position potential project opportunities within the organisation's portfolio of projects. Its primary objectives according to Cova *et al.* (2000) are to:

- *Prioritise* project opportunities and inform internal resource allocation
- *Determine* the optimum mode of entry into the project system and any associated external resource requirements.

Cova *et al.* (2000) present a case study of a d2b opportunity for a shipbuilding project. The process involved an assessment by the prospective bidder of its 'relational position' or 'network position', 'functional position', 'competitive strengths', and the 'attractiveness' of the proposed project. They conclude that, depending upon the competitive position of the company, the process seems to underpin the selling and the offering actions in two different ways:

- *Bidding approach*: Where the supplier submits a specification compliant bid and conforms to the purchaser's decision criteria
- *Solution approach*: Where the supplier develops a non-compliant bid based on the risk approach, which assumes that the purchaser will modify their decision criteria for the project.

Factors influencing the d2b decision

To inform the development of d2b models, surveys by Ahmed and Minkarah (1988), Abdelrazig (1995), Wanous *et al.* (2000), Odusote and Fellows (1992) and Shash (1993) within the construction industry identified and ranked 31, 37, 38, 42 and 55 factors, respectively, perceived to influence the d2b decision. For example, Ahmed and Minkarah (1988) identified: type of job, need for work, owner, historic profit and degree of hazard; Odusote and Fellows (1992): client related factors, such as the ability of the client to pay, and the type of work; Sash (1993): need for work, number of competitors bidding, experience in such projects, current work load, and purchaser (owner/promoter/client) identity; and Wanous *et al.* (2000, 2003) identified fulfilling the to-bid conditions imposed by the client, financial capability of the client, relations with and reputation of the client, project size, and availability of time to bid as the top five factors influencing a supplier's decision to bid for a project.

Taken collectively, these studies have much in common, and reveal the top 15 factors that influence the decision to be: type of job and/or project type; current work load; size of contract and/or project size; need for work; past profit in similar projects; purchaser identity; purchaser's ability to pay; experience in similar projects; competitors – number bidding; location of project; tendering duration; degree of safety and/or hazard; labour environment; overall economy (availability of other projects and/or work); and pre-qualification requirement. Further, they suggest that suppliers are influenced more by project characteristics and company related issues, moderately by the bidding situation, and less so by the economic situation and project documentation. Alternatively, the following reasons why firms might decide not to bid are offered by Hillebrandt and Cannon (1990): the lack of skills within the company to undertake the work, unsatisfactory payment arrangements, too many competitors, inadequate capacity in the estimating department, and unsatisfactory experience in a particular geographical area. However, unsatisfactory past experiences of a particular purchaser or consultant regarding personality or payment, high cost and inadequate information, often resulted in inflation of the bid price rather than a refusal to bid.

While there is some agreement between the studies on the identity of the major factors considered to be significant, it is unlikely that decision makers would consider all these factors for every bid opportunity. Smith (1995) suggests that for each potential project the supplier makes a set of intuitive and subjective judgements in the light of the prevailing circumstances, derived according to Ahmad (1990) from a combination of gut feelings, experience and guesses.

Cova *et al.* (2000) categorise the main criteria used to assess project opportunities under the following headings: attractiveness of the project to the bidder, and competitive strengths of the bidder in relationship to the project. Similarly, Lin and Chen (2004) classify the main d2b criteria under two categories:

- *Company factors*: These include the company's reputation and mission, and its internal resources
- *Bid opportunity factors*: These include the competitive environment, probability of project going ahead, and project risk.

Lowe and Parvar (2004) undertook semi-structured and unstructured interviews with the decision makers from a UK construction company (these included sales, marketing and estimating personnel) followed by functional decomposition to organise and classify the factors perceived to be important in the d2b process. Their model contains 21 items classified under seven headings (see Table 16.1).

A pro-forma to assess these items was used to collect d2b data for 115 historical projects from the organisation. Comparison of the distributions for the responses to each item, between accept and reject an opportunity to bid, reveals a complex decision-making process. Moreover, they illustrate the difficulty in determining a 'kill score' for each item (a score below which a no-bid decision would be triggered). Additionally, Lowe and Parvar (2004) established that the decision makers were only able to discriminate between the outcomes 'bid' and 'no-bid' for eight of these items: strategic and marketing contribution of the project; competitive analysis of the bid environment; competency – project size; competitive advantage – lowest cost; resources to bid for the project; feasibility of alternative design to reduce cost; external resources (implementation); and bidding procedures. Further, they establish that only these eight items had a significant linear relationship with the decision to bid.

Factor analysis of the 21 items of the bid/no-bid pro-forma identified six underlying dimensions: responsiveness to opportunities, project relationships, strategic competitive advantage, project procedures, financial relationships, and project risks. Correlation revealed that only three of these dimensions had a significant linear relationship with the d2b outcome. These are, in order of strength of relationship: strategic competitive advantage, responsiveness to opportunity (both of which have a positive relationship) and project risk (which correlates negatively). The positive relationships are consistent with those established for the primary items, while the negative relationship is perhaps indicative of a risk-aversion strategy adopted by the organisation. Again, tests for differences revealed that the decision makers were only able to discriminate between bid/no-bid for the same three factors.

These results suggest that, when deciding upon whether or not to accept a bid opportunity, decision makers are effectively using far fewer factors to inform this decision than indicated by some sections of the literature (see, for example, Ahmed & Minkarah 1988; Odusote & Fellows 1992; Shash 1993).

Table 16.1 Items considered to influence the bid/no-bid decision.

Opportunities	
(1)	Economic contribution of the project
(2)	Strategic and marketing (non-monetary) contribution of the project
(3)	Competitive analysis of the tender environment
(4)	Feasibility of alternative design to reduce cost
Resources	
(5)	Resources to tender for the project
(6)	Internal resources (managerial and technical) to support the implementation of the project
(7)	Financial resources to support the implementation of the project
(8)	External resources (plant, materials and subcontractors) to support the implementation of the project
Project relationships	
(9)	The current relationship with the client
(10)	The current relationship with the client's professional advisors
Project procedures	
(11)	Form of contract
(12)	Contract conditions
(13)	Tendering procedure
Project characteristics	
(14)	Competency – project type
(15)	Competency – project size
(16)	Competency – location
(17)	Experience
Risks	
(18)	Risks involved due to the nature of the project
(19)	Financial capability of the client
(20)	Speed of payment of the client
Competitive advantage	
(21)	Lowest cost

Source: Lowe and Parvar (2004) and Lowe *et al.* (2004).

The d2b output

Ansoff (1965) suggested the following classification of bid opportunity decision outcomes:

- Reject the opportunity to bid
- Provisionally accept the project and prioritise as follows:
  - Add it to a reserve list
  - Remove another project from the reserve list and replace it with the current project
- Unconditionally accept the ITT.

Alternatively, Fellows and Langford (1980) offer the following five possible outcomes: returning the documents; submitting a 'cover price';[V] providing detailed estimates and bid conversion; preparing a bid based on approximate estimates; or reworking the bid. They suggest that each of these possible 'outcomes' is evaluated, using a mixture of intuition, past data, research information, etc., against each 'utility criterion' to give an assessment of the desirability of a particular outcome. However, according to Skitmore (1989) the usual options are simply acceptance or rejection of the opportunity, although rejection does not mean that the supplier does not submit a bid.

d2b models

Generally, the objectives behind the development of a d2b model are to:

- provide a decision support system for the d2b process by systematising the existing 'hard' and 'soft' knowledge within the organisation held by senior management, marketing and technical personnel
- reduce the overall number of bids submitted, but maintain or increase the number of successful bids and overall profitability.

Further, any improvement in the selection of more profitable projects for which an organisation has the 'best chance' of submitting successful bids would give that organisation a significant competitive advantage. While previous investigators have suggested that suppliers could improve the efficiency and effectiveness of bid preparation by adopting a more systematic approach (Ward & Chapman 1988), as previously stated, this has primarily been directed to the bid mark-up problem. Having said this, the following models of the d2b process have been devised.

Eastham (1987) divided the d2b process into objective and subjective decisions, the relative importance of these factors being based on content analysis of interviews. He developed a model with relevant weightings for different items. A main feature of the d2b model was that it was based on 90% subjective and 10% objective decisions to establish the desirability of the project. Unfortunately, no assessment of its applicability and/or accuracy of prediction was provided.

Ahmad (1990) presented a deterministic worth-evaluation d2b model, where thirteen individual factor 'worths' were weighted and combined additively to generate an overall score based on the subjective assessment of the project and the objectives of the firm. A suggested decision was made by comparing this score with a desired minimum (threshold) score. Although the model was illustrated using a hypothetical bidding decision-making scenario, the model was theoretical, based on a set of attributes obtained from a questionnaire survey of 400 US general contractors. Thirteen attributes (factors), lower-level criteria representing major objectives of a construction firm, are divided into four hierarchical groups: job-related, market-related, firm-related and resource-related.

A parametric approach to modelling the d2b process was provided by Wanous *et al.* (2000), based upon the findings of six semi-structured interviews and a formal questionnaire survey of Syrian contractors. The derived model incorporated thirteen positive and five negative bidding factors. While the model was based upon perception of the importance of the factors, 162 real bidding situations were used to optimise and improve the model. The results of the model were evaluated using 20 real-life bidding situations; it proved to be 85% accurate in simulating the actual decisions. Using the same 18 input variables, Wanous *et al.* (2003) developed a neural network[vi] bid/no-bid model again using 162 projects to train the model with 20 randomly selected cases used as a validation sample. Based on these 20 projects, the model had a reported 90% classification rate, a high correlation between the predicted and actual decisions ($R^2 = 0.8120$) and a low prediction error (RMS = 0.1744).

Logistic regression analysis[vii] was used by Lowe and Parvar (2004) to develop two predictive models of the d2b process, derived from 115 historic projects. While a model based on six factor scores was ultimately rejected, a model using the forward stepwise (likelihood ratio) method and incorporating three of the 21 items (see Table 16.1) as predictor variables (competitive analysis of the bid environment, competency – project type and the non-monetary contribution of the project) demonstrated a high prediction capability. The model accounted for 70.9% of the variance (Nagelkerke r^2) with an overall classification accuracy of 94.8%; this increased to 96.5% when two misallocated cases were removed from the analysis. These results are encouraging: a reliable model of the d2b process has been developed using a relatively small number of key variables. Further, the model has a higher predictive accuracy than previous attempts to model this decision process.

A neural network model of the d2b process was developed by Lowe *et al.* (2004), again derived from 115 historic projects. The optimum network architecture selected for further development, and integration into a decision support system (DSS), consisted of a network with 21 nodes in the input layer (the 21 input variables referred to in Table 16.1), three hidden layers (with 15 neurons in the first hidden layer, 10 neurons in the second hidden layer, 10 neurons in the third hidden layer), and four nodes in the output layer (representing the following four options: accept the opportunity to bid, add to a reserve list, replace another project with the current project in the reserve list, or reject the opportunity to bid). The network captured the principal relationships in the training data well, as indicated by an RMS error of 0.001 and a classification rate of 100%. Likewise, the model had good generalisation capability as indicated by an RMS error of 0.001 and a classification rate of 100% for the test data set. Although it demonstrated an acceptable degree of accuracy in representing the data collected, the model, as with the logistic regression model of Lowe and Parvar (2004), had the limitation of only representing the responses of a single organisation.

A fuzzy linguistic approach to the d2b process was proposed by Lin and Chen (2004), where users subjectively assessed screening criteria in linguistic terms and fuzzy values were used to weight their importance. The developers suggested that this approach provided a natural framework for the representation and manipulation of assessment criteria in a context where incomplete data and vague environments existed. Term values were then approximated by their membership functions, and by applying fuzzy logic arithmetic, values from different decision makers could be combined under different criteria. Thus, these imprecise d2b criteria were allowed to take on exact values. The authors illustrated the principles of their model using a case study: an evaluation of a bid opportunity for an international co-development commercial aeroplane project in Taiwan. The decision problem was subdivided into six main criteria and eleven sub-criteria.

Summary

Whether or not to bid for a project is a strategic decision requiring the consideration of strategic intent, competency acquisition and the long-term aims and objectives of the organisation. Moreover, the decision is extremely important to suppliers; besides the issues of resource allocation, the preparation of a *bona fide* bid commits the organisation to considerable expenditure which is only recovered if the bid is successful. Analysis of the literature has identified some 85 factors considered to be important in the d2b process. Deliberation and assessment of these factors (or subgroups of these factors) facilitates: a systematic approach to the decision-making process, which can improve the quality of the decision making by ensuring that all relevant items are considered and assessed; consistency in the decision process; increased productivity; further, they assist in achieving the strategic objectives of an organisation. Additionally, the application of d2b models in the form of DSSs can enhance these benefits further by providing the analyst with more convincing and reliable results; speeding up the decision process, thereby generating a cost saving, and improving the communication of the rationale adopted by the decision makers within the bidding organisation.

While artificial neural networks, regression analysis and analytical hierarchy process (AHP) techniques have been used to model the d2b decision, as illustrated in the previous section, there are still relatively few models. Those that have been developed are predominantly theoretical and/or based on the perception of the decision makers as to the relationships between and importance of the various decision criteria. Moreover, there is little evidence to suggest that these models have been adopted in practice, despite protestations by some researchers that their industrial collaborators had favourably received the models.

Determining the bid submission (price)

In the following section the topics of purchaser bid evaluation and supplier bid (price) determination are presented together, as they are interlinked: the criteria used by purchasers to evaluate a bid informs and shapes the submitted bid. More importantly, as the majority of bidding models developed to date have concentrated mainly on determining the optimum bid price, the proximity of these two aspects clearly illustrates the potential limitations of these models, as clients move from 'lowest price' to multi-criteria selection.

Supplier selection

It is suggested that competition promotes economy, efficiency and effectiveness in expenditure (Procurement Policy Unit 2000), while contributing to the competitiveness of suppliers in both domestic and overseas markets. Assisting the competitiveness of suppliers will also assist purchasers to obtain future value for money and security of supply in a competitive market. The UK Government's Procurement Policy Guidelines (Procurement Policy Unit 2000) require goods and services to be acquired by competition unless there are compelling reasons to the contrary. They also recommend that whether or not there is any legal requirement for it, advertising proposed contracts could be a useful means of ensuring that the potential of the market is fully tested.

Traditionally, in many market sectors, if suppliers have been pre-selected on their capability to meet the purchasers' procurement objectives, tendered on the same information (detailed design and/or specifications) and submitted a compliant bid, then the bidder with the lowest bid price would usually be awarded the contract: the 'lowest-price wins'. The following statement succinctly illustrates this practice:

> *...for every contract there is the optimum bidder who is not only capable of fulfilling the Clients' requirements in terms of time, quality and risk but also in respect of cost is also willing and able to submit a bid lower than any competitor. A fundamental goal of any competitive bidding system is to reveal the identity of this optimum bidder and determine the bid price.* (Drew 1993)

Lowest price

Generally, while it is recognised that an element of competition is necessary, the application of the low bid criterion as the sole criterion for discriminating between bidders is believed to be unsound in several ways. For example, a recent review of the UK construction industry, which has a long history of appointing suppliers by competitive bidding, concludes:

> *...too many clients are undiscriminating and still equate price with cost, selecting designers and constructors* [suppliers] *almost exclusively on the basis of tendered price. This tendency is widely seen as one of the greatest barriers to improvement. The public sector, because of its need to interpret accountability in a rather narrow sense, is often viewed as a major culprit in this respect.* (Egan 1998)

'Lowest-price' does not guarantee the overall lowest project cost upon project completion, while there is an increased possibility of financial collapse of the supplier, bad performance, delay in completion, time and cost over-runs. Moreover, Hatush and Skitmore (1998) assert that it is a somewhat risky and short-sighted approach, as a supplier may inadvertently submit an unrealistically low, or suicidal, bid – a well documented phenomenon known as the 'winner's curse'. Economist William Vickrey has long held that the low bid criterion is not economically efficient, and in 1961 he was able to show that what has become known as the Vickrey Auction (Vickrey 1961), in which the low bid criterion is retained for supplier selection but at the contract price of the second lowest bidder, is theoretically better, as the price obtained in this way is closer to the market consensus.

Further, this approach is not suitable where suppliers are responsible for elements of design, requiring the evaluation of more than one criterion. At the next level of simplicity, evaluation will be based on two variables, price and quality, while the evaluation of the 'hard issues' (price, delivery, quality and performance, etc.) rapidly becomes difficult, requiring purchasers to develop rating techniques. Principally, therefore, there has been an acceptance that relying on lowest price is not always in the purchaser's best interest and as a result there has been an increasing transition from 'lowest price wins' to 'multi-criteria selection' practices in the supplier selection process, the motivation being to attain best value (for money) for the purchaser (Wong *et al.* 2000).

Best value/value for money

In response to these and other criticisms of competition based solely on price, the Procurement Policy Guidelines (Procurement Policy Unit 2000) stipulate that all public procurement of goods and services, including works, within the UK is to be based on value for money, and that the criteria for determining the award of contracts should rarely rely on price alone. The guidelines define 'value for money' as the optimum combination of whole-life cost and quality (or fitness for purpose) to meet the user's requirement. In most cases, value for money (most economically advantageous offer/tender (MEAT) principles in EC terms) will involve other factors such as whole-life cost, quality and delivery against price. The Procurement Policy Unit recommends that appropriate investment appraisal techniques should be used in assessing which compliant bid offers best value for money.

Bid evaluation criteria

The purpose of bid evaluation is to determine each bid in a manner that permits a meaningful comparison. It is often thought that if the proposal variables can be reduced then like-for-like comparison will be easier. Therefore, despite the previous comments, bid price is a key factor in determining the award of a contract. Other project specific factors, however, affect the success of a project, and therefore should also be considered (Mahdi *et al.* 2002).

The criteria by which purchasers evaluate bids can be categorised as follows:

- *Commercial*: Compliance with the ITT, delivery requirements, tender validity period; acceptance of the purchaser's terms and conditions of contract, and payment arrangements; and length of time that product range/design will be on the market
- *Technical*: Consideration of: performance and productivity standards; quality (fitness for purpose); inspection requirements; operational and maintenance costs; professional competence; technical/professional support; standardisation; after-sales service; cost and availability of spares and/or consumables; provision of manuals and training; and warranties, etc.
- *Financial*: Life cycle costing comparisons; quantifiable financial benefits; fixed or variable pricing; cost of components, spare parts, consumables and servicing; financial qualifications to fulfil the contract; risk analysis and financial appraisal, etc.

The criteria for evaluating bids should be established before an ITT or quotation request is issued in order to ensure effective and transparent bid evaluation. Purchasers are advised, if they are not to adopt the lowest bid principle, to set out in detail the criteria for evaluation in the tender documentation for the information of bidders.[viii]

Increasingly, partnering is used to create long-term and durable relationships (see also Chapters 4 and 8). For example, a recent UK Government initiative is the

Smart Procurement Initiative (SPI),[ix] which has mandated changes in the way the Ministry of Defence (MoD) and industry do business. The SPI advocates the pursuit of partnering arrangements in suitable situations. Partnering has had an impact on supplier selection, moving selection criteria from 'low bid' to 'best value', as the evaluation of so-called 'hard issues' will give only a limited indication of the potential for a supplier to be a durable partner in any long-term commercial relationship. Likewise, longer-term partnering arrangements are unlikely to be fully pre-defined and as a result cannot be fully priced at the outset. In such circumstances, price will be less of a determining factor in contract award, requiring other markers to future performance, such as the commercial and strategic processes within organisations: 'soft issues'. Public accountability requires that the evaluation of these soft issues must be as rigorous and auditable as the more objective assessment of 'hard issues'. A Soft Issues Bid Evaluation Tool (SIBET)[x] has therefore been developed by the MoD for this purpose.

The implications on the bid preparation process of the transition from 'lowest price' to 'multi-criteria' evaluation requires the bidder to signal and substantiate their ability to meet these criteria within their bid documentation.

Multi-criteria selection

As previously mentioned, in order to develop appropriate bidding models, an appreciation of how purchasers evaluate bids is required. Studies by Hatush and Skitmore (1998), Wong *et al.* (2000), Fong and Choi (2000) and Madhi *et al.* (2002) have identified 24, 37, 68 and 127 criteria, respectively, involved in supplier selection. The most common criteria considered by purchasers during the bid evaluation stage have been found to include: bid price, financial capability and/or stability, technical ability (past performance, past experience and resources), management capability (current workload, current capabilities; past purchaser–supplier relationship and work strategy), health and safety performance of suppliers and reputation (Hatush & Skitmore 1997, 1998; Fong & Choi 2000; Madhi *et al.* 2002). However, bid price was found to be the only criterion considered by all purchasers in the bid evaluation phase (Hatush & Skitmore 1997).

The analysis by Wong *et al.* (2000) of the responses of both public and private construction clients revealed much similarity between the two subgroups. Overall, the following were deemed to be important: ability to complete on time, ability to deal with unanticipated problems, maximum resource and/or financial capacity, actual quality achieved for similar works, quality and quantity of managerial staff, site organisation, rules and policies (health and safety, etc.), training or skill level of craftsmen, comparison of client's estimate with tender price, amount of key personnel for the project and quality and quantity of human resources. In addition, they found an increasing use of project-specific criteria (PSC), while 'lowest price' was not necessarily the purchaser's principal selection criterion. They established a growing realisation that cost has to be tempered with evaluation of PSC in any attempt to identify value for money.

They also established that, while PSC are used in bid evaluation, public sector supplier selection was still dominated by the principle of 'lowest bid price'. However, private clients favoured an evaluation strategy where bid price was equally as important as PSC. Wong *et al.* (2000) put forward a possible explanation for this difference: that the public sector behaviour could be attributed to the need for financial accountability and deficiencies in public procurement systems.

Multi-criteria decision support systems

Hatush and Skitmore (1998) address the supplier selection problem using multi-criteria utility theory; Mustafa and Ryan (1990), Fong and Choi (2000) and Madhi *et al.* (2002) propose supplier selection models using the analytical hierarchy process; while Sönmez *et al.* (2001) adopt an evidential reasoning approach for the selection of the most appropriate supplier. While these models were initially developed to assist purchasers to select an appropriate supplier, they can be used to the benefit of the bidder to pre-judge the purchaser's evaluation of their bid documentation and/or proposal.

Commentary

Despite moves towards partnering and open book arrangements, competitive bidding, especially for the appointment of subcontractors, remains prevalent (Fu *et al.* 2002), while current and expected EU legislation promotes the use of competition. Also, there is, perhaps, an indication within the UK that partnering does not deliver 'best value' – as signified by the recent decision by Network Rail to suspend their partnering arrangements and return to traditional competitive bidding.

The objectives of the tendering process and in particular supplier evaluation can be illustrated by the following:

> *The principal aim of the tendering process is to select the goods and/or services which offer best value for money in performing the outputs required. Therefore, it is not appropriate to accept the lowest price without full evaluation of the total offer. Purchase price is only one consideration when selecting a supplier. As the value and/or complexity of products or services increase, it becomes more important to consider whole of life costs. Moreover, meeting user requirements, quality and service are critical and can be as or more important than price.* (Treasury Corporation of Victoria, Australia)

The rationale for using an objective bid evaluation method is that purchasers may accomplish most of the objectives, i.e. reduce *ex-post* cost and minimise contract failure. The advantage of these alternative criteria, from a purchaser's perspective, is that they attempt to safeguard against the acceptance of unrealistically low bid prices and the resulting claims, disputes and adversarial relationships during the project. It can, of course, be argued that the use of non-low bid criteria will result in bidders adjusting their prices upwards to try and find the criterion level, and that the incentive to develop more efficient methods of production will be lost. The counter argument is that, contrary to first impressions, innovation, technology development and cost reduction will not be discouraged as suppliers will bid at what they believe to be the market price, with any such cost savings made by one supplier still producing significantly higher profit margins in contracts won. Of course, when such savings are available throughout the industry, bid prices would be expected to gradually fall and the savings eventually passed on to the purchaser.

The findings show that while competition is still the most common approach to supplier selection, purchasers (particularly those in the construction sector) have been influenced to some extent, either by good practice documentation and/or industry commentators, so that 'bid price' and 'multi-evaluation criteria' have equal status when evaluating bid submissions. This transition towards multi-criteria selection has implications for the development of bidding models.

Bid price determination

The success or failure of a bid is, therefore, significantly influenced by the formulation of the bid price. Competitive bidding, however, unlike traditional sales, does not allow an organisation to market test the price it sets for its products and services.

Pricing policies and systems

The strategy and tactics employed by an organisation to implement its pricing policy should reflect the level of profitability it is trying to achieve. As previously stated, mark-up, in addition to profit, also includes additions for running costs or overheads. Further, the costs for all unsuccessful bids have to be recouped from those that are successful.

Generally, pricing policies contain the following components:

- *Business policy*: For most large organisations, policy is predetermined and documented; examples include lowest price and best value, high quality price leader, or a mid-range policy
- *Business objectives*: The strategic objectives and mission statement of the organisation
- *Pricing strategy*: A set of bid-specific objectives developed by the bid team.

Determine bid price strategy

The bid price should be prepared and structured to benefit from the bidder's strengths and weaknesses, accounting for the competition and the purchaser's assessment criteria. Bid price strategy should consider: *price objectives*, derived from the outcome of previous analysis; an appropriate balance between profitability and the offer price; and the *winning price*. Determining the winning bid price is a challenge: often referred to as the bidder's dilemma – too high a profit margin and the bid will be unsuccessful, too low and despite winning the bid the project could easily return a reduced or no profit. The process is linked to the purchaser's criteria for awarding the contract. However, the prime factor in deciding upon a price strategy is the consideration of the winning price. Strategies include: submitting the lowest price, profit maximisation, submitting a loss-leader bid, gaining market entry, or responding to competitor, purchaser and market factors.

In the construction sector, Drew and Skitmore (1990) established two very successful methods of acquiring contracts:

(1) Consistently bidding very competitively for specific types of work, and thereby having a comparatively low bidding variability relative to other bidders
(2) Being inconsistently competitive and having a comparatively high bidding variability relative to other bidders.

They later ascertained a significant correlation between competitiveness and bidding variability: less competitive bidders were more variable in bidding (Drew & Skitmore 1992), while differences in supplier competitiveness are greater for different contract sizes than for different contract types. The most competitive bidders appear to be those with a preferred contract size range (Drew & Skitmore 1997).

Pricing systems

According to Rathmell (1966), 'economic products lie along a goods–services continuum, with pure goods at one extreme and pure services at the other, but with most of them falling between these two extremes'. Within the context of construction, building and construction work is predominantly subcontracted, with the main supplier primarily providing management services. In terms of pricing, the contractor solicits, selects and compiles, with judgement, quotations from subcontractors into a single bid. Projects are undertaken by temporary coalitions of firms, with successful completion of the project balanced against profit and their long-term interest in survival and growth (Winch 2001). They often form a loosely organised set of subcontractors who work from time to time for a main contractor – a relationship that tends to be essentially long-term and rarely based on price competition. More often than not the subcontractors do not have to bid to win the work, although main contractors often 'test the market' every few years by holding a tender competition between subcontractors (Eccles 1981).

This suggests that main suppliers (contractors), within the construction sector belong to one of the service industries, and service industries are not known for their sophistication in pricing (Hoffman *et al*. 2002).

The following pricing systems exist:

- *Cost-related systems*: These include standard-cost pricing (covers standard variable cost and fixed cost per unit, plus profit, adjusted on the basis of competitor pricing); cost-plus-profit (standard mark-up applied to total cost of each product); break-even analysis or target profit pricing (determines the price that will yield the required profit); and marginal pricing (refers to the marginal cost of manufacturing each unit)
- *Market-related systems*: These include perceived value pricing (pricing based on assumptions of purchaser beliefs of 'value'); psychological pricing (price used as a tool to condition purchaser beliefs, e.g. quality or value); promotional pricing (discounts offered to generate high turnover); and skimming (a high price is bid to 'skim the cream' off the market)

- *Competitor-related systems*: These include competitive pricing (tackling the price leader in a particular segment); discount pricing (set artificially high prices and offer discounts to attract purchasers); and penetration pricing (significantly under-cutting competitors' prices to generate turnover).

The pricing strategies of service industries have been classified as either cost-based or market-orientated (Gabor 1977), with full cost pricing, in practice, being the most popular pricing policy. In general, 'real-world' pricing practices in service industries essentially differ in the emphasis placed on production costs and prevailing prices – which mirrors most manufacturing organisational structures of separate, and often conflicting, production and marketing departments – with 'the ideal pricing policy being simultaneously profit based, cost conscious, market-oriented and in conformity with any other aims the businessman may have' (Gabor 1977, p. 43). Clearly though, the amount of available knowledge concerning costs and prices depends on the products involved, with services costs generally being harder to calculate than those incurred in the production of commodities.

Mochtar and Arditi (2001) confirm the widely held belief that pricing strategy in construction is predominantly cost based. However, Runeson and Skitmore (in press) have examined the tenability of two mutually exclusive accounts of construction bid pricing:

(1) A full-cost pricing policy is used
(2) Classical microeconomic theory holds.

They demonstrate that the nature of construction firms' marketing activity and price movements generally make (1) highly unlikely. Likewise, Runeson (1996, 2000) has also shown that the predominant form of construction contract pricing is unlikely to be an absorption, or full-cost, pricing policy – once again offering support for the tenability of the neo-classical position. An alternative approach has been suggested in a recent 'comment' by Weverbergh (2002), advocating 'the economic theory of auctions' to underpin empirical tests of statistical bidding models with construction contract bidding data.

However, what is apparent is that:

> *Businesses that use price as a strategic tool will profit more than those who simply let costs or the market determine their pricing.* (Kotler 2000, p. 459)

Bidding and pricing tactics

Suppliers may adopt various bidding and pricing tactics; these include:

- *Tactical withdrawal of bid*: However, bidders need to consider the implications and impact that this action may have upon existing and future relations with the purchaser
- *Seeking an extension to the submission date*: Although it is better to request an extension rather than risk missing a submission date or submitting a poor or incomplete bid, Tweedley (1995) cautions against this tactic unless there are compelling reasons to do so
- *Submitting a non-compliant offer*: Where cost saving for both the supplier and the purchaser can be shown, for example, through value engineering, a non-compliant (or innovative) offer may give the bidder a competitive advantage.

Likewise, they may implement pricing tactics; these include:

- *Overpricing*: Alternatively referred to as the submission of a cover price. An extreme instance of this is 'collusion' which corrodes the basis and attacks the rationale of the competitive tendering system by restricting competition. This can lead to a reduction in the number of available bidders, an increase in the average bid price and a reduction in bid variance, potentially resulting in artificially increased prices. Zarkada-Fraser and Skitmore (1999) consider that the decision to participate in some form of collusion is primarily focused on the individual. Their investigation established that there is a minority of decision makers that admit they would consider participating in some form of collusive tendering agreement under certain circumstances.
- *Unbalanced allocation of profit*: Where suppliers are required to submit a detailed breakdown of their bid, the bid team has a variety of methods for distributing the profit element within the bid.
- *Tactical pricing*: Where payment is staged there is an opportunity for manipulation of the bid by the supplier (King & Mercer 1988). Examples include: front-end loading, item spotting, back-end loading, and maximising the net present worth of the cash flow. While there is evidence of the use of tactical pricing or unbalanced bidding, it would seem that front loading and item spotting are by far the most common in market sectors such as construction (Green 1989). This perhaps is not the case in service industries.

There is limited research in this area, both in investigating practice and model development to systematise the process. However, it is apparent that the techniques used in practice are both subjective and opportunistic.

Price formulation

After deciding to submit a bid, there usually follows a two-stage price formulation process consisting of a baseline estimate and mark-up. Formulated at the operational level, the baseline estimate is forwarded to the business strategy level, where senior management decides the appropriate bid level; the baseline estimate is usually combined with a mark-up to form the bid. Different bidders utilise distinct mark-up policies. Bidding strategy involves determining the mark-up level to a value that is likely to provide the best pay-off. Standard bidding models, according to Male (1991), assume that bidders endeavour to maximise their expected profit. The bidder, however, may be attempting to fulfil other objectives, including minimising expected losses, minimising profits of competitors, or obtaining a contract in order to maintain production (Drew & Skitmore 1997).

Cost estimating

All bidders, irrespective of the pricing system they adopt, need to know the likely cost components of a bid, if only as an indication of their bottom line. Cost estimating therefore plays an important part in pricing a bid and is critical to the overall success of the resulting project. Estimating, however, falls outside the scope of this chapter: for more information see Ashworth and Skitmore (1983), Akintoye (2000) and Akintoye and Fitzgerald (2000). However, in contrast with the usual assumptions of auction theory, in construction contract auctions, bidders have both different costs and imperfect estimates of them.

Setting the bid price

Bidding performance is a reflection of the strategic process (Drew & Skitmore 1997) and is concerned with the competitive relationships between the bids submitted to the purchaser. Because a bid is an estimate of the unknown market price, most bidders submitting a genuine bid are attempting to submit a bid that is low enough to win the contract but high enough to make a profit (Park & Chapin 1992).

> *Essentially, the bid price problem is one of optimising the price of the bidder's proposal against the requirements of the client and the performance of competitors' bids in order to maximise the bidder's profits while ensuring that it maintains a reasonable chance of winning the bid.* (Bussey *et al.* 1997)

At the time of submitting the bid, the maximum level of competitiveness can be taken to be the lowest bid, so the bid price will become the optimum bid. The optimum bid has been defined as:

> *...the lowest priced evaluated bid which has undergone a process of assessment to identify and, where necessary, to price the consequences inherent in the submission.* (Merna & Smith 1990)

As previously mentioned, suppliers will often base their bid price on an estimate of the cost of the inputs to the process. The procedure of transforming the estimate into a bid is a management activity sometimes referred to as tender adjudication, and is essentially about submitting a winning bid at the best possible price.

> *Tender adjudication is concerned with getting the job at the best price; securing it in competition by the smallest possible margin; and on the best commercial terms procurable...* (Tassie 1980)

To arrive at the bid price, a mark-up (additions for overheads, project financing costs, required profit and risk margin) is added to the net cost estimate. The technical authority will provide the bid team with an analysis of the net estimate, indicating its key components, together with clarification on how risks identified during the estimating process have been resolved. Generally, an organisation's annual business plan will specify its overheads budget and baseline profit requirement. Therefore, the only item left for consideration is the risk margin: the supplier's assessment of the special risks and/or the commercial attractiveness presented by the project under review. This decision will be made taking into consideration any knowledge of the purchaser's available budget for the project; approaches likely to be adopted by the competition; and an appraisal of the potential (realistic) future revenues and margins obtainable through contract variations.

Mark-up estimation, however, is an amorphous decision problem rendering analysis and formulation of a satisfactory solution mechanism problematic. Identification of all the related factors that inform rational decision making is both protracted and complex, likewise the analysis of their individual influence and the measurement of their collective bearing on the decision. As a result, conventional practice is to form bid decisions on the basis of intuition, derived from a mixture of gut feeling, experience and guesses. According to Moselhi *et al.* (1993), this implies the application of some degree of pattern recognition rather than computation or deep reasoning about the components of the problem. Likewise, Fayek (1998) asserts that the margin-size decision process predominately involves making qualitative and subjective assessments and that, despite the problems involved, a need exists to structure and formalise the process. Many different theoretical approaches to competitive bidding have been proposed and tested with varying results.

Factors influencing the bid price decision

Flanagan and Norman (1982) identified five key factors as influencing contractors' bidding behaviour: market conditions, current and projected workload, the size and complexity of the work, client type and project size. Likewise, research within the construction industry has sought to determine the order of importance of the factors considered by suppliers when setting the bid price. Surveys by Ahmed and Minkarah (1988), Shash and Abdul-Hadi (1992), Dulaimi and Shan (2002), Eastham (1987) and Shash (1993) identified and ranked 31, 37, 40, 52 and 55 factors, respectively. For example, Eastham (1987) identified: subcontract requirements, type and size of job, competitors, client and professionals, and labour requirements; Ahmad and Minkarah (1988): degree of hazard, degree of difficulty, type of job, uncertainty in estimate and historic profit; Teo *et al.* (1991) report: accuracy of supplier's cost estimate, credit worthiness of client, contractual liabilities, type of job and relationship with consultants; and Shash (1993): degree of difficulty of the work, risk owing to the nature of the work, current workload, need for the work and contract conditions as the top five factors influencing a supplier's bid price decision. Further, Teo *et al.* (1991) comment that closer examination of the responses shows that most suppliers tend to place more emphasis on personal relationships when adjudicating (reviewing) their bids. The least important factors were found to be those relating to general market and political conditions. Likewise, factors such as the number of competitors bidding and the perceived competitiveness of competitors were generally ranked comparatively low. Type of job, experience, competitors, owner and strength in the industry are perceived to influence the supplier's confidence in winning a project (Ahmad & Minkarah 1988; Sash 1993), while both these studies found that the need for work and location of the project influenced the suppliers' motivation to obtain a job.

Again, taken collectively, these studies have much in common. They reveal that the top 15 factors that influence the bid price decision are: risk involved in investment; degree of difficulty; size of contract and/or project size; need for work; uncertainty in (reliability of) cost estimate; current workload; past profit in similar projects; owner (private or public); type of contract; type of job or project type; degree of safety and/or hazard; anticipated rate of return on project; overall economy (availability of other projects and/or work); risk involved owing to the nature of the work; and project cash flow. Further, they suggest that bidders are predominantly influenced by project characteristics and company related issues, moderately so by the economic situation and project documentation, and to a lesser extent by the bidding situation.

The supplier's need for work and risk has been shown to significantly affect bid mark-up (De Neufville & King 1991), while bidding decision making has been found to be influenced both by the results of previous bidding attempts and by construction experience derived from undertaking projects (Fu *et al.* 2002, 2003); these authors found that suppliers that bid more frequently were more competitive than those that bid occasionally; further, they established the existence of 'experienced market players' who display the traits of a more competitive and consistent bidding performance.

As with d2b, the bid mark-up decision is again a dynamic one where the factors considered vary over time, from project to project and from organisation to organisation.

Risk and uncertainty

In principle, Chapman *et al.* (2000) assert that risk analysis should be incorporated into the process of bid preparation to assess uncertainty inherent in the obligations required by the contract, and also to assist in the formulation of bids that provide a suitable balance between the risk of not being awarded the contract and the risk associated with potential profit and losses if the contract is acquired.

Risk and uncertainty relating to the bid process can be categorised under the following headings:

- *Factors relating to the influence of the decision makers*: As risk assessment decisions are predominantly made intuitively and subjectively, the decision-making process itself gives rise to additional risks and uncertainties.
- *Factors relating to the bidding process*: According to Smith (1995), these may be divided into three distinct groups: risks relating to the project itself, for example, procurement system and intended contractual arrangements, adequacy of the tender documentation, planned project time-scale and the implications of delayed or non-completion, fixed or fluctuating (variable) price, level of prime cost sums, and the degree of technological difficulty; risks relating to the purchaser and the professional team, for example, uncertainty over whether or not payment will be on time, under payment of payments on account, under payment for variations to the contract, delay in agreeing the final account, and purchaser's advisers' interpretation of the specification; and risks relating to the bidding process, for example the selection of appropriate work rates and price stability.

As cited in several influential contributions in the construction literature (e.g. Raftery 1991; Hillebrandt 2000; Runeson 2000), the extent of the uncertainties involved in forecasting future costs, as well as the behaviour of competitors and the market in general, requires contract bidders to devote a far greater amount of energy and resources to marketing than is currently admitted in the economic theory of auctions.

Bidding models

Over the past 50 years extensive research has been undertaken into the development of analytical bidding models, with the aim of improving a bidder's likelihood of submitting the 'optimum' bid. These mathematical models can be dated from Friedman's innovative work (Friedman 1956) and include statistical bidding models, approaches based on historical data, multiple regression analysis and artificial neural network techniques. Other researchers have taken a less mathematical approach, attempting to analyse the human decision-making process as it applies to forecasting. However, most of the bidding literature is concerned with setting a mark-up, m, so that the probability, $Pr(m)$, of entering the winning bid reaches some desired level. Several models have been proposed for calculating $Pr(m)$ (e.g. Friedman 1956; Gates 1967; Carr 1982; Skitmore & Pemberton 1994).

The problems involved in monitoring the performance of competitors

All organisations monitor, to a degree, the bidding performance of their competitors. The use of these bidding patterns to model and predict bidding performance requires the acceptance of a series of interrelated assumptions which influence the reliability of the results obtained and can produce the following problems:

- *Availability of information*: There is an assumption that there is an adequate supply of information on competitors' bids. However, depending upon the market sector, this may be erroneous; for example, in some sectors relatively few submitted bids are published and the bidder may only be able to allocate the 'winning bid' to the successful bidder. In others, even this information remains confidential. Obtaining information required by quantitative models can therefore be problematic, particularly as Hillebrandt (1974) recommends that data used should not span a period of more than three months. Alternatively, in other sectors, there may be a large number of potential bidders, in which case the collection and assessment of data is both time consuming and expensive to undertake.
- *Continuity of bidding behaviour*: The use of historical data assumes that competitors will continue to bid as in the past: this is held to be unrealistic (Chapman *et al.* 2000).
- *Conformity to statistical independence*: Predominantly, modelling approaches require that, for each competition, every rival bid is statistically independent.
- *Homogeneity of underlying costs*: For organisations to be able to compare their submitted bid (or underlying estimate) with the bids of others, or to treat all rival bidders as one 'average' competitor, necessitates the assumption that there is no significant difference between the competitors' cost estimates. Unfortunately, this is not accurate due to the variability of each supplier's performance.

Taken collectively, these assumptions imply that competitors do not discriminate between contracts, rather they randomly select values from a constant bidding pattern, oblivious to any deviation in attractiveness of contracts. Also, increasingly, as indicated earlier in this chapter, tenders are not always expressed or assessed in terms of the 'lowest' price; in these circumstances where alternative award criteria are applied, it can be difficult to obtain information concerning the purchaser's subjective decision making. Further, these approaches do not attempt to predict the outcome of any particular bid: they simply attempt to predict a trend over a period of time, and it may be that most organisations do not submit enough bids in any given period of time for the trend to become established or useful. Despite these comments, used cautiously, bidding models can give suppliers a competitive advantage.

Classification of bidding models

As stated in the Introduction, there are currently in excess of 1000 papers associated with modelling the bid price decision; generally, these models fall into four main categories.

Models based on probability theory

Most bidding models based on probability theory have been derived from the work of Friedman (1956). He asserted that in the context of a bid auction, by using a probability distribution, it was feasible to model the behaviour of each bidder as a function of the cost estimate. According to Friedman, over time the bidder's principal goal is profit maximisation, which could be measured by multiplying the expected profit by the probability of winning to provide an anticipated percentage profit. This basic probability model requires data of both winning and losing bids. However, while it may not always be possible to obtain information on losing bids, it is more likely that the winning bid will usually be identified. Moreover, it has been claimed that a bidder's performance against the successful bid is of prime importance. Again, this implies lowest price is the most important criterion and ignores the other award criteria discussed earlier.

Numerous models have been developed on this foundation, centred principally on the efforts of Gates (1967), who asserted that a bidding organisation, by evaluating its own bid against the winning bid, could determine the mark-up which would have been required to win the bid opportunity. Others have sought to improve the precision of probability models by incorporating, for example, mechanisms to account for factors such as project complexity, duration, the number of suppliers competing for a specific project, the perceived attractiveness of the project, the likely bidding performance of competitors, project size, and transient movement in market price. Additionally, King and Mercer (1990) investigated the implications of predicting optimum mark-up when bidding with uncertain costs, while Hillebrandt (1974) proposed incorporating professional expertise within an econometric framework called Shackle's Degree of Potential Surprise. This approach seeks to integrate the expert opinion and individual feelings of the decision makers within an organisation's established decision-making process, taking into account the individual's personal risk preference. A detailed synopsis of statistical modelling can be found in Skitmore (1989) or Skitmore (1991a).

Recently, Crowley (2000) compared and contrasted the models developed by Friedman (1956), Gates (1967) and Carr (1982). He concluded that Friedman's model was theoretically correct, yet the bid problem was incorrectly specified; Gates's model was practically correct in recognising historical as well as present variability, yet the formula was incorrectly specified. He recommended the use of Carr's model. Likewise, Skitmore (2004), using real and typical sets of construction bid data, compared Friedman's (1956) model, Gates's (1967) model, Carr's (1982) model and two versions of Skitmore's (1991b) model against pure chance; he established that at best they produced only a marginal improvement on chance. However, there is still more that could be done to develop these models, for example, by grouping similar bidders and by incorporating interdependence.

Multiple regression models

The basic expression of the multiple regression equation is:

$$y = a + b_1x_1 + b_2x_2 + b_3x_3 + \cdots + b_nx_n$$

where y is the dependent variable (the observation to be predicted), a is a constant, x_1, x_2, etc., are independent variables (the various characteristics assumed to influence the prediction), and b_1, b_2, etc., are coefficients which represent the weightings linked to the characteristics. Multiple regression analysis has two functions: to summarise selected data and to analyse them for (statistically) significant trends. In a bidding model, therefore, y could represent the value of the lowest bid, while the independent variables x_1, x_2, etc., might be the bidder's cost estimate, the number of bidders, or measures of complexity and project size, etc. Details of the various regression techniques are clearly described in Skitmore and Patchell (1990). Examples of the application of this approach include Carr and Sandhal (1978) and Seydel and Olson (2001) who have utilised regression analysis to identify and link project and situational characteristics to the combined distribution of bid acceptance probabilities.

Neural network models

Optimal bid price models have been developed using the neural network approach. For example, artificial neural networks (ANN) models have been developed to aid the prediction of winning bids (McKim 1993), predict mark-up (Dozzi *et al.* 1996) and for optimum mark-up estimation (Moselhi *et al.* 1991; Moselhi & Hegazy 1993; Li & Love 1999; Li *et al.* 1999). Dozzi *et al.* (1996) described a bidding model that utilised 21 attributes, of which 11 were used within a neural network model. The network of Moselhi *et al.* (1991) comprised a set of input parameters that described a bid situation and produced an output pattern of three optimum mark-up values that approximated the results of Friedman's (1956), Gates's (1967) and Carr's (1987) bidding strategy models. Li and Love (1999) presented a hybrid system combining both neural networks and rule-based systems, which used current workload, project complexity, number of competitors, market conditions, project size, project type, location, working cash requirements, overhead rate, and labour availability as input variables.

A significant shortcoming of ANN systems is their inability to explain their recommendations. In an attempt to over come this, Li *et al.* (1999) utilised the KT-1 method to automatically extract rules from a trained neural network. The model was trained using 25 successful bidding examples with five cases used for validation purposes; the developers reported an average error rate of 6.7%.

Novel pricing approaches

Most bid price models have attempted either to determine the probability of a bidder winning a bid based only upon the price of its proposal, or to estimate the optimum mark-up. As shown earlier, in practice, however, purchasers usually evaluated bids on the basis of multiple criteria which they perceived to have value. More recently, bid strategy models have been developed to address these mutliple award criteria. For example, Cassaigne and Singh (2001) describe an intelligent tactical decision support system developed to aid pricing decisions within dynamic competitive environments. The system uses non-linear models, optimisation and learning algorithms to solve generic price-setting problems. Likewise, Bussey *et al.* (1997) evaluated two novel pricing approaches which considered the multiple criteria influencing the purchaser's bid selection and which modelled the three main actors within the bid pricing problem: the purchaser, the competitors and the bidder. Both models adopted multi-attribute utility theory to capture the bid selection behaviour of the purchaser, while possibility theory was used to model the performance of competitors in the bid. The resulting models calculated the possibility of the bidder winning a bid at different bid prices, allowing the bidder to optimise their bid price with the intention of improving their overall success rate and profitability.

Further examples of bidding decisions models that use utility theory include those developed by Dozzi *et al.* (1996) and Seydel and Olson (2001). Dozzi *et al.* (1996) developed a bid mark-up model using 21 criteria in the bidding decision, the majority of which were evaluated subjectively. Their model could be customised to take account of variation in the decision makers' attitudes, preferences and bias to risk. While an example was provided to illustrate the technique, no performance data were given. Seydel and Olson (2001) proposed a hybrid multi-criteria method to model the bidding decision, which advocated a pairwise comparison procedure to generate criterion weights, and a linear transformation procedure to calculate relative scores for bidding alternatives. The approach was illustrated and evaluated using a set of 83 construction projects.

Alternatively, the analytical hierarchy process (AHP) has been applied by Seydel and Olson (1990), Cagno *et al.* (2001), and Marzouk and Moselhi (2003) to the bidding process, where competing bids were evaluated on a multiple criteria basis. Seydel and Olson's (1990) model incorporated decision-maker preferences into the bidding process and combined stochastic modelling with AHP; their inputs were cost data, competitor data, and decision-maker preferences, while output was a set of composite weights by which alternative bid mark-up may be ranked by multiple criteria. Similarly, Cagno *et al.* (2001), developed a simulation approach based on the AHP to assess the probability of winning in a competitive bidding process. The model allowed the bidder to define a bidding strategy on the basis of the information currently available regarding the owner, the competitors, and their own bid profile; an example was provided, but no performance data. Marzouk and Moselhi (2003) presented a generic decision support model for construction bidding which utilised multi-attribute utility theory and the AHP; the model provided a decision support environment for both estimating mark-up and bid evaluation. Further, the model could be adapted to take account of the user's business environment and bidding strategy and could account for the decision maker's attitude towards risk.

Fayek (1998) developed a competitive bidding strategy model based on the techniques of fuzzy set theory, which permitted bidders to evaluate their bidding objectives and to consider the influence of the various corporate, commercial, project, purchaser and competitive factors in 'qualitative and approximate' terms appropriate to the subjective nature of the mark-up size decision. Moreover, the model did not assume that the lowest bidder would necessarily be awarded the contract; rather, it considered the effects of other, qualitative criteria that may influence the purchaser's decision and thus the most suitable margin size. Despite addressing some of the issues raised above, the model can be criticised in terms of making 'black-box' decisions. The developer, who recommended the incorporation of an explanation facility to indicate how the output was obtained, has acknowledged this. Finally, while a sample project was analysed using the model to illustrate how it could be used in practice, no generalised performance statistics were provided.

An optimal approach was used by Shen *et al.* (1999) to develop a bid model based on the purchaser's unit time value, for price–time bi-parameter construction contracts for situations where bidders are faced with the problem of deciding which combination of bid price and contract time to submit. Using the same approach, Drew *et al.* (2002) developed an optimal bidding strategy using three popular two-envelope fee bidding formulations; the authors suggested that this approach could be applied to other two-parameter formulations. Finally, Chua *et al.* (2001) have developed a case-based reasoning approach to model bid decision making: the output derived was the optimal mark-up. While illustrative examples were provided for all three models, based on either an individual or several sets of hypothetical cases, their performance has not been tested using real project data.

Adoption of models by practitioners

Despite the wealth of research activity over the past 50 years in developing bidding models, in practice there has been relatively little use made of these models. Moreover, in many organisations the decision on risk margin is a very subjective one. Surveys of the application of bidding models within the construction sector (see Table 16.2) have revealed the limited use of any kind of mathematical or statistical bidding model. Additionally, within the context of Singapore, Chua and Li (2000) found that 80% of their respondents had never used any statistical model to assist their bidding decisions, while almost 95% of Dulaimi and Shan's (2002) respondents were comfortable with the way they made the mark-up decision.

There appears to be an issue concerning the apparent apathy by which these theoretical bidding models have been received in practice. This raises the question: considering the academic interest in this area, why are bidding models not more widely used? Chapman *et al.* (2000) suggested that this may reflect a lack of understanding on the part of practitioners and/or a lack of organisation effort in collecting,

Table 16.2 Usage of probability/mathematical models.

Country	Suppliers using mathematical/statistical bidding models (%)	Researcher(s) (date)
US	11.1	Ahmad & Minkarah (1988)
UK	17.6	Shash (1993)
Australia	12.0	Ting & Mills (1996)
US	14.3	Mochtar & Arditi (2001)
Singapore	0	Dulaimi & Shan (2002)
Syria	3.0	Wanous *et al.* (2003)

collating and interpreting relevant information. They asserted that this was a result of failure on the part of the model developers to persuade practitioners that investing their time and resources is worthwhile, and of theorists to convince practitioners that theoretically sound approaches are practical propositions. Alternatively, Fayek (1998) believed that current models did not correspond with the actual practice of bidders. Likewise, Rothkopf and Harstad (1994) advocated that, generally, the gap between theory and practice was a result of the failure to provide enriched models that took context into consideration and moved towards realism. As a response, Chapman *et al.* (2000) suggested that flexible, easily modifiable models may be more useful than technically accurate models; they asserted that the apparent 'mismatch' could be managed through the learning approach offered by constructively simple models.

eAuctions and game theory

Introduced in Chapter 4, eProcurement refers to the electronic enablement of the purchasing process. Of particular interest to the topic of bidding are electronic reverse auctions (eAuctions), online exercises in which suppliers vie against each other with open bids for the entitlement to provide the purchaser with goods or services. The Office of Government Commerce (2004) claims that during an eAuction the impression of competition is intensified as suppliers compete in real time by submitting lower bids as the auction develops. The process can accommodate both lowest price or MEAT award criteria. Also, in addition to submitting a lower bid, a supplier can introduce new or enhanced value aspects to their bids, which are open to all the competing suppliers. As it is increasingly used in public and private sectors, the Office of Government Commerce (2004) reports efficiency improvements in public sector contracts in the range 20–25%.

Linked to reverse auctions is the application of game theoretic techniques. An explanation of game theory would in itself amount to at least a chapter; it therefore falls outside the scope of this chapter. For more information see, for example, Binmore *et al.* (1993) and Binmore (1994, 1998). Probably the most famous application of Binmore's work is the sale of the British 3G Telecom Licences (Binmore & Klemperer 2002), claimed to be the biggest auction ever.

Summary

This section has shown that, notwithstanding the increased use of partnering and open book arrangements, competitive bidding is still the most commonly used approach to supplier selection. Moreover, existing and anticipated EU legislation advocates the use of competition. However, purchasers (particularly those in the construction sector) have been persuaded, either by good practice documentation or by industry pundits, so that, currently within the UK, 'bid price' and 'multi-evaluation criteria' have equal status when evaluating bid submissions, the motivation being to attain best value. This transition towards multi-criteria selection has implications both on the bid preparation process, requiring bidders to signal and demonstrate their ability to meet these criteria within their bid documentation, and on the development of bidding models.

As with the assessment of whether or not to bid for a project, determining the bid submission (the interplay between bid price and value aspects of the submission) is a strategic decision. Likewise, bidding performance is an expression of a strategic process. A review of the literature revealed 93 factors believed to influence the bidding decision. In particular, risk involved in investment, degree of difficulty, size of contract and/or project size, need for work, and uncertainty in (reliability of) the cost estimate were held to be important, suggesting that bidders were primarily influenced by project characteristics and company related issues. The prominence of risk implies that risk analysis should be central to the decision process.

Over the past 50 years extensive research has been undertaken into the development of analytical bidding models, with the aim of improving a bidder's likelihood of submitting the 'optimum' bid price. Originating from the work of Friedman (1956) they include: statistical bidding models, approaches based on historical data, multiple regression analysis and artificial neural network techniques. More recently, bid strategy models have been produced to take account of mutli-award criteria, employing utility theory, analytical hierarchy process (AHP), fuzzy set theory, and case-based reasoning approaches. Regardless of this wealth of research activity, in practice these models have not been widely adopted. Furthermore, in many organisations conventional practice is to form bid decisions on the basis of intuition, derived from a mixture of gut feeling, experience and guesses.

Research agenda

Despite the longevity of the research into developing bidding (price optimisation) models, further work within the wider area of bidding is still required. To underpin the development of decision support tools for both the d2b and bid submission, research should be concentrated on developing a suitable underlying theoretical framework (Runeson & Skitmore 1999). This research could, for example, include:

- further investigating and modelling the competitive behaviour of suppliers and, in particular, the impact of the competitiveness of their subcontractors and the implications of globalisation and joint ventures on bidding decisions
- further investigating the implications of on-line reverse auctions and eProcurement generally on the competitiveness and bidding behaviour of suppliers.

In terms of the development of decision support tools for both the d2b and bid submission, the reticence of decision makers to employ existing models implies that model developers, as well as the considering the academic rigour of the modelling approaches adopted, need to consider the practical application of the models. In particular, this lack of acceptance of decision support models needs to be investigated, hopefully leading to the development of innovative models that are both suitably robust and theoretically sound to accurately model the complexities of the decision domain, but which are practical enough to be accepted by practitioners. To achieve this, Fayek (1998) has recommended the incorporation of qualitative and subjective judgements and heuristics, instead of involved mathematical or statistical methods, with less dependence on historical project and competitor data. Additionally, Chapman *et al.* (2000) suggest incorporating transparency into the modelling process, acknowledging and addressing the impact of the inevitable uncertainty inherent in cost estimates and in the estimates of the probability of winning related to various levels of bids, and making simple models usable, by dealing directly with the complexities of reality.

Specifically, in terms of project selection (d2b) models, development should be expanded to include objective data, such as the profit and non-monetary objectives achieved by specific project types, size, etc. This would necessitate the incorporation of data generated via cost value reconciliation as described in Chapter 18. With regard to the development of bid submission support models, the increasing use of 'best value' and alternative award criteria should be addressed, in addition to profit maximisation, and incorporated to allow the assessment of alternative value–price combinations, rather than solely generating an optimum bid price or mark-up.

References

Abdelrazig, A.A. (1995) MSc dissertation (unpublished), King Fahd University of Petroleum and Minerals, Saudi Arabia.

Ahmad, I. (1990) Decision-support systems for modelling bid/no-bid decision problem. *ASCE Journal of Construction Engineering and Management*, **116**(4), 595–608.

Ahmad, I. & Minkarah, I. (1988) Questionnaire survey on bidding in construction. *ASCE Journal of Management in Engineering*, **4**(3), 229–243.

Akintoye, A. (2000) Analysis of factors influencing project cost estimating practice. *Construction Management and Economics*, **18**, 77–89.

Akintoye, A. & Fitzgerald, E. (2000) A survey of current estimating practices in the UK. *Construction Management and Economics*, **18**, 161–172.

Ansoff, H.I. (1965) *Corporate Strategy*. Penguin, London.

Ashworth, A. & Skitmore, R.M. (1983) *Accuracy in estimating*. CIOB Occasional Paper 27. Englemere.

Beeston, D.T. (1983) *Statistical Methods for Building Price Data*. E. & F.N. Spon, London.

Binmore, K. (1994) *Game Theory and the Social Contract, Vol. 1: Playing Fair*. MIT Press, Cambridge, MA.

Binmore, K. (1998) *Game Theory and the Social Contract, Vol. 2: Just Playing*. MIT Press, Cambridge, MA.

Binmore, K., Kirman, A. & Tani, P. (eds) (1993) *Frontiers of Game Theory*. MIT Press, Cambridge, MA.

Binmore, K. & Klemperer, P. (2002) The biggest auction ever: the sale of the British 3G Telecom licences. *Economics Journal*, **112**, C1–C23.

Bussey, P., Cassaigne, N. & Singh, M. (1997) Bid pricing – calculating the possibility of winning. *IEEE International Conference*, **4**, 3615–3620.

Cagno, E., Caron, F. & Perego, A. (2001) A multi-criteria assessment of the probability of winning in the competitive bidding process. *International Journal of Project Management*, **19**, 313–324.

Carr, R.I. (1982) General bidding model. *ASCE Journal of Construction Division*, **108**(4), 639–650.

Carr, R.I. (1987) Optimum markup by direct solution. *ASCE Journal of Construction Engineering and Management*, **113**(1), 138–150.

Carr, R. & Sandahl, J. (1978) Competitive bidding strategy using multiple regression. *ASCE Journal of Construction Engineering and Management*, **104**, 15–26.

Cassaigne, N. & Singh, M.G. (2001) Intelligent decision support for the pricing of products and services in competitive markets. *IEEE Transactions on Systems, Man, and Cybernetics, Part C*, **31**(1), 96–106.

Chapman, C.B., Ward, S.C. & Bennet, J.A. (2000) Incorporating uncertainty in competitive bidding. *International Journal of Project Management*, **18**, 337–347.

Chartered Institute of Building (1997) *Code of Estimating Practice* (6th edn). Addison Wesley Longman, Harlow.

Chua, D.K.H. and Li, D. (2000) Key factors in bid reasoning model. *ASCE Journal of Construction Engineering and Management*, **126**(5), 349–357.

Chua, D.K.H. Li, D. & Chan, W.T. (2001) Case-based reasoning approach in bid decision making. *ASCE Journal of Construction Engineering and Management*, **127**(1), 35–45.

Cova, B, Salle, R. & Vincent, R. (2000) To bid or not to bid: Screen the Whorcop Project. *European Management Journal* **18**(5), 551–560.

Crowley, L.G. (2000) Friedman and Gates – another look. *ASCE Journal of Construction Engineering and Management*, **126**(4), 306–312.

Dozzi, S.P., AbouRizk, S.M. & Schroeder, S.L. (1996) Utility-theory model for bid markup decisions. *ASCE Journal of Construction Engineering and Management*, **122**(2), 119–124.

Drew, D. (1993) A critical assessment of bid evaluation procedures in Hong Kong. *Professional Builder*, June. Chartered Institute of Building (Hong Kong Branch).

Drew, D.S. & Skitmore, M. (1990) Analysing bidding performance; measuring the influence of contract size and type. In: *Building Economics and Construction Management: Management of the Building Firm*, CIB W-65. International Council of Construction Research Studies and Documentation, Sydney, Australia; pp. 129–139.

Drew, D.S. & Skitmore, R.M. (1992) Competitiveness in bidding: a consultant's perspective. *Construction Management and Economics*, **10**(3), 227–247.

Drew, D. & Skitmore, M. (1997) The effect of contract type and size on competitiveness in bidding. *Construction Management and Economics*, **15**, 469–489.

Drew, D. Shen, L.Y. & Zou, P.X.W. (2002) Developing an optimal bidding strategy in two-envelope fee bidding. *Construction Management and Economics*, **20**, 611–620.

Dulaimi, M.F. & Shan H.G. (2002) The factors influencing bid mark-up decisions of large and medium-size contractors in Singapore. *Construction Management and Economics*, **20**, 601–610.

Eastham, R.A. (1987) The use of content analysis to determine a weighted model of the contractor's tendering process. In: *Building Cost Modelling and Computers* (ed. P.S. Brandon). E. & F.N. Spon, London; pp. 351–363.

Eccles, R.G. (1981) The quasifirm in the construction industry. *Journal of Economic Behaviour and Organisation*, **2**, 356–357.

Egan, J. (1998) *Rethinking Construction*. HMSO, London.

Fayek, A. (1998) Competitive bidding strategy model and software system for bid preparation. *ASCE Journal of Construction Engineering and Management*, **124**(1), 1–10.

Fellows, R.F. & Langford, D.A. (1980) Decision theory and tendering. *Building Technology and Management*, **18**(9), 36–37.

Flanagan, R. & Norman, G. (1982) Making use of low bids. *Chartered Quantity Surveyor*, **14**, 226–227.

Fong, P.S. & Choi, S.K. (2000) Final contractor selection using the analytical hierarchy process. *Construction Management and Economics*, **18**, 547–557.

Friedman, L. (1956) A competitive-bidding strategy. *Operations Research*, **4**, 104–112.

Fu, W.K., Drew, D. & Lo, H.P. (2002) Competitiveness of inexperienced and experienced contractors in bidding. *Construction Management and Economics*, **20**, 655–666.

Fu, W.K., Drew, D. & Lo, H.P. (2003) The effect of experience on contractors' competitiveness in recurrent bidding. *ASCE Journal of Construction Engineering and Management*, **129**(4), 388–395.

Gabor, A. (1977) *Pricing: Principles and Practices*. Gower, Aldershot.

Gates, M. (1967) Bidding strategies and probabilities. *ASCE Journal of Construction Division*, **93**(CO1), 75–107.

Green, S.D. (1989) Tendering: optimisation and rationality. *Construction Management and Economics*, **7**, 53–63.

Hatush Z. & Skitmore, M. (1997) Criteria for contractor selection. *Construction Management and Economics*, **15**(1), 19–38.

Hatush Z. & Skitmore, M. (1998) Contractor selection using multicriteria utlity theory: an additive model. *Building and Environment*, **33**(2–3), 105–115.

Hillebrandt, P.M. (1974) *Economic Theory and the Construction Industry*. Macmillan, Basingstoke.

Hillebrandt, P.M. (2000) *Economic Theory and the Construction Industry* (3rd edn). Macmillan, Basingstoke.

Hillebrandt, P.M. & Cannon, J. (1990) *The Modern Construction Firm*. Macmillan, Basingstoke.

Hoffman, K.D., Turley, L.W. & Scott, W.K. (2002) Pricing retail services. *Journal of Business Research*, **55**, 1015–1023.

Kennedy, C. & O'Connor, M. (1997) *Winning Major Bids: The Critical Success Factors*. Policy Publications, Bedford.

King, M. & Mercer, A. (1988) Recurrent competitive bidding. *European Journal of Operational Research*, **33**, 2–16.

King, M. & Mercer, A. (1990) The optimum markup when bidding with uncertain costs. *European Journal of Operational Research*, **47**, 348–363.

Kotler, P. (2000) *Marketing Management*. Prentice Hall, Englewood Cliffs, NJ.

Kwakye, A.A. (1994) *Understanding Tendering and Estimating*. Gower, Aldershot.

Li, H. & Love, P.E.D. (1999) Combining rule-based expert systems and artificial neural networks for mark-up estimation. *Construction Management and Economics*, **17**, 169–176.

Li, H. Shen, L.Y. & Love, P.E.D. (1999) ANN-based mark-up estimation system with self-explanation capacities. *ASCE Journal of Construction Engineering and Management*, **125**(3), 185–189.

Lin, C.T. & Chen, Y.T. (2004) Bid/no-bid decision-making – a fuzzy linguistic approach. *International Journal of Project Management*, **22**, 585–593.

Lowe, D.J. & Parvar, J. (2004) A logistic regression approach to modeling the contractor's decision to bid. *Construction Management and Economics*, **22**(6), 643–653.

Lowe, D.J., Parvar, J. & Emsley, M.W. (2004) Development of a decision support system (DSS) for the contractor's decision to bid: regression and neural networks solutions. *Journal of Financial Management of Property and Construction*, **9**(1), 27–42.

Mahdi, I.M., Riley, M. Fereig, S.M. & Alex, A.P. (2002) A multi-criteria approach to contractor selection. *Engineering, Construction and Architectural Management*, **9**(1), 29–37.

Male, S. (1991) Strategic management for competitive strategy and advantage. In: *Competitive Advantage in Construction* (eds S.P. Male & R.K. Stock), Butterworth-Heinemann, Oxford; pp. 1–4.

Marzouk, M. & Moselhi, O. (2003) A decision support tool for construction bidding. *Construction Innovation*, **3**(2), 111–124.

Mattson, M.R. (1988) How to determine the composition and influene of the buying centre. *Industrial Marketing Management*, **17**(3), 205–214.

Merna, A. & Smith, N.J. (1990) Bid evaluation for UK public sector construction contracts. *Proceedings of the Institution of Civil Engineers, Part 1*, **88**, 91–105.

McKim, R.A. (1993) Neural network applications for project management: three case studies. *Project Management Journal*, **24**(4), 28–33.

Mochtar, K. & Arditi, D. (2001) Pricing strategy in the US construction industry. *Construction Management and Economics*, **19**, 405–415.

Moselhi, O. & Hegazy, T. (1993) Markup estimation using neural network methodology. *Computing Systems in Engineering*, **4**, 135–145.

Moselhi, O., Hegazy, T. & Fazio, P. (1991) Neural networks as tools in construction. *ASCE Journal of Construction Engineering and Management*, **117**(4), 606–625.

Moselhi, O., Hegazy, T. & Fazio, P. (1993) DBID: analogy-based DSS for bidding in construction. *ASCE Journal of Construction Engineering and Management*, **119**(3), 446–479.

Mustafa, M. & Ryan, T. (1990) Decision support for bid evaluation. *International Journal of Project Management*, **8**, 230–235.

Neufville, R. de & King, D. (1991) Risk and need-for-work premiums in contractor bidding. *ASCE Journal of Construction Engineering and Management*, **117**(4), 659–673.

Office of Government Commerce (2004) *eAuctions*. www.ogc.gov.uk/index.asp?docid = 1001034 (accessed 17 April 2005).

Odusote O. & Fellows R.F. (1992) An examination of the importance of resource considerations when contractors make project selection decisions. *Construction Management and Economics*, **10**, 137–151.

Park, W.R. & Chapin W.B. (1992) *Construction Bidding: Strategic Pricing for Profit*. John Wiley & Sons, New York.

Procurement Policy Unit (2000) *Procurement Policy Guidelines*. www.ogc.gov.uk (accessed 27 March 2001).

Raftery, J. (1991) *Principles of Building Economics*. BSP Professional Books, Oxford

Rathmell, J.M. (1966) What is meant by services? *Journal of Marketing*, **30**, 33–34.

Rothkopf, M.H. & Harstad, R.M. (1994) Modeling competitive bidding: a critical essay. *Management Science*, **40**(3), 364–384.

Runeson, G. (1996) *The tenability of neoclassical microeconomic theory and tendering theory*. PhD thesis, School of Construction Management and Property, Queensland University of Technology, Australia.

Runeson, G. (2000) *Building Economics*. Deakin University Press, Geelong, Australia.

Runeson, G. & Skitmore, M. (1999) Tendering theory revisited. *Construction Management and Economics*, **17**, 285–296.

Runeson, G. & Skitmore, M. (in press) Construction price formation: full cost pricing or neoclassical microeconomic theory? *Construction Management and Economics*.

Seydel J. (2003) Evaluating and comparing bid optimisation effectives. *ASCE Journal of Construction Engineering and Management*, **129**(3), 285–292.

Seydel, J. & Olson, D.L. (1990) Bids considering multiple criteria. *ASCE Journal of Construction Engineering and Management*, **116**(4), 609–623.

Seydel, J. & Olson, D.L. (2001) Multicriteria support for construction bidding. *Mathematical and Computer Modelling*, **34**, 677–702.

Shash, A.A. (1993) Factors considered in tendering decisions by top UK contractors. *Construction Management and Economics*, **11**(2), 111–118.

Shash, A.A. (1998) Bidding practices of subcontractors in Colorado. *ASCE Journal of Construction Engineering and Management*, **124**(3), 219–225.

Shash, A.A. & Abdul-Hadi, N.H. (1992) Factors affecting a contractor's mark-up size decision in Saudi Arabia. *Construction Management and Economics*, **10**, 415–429.

Shen, L., Drew, D. & Zhang, Z. (1999) Optimal bid model for price–time biparameter construction contracts. *ASCE Journal of Construction Engineering and Management*, **125**(3), 204–209.

Skitmore, M. (1989) *Contract Bidding in Construction: Strategic Management and Modelling*. Longman Scientific and Technical, Harlow.

Skitmore, M. (1991a) An introduction to bidding strategy. In: *Competitive Advantage in Construction* (eds S.P. Male & R.K. Stock), Butterworth-Heinemann, Oxford; pp. 139–162.

Skitmore, R.M. (1991b) The contract bidder homogeneity assumption: an empirical analysis. *Construction Management and Economics*, **9**(5), 403–429.

Skitmore, R.M. (2004) Predicting the probability of winning sealed bid auctions: the effects of outliers on bidding models. *Construction Management and Economics*, **22**(1), 101–109.

Skitmore, R.M. & Patchell, B. (1990) Developments in contract price forecasting and bidding techniques. In: *Quantity Surveying Techniques: New Directions* (ed. P.S. Brandon), BSP Professional Books, Oxford; pp. 75–120.

Skitmore, M. & Pemberton, J. (1994) A multivariate approach to construction contract bidding mark-up strategies. *Journal of Operational Research Society*, **45**(11), 1263–1272.

Smith, A.J. (1995) *Estimating, Tendering and Bidding for Construction*. Macmillan, Basingstoke.

Sönmez, M., Yang, J.B. & Holt, G.D. (2001) Addressing the contractor selection problem using an evidential reasoning approach. *Engineering, Construction and Architectural Management*, **8**(3), 198–210.

Stark, R.M. & Mayer, R.H. (1971) Some multi-contract decision-theoretic competitive bidding models. *Operations Research*, **19**(2), 469–483.

Stark, R.M. & Rothkopf, M.H. (1979) Competitive bidding: a comprehensive bibliography. *Operations Research*, **27**(2), 364–390.

Tassie, C.R. (1980) Aspects of tendering: converting a net estimate into a tender. In: *The Practice of Estimating*. Chartered Institute of Building, London.

Teo, D.H.P., Quah, L.K., Torrance, V.B. & Okoro, M.I. (1991) Risk evaluation and decision support system for tendering and refurbishment contracts. In: *Management, Quality and Economics in Building* (eds A. Bezalga & P.S. Brandon), E. & F.N. Spon, London; pp. 301–319.

Thorpe, T. & McCaffer, R. (1991) Competitive bidding and tendering policies. In: *Competitive Advantage in Construction* (eds S.P. Male & R.K. Stock), Butterworth-Heinemann, Oxford; pp. 163–194.

Ting, S.C. & Mills, A. (1996) Analysis of contractors' bidding decisions. In: *Proceedings of CIB W92 Procurement Systems Symposium, North Meets South: Developing Ideas* (ed. R. Taylor). Durban, South Africa; pp. 53–65.

Treasury Corporation of Victoria, Australia (2002) *Bid Analysis Evaluation Policy*. www.vic.gov.au/treasury/treasury.html (accessed 10 January 2002).

Tweedley, N. (1995) *Winning the Bid: A Manager's Guide to Competitive Bidding*. Pitman, London.

Tucker, S.N., Love, P.E.D., Tilley, P.A., Salomonsson, G.S. MacSporran, C. & Mohamed, S. (1996) *Perspectives of Construction Contractors' Communication and Performance Practices, Pilot Survey*. DOC 96/29 (M), Division of Building, Construction and Engineering (DBCE). CSIRO Press, Dickson, Australia.

Vickrey, W. (1961) Counterspeculation, auctions, and competitive sealed tenders. *Journal of Finance*, **16**, 8–37.

Wanous, M., Boussabaine, A.H. & Lewis, J. (2000) To bid or not to bid: a parametric solution. *Construction Management and Economics*, **18**, 457–466.

Wanous, M., Boussabaine, A.H. & Lewis, J. (2003) A neural network bid/no bid model: the case for contractors in Syria. *Construction Management and Economics*, **21**, 737–744.
Ward, S.C. & Chapman, C.B. (1988) Developing competitive bids: a framework for information processing. *Journal of the Operational Research Society*, **39**(2), 123–134.
Weverbergh, M. (2002) A comment on 'Predicting the probability of winning sealed bid auctions: a comparison of models'. *Journal of the Operational Research Society*, **53**(10), 1156–1158.
Winch, G.M. (2001) Governing the project process: a conceptual framework. *Construction Management and Economics*, **19**, 331–335.
Wong, C.H., Holt, G.D. & Cooper, P.A. (2000) Lowest price or value? Investigation of UK construction clients' tender selection process. *Construction Management and Economics*, **18**, 767–774.
Zarkada-Fraser, A. & Skitmore, M. (2000) Decisions with moral content: collusion. *Construction Management and Economics*, **18**, 101–111.

Endnotes

[i] A competition for the right to perform services or to acquire property (Stark & Mayer 1971).
[ii] Mark-up, in addition to profit, also includes additions for running costs or overheads.
[iii] 'Non-bona fide tenders' are a commercial reality derived from a perception that suppliers will be penalised in the future for not responding positively to an invitation to tender.
[iv] Due to the commercial nature of this information, it is difficult to obtain precise figures on the costs incurred by suppliers when responding to a bid opportunity. However, despite the age of this reference there is no evidence to suggest that these costs have significantly reduced.
[v] A non-*bona fide* tender (Beeston 1983, p. 112).
[vi] Neural networks can be defined as a non-linear function-mapping tool, which maps the relationships between a set of input variables (input vector) to a set of output variables (output vector). The prominent feature of neural networks is their ability to learn from samples or historical data. Moreover, no explicit programming is required, and it is feasible to model the relationships between a set of inputs and a set of outputs, and develop a generalisation of the relationships. The advantages of neural networks are as follows:

- Unlike linear regression, they are able to model interdependencies between input data which will inevitably occur when considering cost and/or price significant variables.
- They can deal more readily with non-linear relationships.
- They can, more effectively than regression, handle incomplete data sets.
- Theoretically, they can model any function, provided they are presented with enough data and given enough time for training.

However, as with other modelling techniques, the requirement to collect large amounts of data to populate the model can be difficult at best and impossible at worst.
[vii] Logistic regression is a conditional probability approach used to estimate the probability of occurrence of an outcome or choice. The technique is distribution free, while probability models have the advantage that significant coefficients can be interpreted in terms of the relationship with the dependent variable.
[viii] *Harmon CFEM Facades (UK) Limited* v. *The Corporate Officer of the House of Commons* (2000) 67 Con LR 1.
[ix] www.dgcom.mod.uk/dgcom/guides/sect18.doc (accessed 10 January 2002).
[x] For more information see *A Soft Issues Bid Evaluation Tool (SIBET) User Manual*. www.ams.mod.uk/ams/content/docs/toolkit/gateway/guidance/linkdocs/sibet.rtf (accessed 10 January 2002).

17 Strategic Purchasing and Supply Chain Management in the Project Environment – Theory and Practice

Andrew Cox and Paul Ireland

Introduction

This chapter provides an introduction to the main theoretical approaches to procurement and supply chain management, and contends that many of the more recent tools and techniques that have been developed within project-based industries to allow practitioners to manage project relationships effectively may not always deliver the benefits expected. Using evidence from the UK construction industry, it is argued that a failure to understand the objective power and leverage circumstances that face industry players prevents construction clients, contractors and suppliers from achieving their operational goals and/or causes them to have unachievable objectives commercially. It is also argued that these problems are compounded by government-sponsored industry reports that unthinkingly advocate the adoption of approaches that appear to have worked in one industry, without proper regard to the dissimilar demand and supply and power circumstances operating in construction.

The chapter, therefore, provides practitioners with a framework for understanding an industry and its supply chains that have a primarily project-based demand and supply structure. It also provides practitioners with a way of thinking about the attributes of buyer and supplier power and leverage in construction, and the appropriateness of different relationship management approaches. Finally, the chapter explains the limited and unique circumstances under which supply chain management approaches, uncritically recommended by recent government and industry reports, can be implemented successfully. This framework may be applied across other project-based industries.

Strategic purchasing, procurement and supply chain management: an introduction

In the 1950s, Forrester (1958) recognised the importance of relationships between organisations, not least because externally purchased goods and services can account for between 50% and 80% of a company's overall expenditure, and effective purchasing can contribute significantly to reduced costs and therefore directly impact profit margins. It appears that Forrester identified a number of key contractual and commercial management issues that today are often referred to as 'strategic purchasing', 'strategic procurement' or 'supply chain management'.

Since then there has been a debate about what is the best way for buyers to manage their commercial relationships with suppliers. This is demonstrated by the multiplicity of approaches that have been developed to explain best practice. These include: integrated purchasing strategy (Burt 1984); supplier integration (Dyer *et al.* 1998); buyer–supplier partnership (Lamming 1993); strategic purchasing (Saunders 1994); strategic supplier alliances (Lewis 1995); supply chain synchronisation (Tan *et al.* 1998); lean manufacturing and supply (Ohno 1988; Womack *et al.* 1990; Hines 1996; Womack & Jones 1996), supply pipeline management (Farmer & Ploos van Amstel 1991); supply network (Nishiguchi 1994); value stream management (Hines *et al.* 2000); agile supply (Mason-Jones & Towill 1999; Naylor *et al.* 1999; Christopher 2000; Mason-Jones *et al.* 2000); and power regimes and supply chain management (Cox *et al.* 2000, 2001, 2002, 2004a,b; Cox 2004a,b,c).

Interestingly, while the subject of commercial management has historically been considered to be quite separate from the purchasing activity, the twin developments of outsourcing and supply chain management have seen a blurring of the distinction. The decisions made about external resource management now have a considerable impact on overall commercial performance. Effective supply chain management now encompasses a wide range of key commercial issues including:

- The identification of desired value propositions
- The extent of outsourcing
- The nature of supply relationships
- The length of supply relationships
- The nature of contract
- The length of contract
- The extent of ethical sourcing and impact of environmental issues
- The pre-contractual selection of competent and congruent suppliers
- The extent of involvement within the supply chain
- The post-contractual management of suppliers
- The management of commercial risk and uncertainty in supply.

Furthermore, the recent emergence of the *supply chain management* concept has represented a significant paradigm shift for procurement strategies and buyer–supplier relationships. This shift has typically focused on the rejection of short-term and highly adversarial approaches in favour of more transparent, equity-based, long-term and collaborative ways of working. However, while the term supply chain management is widely used, the lack of a universally accepted definition leads to considerable confusion for those attempting to develop a supply chain strategy. To understand the appropriateness of supply chain management thinking for project-based industries, it is clearly essential to explain what the concept means and how it differs from other supply management approaches.

Theories of business relationship and supply chain management

A major problem with business relationship and supply chain management thinking has been a failure by buyers and suppliers to think carefully about what business involves commercially and operationally (Cox *et al.* 2004a). Buyers and suppliers have their own distinct commercial goals which define their strategic ends and are very frequently in direct conflict. For example, as shown in Fig. 17.1, the buyer will be attempting to develop effective operational approaches to maximise value for money (improve functionality and/or reduce costs of ownership) while the supplier is attempting to maximise revenues and returns from the exchange.

This conflict may be expressed in another way. Ultimately, business success for a firm or entrepreneur relates to the ability to purchase something cheaply and then sell it dear (i.e. at a profit allowing for any operating costs). This means that the ability to buy and sell effectively must be at the heart of successful business management.

The problem is, however, that although there has always been recognition that both of these competencies (buying and selling) are important, there has been a tendency to address competence in these two areas from a one-dimensional functional perspective. This is not surprising given that most companies create a procurement and supply function to oversee the buying activity and a separate marketing and sales function to manage the selling activity. Therefore, rather than focusing holistically on the dyadic relationship within exchange transactions between buyers and suppliers, the majority of current supply chain thinking focuses on how one side can achieve their goals without necessarily understanding how this impacts on the other party in the exchange.

As a result, most thinking about business relationship management, whether on the buyer or supplier side, is subject to a number of weaknesses in that the approaches tend to be (Cox *et al.* 2004a):

- Descriptive rather than analytical
- Focused specifically on one side of the relationship, emphasising the buyer or supplier perspective, without considering the transaction and interaction between the two parties
- Prescriptive rather than predictive
- Focused on operational management issues, without fully explaining the complex interconnections between the commercial and operational preferences of both the buyer and the supplier in any exchange relationship.

Not all current approaches to business relationship management have all of these four weaknesses, and all of them in different ways provide major contributions to our understanding of the structure and purpose of business relationships. However, in order to outline the strengths and weaknesses of the current thinking within the procurement and supply management literature, the discussion that follows will focus on three broad schools of thought.

The first two to be discussed are supply chain management (encompassing lean and agile) and the interaction approach of the International Marketing and Purchasing (IMP) Group. It is argued that while each of these schools has specific nuances, they also have a great deal in common. In particular, there is a tendency to prescriptively encourage relationship management styles that favour transparent, collaborative and trusting (partnering) ways of working. Furthermore, all of these approaches tend to focus one dimensionally on the commercial and operational goals of the buyer rather than of the supplier. After the discussion of these two schools of thought, the power perspective on business relationship alignment and misalignment is discussed.

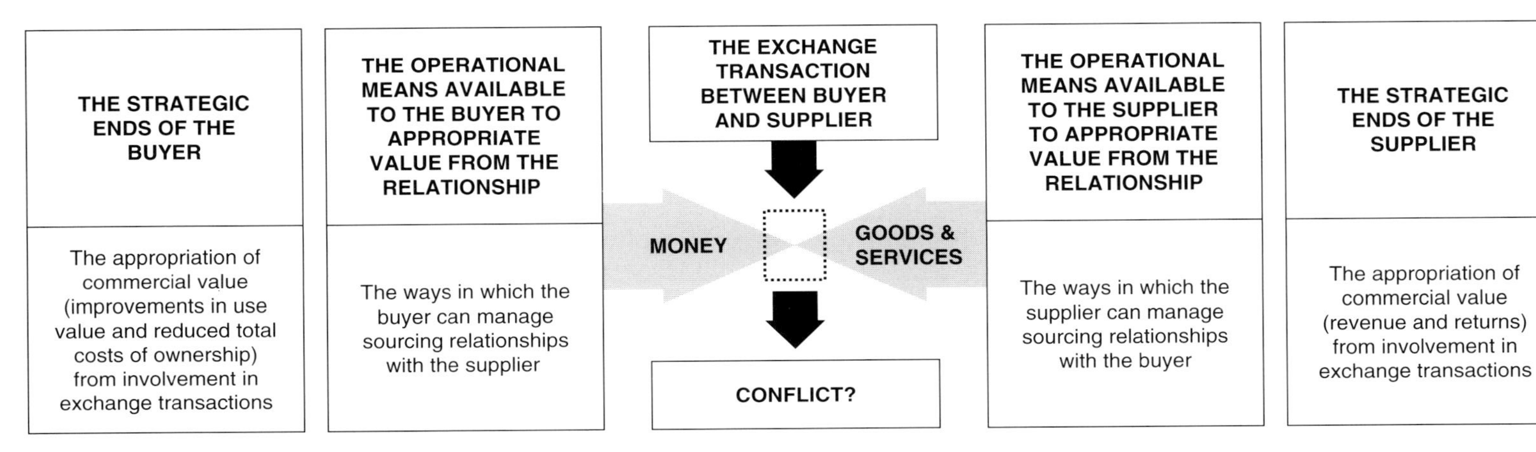

Figure 17.1 Conflict in buyer–supplier relationships (adapted from: Cox *et al.* 2004a, p. 34).

Lean supply chain management

In recent years, a prescriptive approach to procurement and supply improvement has developed, and it is this approach that has dominated recent thinking from government and industry reports in construction (Latham 1994; Egan 1998). Most writers who advocate this approach, known as supply chain management and encompassing the lean and agile schools of thought, tend to prescribe the use of a highly transparent, trusting and long-term relationship between the buyer and the supplier to create a physically efficient supply chain through the reduction of waste in processes or increase in responsiveness in product delivery.

Although the term *lean* was initially coined by Krafcik (1988), the first comprehensive account of the principles underlying lean production can be traced to Womack *et al.* (1990) who described the pioneering work of the Japanese Toyota Company in managing their production processes and supply networks. A key message within the lean literature is the need to customise high volume production to provide end customers with exactly what they want at the time they want it. To achieve this end, the lean approach seeks to improve flexibility, narrow sources of supply, reduce waste and improve flow in the supply network through the adoption of long-term and highly collaborative supply relationships.

A major element of the lean school surrounds the waste reduction tools and techniques developed to improve both internal and external process efficiency. Many authors (Ohno 1988; Womack & Jones 1996; Hines *et al.* 2000) argue that the starting point in understanding how to improve supply chain process efficiency is to understand the different types of waste that do not add value to the final product from the point of view of the client and need to be removed. This requires recognition of the seven key wastes (Ohno 1988):

- overproduction
- defects
- unnecessary inventory
- inappropriate processing
- excessive transportation
- waiting
- unnecessary motion.

Having identified the different types of waste that can exist within production processes and supply chains, the lean literature argues that it is necessary to design a strategy to eliminate it based upon the principles of lean thinking (Womack & Jones 1996):

- Specify value from the customer perspective, not the perspective of any of the supply chain participants
- Identify the value stream that delivers the value proposition to the customer
- Identify those activities within the value stream that add value, are non-value-adding but necessary, or are non-value-adding
- Make those actions that create value flow without interruptions or impediments
- Only make what is pulled by the customer. The value stream should be operating a just-in-time pull system, rather than a producer-focused push system

- Create transparency of strategies and costs within the supply chain. The issue of competitive advantage must be understood at a supply chain level rather than at the individual firm level
- Constantly pursue perfection – the process of reducing waste never ends.

Lean supply chain management in the project environment: is it best practice?

The lean supply chain management approach has become something of a dominant way of thinking about 'best practice' amongst buyers. This is predominantly because the approach, originally developed in the automotive sector in Japan, has also been shown to provide significant benefits in other sectors, such as supermarket retailing, electrical industries, aluminium and computer manufacturing, where demand and supply and power variables have been conducive (Cox *et al.* 2003). However, while it may be highly appropriate for these process-based industries, its universal applicability for practitioners within project-based industries such as construction can be questioned (Cox & Townsend 1998; Cox & Ireland 2001, 2002).

Many studies have shown that waste is omnipresent in construction supply chains, occurring at the interface between processes, disciplines and organisations (Vrijhoef & Koskela 2000). Given the magnitude of non-value adding activities in construction, as indicated in Table 17.1, there is an obvious need for the development of *appropriate* waste reduction strategies in the industry.

As a result, *Rethinking Construction* (Egan 1998) contained specific, measurable challenges for UK construction and suggested a number of ways to improve the way the construction industry performs. Based on a limited number of case studies from around the world, where construction was achieving improvement using these techniques, conclusions were drawn that lean thinking should be applied to the UK construction industry as a whole:

> *Lean thinking presents a powerful and coherent synthesis of the most effective techniques for eliminating waste and delivering significant sustained improvements in efficiency and quality... the UK construction industry should also adopt lean thinking as a means of sustaining performance improvement.* (Egan 1998, pp. 25–26)

Table 17.1 Extent of waste in construction (source: Koskela 1997, p. 6).

Waste	Cost	Country
Quality costs (non-conformance)	12% of total project costs	USA
External quality cost (during facility use)	4% of total project costs	Sweden
Lack of constructability	6–10% of total project costs	USA
Poor materials management	10–12% of labour costs	USA
Excess consumption of materials on site	10% on average	Sweden
Time used for non-value adding activities on site	Approx. 2/3 of total time	USA
Lack of safety	6% of total project costs	USA

Prior to this endorsement, a first conference of the International Group for Lean Construction in 1993 had already developed an extensive lean construction literature. However, the majority of the literature has applied the lean concept without an empirical exploration of the market and supply chain structures in the construction industry. This contrasts with the contextualisation of the lean production literature through the descriptions of the automotive and electrical industries (Lamming 1993; Hines 1994; Nishiguchi 1994).

Despite the popularity of the lean approach in construction there is an 'alternative' approach that is now receiving increasing attention. Although having a similar managerial philosophy based on collaboration, an agile construction approach is argued to be more appropriate given the characteristics of the industry: unpredictable, volatile, highly customised and low volume markets. The proponents of the agile school, therefore, support the contention that lean approaches cannot be considered to be appropriate in all supply chain circumstances, as buyers have to manage in many different demand and supply circumstances.

Agile supply chain management

The 'agile school' argues that while a lean approach makes sense under conditions where demand is predictable, the requirement for variety is low and volume is high (as in typical process manufacturing), an alternative is needed in unpredictable, volatile, highly customised and low volume markets (as in construction). A leading proponent of the agile school supports this view:

> *There are certain conditions where a lean approach makes sense... The problems arise when we attempt to implant that philosophy into situations where demand is less predictable* [and] *requirement for variety is high...* [As a result, many] *firms have been misguided in their attempts to adopt a lean model in conditions where it is not suited.* (Christopher 2000, p. 38)

In market and supply chain situations, characterised by volatile demand and a high customer requirement for variety, the elimination of waste is a lower priority than responding to the dynamic marketplace. In these circumstances, an innovative and market-responsive (agile) supply management strategy is demanded (Stalk & Hout 1990; Mason-Jones & Towill 1999; Harrison *et al.* 1999; Christopher 2000). Table 17.2 summarises the key attributes of a lean and agile supply management approach.

The agile literature also recognises that 'hybrid' or 'leagile' strategies that combine lean and/or agile approaches may be appropriate according to the characteristics of demand (Naylor *et al.* 1999; Christopher 2000, Christopher & Towill 2000; Mason-Jones *et al.* 2000).

However, although the lean and agile supply schools have a difference of focus, there are considerable similarities between the two approaches, particularly in the advice offered to buyers for the management of supply networks. The main focus of both approaches is on the development of collaborative relationships between buyers and suppliers based on trust and a requirement for transparent, co-ordinated and integrated supply chain activity. The main difference between the lean and agile

Table 17.2 Comparison of lean supply with agile supply (adapted from: Christopher & Towill 2000, p. 208).

Distinguishing attributes	Lean supply	Agile supply
Marketplace demand	Stable and predictable	Volatile and unpredictable
Nature of product	Standardised and commoditised	Customised
Product variety	Low	High
Product life cycle	Long	Short
Customer drivers	Cost	Availability
Profit margin	Low	High
Dominant costs	Physical costs	Marketability costs
Stockout penalties	Long term contractual	Immediate and volatile
Purchasing policy	Buy goods	Assign capacity
Information enrichment	Highly desirable	Obligatory
Forecasting mechanism	Algorithmic	Consultative

prescriptions relates to the nature of their operational techniques and how they deal with different demand conditions, rather than this general management philosophy (Watson *et al.* 2003).

When considering *inter-organisational relationships,* however, a consensus about 'best practice' in procurement and supply management has developed in the literature. This is based on the rejection of adversarial buyer relationships with suppliers in favour of the development of 'value-adding' collaborative relationships when integrating the supply chain. However, whether integrated collaborative relationships are always necessary, or possible, is open to debate. Academics outside the lean and agile schools contend that they are not. It is argued that such relationships incur significant costs, either directly through investments, or indirectly through the costs of management resource, and need to be justified by a certain level of return on that investment (Watson *et al.* 2003). With this need in mind it can be argued that a key problem with the lean and agile theories is that they do not make sufficiently clear what the mechanism is that provides the driver for collaboration throughout the supply chain. This problem is shared by the International Marketing and Purchasing (IMP) interaction approach.

The International Marketing and Purchasing (IMP) Group

The IMP Group is a broad literature that has developed its detailed insights into buyer and supplier relationships over 20 years (Ford 1980; Håkansson 1982). The major benefit of the IMP approach has been its unequivocal focus on the need to understand the complex inter-relationships between the buyer and the supplier from the perspective of both parties when they interact. Thus, the IMP approach has provided an extremely comprehensive descriptive account of the nature of business relationships, as outlined below.

As Fig. 17.2 demonstrates, this approach provides a detailed overview of the factors that impact upon buyer and supplier transactions. It is argued that the interactions between buyers and suppliers are shaped by 'environmental' factors, which neither

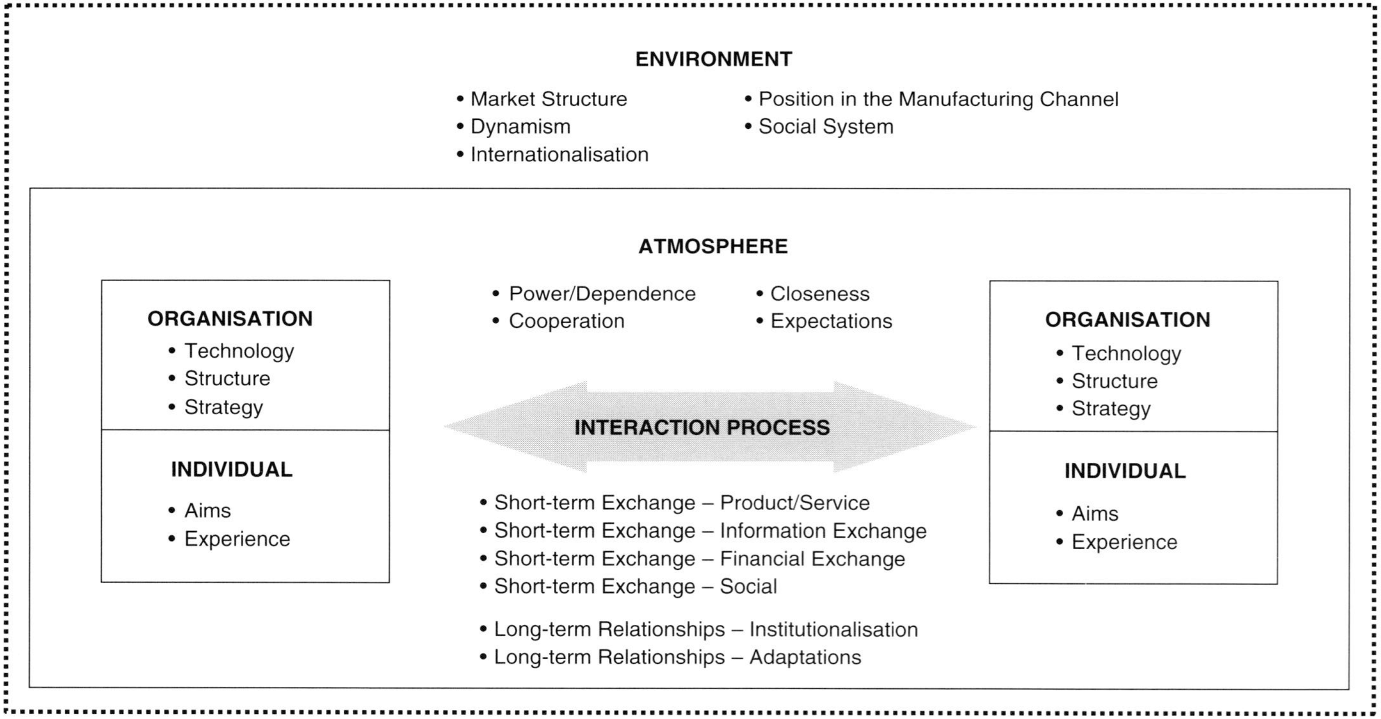

ENVIRONMENT
• Market Structure
• Dynamism
• Internationalisation
• Position in the Manufacturing Channel
• Social System

ATMOSPHERE
• Power/Dependence
• Cooperation
• Closeness
• Expectations

ORGANISATION
• Technology
• Structure
• Strategy

INDIVIDUAL
• Aims
• Experience

INTERACTION PROCESS
• Short-term Exchange – Product/Service
• Short-term Exchange – Information Exchange
• Short-term Exchange – Financial Exchange
• Short-term Exchange – Social

• Long-term Relationships – Institutionalisation
• Long-term Relationships – Adaptations

ORGANISATION
• Technology
• Structure
• Strategy

INDIVIDUAL
• Aims
• Experience

Figure 17.2 The IMP interaction model (adapted from: IMP Group 1982, p. 24).

party to the exchange can directly control (*limiting factors*). This involves such aspects as market structure, dynamism, internationalisation, channel position and social system. There are some elements of the interaction, it is argued, that buyers and suppliers can influence (*handling factors*). These aspects are referred to as 'atmosphere' and include such elements as power/dependence, conflict and co-operation, and expectations (IMP group 2002).

A major benefit of this approach is its inclusiveness – it is argued that there is a need for buyers and suppliers to think about the strategies they should adopt under different types of buyer and supplier interaction, and that sometimes these may create a match or mismatch in a relationship (Campbell 1985). A major weakness of the IMP approach, however, is its lack of predictive focus on the relative importance of the variables specified in an interaction between a buyer and supplier. Therefore, while the IMP approach is comprehensive, the fact that it is primarily descriptive provides managers with limited guidance as to which particular relationships are suitable in specific circumstances. Despite these criticisms, the IMP approach, while remaining essentially pessimistic about the ability of managers to control many of the environmental factors that impinge on relationships (Ford 2001), has provided considerable insight into the factors that impinge on buyer and supplier choices.

The power perspective

The power perspective on buyer and supplier relationship management has a long tradition of its own and has also been supported by the work of the IMP Group. However, unlike the IMP's undifferentiated approach, the power perspective has always asserted that the relative power of the buyer and supplier is the determining factor in the operational and commercial outcome in any transaction. The background to the power perspective is detailed elsewhere (Cox 1997, 2004a; Cox *et al.* 2002, 2003, 2004a).

The *power perspective* contends that the current consensus about what constitutes 'best practice' in procurement and supply management – namely, the rejection of adversarial buyer relationships with suppliers in favour of a more long-term collaborative approach based on trust – contradicts the common-sense logic of economic theory that the best defence of the buyer's position, and the one that ensures that suppliers innovate and pass value to buyers, is the maintenance of perfectly competitive (or highly contested) supply markets, with low barriers to entry, low switching costs and limited information asymmetries (Cox 2001).

The fact that buyers and suppliers can pursue symmetrical or asymmetrical relationship management styles with one another to achieve their business objectives supports the argument that there can be no one 'best practice'. Therefore, competence for buyers and suppliers must reside in their respective abilities to make appropriate choices about how they conduct themselves in very different exchange relationships (Cox 1997). A supply chain strategy based on a generic approach may, therefore, be wholly unsuitable (or even impossible) under particular circumstances.

Following this argument, it is important to stress that, while supply chain management can be regarded as a 'best practice' approach for organisations under *some* circumstances, it is highly unlikely to be 'best practice' in most or all of their external sourcing circumstances. Given this, the first issue to be addressed is not what is the most effective way to undertake supply chain management, but if it is not always appropriate, what is the full range of sourcing choices that managers have to choose from?

The best way to address this issue is by first understanding, logically, the ways in which buyers can work with any supplier (in a reactive or proactive manner) and the scope of their activities within a supply chain (at the level of the first-tier or entire supply chain). By doing this, it is possible to define four basic sourcing approaches that are always available for buyers to select from when they seek to manage their supply relationships. The four options: *supplier selection*, *supply chain sourcing*, *supplier development* and *supply chain management* are summarised in Fig. 17.3.

Supplier selection is arguably the approach that is most commonly used by buyers in all types of organisation, as it places the lowest demands on scarce internal resources and requires the lowest levels of commitment to long-term collaborative external sourcing relationships. Supplier selection implies that the buyer's role is confined primarily to reactive sourcing at the first-tier; therefore, the buyer selects products and/or services from the supply offerings made by suppliers currently

Focus of Buyer Relationship with the Supplier	FIRST-TIER	SUPPLY CHAIN
PROACTIVE	**SUPPLIER DEVELOPMENT** **SD** A process through which the buyer works on a continuous and **proactive** basis with the **first-tier** supplier to transform the current trade-off between product or service functionality and the overall cost of ownership.	**SUPPLY CHAIN MANAGEMENT** **SCM** A process through which the buyer works on a continuous and **proactive** basis with all **supply chain** players, to transform the current trade-off between product or service functionality, and the overall cost of ownership.
REACTIVE	A process through which the buyer **reactively** selects products and/or services from the existing supply offerings made by suppliers currently operating at the **first-tier**. **SS** **SUPPLIER SELECTION**	A process through which the buyer **reactively** selects products and/or services from the existing supply offerings made by suppliers currently operating at the first-tier but having understood the structure of the **supply chain** through which products and services are created and delivered. **SCS** **SUPPLY CHAIN SOURCING**

Level of Work Scope with Supplier and Supply Chain

Figure 17.3 The sourcing options matrix (adapted from: Cox *et al.* 2003, p. 5).

operating in the market. Relationships tend to be relatively short term and arm's length in nature. This is true when a similar reactive approach is used to source supply inputs throughout the supply chain (*supply chain sourcing*).

Supply chain management is arguably the most demanding and challenging of the four options for buyers. A proactive buyer assesses the scope to undertake *supplier development* work (which means improving the competence of the supplier operationally and commercially by working with them on improvement initiatives) at the first tier. If this approach is extended throughout the chain, or in significant parts of it, then it is a *supply chain management* approach. This requires extensive co-ordination of internal business functions and a linking together of all the buyers and suppliers in the chain so that they are able to focus their procurement and supply strategies on the delivery of improvement in functionality and lower costs of ownership for the ultimate buyer in the chain. Fundamental to achieving this increase in value for money, profitability and long-term performance improvement is collaboration at a supply chain level.

Given the four choices, it is argued in this perspective that all four options may be appropriate for buyers under particular circumstances and, furthermore, that not all approaches are feasible or desirable for buyers or suppliers all of the time. To determine which of these approaches is feasible and/or desirable, it is argued that buyers need to consider the following key questions (Cox *et al.* 2003):

(1) What will be the cost to the firm of implementing the different supply management options? To answer this, the buyer has to understand the costs of adopting the reactive and proactive (collaborative) approaches, and the cost of implementing these at the first tier or extending them through the supply chain.
(2) What are the likely benefits from adopting the different supply management options?
(3) What is the probability of the different sourcing options being successfully executed and attaining the anticipated benefits?
(4) What is the power relationship with the supplier? How will this power structure impact:
    (a) Whether the supplier is interested in a proactive approach (i.e. collaborating) or a reactive approach
    (b) Whether the approach can be extended beyond the first tier, and for each of the four options
    (c) The way in which the surplus value is created and distributed.

Having considered these questions, practitioners are able to assess not only which options are desirable and/or feasible, but also whether the buyer or supplier will dominate in the exchange relationship. In addressing these four questions the buyer is effectively conducting an investment appraisal. This is because all supply management decisions carry with them a cost that is expected to yield a return, with risk and power as intervening variables. A key criticism of the lean and agile schools is that they have failed to recognise that intra- and interorganisational power may facilitate and inhibit efforts to plan, control and manage supply networks (Cox *et al.* 2003). In other words, recent government and industry reports may be guilty of ignoring the objective reality of power and leverage in the construction industry in their misguided search for a simple answer to a complex problem.

Firms aiming for superior performance in procurement and supply chain management must therefore understand the power and leverage situation in which they are operating, or which they can create in the future, before they decide on which sourcing strategies are appropriate. The power matrix, as illustrated in Fig. 17.4, provides a way of thinking about buyer and supplier power. The power circumstances in which buyers and suppliers find themselves can be defined either as *buyer dominance, interdependence, independence* or *supplier dominance*. The matrix is constructed based on the idea that all buyer and supplier relationships are predicated on the relative utility and the relative scarcity of the resources that are exchanged between the two parties (Cox *et al.* 2002).

The use of the matrix requires the practitioner to understand what are the key questions that must be asked in order to understand the relative attributes that provide power and leverage to buyers or suppliers. Some of the key questions that

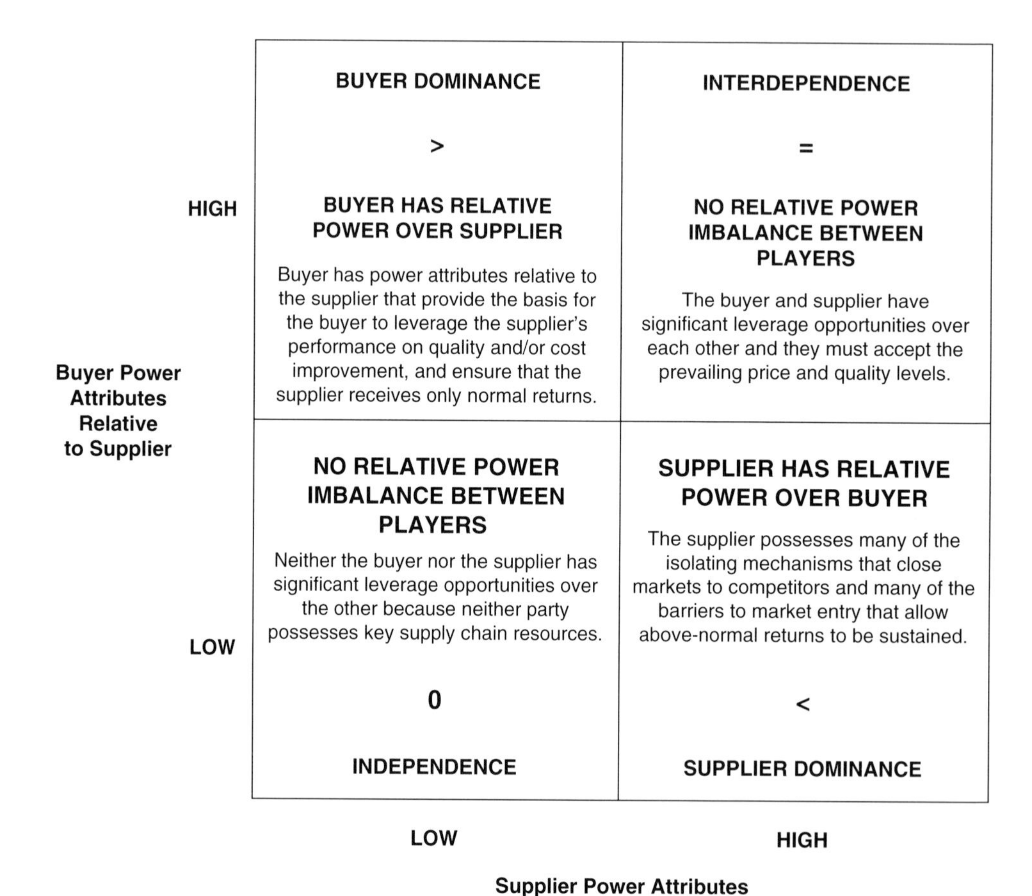

Figure 17.4 The power matrix (adapted from: Cox *et al.* 2000, p. 18).

need to be addressed before a buyer or supplier can locate their own position and that of their current adversaries in the matrix are:

- The balance between the number of buyers and suppliers
- The salience of the buyer's expenditure to the supplier
- The number of available alternative purchasers to the supplier
- The extent of buyer and supplier switching costs
- The extent to which the product or service is commoditised or standardised
- The level of buyer search cost
- The level of information asymmetry advantage that one party has over the other.

The discussion of the power resources in the dyadic buyer–supplier relationship is, of course, only of limited value for practitioners, as one has to consider the entire supply chain in which the firm is operating. Within the construction industry, the dyad between the client and the main contractor is also affected by the relationship that the main contractor has with its subcontractors and their relationship with component, equipment and labour suppliers. Therefore, one has to understand the extended network of dyadic power relationships so that appropriate relationship management strategies can be developed. This is referred to as the *power regime* within which all supply chain relationships must operate. Only by understanding the structure of power within the supply chain and/or network as a whole is it possible to understand the feasibility and/or desirability of introducing *supplier selection*, *supply chain sourcing*, *supplier development* or *supply chain management* strategies (Cox *et al*. 2000, 2002, 2003, 2004a,b).

Construction sourcing in practice: on the appropriate management of procurement in the project-based environment

It should by now be self-evident that proactive sourcing approaches are not always possible in the construction industry and its project related supply chains. This does not mean that *supplier development* and *supply chain management* are impossible to implement. Rather, it means that they are likely to be of benefit only to those lucky few clients, contractors and suppliers who have power resources of *buyer dominance* or *interdependence*, because they are able to manage a regular high volume and highly standardised spend with the same suppliers over a long period of time. Unfortunately, given the structure of demand and supply in construction, this is never likely to be the case for most participants in the industry as the discussion below demonstrates.

Supply chain characteristics in a typical project environment

The UK construction industry, like many project-based industries, has undergone considerable change recently due to increasing pressure on already low margins. In order to assist the industry there have been a number of government- and industry-sponsored reports aimed at raising competitiveness by focusing on improvements in supplier performance and customer satisfaction (Latham 1994; Egan 1998). However, despite the thousands of words written on what the industry should do to improve, these initiatives appear, by and large, to have failed most actors within it. This is because construction supply chains have remained contested, fragmented and highly adversarial with low margins.

Given this, many of those who called for the adoption of supply chain management tools and techniques that have proven successful in the automotive industry would probably argue that the problem is in poor understanding and implementation by practitioners, and that things would be better for everyone if only practitioners would take their medicine as prescribed. It may well be, however, that it is not the invalid that is at fault but the medicine. It could be that the physicians need to heal themselves because they may be peddling snake oil.

What does this mean? As Fig. 17.5 indicates, in the project environment there are a myriad of supply chains for materials, labour, equipment and professional services that have to be integrated by construction firms in the delivery of finished solutions to the end customer. Each of these supply chains has different demand and supply as well as operational and commercial properties that need to be understood so that appropriate sourcing strategies can be developed for the specific products and services involved.

Figure 17.5 suggests that the key generic supply chains that are required for a typical construction solution are rather simple and linear, but the reality is quite different. The ultimate level of complexity involved in the management of a construction project (and the exact number of constituent materials, equipment and labour supply chains that have to be integrated) will be determined by the extensive requirements of the end customer as defined in the original design and specification.

During the construction process, the end customer normally appoints a construction firm and professional services. Within the generic supply chain the construction firm normally fulfils the major 'integrating' role for all upstream supply chains.[i] Surprisingly, throughout the project procurement process, little control or management of the entire supply chain is taken up by the focal organisation (the client) (London & Kenley 2000, p. 10). As a result, each tier is able to manage their supply relationships in such a way that they can effectively act as a procurement gatekeeper. The first-tier organisation typically acts as a gatekeeper to the subcontractors' tiers of suppliers, and each trade subcontractor subsequently acts as a gatekeeper to the materials suppliers operating at the third tier.

Regardless of whether it is the client or construction firm undertaking the purchasing activity, this limited access to upstream supply chain stages creates considerable challenges for those attempting to undertake effective procurement and supply chain management. However, there are a number of other characteristics of construction supply chains that may create further significant difficulties for practitioners responsible for construction procurement. These characteristics are also evident within other project-based industries.

'Manufacture-to-order' one-off supply chain

It is widely stated that the construction industry is a relatively unique industry in the way that it establishes projects to deliver one-off products (Cox & Townsend 1998; Cox & Ireland 2001, 2002). However, it may be contended that other industries display very similar characteristics. Within the IT industry, systems integrators

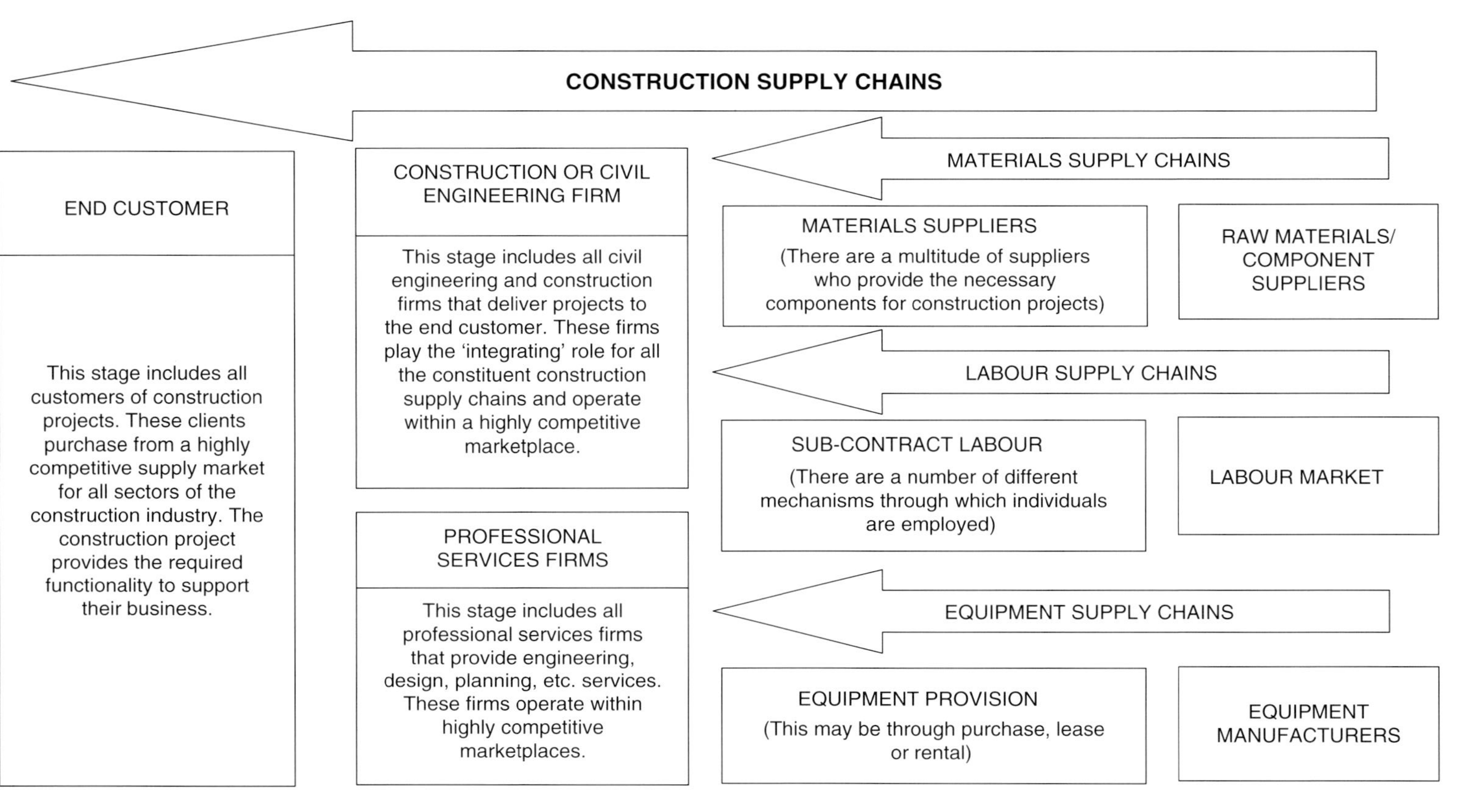

Figure 17.5 The Myriad of construction supply chains (source: Cox & Ireland 2002, p. 411).

develop complex one-off IT systems for clients, and pipeline contractors within the oil and gas industry also develop unique solutions.

In construction, like these industries, it is the client who takes the initiative to start a construction project, and therefore construction is often a customer-driven one-off production process. This leads to the frequent conceptualisation of the construction supply chain as a process explicitly starting and ending with the end user. A rather common representation of a construction process is shown in Fig. 17.6.

Converging supply chain

The final construction supply chain represents a system of multiple supply chains delivering all raw materials, human resources and information, required for the successful completion of a project to the place where the specific end product must arise. With limited pre-fabrication, construction is largely a site operation, confined to the specific location where the final assembly takes place (Nam & Tatum 1988). Therefore, construction takes place at the place of consumption as opposed to the wide and less-specific end market of manufacturing industry. For this reason, the construction supply chain can be considered to be relatively converging in comparison with a manufacturing supply chain.

Temporary supply chain

Construction projects tend to be considered as unique and temporary (Turner 1993; Morris 1997; Turner & Müller 2003). In contrast to manufacturing, this implies a temporary organisation of production for each project characterised by a short-term coalition of participants with frequent changes of membership, often termed 'temporary multi-organisation' (Cherns & Bryant 1984; Luck & Newcombe 1996; Koskela 1997). The participants are totally interdependent on each other, but operate through a variety of contractual arrangements and specific procedures with considerable fluctuations in productivity (Hellard 1995).

While the above literature argues that a construction project is unique and temporary, there is a literature that argues that this is not the case for all construction projects (Cox & Thompson 1998; Cox & Ireland 2002). This literature highlights that while unique, temporary and one-off projects constitute a large percentage[ii] of the market, there are a number of major construction clients, such as McDonald's, whose construction portfolios are often classified as regular process expenditures (Cox & Townsend 1998). It is contended that the regular nature of their construction demand means it is possible for these clients to standardise the design and specification of key elements of their construction projects and manage them in a similar manner to a manufacturing process.

Fragmented supply chain

The literature that examines the structure of the construction industry largely focuses on the first element in structural analysis of industries: the number and relative size of firms in the industry (London *et al.* 1998). In market-based economies, the fundamental assumption is that firms exist to maximise profits and that competition with

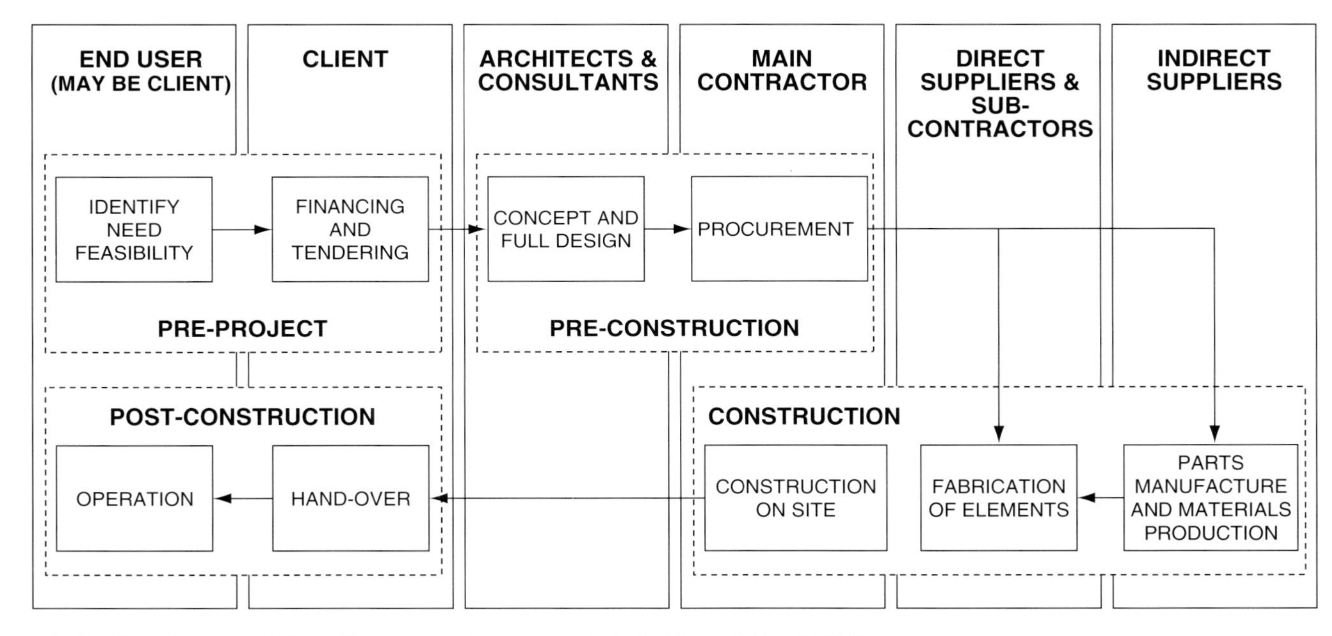

Figure 17.6 Typical representation of the construction process (adapted from: Vrijhoef 1998).

other firms is one of the forces that affects this. If this view is accepted, then a fragmented industry is characterised by the following two criteria: no firm has a significant market share to strongly influence the industry outcome, and fragmented industries are populated by a large number of small and medium-sized companies (Cox & Thompson 1998; Cox & Townsend 1998).

In summary, the construction industry's fragmented, temporary and one-off nature, together with the existence of adversarial relationships and the tendency for self-interest rather than common objective, affects the efficient and effective satisfaction of customer demands (Cutting-Decelle 1997; Nam & Tatum 1988). In particular, it creates power situations that do not very often support the development of long-term collaborative relationships based on non-adversarial trust. Despite these unique characteristics, it is still widely believed that the key to the solution of many of the inherent problems of the construction industry is to be found in promoting commitment, trust and openness among project participants, and to adopt an integrated collaborative project organisation (Latham 1994; Egan 1998). This belief, also widely promoted within other project-based industries, will be fundamentally questioned in the remainder of this chapter.

The implementation of supply chain management in the project environment

As a result of the primarily project-specific demand and supply structures outlined above, the majority of construction procurement buyers have typically adopted *supplier selection* approaches to sourcing in the past. This is because these demand and supply structures create power structures that make more proactive approaches unfeasible and/or undesirable, both operationally and commercially. This is why construction relationships tend to be short-term and often adversarial. It can be argued that, rather than this being perceived as an error, this may in fact be simply a common-sense approach for buyers to adopt in a demand and supply environment that is one-off rather than long-term. However, while the highly competitive nature of the construction industry has historically 'guarded' against pre-contractual opportunism, buyers have to be aware that suppliers may regard a post-contractual opportunistic approach as highly appropriate in one-off rather than repeat games.

On the other hand, there are buyers who require construction on a more regular basis, who have the capability to standardise their requirements across multiple projects, for whom the feasibility and/or desirability of a more proactive sourcing approach may be much higher. It is important to understand that whether or not a buyer can develop such a proactive approach is not just a matter of choice. Proactive supply management (*supplier development* or *supply chain management*) requires that the buyer has both the internal resources and capabilities to undertake supplier development at the first tier and/or throughout the chain, and also a conducive external power structure to provide sufficiently willing partners who are prepared to be transparent and open to operational and commercial leverage. In short, the internal and external risk and incentives structures must support proactivity.

For this to occur the external power structures at the first tier and throughout the supply chain (in the power regime) have to be buyer dominant or interdependent, and they must persist over time. In the absence of these power structures there is little point in buyers attempting proactive sourcing initiatives. In other words, unless the demand and supply structures can be constructed such that a repeat game rather than a one-off game is being played, proactive sourcing is illusory. This, in part, explains why attempts at project partnering are bound to fail in practice (Bennett & Jayes 1995, 1998). This is because the project-specific learning is difficult to replicate or transfer across projects with dissimilar partners.

An important power variable in project procurement: regularity of demand

While it is evident that the possession of a conducive position of relative power is of high importance, there are some variables that are critical within a project-based procurement environment. Within the construction industry, the majority of clients are not in position of consistent dominance over the supply base because of the nature of their *ad hoc* construction profiles. Those clients who possess a high level of *regular* construction spend and who can standardise their requirements are clearly in a better position to be able to leverage the supply chain effectively and implement supply chain management approaches successfully. It is interesting to note that the client members of the Construction Task Force for *Rethinking Construction* (Egan 1998) all had regular spend of this type, which made proactive sourcing possible for them. It is an irony that these individuals did not consider the fact that just because something was possible for them it might not be feasible and/or desirable for everyone else in the construction industry.

In the power literature, the belief that power relationships in supply chains are highly transaction-specific, and that there are variations between firms and transactions in relation to the dimensions of utility and scarcity, is supported by the following two statements (Cox *et al.* 2002):

- The resources that give a buyer power over a specific supplier in the context of a particular transaction will not necessarily give it power over other suppliers in the context of an equivalent transaction
- The resources that give a buyer power over a specific supplier in the context of a particular transaction will not necessarily give it power over the same supplier in the context of other transactions.

What is not made explicit in this argument is the variable of *time* and how this may impact on the power resources in buyer–supplier relationships. This variable is of particular importance when developing appropriate buyer–supplier relationships in the project environment – especially in construction supply chains because of the project-specific and one-off nature of construction procurement for the majority of clients and construction firms. Therefore, before considering the issue of time (in the context of regularity), it is necessary to clarify the definition of a project and the distinction between this and a regular construction process spend.

In a definitive work on project management, Turner (1993) defined a project as

> *. . . an endeavour in which human, material and financial resources are organized in a novel way, to undertake a unique scope of work, of given specification, within constraints of cost and time, so as to achieve beneficial change defined by quantitative and qualitative objectives.* (Turner 1993, p. 5)

This definition is strengthened by Turner and Müller (2003) who argue that a project should be viewed as

> *...a temporary organisation to which resources are assigned to undertake a unique, novel and transient endeavour managing the inherent uncertainty and need for integration in order to deliver beneficial objectives of change.* (Turner & Müller p. 7)

This second definition emphasises that projects are unique, require novel processes and are transient with a clear start and end. However, although the majority of clients procure discrete one-off projects, there are those with an ongoing demand that source construction in a relatively regular manner. This regularity (process spend) may lead to a construction portfolio that either involves:

- Projects that are similar because of a standardised design and specification, and are undertaken either in a consecutive or overlapping manner. The client may wish to select a number of preferred suppliers through a long-term collaborative arrangement to deliver their requirements.
- Projects that have distinct design and specifications, but form part of an ongoing construction requirement. In the absence of a standardised specification, whether this option involves the selection of preferred suppliers and long-term relationships will depend on the procurement competence of the client and the capabilities of the supplier to deliver the range of requirements demanded by the client.

However, irrespective of whether the design and specification are standardised or not, it is evident that the nature of regularity involves two separate variables: *regularity of workload* and *regularity of relationship*. These two variables related to regularity are shown in Fig. 17.7.

Given this distinction, it is clear that the appropriate management of construction spend and supplier relationships may vary based on the nature of the regularity of the demand and supply variables that have to be managed:

(1) For *one-off projects with different suppliers* it is likely that *reactive sourcing* approaches (such as *supplier selection*) based on adversarial and opportunistic short-term relationships are likely to persist
(2) When there are *one-off projects with preferred suppliers* it is possible for opportunistic and adversarial *reactive sourcing* approaches to be adopted that could include *supplier selection* and *supply chain sourcing*
(3) For those with an *ongoing portfolio with different suppliers* it is possible to be both *reactive and proactive*, but the proactive sourcing approach would probably only extend to *supplier development* activities
(4) When there is an *ongoing portfolio with preferred suppliers* it is also possible to pursue both reactive and proactive approaches, but there may be scope for both *supplier development* and *supply chain management* approaches.

Table 17.3 summarises the limited and unique circumstances, in relation to regularity, under which particular supply management approaches may be feasible. However, while these different relationship management approaches may be

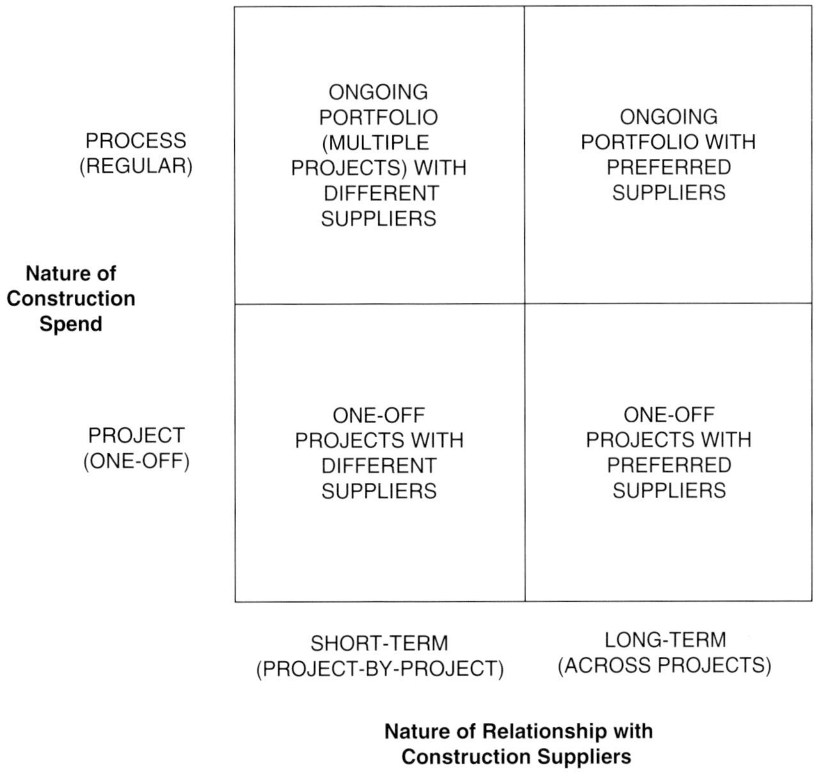

Nature of Construction Spend	SHORT-TERM (PROJECT-BY-PROJECT)	LONG-TERM (ACROSS PROJECTS)
PROCESS (REGULAR)	ONGOING PORTFOLIO (MULTIPLE PROJECTS) WITH DIFFERENT SUPPLIERS	ONGOING PORTFOLIO WITH PREFERRED SUPPLIERS
PROJECT (ONE-OFF)	ONE-OFF PROJECTS WITH DIFFERENT SUPPLIERS	ONE-OFF PROJECTS WITH PREFERRED SUPPLIERS

Nature of Relationship with Construction Suppliers

Figure 17.7 The nature of regularity in construction procurement.

Table 17.3 The implementation of sourcing options in different conditions of regularity.

Nature of construction spend	Nature of relationship with construction suppliers	Supply management option			
		Supplier selection	Supply chain sourcing	Supplier development	Supply chain management
Process (regular and ongoing)	Long term (ongoing)	Possible	Possible	Possible	Possible
	Short term (by project)	Possible	Possible	Possible	Not possible
Project (one-off)	Long term (ongoing)	Possible	Possible	Not possible	Not possible
	Short term (by project)	Possible	Not possible	Not possible	Not possible

possible, it is important to understand their appropriateness given the firm's power position within the supply chain.

Conclusions

It follows from this discussion that, while government and industry reports may call on everyone in construction to undertake proactive sourcing approaches, such generic recommendations provide evidence of poor thinking by the authors of these reports. This is because it is impossible to generalise about the way that particular clients, contractors and suppliers should operationalise supply management under particular power and leverage circumstances. In reality, when a construction project is designed and specified it results in the creation of a particular (and often unique) power regime that has to be managed. Thus, whatever some may wish, the reality in construction projects is that sometimes it is impossible to create power regimes that will allow the effective development of proactive sourcing approaches. On the contrary, one might conclude, given the typical client's *ad hoc* and one-off demand and the high levels of fragmentation and contestation in the majority of construction supply markets, that conducive power regimes for proactive sourcing are highly unlikely. This argument is applicable across all project-based industries.

Despite this there is some hope for those who would like to see the construction industry move towards proactive sourcing. For those clients, contractors and suppliers who have a regular demand that is also of high volume and capable of being standardised, there may well be opportunities for the creation of supplier development and supply chain management approaches. Unfortunately, this is unlikely to be possible for the vast majority of participants who struggle to make acceptable returns from participating in a highly fragmented and opportunistic industry. Reactive sourcing will, therefore, continue to be, as it has in the past, the primary tool of procurement and supply management in the project environment, whatever the authors of government and industry reports believe.

Future research

This previous discussion implies that academic research in the future needs to consider a number of key conditions that must be in place so that effective supply management can be implemented in the project environment (Cox 2004c; Cox *et al.* 2004b). These conditions highlight the need for academics and practitioners to understand both the commercial and operational circumstances that exist between buyers and sellers in transactional exchange, the distinct and often conflicting goals of both parties when they enter into business relationships, and the appropriate supply management approaches given these circumstances. This fundamental understanding of dyadic buyer–supplier exchange and relationship alignment needs to be extended across all supply chain relationships and take into account whether project demand is one-off or serial (Ireland 2004).

The first condition for successful relationship management is that the buyer and supplier in any relationship understand the power and leverage circumstances that they are currently in, and how these can be modified in the future. If either party is uncomfortable with their relative power position, conflict is highly likely, as they will inevitably attempt to re-engineer a more favourable balance. This condition requires that both parties understand the power matrix, as introduced in Fig. 17.4 (Cox *et al.* 2000).

The second condition requires the buyer and supplier to have commensurable commercial strategies. The need for an acceptable 'trade-off' is critical, as buyers and suppliers have their own distinct commercial goals, which are very frequently in direct conflict. For example, buyers normally attempt to develop effective operational approaches to maximise value for money (improve functionality and/or reduce costs of ownership), while suppliers normally attempt to maximise their revenue and returns from the exchange. It is also important that these choices are linked to an acceptable supplier competence and congruence for the buyer, and an appropriate customer account management strategy for the supplier (Cox *et al.* 2004a).

The third condition is that buyers and suppliers must have commensurable operational strategies. By this we mean that the buyer selects an appropriate supply management strategy (as introduced in Fig. 17.3) that is commensurable with the competitive market strategy being pursued by the supplier. The operational strategy of the supplier may be one of cost leadership, differentiation, or a combination of these two.

The fourth condition relates to the ability of the buyer and supplier to differentiate between alternative relationship management styles and select the one that is the most appropriate given the power circumstance, commercial goals and operational strategies of each party in the transaction. Relationship management styles may be considered to consist of two main dimensions. These are the way of working between the buyer and supplier (arm's length or collaborative) and the way in which the surplus value created in the relationship is shared. This distinction is explained in detail elsewhere (Cox *et al.* 2003).

The final condition of successful relationship management is the need for there to be top management support and internal buy-in from within both the buyer's and the supplier's organisation. This buy-in and the requisite internal environment may also be determined by the presence of other factors such as robust IT systems that provide supporting information and a willingness to invest in training and development.

In summary, it is important that future research enables academics to assist practitioners by helping them to understand the most appropriate relationship management strategies available given the power and leverage circumstances they are in. With the five conditions outlined previously in place, business relationships can be made to work effectively, as both the buyer and supplier will have a full understanding of the need to align their commercial and operational goals both internally and externally, and the appropriateness of particular actions in specific internal and external power and leverage circumstances. Without such an understanding, practitioners could well blindly pursue particular sourcing and relationship management approaches, such as 'win–win' partnering arrangements or integrated supply chain management, which may not always be appropriate given the power and leverage circumstances possible in the project environment (Cox 2004b).

References

Bennett, J. & Jayes, S. (1995) *Trusting the Team: The Best Practice Guide to Partnering in Construction*. Reading Construction Forum, Reading.

Bennett, J. & Jayes, S. (1998) *The Seven Pillars of Partnering*. Thomas Telford, London.

Blismas, N.G. (2001) *Multi-project environments of construction clients*. PhD thesis (unpublished), Loughborough University, Loughborough.

Burt, D.N. (1984) *Proactive Procurement*. Prentice-Hall, Englewood Cliffs, NJ.

Campbell, N.C.G. (1985) An interaction approach to organisational buying behaviour. *Journal of Business Research*, **13**, 35–48.

Cherns, A.B. & Bryant, D.T. (1984) Studying the client's role in construction management. *Construction Management and Economics*, **2**(2), 177–184.

Christopher, M. (2000) The agile supply chain: competing in volatile markets. *Industrial Marketing Management*, **29**(1), 37–44.

Christopher, M. & Towill, D.R. (2000) Supply chain migration from lean and functional to agile and customised. *Supply Chain Management*, **5**(4), 206–215.

Cox, A. (1997) *Business Success*. Earlsgate Press, Boston.

Cox, A. (2001) Understanding buyer and supplier power: a framework for procurement and supply competence, *Journal of Supply Chain Management*, **37**(2), 8–15.

Cox, A. (2004a) The art of the possible: relationship management in power regimes and supply chains. *Supply Chain Management*, **9**(5), 346–356.

Cox, A. (2004b) Business relationship alignment: on the commensurability of value capture and mutuality in buyer and supplier exchange. *Supply Chain Management*, **9**(5), 410–420.

Cox, A. (2004c) *Win–Win? The Paradox of Value and Interests in Business Relationships*. Earlsgate, Stratford-upon-Avon.

Cox, A. & Ireland, P. (2001) Managing construction supply chains: the common-sense approach for project-based procurement, In: *Proceedings of the 10th International Annual IPSERA Conference*, 9–11 April, Jönkoping, Sweden; pp. 201–214.

Cox, A. & Ireland, P. (2002) Managing construction supply chains: a common sense approach. *Engineering, Construction and Architectural Management*, **9**(5/6), 409–418.

Cox, A. & Thompson, I. (1998) *Contracting for Business Success*. Thomas Telford, London.

Cox, A. & Townsend, M. (1998) *Strategic Procurement in Construction*. Thomas Telford, London.

Cox, A., Sanderson, J. & Watson, G. (2000) *Power Regimes: Mapping the DNA of Business and Supply Chain Relationships*. Earlsgate Press, Boston.

Cox, A., Sanderson, J. & Watson, G. (2001) Supply chains and power regimes: toward an analytical framework for managing extended networks of buyer and supplier relationships. *Journal of Supply Chain Management*, **37**(2), 28–35.

Cox, A., Ireland, P., Lonsdale, C., Sanderson, J. & Watson, G. (2002) *Supply Chains, Markets and Power: Mapping Buyer and Supplier Power Regimes*. Routledge, London.

Cox, A., Ireland, P., Lonsdale, C., Sanderson, J. & Watson, G. (2003) *Supply Chain Management: A Guide to Best Practice*. Financial Times/Prentice Hall, London.

Cox, A., Lonsdale, C., Sanderson, J. & Watson, G. (2004a) *Business Relationships for Competitive Advantage: Managing Alignment and Misalignment in Buyer and Supplier Transactions*. Palgrave Macmillan, London.

Cox A, Watson, G., Lonsdale, C. & Sanderson, J. (2004b) Managing appropriately in power regimes: relationship and performance management in 12 supply chain cases. *Supply Chain Management*, **9**(5), 357–371.

Cutting-Decelle, A.F. (1997) The use of industrial management methods and tools in the construction industry: application to the construction process. In: *Proceedings of First International Conference on Concurrent Engineering in Construction* (eds C.J. Annumba & N.O.F. Evbuomwan). Institution of Structural Engineers, London; pp. 306–318.

Dyer, J.H., Cho, D.S. & Chu, W. (1998) Strategic supplier segmentation: the next best practice in supply chain management. *California Management Review*, **40**(2), 57–77.

Egan, J. (1998) *Rethinking Construction*. HMSO, London.

Farmer, D.H. & Ploos van Amstel, R. (1991) *Effective Pipeline Management: How to Manage Integrated Logistics*. Gower, Aldershot.

Ford, D. (1980) The development of buyer–seller relationships in industrial markets. *European Journal of Marketing*, **14**(5/6), 339–354.

Ford, D. (2001) *Understanding Business Marketing and Purchasing*. International Thomson Business Press, London.

Forrester, J. (1958) Industrial dynamics: a major breakthrough for decision makers. *Harvard Business Review*, **36**(4), 37–66.

Håkansson, H. (ed.) (1982) *International Marketing and Purchasing of Industrial Goods: An Interaction Approach*. John Wiley & Sons, Chichester.

Harrison, A., Christopher, M. & van Hoek, R. (1999) Creating the agile supply chain. *School of Management Working Paper*, Cranfield University, Cranfield.

Hellard, R.B. (1995) *Project Partnering: Principles and Practice*. Thomas Telford, London.

Hines, P. (1994) *Creating World Class Suppliers: Unlocking Mutual Competitive Advantage*. Pitman, London.

Hines, P. (1996) Purchasing for lean production: the new strategic agenda. *International Journal of Purchasing and Materials Management*, **32**(1), 2–10.

Hines, P., Lamming, R., Jones, D., Cousins, P. & Rich, N. (2000), *Value Stream Management: Strategy and Excellence in the Supply Chain*. Financial Times/Prentice Hall, London.

Ireland, P. (2004) Managing appropriately in construction power regimes: understanding the impact of regularity in the project environment. *Supply Chain Management*, **9**(5), 372–383.

IMP Group (1982) An interaction approach. In: *International Marketing and Purchasing of Industrial Goods: An Interaction Approach* (ed. H. Håkansson), John Wiley & Sons, Chichester; pp. 10–27.

Koskela, L. (1997) Lean production in construction. In: *Lean Construction* (ed. L.F. Alarcón). A.A. Balkema, Rotterdam; pp. 1–9.

Krafcik, J. (1988) Triumph of the lean production system. *Sloan Management Review*, **30**(1), 41–52.

Lamming, R. (1993) *Beyond Partnership*. Prentice Hall, New York.

Latham, M. (1994) *Constructing the Team*. HMSO, London.

Lewis, J.D. (1995) *The Connected Corporation*. Free Press, New York.

London, K. & Kenley, R. (2000) Mapping construction supply chains: widening the traditional perspective of the industry. In: *Proceedings of the 27th Annual European Association of Research in Industrial Economics (EARIE) Conference*, 7–10 September, Lausanne, Switzerland.

London, K., Kenley, R. & Agapiou, A. (1998) The impact of construction industry structure on supply chain network modelling. Paper presented at *Logistics Research Network Annual Conference*, September, Cranfield School of Management, Cranfield.

Luck, R. & Newcombe, R. (1996) The case for integration of the project participants' activities within a construction project environment. In: *The Organisation and Management of Construction: Shaping Theory and Practice* (eds D.A. Langford & A. Retik). E. & F.N. Spon, Glasgow; pp. 458–470.

Mason-Jones, R. & Towill, D.R. (1999) Total cycle time compression and the agile supply chain. *International Journal of Production Economics*, **62**(1/2), 61–73.

Mason-Jones, R., Naylor, B. & Towill, D.R. (2000) Lean, agile or leagile? Matching your supply chain to the marketplace. *International Journal of Production Research*, **38**(17), 4061–4070.

Morris, P.W.G. (1997) *The Management of Projects*. Thomas Telford, London.

Nam, C.H. & Tatum, C.B. (1988) Major characteristic of constructed products and resulting limitations of construction technology. *Construction Management and Economics*, **6**(2), 133–148.

Naylor, J., Naim, M.M. & Berry, D. (1999) Leagility: integrating the lean and agile supply chain. *International Journal of Production Economics*, **62**(1/2), 107–118.

Nishiguchi, T. (1994) *Strategic Industrial Sourcing*. Oxford University Press, New York.

Ohno, T. (1988) *Toyota Production System: Beyond Large-Scale Production*. Productivity Press, Cambridge, MA.

Saunders, M.J. (1994) *Strategic Purchasing and Supply Chain Management*. Pitman, London.

Stalk, G. & Hout, T. M. (1990) *Competing Against Time: How Time Based Strategies Deliver Superior Performance*. Free Press, New York.

Tan, K.C., Kannan, V.R. & Handfield, R.B. (1998) Supply chain management: supplier performance and firm performance. *International Journal of Purchasing and Material Management*, **34**(3), 2–9.

Turner, J.R. (1993) *The Handbook of Project Based Management*. McGraw-Hill, London.

Turner, J.R. & Müller, R. (2003) On the nature of the project as a temporary organisation. *International Journal of Project Management*, **21**(1), 1–8.

Vrijhoef, R. (1998) *Co-makership in construction: towards construction supply chain management*. Thesis of Graduate Studies, Delft University of Technology/VTT Building Technology, Espoo.

Vrijhoef, R. & Koskela, L. (2000) The four roles of supply chain management in construction. *European Journal of Purchasing and Supply Management*, **6**(3/4), 169–178.

Watson, G., Cox, A., Lonsdale, C. & Sanderson, J. (2003) Thinking strategically about supply chain relationship management: the issue of incentives. In: *Global Logistics and Distribution Planning* (ed. D. Waters), Kogan Page, London; pp. 104–116.

Womack, J.P. & Jones, D.T. (1996), *Lean Thinking: Banish Waste and Create Wealth in Your Organisation*. Simon Schuster, New York.

Womack, J.P., Jones, D.T. & Roos, D. (1990) *The Machine that Changed the World*. Rawson Associates, New York.

Endnotes

[i] However, it should be noted that there is a high degree of subcontracting within the industry, with main contractors appointing third parties to deliver 'packages' that can be integrated within the solution. It has been argued in the construction management literature that the unmanaged use of subcontracting is a major factor that leads to the high level of adversarialism within the industry, as there is another player attempting to earn margins to the detriment of other firms. This is discussed in detail elsewhere (Cox & Thompson 1998).

[ii] The nature of demand for construction can be segmented in a number of ways. Cox and Townsend (1998) differentiate between clients who have a regular requirement for construction work of similar value and content (process spenders), and infrequent purchase clients (commodity spenders). Cox and Thompson (1998) contend that clients that possess regular process spends are unlikely to constitute more than 25% of the total UK market, while Blismas (2001) contends that in 1999 multi-projects accounted for 10% of the entire industry's output and as much as 30% of contractor's output.

18 Signals from Site – Embodied Logic and Management Accounting on Construction Projects

Andrew Ross and David Hugill

Introduction

This chapter examines how commercially relevant information is provided, with the purpose of reporting a company's financial position, by enterprises who engage in construction projects. It considers how the conflicting objectives of the project team and senior management have created a range of interorganisational communication processes that seek to both obscure and uncover the 'true' financial position. It reports upon an empirical analysis of a sample of cost–value reconciliation (CVR) forms that illustrate the variability of interorganisational communication processes for the reporting of project financial information.

Construction projects are undertaken in a commercial environment constrained by complex legal agreements. Having contracted to perform their obligations, each party must then manage resources, aiming to achieve a satisfactory commercial outcome – minimally an income in excess of expenditure. The tools and techniques for managing this process require information concerning the financial consequences of resource usage, not only for work undertaken, but also future resource commitments. Thus, the simple though stark reality of a contract sum increasingly constrains the scope of management, as monies are expended and both the time and finance available diminish. Organisations are also required to collect and codify cost and value data in order to satisfy audit requirements. We suggest that the data used in the CVR process are a potentially valuable source of information that can be used to model project value and cost profiles, production outputs and resource inputs, which so far has been overlooked. The chapter commences with an overview of the governance requirements of organisations, considers their application at a project level and also discusses some of the organisational barriers to the diffusion of reliable information. The empirical work on which the chapter is based is a documentary analysis of a selection of contractors' cost–value reconciliation reports.

Control accounting

The tools used by management in this circumstance fall under the general description of accounting, of which there are three main functions: stewardship, decision accounting, and control accounting (Sizer 1986, p. 18). Whilst the first is minimally distinguishable from the others in terms of the principal end users (shareholders), decision and control accounting can, in the first instance, be combined into management accounting. However, for the purposes of this chapter, our concern is limited to control accounting.

The chapter aims to raise awareness of the potential for data collection and analysis, and identifies that there is a lack of empirical work in this area of construction management. Walker and Wilkie (2002) devote a chapter and two appendices to 'cost and value comparison', which reveals more of the detailed features of the process than Upson's earlier text (Upson 1987). Cooke and Williams (1998) include a brief outline of the process. The most thorough exposition to date, however, is that produced by Barrett (1992) for the Chartered Institute of Building. The primary limitation of these treatments is that they report upon standardised practice, failing to reflect the actual praxis of CVR reporting. Thus, following Anderson *et al.* (1989), we shall be performing a detailed examination of certain features of the management accounting process, since: 'Praxis only becomes available for analytic description in the detail' (Anderson *et al.* 1989, p. 126).

The preparation and instrumentality of management accounts is furnished through the data collected to *account* for construction activity, at a particular point in time. The format in which the data is assembled is usually prescribed by the organisation using generic protocols; it is to be read by other senior management personnel, who are often in a hierarchically senior position to the site team. The decisions made by these personnel have an impact upon the autonomy of the site team, and their judgements are formulated on the basis of their interpretation. To this end, the current work is limited to one aspect of the overall project, that is, the format of management accounts produced from site, typically using pro forma reports, and speculates upon the reasoning behind their design.

Management accounting's origins

The origins of management accounting can be traced in both American and British industry, although Professor John Sizer's account leads to the conclusion that the often-overlooked activities of the Anglo-American Council of Productivity provided a influential conduit in the immediate post-war period, making management accounting 'fashionable' (Sizer 1986, p. 15). Before this, management accounting emerged out of 'cost and works' accounting in the 1950s and an examination in 'management accountancy' was used to award Fellowship status to members of the Institute of Cost and Works Accountants (FCWA). In the mid-1960s, the name of the Institute's journal was changed from *Cost Accounting* to *Management Accounting*, and the status of 'management accounting' was elevated further in 1972 when the Institute modified its name to the Institute of Cost and Management Accountants. The Institute became the Chartered Institute of Management Accounting when a Royal Charter was granted in 1975.[i]

In the USA, management accounting emerged when firms started to rely more on internal operations than the market to conduct business. Beginning in the nineteenth century in textiles, railroads and the manufacture of iron and steel, 'entrepreneurs linked processes in the new single-activity organizations because they believed greater value could be achieved by managing the processes in a centrally controlled organization than by exchanging outputs from the processes in the market' (Johnson & Kaplan 1987, pp. 19–20).

Whilst cost and works accountants in the UK were engaging in greater managerial intervention, it has been argued that their US counterparts were failing in their duty adequately to serve the needs of American management. Despite a panoply of accounting practices,[ii] it was claimed that little progress had been made since the mid-1920s, leading to:

> *...distorted product costs, delayed and overly aggregated process control information, and short-term performance measures that do not reflect the increases or decreases in the organization's economic position.* (Johnson & Kaplan 1987, p. 13)

Johnson and Kaplan attributed these shortcomings to a re-orientation by management in their direction of the affairs of corporations:

> *After 1925 a subtle change occurred in the information used by managers to direct the affairs of complex hierarchies.... Until the 1920s, managers invariably relied on information about the* ***underlying processes, transactions and events*** *that produce the financial numbers. By the 1960s and 1970s, however, managers commonly relied on financial numbers alone.* (Johnson & Kaplan 1987, pp. 125–126; emphasis added)

Johnson and Kaplan argued that management had become uninterested in direct measures of activity in the firm, relying instead on single surrogate 'financial numbers'. Referring to the 'transition' in managerial thinking, they were unable to explain the complex forces that brought about this change.

Johnson and Kaplan (1987) used the provocative title of their book *Relevance Lost. The Rise and Fall of Management Accounting* to draw attention to shortcomings in management accounting, taking as their start point the instrumental value of information derived through management accounting practice to provide management with the wherewithal better to manage the business.

Accounting regulations

Whilst the primary concern of this chapter is *management* accounting, various connection(s) with *financial* accounting can be observed. According to Sizer (1986, p. 18), financial accounting is concerned with stewardship – the preparation of periodic statements in the form of profit and loss accounts and balance sheets that provide a true and fair view to shareholders and meet the requirements of the law, whereas management accounting combines decision and control accounting. The former includes estimates of costs and revenues associated with particular alternatives; the latter information assists management to plan and control effectively.

Accounting information is prepared to certain 'Standards': Statements of Standard Accounting Practice (SSAP) and Financial Reporting Standard (FRS). There are currently eleven SSAP (1–25) and fifteen FRS (1–15).[iii] Following their introduction in the early 1970s, the status of SSAPs was not properly established in law until the Companies Act (HMSO 1989) required firms to state whether or not their financial statements had been prepared in accordance with Accounting Standards. Moreover, firms are required to explain any material departures from them (Gowthorpe 2000, p. 52).

The Accounting Standards Board (ASB) develops standards to be applied by accountants, which are then issued by six principal accounting bodies,[iv] who are also represented on a Consultative Committee of Accountancy Bodies (CCAB).

Of particular interest in this discussion is SSAP No. 9 (SSAP9) – 'stocks and work in progress', which succinctly identifies its genesis in its opening paragraph:

> *No area of accounting has produced wider differences in practice than the computation of the amount at which stocks and* ***work in progress*** *are stated in financial statements.* (SSAP9, quoted in Barrett 1992, p. 4; emphasis added)

Thus, it is intended that amounts included in financial statements for work in progress (WIP) will not vary because of widely different practices. The statement is therefore intended to eradicate the variety of practices, lending assurance to those reliant on the financial report that WIP (a contribution to sales value) has been calculated consistently (as between companies, and years) and appropriately – having regard to the detailed requirements of the statement. In short, it serves to reassure readers of the accounts that sales (and hence profits, since costs are unaffected) have not been inflated.

Construction organisations adopt a framework for the financial management, which imposes obligations on its directors to produce annual reports that are an accurate and true reflection of the organisation's financial position. Company accounts and an operating statement that are audited within a stated accounting policy and governance framework, support the published annual report. The governance framework seeks to ensure that records are available to demonstrate how the stated financial position has been arrived at, that prudence has been exercised in decision making, and that wherever uncertainty exists regarding revenue or costs, conservative estimates have been made. These three principles of records, prudence and conservatism are in place and embodied within an accounting policy that is published within the annual statement. The accounting policy generally follows the guidelines suggested by SSAP9 with exceptions that take into account the project nature of construction organisations, and the published accounts are audited to ensure that the policy has been applied. It relates to the published accounts of companies rather than the management accounts and, although there is no legal requirement to comply with its provisions, auditors can decide to qualify the accounts if they are not convinced that the company has alternative robust arrangements.

The organisation's accounting policies consequently influence the reporting of aggregated business units within organisations and also the reporting of a project's financial position. In reporting their financial position, construction organisations are faced with unique difficulties in comparison with other sectors, allied to the idiosyncratic nature of the commercial management of projects. External factors that influence the reporting processes adopted by organisations include the client's procurement arrangement that in general prescribes processes for the valuation of completed work and WIP. Internal factors that influence the processes will be discussed later. SSAP9 identifies the following precepts that need to be considered by organisations:

- Timing
- Cost definition
- Attributable profit
- Losses
- Reconciliation
- Overheads
- Net realisable value
- The treatment of claims

The following section introduces these precepts and demonstrates their application in practice by analysing a sample of organisations' forms.

Timing

An accounting period is defined as a year, and although the construction industry's clients are influenced by the accounting cycle, the projects they commission obviously do not neatly fit within the annual reporting cycle. As a result, an organisation's reporting processes need to take into account the value, cost and profits that are generated by projects that:

- Commence before, and are being undertaken within, the accounting period
- Have been physically completed before the accounting period but remain financially active
- Commence and finish within that accounting period.

Organisations adopt a reporting framework that is usually monthly for the production of project financial reports which are supported by costs statements and statements of value. The dates for the reporting of costs are generally common to all projects and are typically at the end of each calendar month. In the examples considered, the organisations did not refer to the current industry standards such as CI/SfB or Uniclass for the reporting of construction project costs, and had developed unique taxonomies that tended to reflect the nature of their organisation. In some cases costs were aggregated into broad cost heads, such as labour or plant; in other instances, costs were reported at a detailed level, such as different types of materials. The cost statement was produced from records of production, which included materials received by the site, labour and plant returns, subcontractor liabilities, and schedules of overhead costs. A range of systems of accruals were used to take account of the delay between receipt and payment for resources.

Form analysis

The following is an account of a sample of reporting formats (pro formas), drawing attention to their *instrumental* features to illustrate how, in Baccus's words, the 'embedded logic' (Baccus 1986) of the forms can be witnessed. It makes explicit how the design of the forms furnishes their instrumentality in the ongoing reporting of the commercial progress of each construction project, on each and every occasion of the reports' production.

A sample of pro formas has been obtained from a range of construction (building and civil engineering) companies. No attempt has been made to design the 'sample'; it is simply what has been obtainable utilising a variety of contacts with practitioners. A cursory characterisation of the companies reveals a number of multinationals; others are companies operating in their immediate locale.

Given the limited amount of literature concerning this aspect of commercial management in construction, and the absence of any research on the design of management accounting pro formas, it is necessary to warrant the approach taken in providing what follows as analysis-through-description.

Initially, some rudimentary views of the forms will be provided, covering such matters as the overall structure of the reports – title, size (number of pages), attestation (if any), and managerial function. More detailed gloss noticings will then be furnished of subdivisions, and a discussion of the imperatives for form design and use will be provided. Since a closer examination of the data is to follow, the attribution of 'size' on this basis will overlook certain other features of the data that will be discussed later, and used to delineate categories. Thus, in this preliminary account, data categories are overlooked.

The examples within this chapter tabulate summarisable[v] features of the pro formas. From this presentation of the data, we can see that accounting for construction activity on a periodic basis (all of the reports are for a 'month'[vi]) can take up a single page, several pages, and even up to ten pages.[vii]

The baptism of a document furnished for managerial purposes (its *naming*):

> *Naming appears as a queer connexion of a word with an object. . . . as it were a baptism of an object.* (Wittgenstein 1967, p. 38)

The *reports* (their status is at least minimally distinguishable as having this format) are presented under a number of guises. In two instances, there is no obvious title; of the remainder 'cost–value reconciliation' is commonly used. The array of titles is reproduced below:

- Contract analysis sheet
- Cost–value reconciliation
- Contract review
- Monthly contract cost/value report
- Internal valuation
- Monthly reconciliation.

Report size and identifiable sections

Summarising features or sections of the reports is a more difficult undertaking, most obviously because of the wide range in *size* (pages are used as the simplest available indicator here), but also due to the extent to which the whole (report) is logically capable of disaggregation. That this is a necessary undertaking is warranted by the wide range of report sizes encountered in practice, and to furnish an answer to the obvious question of how some firms are able to achieve the object in one page, whereas another requires ten.[vii] This problem is also entwined with the demonstration that reports are presented in sections, and how it is that different parts are distinguished from one another (the design of reports). The format of the reports, its contribution to *signalling* from site, is taken up in a later section of this work.

Further, it is interesting to pause and consider what is accomplished in a more disaggregated report, as opposed to a lesser one. Can the objectives of a one-page report be the same as one in which ten pages are produced (these examples are taken from the current dataset)? What do the titles of the sections and subsections announce? Can their object be detected from this signal? On inspection, the more comprehensive forms appeared to require more information to support the summary; the implications of this are presumed to relate to a company requirement for auditability at project level and to ensure that the site team conforms to standardised processes.

Describing the pro formas collected, to provide an understanding of management accounting in construction, presents difficulties because of the wide range of formats in use. Since the principal object of this discussion is to comment on events *in practice*, it is clear that the presentation of this information must remain a paramount feature.

Barrett (1992) notes that such statements are couched in terms which draw directly upon SSAP9, typically of the form:

> *The work in progress is stated at* ***cost*** *plus* ***attributable profit*** *(if any), less* ***foreseeable losses*** *(if any), and* ***progress payments*** *received and receivable.* (Barrett 1992, p. 9; emphasis added)

Before examining pro formas in the light of this statement it is necessary for it to be understood. Initially we can see that the WIP comprises four components (these have been emboldened), and associated with these are what we shall initially refer to as an instruction forming part of a calculation ('plus' or 'less'). Before looking at the calculation, we shall examine each of the components.

Cost

Cost has been defined as 'that expenditure which has been incurred in the normal course of business in bringing the product or service to its present location and condition' (SSAP9). This simple definition belies the processes that are required to report upon a project's cost. Management have to apply their judgement to arrive at a fair approximation of the costs that have been incurred on a project; in order to inform this judgement organisations systematically collect information to support the reporting of costs. The cost report is generally produced by an accounting department and takes account of both direct costs (costs that have been paid for) and accrued costs (costs that have been incurred for which a judgement was made about their scale); an example is provided (Fig. 18.1). The information used to report the costs in the sample of forms considered originated from three sources: the site, head office, and materials and plant suppliers.

B. Cost reconciliation	Previous	Current
18. Cost per printout **at**		
18(a). Insurance levy for month @%		
19. Damages for delay (to be accounted)		
(a) Claim main contractor		
(b) Liquidated damages		
Total item 19		
20. Late costs (to be accounted)		
(a) Late materials		
(b) Late equipment		
(c) Late subcontractor		
(d) Contra charges not invoiced		
(e) Late other		
(f)		
Total item 20		
21. Total costs including provisions and late costs (18+19+20)		

Figure 18.1 Summary cost report.

Site information

The sites produce information that is related to the amount of resource used in 'bringing the product to its present condition' (SSAP9). The resources include those that are internal to the organisation, such as labour or supervisory staff, and also those resources supplied by other organisations, such as material, plant, or subcontractors. The information provided by the site allows for the computation of accrual information which, in the case of materials suppliers, is usually based upon the physical deliveries of materials to site, supported by documentation, order values and a form of recording, typically, a materials-received report. Standardised forms used to collect information on supplier details, initial order values and deliveries, or plant hired are regularly compiled and passed to a central department, and are common across the industry. The accounting department uses the information received from the sites, and the procurement department (in some organisations referred to as the buying department) calculates accrual values and process payments. This process is relatively robust and transparent; the main difficulties that were considered to arise on inspection of the forms were either from missing data (material delivery tickets or plant hire reports missing) or from incorrect calculation of accrual values (caused by missing order values or data input error). The sites also produce a schedule of materials on site which were identified in some of the cost statements (often as a negative accrual) and were excluded from the total costs (these materials were not considered as being incorporated into the works and therefore do not meet the accounting policies' requirements of 'bringing the product to its present condition').

The process for producing information to support the reporting of the costs of subcontractors' work was broadly similar to materials and plant; however, it was subject to more judgement and, accordingly, could be considered to be less transparent. This provides the site team with an opportunity to protect some of their decision-making autonomy. Subcontractors are procured under a wide range of contractual arrangements; however, they are usually reported under three categories: domestic (this includes labour only, labour and plant, and supply and fix), nominated, or named. Figure 18.2 illustrates an accrual adjustment form used by one organisation. The costs heads in use for these are often separate, as the profits and/or losses reported are often related to the nature of the relationship that exists between the reporting organisation and its subcontractors. The process for the completion of a cost accrual for a subcontractor typically refers to the subcontractor's application for payment for work complete, materials delivered to site and variations and claims. The site team assess the valuation and compile a statement of accrual based upon their judgement as to the entitlement of the subcontractor. The summary accrual figure would then be entered onto a standardised form, which included headings for order value, discount, contra charges and last payment details. This is then passed to the accounting department for aggregation into summaries of costs for each of the cost heads.

COST–VALUE RECONCILIATION – COST ADJUSTMENT SHEET			
MONTH			
COMPUTER PRINT OUT			
ADJUSTMENTS			
ACCRUALS (UNDER/OVER COSTING)	+	–	
1.0 Subcontractors – Domestic			
1.1 Subcontractors – Nominated			
1.2 Subcontractors – Labour only			
1.3 Consultants/Professional Fees			
INTERNAL ACCRUALS			
2.0 Insurance pre-payments			
2.1 Staff cars			
2.2 Plant internal			
3.0 COST CORRECTIONS – Transfers			
4.0 COST CORRECTIONS – Incorrect invoices			
5.0 SALARY ADJUSTMENTS – To end of month			
6.0 WAGES – Ditto			
7.0 CONTRA CHARGES to F/P – Not yet costed			
8.0 MAINTENANCE/SNAGGING			
9.0 A.N.OTHER			
TOTALS			
TRUE COST			

Figure 18.2 Cost adjustment summary.

The head office's role in the production of cost information tended to relate to the allocation of costs of company and project overheads. At tender stage the organisations would have predicted the required project overheads; these would normally include supervisory staff, such as site agents, engineers and commercial managers. These allowances were generally time related and based on the expected contribution that the staff were to make on the project. During project execution, the staff costs allocated to the project tend to be for those who are directly employed, together with a percentage contribution for staff who are involved with a number of projects.

The interim valuation of construction projects and the uncertainty that surrounds the evaluation of work is a central issue to the variance of approaches taken by construction companies. Projects often incur costs that may be paid for by another party at a future date; the uncertainty about whether recovery will take place is identified by the accounting policy which states that 'if it is not likely that any reasonable expectation of future revenue to cover total cost is forthcoming, the total cost should be charged against the accounting period in which the cost is incurred' (SAPP9). Consequently, a judgement has to be made regarding the likelihood of future revenue when deciding whether to include a cost within an accounting period.

The page and section headings shown above (Fig. 18.2) unsurprisingly reveal that 'cost' is a common and recurring element in pro formas. Firms frequently take a figure from a computer printout, and then make adjustments to this figure under a series of headings and/or criteria. This is understood to be a part of a process recognised as 'reconciliation'. In making adjustments, cost data are aligned with the 'value' of work at a common point in time. Thus, when profit is subsequently calculated, it is clear that the straightforward arithmetic is being performed on figures that reflect a partially executed project where all of the resources used to create the value of the works, at that point in time, have been included.

Barrett's comprehensive publication for the Chartered Institute of Building lists what is referred to as 'basic cost headings': 'materials', 'plant', 'staff and overheads', 'labour', 'subcontractors', 'residual credits' and 'cost transfer to match revenue', the last two items being deductions (Barrett 1992, p. 49). For each of these categories, three data categories are provided: 'up to current period', 'current period' and 'total'. We can see that a figure for the current period (month), when added to the previous, will provide a 'total'.

The analysis of the costs to arrive at a 'true cost' shown in Fig. 18.2 commenced with computer printout; thus the adjustments are to the figure reported from this printout. In Fig. 18.3, there are eight headings from which we may conclude that these matters warrant more attention than others, but the sheet on its own does not inform us why this is the case.

Other pro formas repeat this design for calculating and presenting cost for managerial and accounting scrutiny: cost is presented as a compound item, revealing an array of categories which either 'build-up' or 'build-up and adjust' a cost both at this point in time and (by arithmetic) in a given period. One pro forma adopts a simpler approach with single line entries (across columns: 'This month', 'Previous month', and 'Difference') for 'Cost printout' and 'Adjustments'. Thus we can conclude that most

	Valuation details			Cost details			Earnings[1]	
	Total of previous month's gross value	Gross valuation to date	Current month's value	Total of previous month's prime cost	Total of prime cost reconciled	Current month's prime cost	Total of earnings level 1	Current month's earnings
A								
B								
C								
D								

[1]The separation of earnings into two categories implies that alternative ways of accounting for the surplus (or deficit) of value over cost are required. Given that the current work is solely reliant on pro formas, it is not known what purpose is being served by these feature on the basis of these data alone. Nevertheless, attention is drawn to this feature, and subsequent enquiries can be conducted to explicate both the signals and the basis for their separateness.

A Prelims and direct work

B Direct subcontractors

C Nominated subcontractors

D Nominated suppliers

Figure 18.3 Reconciliation of value and cost.

companies prefer to disaggregate costs into various categories, as illustrated in the two examples above.

In the examples described so far, 'cost' is treated as a categorical entity capable of disaggregation and adjustment by performing calculations to assess its magnitude at a specified point in time. Other reports feature monitoring of costs and their categories against value as within a budgetary regime.

Attributable profits

Our understanding of the term attributable profit is aided by the following definition:

> *That part of the total profit currently estimated to arise over the duration of the contract (after allowing for likely increases in costs so far as not recoverable under the terms of the contract), which fairly reflects the profit attributable to that part of the work performed at the accounting date.* (SSAP9, Part 2, para. 23)

Thus, attributable profit is *part* of total profit, and thereby constitutes a reduction in the whole. In assessing the magnitude of the part, the relevant portion of construction activity must be computed on the accounting date – a temporal component is introduced and becomes warrantably important. Thus a marker is put down reflecting the partially completed project; in assessing the part, consideration of total profit is required, because events since the estimate was prepared, the bargain was struck between the parties and work commenced may have altered the total profit expected.

The release of the profits generated by projects is an area where most companies diverged from the recommendation of the SSAP9 guidelines, which categorise projects as short- and long-term. A long-term contract 'is entered into . . . manufacture or provision of a service . . . for a period exceeding a year'. Profit can be taken at interim stages, only if the outcome can be assessed with reasonable certainty. It also suggests that if the contract is short-term, no profit can be taken until completion. Due to the scale of projects within construction, most companies release profit from their projects at an interim stage. They comply with the guidelines in preparing a statement of assessment of the outcome. A 'cost and value to complete exercise' is generally completed as part of the process of producing the reconciliation. The relative certainty of outcome obviously changes as the project progresses, and this is often reflected in the documentation supporting the cost and value to complete exercise. At the early stages, the estimate and allowances are used; as the project progresses the more deterministic resourced programmes and subcontract procurement schedules tend to be used.

Foreseeable losses

The third component (the second adjustment to 'cost') is *foreseeable losses*, defined as:

> *Where a loss on a contract as a whole is foreseen, a proportion of the overall loss, calculated either by reference to time . . . . or to the expenditure incurred may normally be taken into account . . .* (SSAP9)

This must be incorporated as a deduction, and self-evidently requires an assessment of events in the future (both their occurrence and their consequences) expressed in economic terms. This assessment therefore entails the formulation of probable events, the resource implications, and the translation of the resources into some economic measure, all of which is clearly and manifestly problematic. If an organisation were likely to make a procurement loss which has been identified during the project, a cost provision is normally made within the cost statement; the extent of the provision would usually relate to expenditure. The application of judgement, as to the scale of the loss, and when to declare the loss within the accounts, is a further area of uncertainty in cost accounting.

Claims and variations

Due to the uncertainty of the construction process, variations and claims arise on virtually every project. The evaluation of a variation or a claim is often prescribed by the contract and can vary from being based on the contract estimate to quantum merit. The difficulty in accounting for the costs and value for variations relates primarily to the likelihood of recovering costs. SSAP9 identifies that if the variation is 'approved' (there is a judgement to be made regarding the definition of approved) and an amount is to be received, but settlement has yet to be reached, then a conservative estimate should be made. Some organisations require that commercial managers identify the extent of net realisable value (NRV) that is derived from conservative estimates of claims and variation; others required the amount to be identified separately, with senior management taking the decision as to the 'safety' of taking the claim or variation to value. For an allowance to be made for value of a claim, SSAP9 identifies that:

> *Provision can only be made when negotiations have reached an advanced stage and there is evidence in writing of the acceptance of the claim and on indications to the sum entitled to. (SSAP9)*

However, the reality of construction contract administration means that often claims and variations are aggregated in order to settle accounts at the end of a project, and parties tend to be reluctant to agree to the acceptance, let alone the sum entitled to. Therefore, senior management must take account of the relationships that exist between organisations before making a judgement concerning the value of claims and variations.

The process of reconciliation

In calculating attributable profit, an accurate assessment of the value of the work at the accounting date is required, and this must also be aligned in as precise a way as possible with costs at the same date. Comparisons of cost and net realisable value must be made on a regular basis for each contract, if they are of produce accurate management accounts. Net realisable value can be defined as:

> *...the amount of which...items of stock and work in progress can be disposed of without creating a profit or loss in the year of sale* (SAPP9)

or the contractor net valuation of the works. As can be seen in the empirical section of this chapter, NRV is sometimes called net sales value (NSV). The timing of the comparisons is usually monthly and is carried out by the commercial team on the site and reported in a monthly cost meeting with senior management. The cost–value reconciliation (CVR) forms the primary means of financial communication from the site to head office. The calculation of the NRV, which is eventually aggregated into the company's accounts, commences with the contractor's external evaluation of the works as submitted to the client or client's representative, which provides an independent reference point for its eventual computation. The information supporting the CVR is often scrutinised by senior management to check that the NRV is a fair reflection of the site's performance.

In examining the data set, our attention is required where the elements listed above are present. This is not such a straightforward matter because the selected pro formas share few *common features* through which the array and detail of the instruments can be made visible. This part of the management accounting process has therefore taken guidance from regulations governing the preparation of financial accounts, in particular the valuation of work in progress.

Of secondary *interest* (not importance, since this is not yet clear) are the other data collected for presentation in the pro formas, and what they might reveal in terms of relevant indicators of performance, accomplishments and achievements on a construction project, at some specified point in time, and on completion of the works undertaken in accordance with the contract.

In Figure 18.3, the first column contains four categories/headings:

(1) Prelims and direct work
(2) Direct subcontractors
(3) Nominated subcontractors
(4) Nominated suppliers.

Each of these categories (if utilised on the particular project) will have an entry in each of the columns. Both the Valuation details and the Cost details (specifically the 'to date' figures) are calculated on other sheets in the report to make various adjustments, as previously discussed. The value and cost in the 'month' are then calculated by simple arithmetic: subtraction (e.g. gross valuation to date *less* total previous month's gross value).

A feature of the pro formas that appears in the cost section of only one example is 'provisions'. Provisions are further adjustments to account for future, foreseen difficulties on a project. Most of the pro formas, where this is a feature of practice that is reported, address this as an adjustment to 'value' and this will be discussed in the next section. For the moment, it is simply noted that some firms chose to highlight this matter as a cost item.

Cost acts as a common denominator to represent the deployment of resources in a way that permits their accounting. Cost sits on one side of the simplest equation whose object is to identify the profit from the activities of the enterprise. Because accounting regulations require the identification of attributable profit, it is necessary to present a sufficiently detailed account of the project's accomplishments for profit to be identified in such a way that the accountancy criteria will be satisfied.

The approach advocated by Barrett (1992) was to commence a calculation starting with the contractor's 'application', and then proceed through a series of adjustments (both additions and deductions) to arrive at an 'internal valuation' (Barrett 1992, pp. 25–48). Barrett's CVR pro forma completed this journey by travelling through four main sections:

(1) Application and certificate
(2) Adjustments
(3) Over valuation
(4) Provisions (Barrett 1992, p. 18).

Thus, we see here a *reconciliation* taking the form of two sets of figures, possessing authority for two parties: contractor and client (via their representative). The contractor applies for monies and the client certifies monies (following a clearly specified contractual mechanism); that these sums are not equal requires a reconciliation which spells out in clear terms how the difference has arisen. In the case of Barrett's pro forma, this difference is manifestly expected to take the form, and be capable of achieving practitioner logic, through the four sections. The interest here is to see whether such a process is undertaken in practice, and what sections are specified.

The main precept for evaluation is to assess the net value of the works that are complete at the same date as the costs are reported. As noted earlier, organisations apply a common date, sometimes referred to as the cost cut-off date, for the production of a cost statement. The other process relevant to the assessment of net value relates to the timing of receipt of income. To minimise an organisation's cash requirement, it will identify a date by which income is expected to be received, a cash cut-off date. The cash cut-off date often determines the date of the external valuation of the works, with time allowed for negotiation, certificate issue and client payment of certificates. As a result, the date of the external valuation and cost cut-off date are often misaligned and there is a need for adjustment for WIP to ensure an equitable comparison is made between NRV and costs. The two other adjustments to the external valuation, in order to arrive at the NRV, are for over valuation and under valuation of the works. The evaluation of work in progress will be considered as an over valuation or under valuation in this context.

Internal valuation

The internal valuation of a project can be considered as an evaluation of the resource allowances that the estimator made for production of the project. The allowances are usually 'internal' to the reporting organisation, and typically are for labour, plant, materials, subcontractors and on cost. It is rare for the allowances to be disaggregated to a finer detail, despite this information being collected during the estimating process. Typically, the documentation available to the commercial manager to develop the internal valuation will be a 'bill' or schedule of allowances under the above headings. The external valuation is the basis for the calculation of the project resource allowances that will be adjusted and then reconciled with the costs.

The internal valuation commences almost immediately after the external valuation has been submitted; the time taken for its compilation is dependent of the complexity and scope of the project, and two weeks for its calculation is not unusual. Often, allowances will be reallocated, as the structure of the project's organisation is not as was originally envisaged at tender stage. If variations or claims have been included within the external valuation, the commercial manager will use their judgement as to where to allocate the appropriate allowances: the principle of matching expenditure to income is generally followed; however, there are few explicit rules on this. The completed internal valuation should correspond to the external application for payment. Next adjustments are undertaken in order to arrive at the net valuation of the works.

In the example illustrated in Fig. 18.4, the CVR is presented in sections A–D and 'value' appears within section A. This part of the report is the largest of the four sections comprising 17 items plus eight sub-items. 'Internal valuation' is the sum of five items. 'Internal valuation' is a modified version of another category of valuation, and the category 'internal' operates for readers to signify its special property, aligning it with the corporate entity, as opposed to external parties (architects, clients, etc.). This example presents a complex arrangement for determining the appropriate value, the reconciled value, for use in arriving at attributable profit.

The second example contains fewer items. A clue to the direction of its calculations is provided through the incorporation of design features such as capital letters, bold font and single and double lines. Thus, important and/or prominent figures are signified through the deployment of such adornment and/or decoration, primarily to the category identifiers (the naming alongside the numerals) and also to the figures themselves. From this analysis, the following matters are capable of separation (they stand out) from the entire section:

- Gross submitted
- Total net income
- (A) Total gross income
- Work in progress
- (B) Net WIP
- (C) Total WIP
- True value (A + C).

We can immediately see that True value is the sum of two items, leaving (B) to fulfil some other purpose which is not apparent from the remaining content of the pro forma.

The management accounting area of 'Value' was examined in two pro formas, as part of the examination to explicate cost and value reconciliation pro formas, particularly in relation to 'attributable profit' (itself an agglomeration of actual and forecast profitability). In the third example, an extract is taken from a CVR comprising five pages, three of which are devoted to 'Internal valuation'. In common with the earlier examples, the sequencing of information commences with monies applied for '1. External value of production' and culminates in '10. Value of production to Board Report 1'. Value is bifurcated into 'External' and another category which is unspecified but

A. Work in progress – reconciliation	Previous	Current
1. Final invoice/applic. net of retention (NOR) and claims		
2. Paid claims (NOR)		
3. Less adjustments		
Payments not anticipated		
(a) Contract Work		
(b) Variations d/wk fluctuations		
4. Net external valuation		
Cash receivable (1 + 2 + 3)		
5. Less cash received at		
6. Work in progress receivable (4 – 5)		
7. Add? anticipated retention on 4		
8. Add work to be applied for (including retention)		
(a) Contract work		
(b) Variations d/wk fluctuations		
Total item 8		
9. Add work completed not payable		
(a) Contract work		
(b) Variations d/wk fluctuations		
(c) Off site materials		
(d) Preliminaries		
Total item 9		
10. Work: in progress (6 + 8 + 9)		
External valuations		
Less provisions		
11. Preliminaries		
12. Over value		
13. Payments certified on account		
14. Maintenance known problems		
(a) Maintenance retention		
(b) Remedial maintenance		
(c) Future losses		
15. Other		
(a) Redundancy/severance		
(b) General services		
16. Total provisions (11 + 12 + 13 + 14 + 15)		
17. Internal valuation (4 + 7 + 8 + 9 – 16)		

Figure 18.4 Work in progress reconciliation.

consists of a figure submitted in a report to the Board of Directors; thus a distinction is seeable but unclear.

A number of forms identified other adjustments to value; the main headings were over measure and under measure which were categorised into on cost in advance and/or arrears, measured works and items in dispute.

Over measure

The commercial benefit of over valuing the works is primarily to enhance cash flow, and the information asymmetry that exists between the parties in most procurement arrangements serves to support its practice. The client's representatives are aware of the practice and often apply hidden retentions to balance against over valuation. Over valuation can take place due to unbalancing of tenders, recovery of enhanced margin due to variations, claiming for works before their actual completion, and claiming for questionable entitlements. The over valuation of the works can be defined as application for monies in advance of their eligibility. The allowances for these items are deducted from the gross internal valuation of the works to give the net sales value. In one instance these were categorised as on cost in advance, measured work, and disputed items.

On costs in advance

On cost can be defined as project contribution, and can include profit as well as overhead contribution. All evaluation of work can include an element of on cost; therefore, if works that have been claimed for that have not yet been carried out there is an element of on cost generated in advance. If the valuation of the works is not adjusted for on cost generated in advance, an over-reporting of the allowances for overheads and profit will occur, giving an over optimistic assessment of the project's position. A range of techniques for the evaluation of on cost in advance were observed in the forms. Usually the on cost associated with the net valuation of the works was deducted from the gross on cost which will include on cost associated with variations, claims, over measure and preliminaries. The residual amount will then be the on cost in advance of entitlement. This was considered as an over measure and was shown as such.

Measured work

The complexity of valuation of measured works, which will include work completed by the reporting organisation and a range of subcontractors' works, means that a judgement has to be applied by all parties. The information asymmetry that exists in this case (the constructing organisation often has more information on the extent of completeness of its works than the party valuing the work), means that there is an element of performance ambiguity, i.e. the party valuing the works has imperfect information and does not have the resources available to check the claiming organisation's valuation. Consequently, an over application of measured works can take place.

Disputed items

Almost all contracts have explicit procedures for managing the uncertainties and variations that arise on contracts. The definition of a claim and variation is usually clear; notices or instructions usually precede the events and they are valued using a set of explicit processes which refer to either the original priced documentation or market rates of quantum merit. The evaluation variations and claims are problematic; as stated earlier, the internal allowances associated with a variation or claim are allocated as either labour, plant, materials, subcontractor or on cost, while the senior management have to decide when and how much of a claim and the associated allowances to take to value. Some organisations net the internal valuation of all variations and claims that are in dispute, and in one instance ranked them as likely, probable and unlikely.

Under valuation of the works

In a similar fashion to over measure, an under measure schedule was produced by some organisations which identified those items where work was completed and costs incurred but which had not been included in the valuation application. The inclusion of a large element of under measure is a signal of failure in the commercial management system to identify work completed or an entitlement. The allowances that are derived from the under measure calculation process are added to the internal valuation to arrive at the net internal valuation of the works. Work in progress is defined as: work that has been carried out in the period between the cost cut-off date and the valuation date, if the cost-cut off date is after the date of the external valuation. The resultant WIP is treated as an under measure. This is the usual position on the reporting of projects, and can cause the commercial manager a lot of administrative effort as it entails two evaluations of the work, one at the date of external valuation and another at the date of the cost cut-off. Normally the WIP is only adjusted for internal labour and materials, as subcontractor liabilities are taken as costs and are aligned with the external valuation.

The CVR documentation used by one of the organisations was particularly comprehensive. The following pro formas were found in the most comprehensive reports examined:

- Internal valuation broken down into a set of allowances
- Over measure schedule
- Under measure schedule
- Cost statement
- Subcontractor liabilities
- Buying reconciliation
- Materials reconciliation
- Cost to complete exercise
- Estimated final account/profit margin
- Architect's/engineer's certificate
- Narrative explaining variances.

The reconciliation was normally undertaken after the net internal valuation had been completed and the cost statement produced. The commercial manager assessed the completeness of the costs statement to identify and correct obvious errors, which may arise from misallocation of costs or over- or under-assessment of accrual values. Then a preliminary reconciliation was generally carried out to identify any obvious misallocation of allowances. Some internal valuations were very complex, as the original documented allowances were significantly different from the project costs, and re-allocation of allowances was often carried out to enhance commercial return by the estimating department. If, after this initial adjustment, large variances were identified this initiated a requirement for a detailed reconciliation at a resource level of a particularly problematic activity.

The NRV may be less than cost if there has been an increase in cost, physical deterioration, errors in production or errors in purchasing. The buying or procurement reconciliation would identify those allowances that may yield a profit or loss due to purchasing. A materials or resource reconciliation would identify whether production output from the resources was within the standards identified within the estimate. The narrative, accordingly, should refer to these other sources of information to triangulate the CVR. In practice, due to the imperfections of the information supporting the processes, the lack of time available and the manipulation of the information, this triangulation rarely took place.

The construction commercial manager will have more information to make judgements upon the areas of commercial uncertainty than the senior management, who will be reviewing the CVR statements and may wish to retain decision-making autonomy and to maintain a contingency to allow for some future flexibility. The extent of autonomy and the size of contingency will vary from project to project. However, some of the factors that may influence the size of contingency would be: extent of future losses, robustness of cases for claims or variations, client team's approach to the valuation of work, the extent of exposure to claims from internal subcontractors, the ability to contra charge costs.

The ethnography of project financial performance reporting is an area of construction management that has, as yet, been under researched. Senior management may well be of the view that they would prefer to be told the full picture and that they would use their experience to make decisions on the amount of value to take for items in dispute. These conflicting views of the CVR process lead to one party obscuring aspects of this picture to retain autonomy and flexibility in decision making, and the other party attempting to uncover the true picture.

Obscuring the picture

The commercial manager can choose from two options to obscure the financial performance of the project, either enhancing the project's costs or adjusting the project's valuation. To enhance a project's costs, the site reporting team generally choose a cost area which they can adjust without monies flowing to another party; in many cases, this tends to be subcontractor liabilities. The subcontract accruals which form the basis for the subcontract cost headings can be artificially increased, thus hiding a contingency that can be used at a later date. A corresponding increase in the subcontract allowance will ensure that there will be no unexplainable subcontract variances. Another manipulation of the cost statement may be to include a cost provision for future losses. This, however, is more observable and more likely to raise senior management's attention to potentially available additional on cost. The adjustment of value may be another tactic used by the site team: for example, increasing the extent of over measure, reducing the allowances that are taken to value, and then under reporting the profitability of projects. The most comprehensive cost reports reviewed required the date and amount of the last subcontract certificate to be included. This suggested an organisation that was aware of the potential for misreporting and, as a result, a corresponding additional reporting burden was placed on the site team.

Uncovering the true position

Obviously, senior commercial managers of organisations who have had extensive experience of obscuring the picture when they were site-based, are aware of the process and in some cases tacitly concur with the underlying reasons for the practice. This concurrence signals an acceptance that site teams require an element of flexibility and autonomy in decision making; however, most senior managers would require an indication of the size of the contingency developed. The forms used for the reporting of costs and value are therefore designed to provide an accurate representation of financial performance, a means of auditing the site team's decision making and also the provision of records for company accounts.

Conclusion

This chapter has shown that organisations vary widely in their practice of project financial reporting in terms of the extent, content and use of reports. As a result, the administration costs are variable. Most of the information generated by the organisations studied was reported within a monthly cost report, while little analysis of the data to identify trends of cost flow, cash flow or asset specificity of subcontracting resources was carried out. When analysing information systems within contemporary organisations one should be aware of the divergent interests and perspectives that underlie such systems. The exercise of managerial power and the political action that occurs during the CVR process is highly complex, and attempts to observe it in action are often thwarted by individuals attempting to conceal their motives, as well as the commercial interest of organisations.

Walsham (1993) suggested that accounting information systems are only one way of looking at the world and that they institutionalise certain boundaries and ascribe a privileged position to numeric data, which can exert pressures that may be to the detriment of communities of practice. A research agenda that seeks to define these boundaries and describes how site communities protect their autonomy by signalling, using codified systems and how management use and respond to these signals, may help uncover the richer pictures that exist behind the forms. It could be useful to start by attempting to identify the factors that influence the design of systems, establish the compliance with organisational processes and investigating the opportunities to use the data generated to develop data sets that may be used for future predictive purposes.

The effectiveness of modelling construction costs is reduced due to the entanglement of cost and price data. The development of processes for data capture and reporting that uncover resource costs, which are based on greater transparency, will increase the portability of information from sites and organisations, and allow construction design and production economics to benefit from the massive improvements in information analysis. Until these processes are developed, the systems used for the design cost modelling and resource prediction will remain largely judgemental and based on information that has been manipulated to enhance the asymmetry between parties.

References

Accounting Standards Board (1975) *Statement of Standard Accounting Practice, No. 9*. (Amended December 1998.) The Accounting Standards Board, London.

Anderson, R.J., Hughes, J.A. & Sharrock, W.W. (1989) *Working For Profit: The Social Organisation of Calculation in an Entrepreneurial Firm*. Avebury, Gower, Aldershot.

Baccus, M.D. (1986) Multiple truck wheel accidents and their regulations. In: *Ethnomethodological Studies of Work* (ed H. Garfinkel), Routledge & Kegan Paul, London, pp. 20–59.

Barrett, F.R. (1992) *Cost Value Reconciliation* (2nd edn). Chartered Institute of Building, Ascot.

Cooke, B. & Williams, P.N. (1998) *Construction Planning, Programming and Control*. Macmillan, London.

Dennis, A. (2003) Skepticist philosophy as ethnomethodology. *Philosophy of the Social Sciences*, **33**(2), 151–173.

Garfinkel, H. & Sacks, H. (1986) On formal structures of practical actions. In: *Ethnomethodological Studies of Work* (ed H. Garfinkel), Routledge & Kegan Paul, London; pp. 160–192.

Gowthorpe, C. (2000) Keeping up with the standards. *Management Accounting*. Chartered Institute of Management Accountants, London.

HMSO (1989) *Companies Act*. HMSO, London.

Johnson, H.T. & Kaplan, R.S. (1987) *Relevance Lost. The Rise and Fall of Management Accounting*. Harvard Business School Press, Boston, MA.

Sizer, J. (1986) *An Insight into Management Accounting*. Penguin, London.

Upson, A. (1987) *Financial Management for Contractors*. BSP Professional Books, Oxford.

Walsham, G. (1993) *Organizational Metaphors: Interpreting Information Systems in Organizations*. John Wiley and Sons, Chichester.

Walker, I. & Wilkie, R. (2002) *Commercial Management in Construction*. Blackwell Science, Oxford.

Wittgenstein, L. (1967) *Philosophical Investigations* (3rd edn), Blackwell, Oxford.

Endnotes

[i] The origins of the modern-day Chartered Institute of Management Accounting are traced by John Sizer (1986, pp. 15–17).

[ii] Johnson and Kaplan (1987, p. 12) identify eight main practices in the management accountant's tool-kit: *cost accounts* for labour, materials and overhead; *budgets* for cash, income, and capital; *flexible budgets, sales forecasts, standard costs, variance analysis, transfer prices*, and *divisional performance measures*.

[iii] The first Financial Reporting Standard was issued in 1991. They are replacing Statements of Standard Accounting Practice, which were first introduced in the early 1970s.

[iv] Membership of the Consultative Committee of Accountancy Bodies (CCAB): Institute of Chartered Accountants in England and Wales (ICAEW); Institute of Chartered Accountants in Scotland (ICAS); Institute of Chartered Accountants in Ireland (ICAI); Association of

Chartered Certified Accountants (ACCA); Chartered Institute of Management Accountants (CIMA); Chartered Institute of Public Finance and Accountancy (CIPFA).

[v] 'Summarisable' here is intended to capture the simplest features of the twelve reports where the myriad array of presentations makes such summarising problematic. Such features are the first, *gloss* noticings on examining the reports, and furnish a starting point for the analysis. They are in some sense (and it is their *sense* which is the object of the current investigation) comparable with the 'glosses' of Garfinkel and Sacks (1986), elegantly described as the 'excavation and description' of interactional work by Dennis (2003, p. 155) who refers to members' formal accounts as 'reports, descriptions, scientific findings, and so on'; here the study of management accounting formats and their *interactional* work, falls neatly within this class of eminently suitable materials for a praxiological study of construction management.

[vi] The degree to which reports reflect events contemporaneously is referred to by Anderson *et al.* (1989, p. 129) as the 'lagged sense of now'. Their study of a firm operating fast-food outlets revealed that 'to those engaged in measuring the company's profitability, there seems no other sensible way of dealing with their operational and organisational contingencies than freezing activities in fortnightly blocks. It is only by doing this that "realistic" measures of sales, purchases, labour costs, overheads, and so on can be obtained.'

[vii] Various sizes (number of pages) were found amongst the current sample: 1 page, four reports; 2 pages, one report; 3 pages, one report; 4 pages, three reports; 5 pages, two reports; 10 pages, one report.

19 Afterword – Identifying and Defining a New Interdisciplinary Research Agenda for Commercial Management

David Lowe

Introduction

As indicated in Chapter 1, the past two decades have seen an increase in the number of individuals within organisations assuming the title Commercial Manager, so that commercial managers can now be found across a spectrum of industries, especially those that are predominantly project based. In the UK construction industry, for example, a clear trend can be observed of the role of the traditional contractor's quantity surveyor evolving into that of a commercial manager.

Although Chapter 1 established some differences in approach and application both between and within industry sectors, there was still sufficient similarity and synergy in practice to identify a specific role of commercial management in project-based organisations. Commercial management was defined as:

> *The management of contractual and commercial issues relating to projects, from project inception to completion.*

This book consists of 18 original contributions, all addressing specific topics pertinent to commercial management, as defined above, within the context of project orientated and predominantly business-to business (b2b) centric organisations: the commercial management of projects. The contributors have, from their own perspective and area of expertise, presented key issues relevant to the development of commercial management as a discipline, as well as to the practice of individual commercial managers. The topics cover the whole spectrum, from describing fundamental theoretical explanations to providing normative advice to the practitioner. Each chapter concludes with a section highlighting areas for further research within the particular subject area, and all are valuable contributions in their own right. This concluding chapter, therefore, should be seen as an attempt to summarise and collate these disparate research areas. Also, it seeks to present a justification for the establishment of a research network and an associated research agenda to support both the practice and the professionalisation of commercial management. This research is seen as a first step in developing a body of knowledge for commercial management, and parallels are drawn with the emergence of project management as a widely recognised core competency, and its moves towards professional status.

Professionalisation of commercial management

Current professional standing of commercial managers

Apart from the construction sector, where there is a tendency for commercial managers to be corporate members of the Royal Institution of Chartered Surveyors (RICS) and/or the Chartered Institute of Building, there is a distinct lack of professional representation and recognition of the commercial management function. Increasingly, the lack of status for commercial management has led both individuals and the organisations that employ commercial managers to explore ways of addressing this issue, for example, through the development of an MSc programme in Commercial Management and more recently an MBA for Commercial Executives. This impetus for raising the status of commercial management in many ways parallels the emergence of project management in the latter part of the twentieth century. Consequently, it would appear that we are observing the embryonic stages of the professionalisation of the commercial management function.

Foundations of a profession

Whilst there is some debate over what constitutes a profession, the importance of a profession's knowledge base is widely acknowledged (Eraut 1994). Moreover, as Eraut goes on to state, 'the power and status of professional personnel is considerably influenced by the degree to which they can lay claim to unique forms of expertise and the value placed on that expertise'. Unlike medicine or law, commercial management has no discrete knowledge base; as with project management and quantity surveying, for example, it is an amalgam of topics, drawing upon research and emerging practice from a variety of subject areas. To appropriate Morris (2004), if there really is a discipline of commercial management that can be delineated, 'then there needs to be some knowledge about it that can be articulated with a reasonable degree of robustness'.

Similarly, Crawford (2004a) asserts, 'a body of knowledge, standards and related assessment and qualification processes can therefore be seen as essential building blocks in the formation and recognition of a profession'. Further she suggests that: the definition of a distinct body of knowledge and the development of standards, providing a baseline for competence, are important building blocks in professional recognition for individuals and for an emerging profession. In Crawford's model, research forms the foundations of a profession, on which a body of knowledge (the plan and structure) is developed; this in turn informs the development of standards, on which education and training can be developed and lead to qualification. Bodies of knowledge are therefore used to provide a consistent structure for professional development programmes, the certification of professionals and accreditation of education programmes. Moreover, within the field of project management, they have been a significant feature as the discipline aspires to recognition as a profession.

Bodies of knowledge

Frequently, mature professions, to protect their position, seek to define themselves in terms of their core skills/competencies and/or knowledge base (see, for example, RICS 1992: which sought to establish the core skills and knowledge base of the Chartered Quantity Surveyor). Likewise, project management associations, for example, the International Project Management Association (Caupin *et al.* 1998), the Project Management Institute (PMI 2000) and the Association for Project Management (APM) (Dixon 2000), have defined bodies of knowledge (BOKs), in order to establish in what a project management practitioner should be knowledgeable (i.e. a set of professional competences) to underpin professional certification. Boyatzis (1982) suggested that a discipline's BOK is one of the fundamental elements of competencies (competence, being role-specific and comprising the knowledge, skills and behaviours needed to perform the role, are therefore linked to individual behaviour and task performance). According to Gale (2004), standards relating to competence fall into two main areas: those describing what a practitioner is expected to know and those describing what practitioners are expected to be able to do. The latter primarily takes the form of performance-based or occupational competence standards. He states that the assumption behind their development and use (in the case of project management) is that they describe the requirements for effective performance in the workplace. However, the purpose of BOKs extends beyond providing a basis for establishing competence to practice. According to Morris (2001), they should reflect the purpose and provide a set of topics, relationships and definitions of a particular domain, and hence a distinct professional discipline.

Arising from the development of the APM BOK, Morris (2004) refers to the transition of project management to the 'management of projects' (Morris 1994), which takes into account the projectisation of organisations. Within the management of projects model, he sees project managers as the vehicle through which not only projects and portfolios and programmes of projects are delivered, but also strategy. Morris's stance is somewhat expansionist, positioning project managers at the centre of the management of projects. Further, in Morris's definition of project management and the APM BOK which he helped develop, there is an attempt to claim the commercial function as part of or subservient to project management. Our observations of the practice of commercial management across several industry sectors (Chapter 1) leads us to the conclusion that, far from being subservient, commercial management is a distinct although parallel process within commercial, b2b organisations.

Certification of professionals

Regarding the certification of professionals, the RICS, for example, to support their practice qualifications, have published requirements and competencies to underpin the assessment of professional and technical competence (RICS 2002). These generic competencies are used to assess a candidate's competence to carry out the work of a qualified surveyor, and are classified under three distinct categories: mandatory, core and optional, and are defined at three levels of attainment:

- *Level 1*: Knowledge and understanding
- *Level 2*: Application of knowledge/understanding ('doing it')
- *Level 3*: Reasoned advice and/or depth of technical knowledge.

Developing a body of knowledge for commercial management

The research base and BOK of a discipline are therefore intrinsically linked: the establishment of a research agenda and BOK is fundamental to the professionalisation of the commercial management function. In proposing the development of a BOK for commercial management, the following aims, adapted from those set by UMIST's Centre for Research into the Management of Projects (Morris *et al.* 2000; Morris 2001), when undertaking the development work for the Fourth Edition of the Association for Project Management Body of Knowledge, are suggested:

- Ascertaining and classifying the topics that commercial management professionals deem necessary to be known and understood by those claiming to be competent in commercial management
- Defining what is meant by those topics at a generically useful level
- Developing the body of literature that supports these topic areas
- Developing a body of knowledge that best represents commercial management.

The International Project Management Association (IPMA) classifies project management under four levels:

- Project director
- Project manager
- Project management professional
- Project management practitioner.

Similarly, in commercial management, several organisations (for example, Rolls Royce and BT) have identified and defined generic role and competencies at: commercial director, commercial executives, commercial manager, commercial adviser and commercial officer level. This initial work may inform the development of a commercial management body of knowledge.

A community of practice for commercial management

Returning to the need to raise the status of commercial mangers, as suggested in Chapter 1, an obvious development would be the establishment of a generic institute or association to represent commercial managers, disseminate good practice and certify competency to practice. However, these formal structures might perhaps represent nineteenth century ideals more than the current aspirations of commercial management practitioners. Possibly a more progressive response would be to utilise the Internet to establish loosely based 'communities of practice' (Lave & Wenger 1991) as a vehicle to disseminate good practice and cutting edge research pertinent to commercial managers. In support of this proposition, Crawford (2004b) notes that, between the members of the global community of project management practice, the majority of activity promoting global communication and co-operation (the free sharing and exchange of ideas) is happening outside recognised organisations, through online discussion groups and unofficial initiatives. However, in order to support a 'community of practice' there is still a requirement to establish a research foundation by co-ordinating the disparate strands of research currently being undertaken and to produce an associated BOK to underpin the development of commercial management as a distinct discipline.

Research agenda

The following section provides a synopsis of the research topics proposed in the preceding chapters. While they cover diverse areas, there is much synergy in the research avenues suggested, each warranting the consideration of both academics and practitioners. As outlined above, research is seen as the foundation of a profession; therefore, this section seeks to classify and collate research pertinent to commercial management and to develop, based on leading edge practice, a research agenda to support the development of the professionalisation of the discipline. When preparing a research agenda for commercial management, all the proposed issues should be considered; however, it is apparent that certain topics are more pertinent than others. A summary of these main issues is provided later in the chapter.

External milieu

Chapter 2: Competition

Hedley Smyth outlines a number of specific research topics pertinent to the area of competition, these include:

- The analysis of competitive processes, taking into account the size of suppliers and project complexity, specifically addressing the dynamics of moving between levels in a particular market. This would necessitate defining the levels and thresholds more precisely and identifying the barriers of entry at each level.
- An investment analysis of structural solutions into market divisions based upon procurement option.
- The analysis of outworking of market restructuring of firms into asset, income and mega-subcontracting, including comparisons of the features and experiences across different project industries.
- Investment analysis, at a detailed level, of performance improvement, including service and/or product differentiation, management of leanness, agility and supply chains, and management of other competencies such as organisational learning/knowledge management (OL/KM), relationship marketing and management (RM) and emotional intelligence (EI).
- The analysis of continuous improvement areas, particularly in relation to concepts such as OL, KM, RM and EI, and the area of human systems

within and across projects, covering the main office–project interface and the client–contractor interface.

- An investigation into the dimensions to change the image of complex project sectors as working environments in order to improve capacity to recruit in the future.
- Research into the dimensions to change the culture of complex projects from a task orientated culture that is based upon personalities and blame to one based upon investment in systems, competencies and technologies for performance improvement.

As Smyth states, there are many detailed issues within each of these main areas, some of which may be elevated from the tactical to the strategic over the next decade, so awareness will be needed to adjust research to emergent trends. The strategic concern is to be as risk averse as possible, and so structural and process shifts recommended from research output must embrace this reality, yet demonstrate clearly the necessity and inevitability of considerable change.

Chapter 3: Culture

Richard Fellows considers that the agenda for researching culture and its effects remains extensive and diverse, reflecting the all-pervading nature of the subject and the inherent dynamism of people. At a fundamental level, research into culture remains differentiated into two primary approaches: anthropological and sociological. In construction, the International Council for Research and Innovation in Building and Construction (CIB) has established a task group (TG-23) to study culture in the industry globally. The research agenda of the task group for 2003–2005 is to:

- Extend the studies of TG-23 to research cultures existing in the construction industry worldwide from societal and organisational perspectives.
- Stimulate and facilitate international exchange and collaboration for studies into re-engineering processes adopted by construction organisations to examine their culturally dependent consequences.
- Research cultural facets of the construction industry to determine feasibilities of changes to improve performance.
- Assemble a 'cultural inventory' to assist organisations in adapting to new cultural situations across construction industry sectors and countries.

This would entail two aspects – reviewing the current applicability of research profiling national cultures, and by using the common instrument of the organisational culture assessment instrument (OCAI) questionnaire to research cultural profiles of construction organisations and sectors in various countries.

Chapter 4: Procurement in the context of commercial management

David Langford and Mike Murray maintain that the issue of procurement has been at the forefront of commercial management in many industries, the central issue being: how do we procure large capital projects on time, on cost and with an appropriate level of performance? Understanding the supply chain process, they suggest, is one of the key competences in the commercial management of most project-based industries.

They utilise Morgan's (1986) idea of metaphors to explain the movement from traditional procurement to partnering and performance based contracts: a transition from the machine metaphor, via the organic metaphor and the political metaphor, to the cultural metaphor. Future procurement practice, they speculate, will embrace the growing concern for corporate social responsibility, the issue of environmental audits and the incorporation of organisational learning and knowledge management. Langford and Murray view this latter development as the catalyst for a new professional within the project team: the 'knowledge manager'. Research, therefore, should be undertaken to address these emerging themes.

Corporate milieu

Chapter 5: Corporate governance and shareholder value

Marc Goergen and Luc Renneboog suggest that future research into corporate governance (an intricate amalgam of internal, external and regulatory mechanisms) should:

- Take into account that the efficiency of these corporate governance mechanisms should not be studied in isolation, as some mechanisms are complements or substitutes to one another.
- Bear in mind the regulatory framework (in terms of corporate law, stock exchange regulation, codes of corporate governance) as well as the enforceability of existing rules because they determine the effectiveness of corporations in generating shareholder value.

Chapter 6: Strategies for solutions

Andrew Davies and Michael Hobday suggest that the rise of integrated solutions calls for more in-depth research on the implications of this new activity for project activities and project business organisations. For example, the traditional project life cycle now extends over many years or even decades and includes important pre-bid and post-project phases of activities. Their research points to four key phases in an integrated solutions life cycle:

(1) Engaging with the customer in high-level strategic negotiations, often before an invitation to tender has been issued.
(2) Working closely with the customer to develop a value proposition during a bid or offer phase.
(3) Project managing the systems integration process.
(4) Operating a product or system during a specified contractual period.

Future research, they suggest, is required to analyse and explain how suppliers and customers are working together in strategic partnerships to identify, capture and share in the added value created over such an extended life cycle.

Chapter 7: The effectiveness of marketing spend

Christopher Preece, Krisen Moodley and Michael Brown hold that the future direction for marketing, within the construction sector, is for firms to take a more planned approach, being able to identify objectives early and evaluating the effectiveness of marketing strategies in generating business for the organisation. They recommend that companies should attempt to model or develop frameworks to identify the most appropriate marketing tool for a given situation. They suggest that researchers should:

- Develop marketing metrics to address the issue of value for money, incorporating a series of performance indicators and the application of benchmarking to cover the whole range of marketing activities.
- Investigate the measurement of marketing spend and its effectiveness, addressing factors such as customer service and client satisfaction, and the impact of these areas on marketing and organisational performance.
- Investigate the marketing and promotional mixes used by companies, including the role of marketing in relation to other related functions such as corporate communications, and their impact on the brand reputation of the firm.
- Investigate brand building and communication strategies, brand protection and reputation management.

Chapter 8: Trust and commercial managers: influences and impacts

Will Swan, Peter McDermott and Malik Khalfan advocate further in-depth research in order to develop more effective trust building relationships within the construction industry as a whole. Simultaneously, there is also a strong need to define clearly the roles and responsibilities of commercial managers in the different organisations involved within a construction project to achieve a no blame, truthful, reliable and responsive culture within the construction industry. An agenda for further research could include investigating how:

- The experience of commercial mangers in interacting with one another can contribute towards better trust building.
- The high levels of uncertainty are addressed in a new relationship in order to effectively manage risk.
- Procurement mechanisms can be a key driver of trust within the construction industry.
- Organisational complexity can create issues for the management of trust and how they could be resolved.
- The role of the commercial manager influences the development of trust.

Chapter 9. Outsourcing

Jan Bröchner presents the rudiments of a simple theory for outsourcing: information costs and distance have implications for the cost of enforcing contractual obligations. What this theory lacks, however, are tools to distinguish how markets for goods and services economise specialised knowledge. He considers that we need a more straightforward understanding of the emerging global patterns of accumulated knowledge and how it is bundled for a heterogeneous set of customers. Government regulation, not least in the fields of competition and the protection of intellectual property, is a strong formative influence for this international division of human skills.

The strongest driving force for outsourcing remains the development of information and communication technologies that shrink distances. Likewise, the new global patterns of production can be expected to mirror technology, markets and how both new and old welfare societies are transformed. However, we should not exclude the possibility that the trend to outsource will be weakened, not because new barriers are erected, but because old ones are torn down. He recommends further research to:

- Investigate why external suppliers have the potential to be better producers, and the possibility that actual patterns of outsourcing result from both features of the companies that outsource and conditions in the external supplier market.
- Examine how the business cycle is related to a willingness to outsource.
- Examine the transfer situation and investigate how to retain and develop an efficient procurement function after the majority of those with the appropriate skills have left the outsourcing organisation.
- Analyse the fundamental differences between the various types of services and types of industries, so that the best objects of comparison can be more readily identified. This avenue of research should provide a sounder basis on which to design service level agreements and other contractual documents and practices.

Projects milieu

Chapter 10: Management of uncertainty

Kalle Kähkönen presents six important areas of risk and opportunity management, which demonstrate missing knowledge, and are thus also subjects for further research and development. While he presents some solutions for filling these knowledge gaps, he suggests that more attention should be directed to these areas to create a better understanding of the nature of managing risks and opportunities. These six areas are:

(1) Developing a conceptual foundation based on localised definitions of risk and opportunity.
(2) Developing a holistic paradigm for enterprise-wide risk and opportunity management by scaling risk and opportunity management vertically in organisations.
(3) Developing core and accessory processes for risk and opportunity management.
(4) Developing the entity of principles for risk and opportunity identification.
(5) Investigating the causes of risk and opportunity modelling inconsistency.
(6) Developing methods of capturing risk knowledge and creating risk knowledge models.

Chapter 11: Conflict management and dispute resolution

Peter Fenn holds that if we seek to avoid disputes it is self-evident that we seek to predict their occurrence. He draws a parallel with the concept of preventative medical research, in particular, aetiology (the study of causes). The adoption of an aetiological approach, he suggests, might throw new light on disputes and lead to the development of predictive tools. Therefore, research should seek to:

- Develop a list of causal criteria associated with dispute occurrence by adopting an aetiological approach.
- Investigate the relationships between these causal criteria and dispute occurrence.
- Develop a predictive model for the occurrence of disputes to facilitate their avoidance based on these causal criteria.

Chapter 12: Performance measurement

Malcolm Horner lists, as he discerns it, the challenges for the research community, in the form of a series of interdependent questions, in the hope of stimulating the design of suitable research programmes. Collectively, these questions generate the following research topics to:

- Investigate how a measurement-averse industry may be encouraged to develop, use and preferably share, a consistent set of performance indicators.
- Develop project models which are sufficiently simple to encourage data collection, yet sufficiently accurate to produce data of real value.
- Develop a hierarchical system of key performance indicators (KPIs) that permit measurements at the project level to be pooled into national statistics.
- Investigate the design of specific measures for 'lean' performance, stakeholder and user values, value-stream mapping, performance of members of integrated supply chains, the influence of the design and/or product and/or service on its users, whole life performance, operating and maintenance environments and their effect on functionality, motivation and skill of operatives and mangers, and buildability.
- Examine the possibility of creating single units of measurement for the three dimensions of sustainability: economic, environmental and social performance. Likewise, develop a single organisational business measurement system to account for finance, cost, time and productivity, linked to bonus, wages, planning, estimating and risk management, and which addresses issues of health and safety, the environment and sustainability.

Chapter 13: Value management of complex projects

Interestingly, John Kelly commences his chapter by asserting that commercial management is an ideal vehicle to facilitate the introduction of value management and value engineering into current construction processes. He considers that research work in value management and value engineering is required to:

- Investigate the application of value management and engineering in the manufacturing areas of customer orientated product innovation and in the performance specification of production space.
- In construction, further investigate the application of value management and engineering in the areas of strategic and project briefing.

- Investigate the utilisation of value management in the public sector via the theories and applications that lie behind Best Value legislation.

Project milieu

Chapter 14: The governance of project coalitions – towards a research agenda

Graham Winch presents an approach to the commercial management of complex projects that represents a break from the traditional analysis of standard forms of contract and procurement routes which currently predominates. Instead it takes a transaction cost perspective and links the analysis of procurement routes to organisation theory. He hopes that the proposed framework will help crystallise this change, and that the propositions suggested will provide an initial stimulus for further research in commercial management. Specifically, he suggests that research should be focused to:

- Analyse transaction costs between discrete alternatives; this would require obtaining good quality data on the 'costs of doing business'.
- Understand the drivers of transaction costs, bearing in mind that in a project context these are not necessarily independent, and that appropriate governance arrangements can motivate the search for reduced production costs.
- Analyse alternative governance modes in project-based industries to include both transaction and comparative production.

He concludes that much more rigorous data collection and formalisation of the propositions are required before we can be confident of the benefits of the proposed framework for understanding the governance of project transactions.

Chapter 15: Contract management

Will Hughes suggests that a future research agenda, within the context of construction contract management, should:

- Examine the changing nature of professional roles, new financial structures for companies in the supply chain, and new contractual mechanisms to give effect to the radical new business models that are emerging at the leading edge of practice.
- Investigate the way that institutional forces might be balanced with the wider interests of buyers and suppliers, for example, by examining the changing role of architects and the liability for the design and building process within the construction sector.
- Analyse the financial structures of suppliers to develop the required business models to support this form of performance liability.
- Investigate the relationship between marketing and quality management to promote more appropriate management techniques in the construction sector.
- Investigate the extent to which more effective models of client satisfaction and performance guarantees can be adopted, rather than rely on documents that enable suppliers to demonstrate that they have discharged their contractual obligations with little residual liability.

Commercial management, he believes, brings a much needed reality check to the process of developing effective business models for a future construction sector.

Chapter 16: Bidding

David Lowe and Martin Skitmore maintain that despite the longevity of the research into developing bidding (price optimisation) models, further work within the wider area of bidding is still required. To underpin the development of decision support tools for both the decision to bid (d2b) and bid submission, they state that research should be concentrated on developing a suitable underlying theoretical framework (as suggested by Runeson & Skitmore 1999). In terms of the development of decision support tools for both the d2b and bid submission, they propose that model developers should:

- Consider the practical application of bidding models as well as the academic rigour of the modelling approaches adopted.
- Investigate the reticence of decision makers to utilise existing decision support models, in the hope that innovative models will be developed that are both suitably robust and theoretically sound to model accurately the complexities of the decision domain, but which are practical enough to be accepted by practitioners.
- Develop project selection (d2b) models to include objective data, such as the profit and non-monetary objectives achieved by specific project type, size, etc. This would necessitate the incorporation of data generated via cost value reconciliation.
- Develop bid submission support models that incorporate 'best value' and alternative award criteria, in addition to profit maximisation, and allow the assessment of alternative value and price combinations, rather than solely generating an optimum bid price or mark-up.

Chapter 17: Strategic purchasing and supply chain management in the project environment – theory and practice

Andrew Cox and Paul Ireland suggest that, in order to implement effective supply management in the project environment, future academic research needs to consider a number of key conditions. They hold that successful relationship management requires the buyer and supplier to:

- Understand the power and leverage circumstances that they are currently in, and how these can be modified in the future.
- Have commensurable commercial and operational strategies.
- Differentiate between alternative relationship management styles and select the one that is the most appropriate given the power circumstance, commercial goals and operational strategies of each party in the transaction.
- Have top management support and internal buy-in.

These conditions, they propose, highlight the need for academics and practitioners to understand the commercial and operational circumstances that exist between buyers and sellers in transactional exchange, the distinct and often conflicting goals of both parties when they enter into business relationships, and the appropriate supply management approaches given these circumstances. This fundamental understanding of the dyadic buyer–supplier exchange and relationship alignment needs to be extended across all supply chain relationships and take into account whether project demand is one-off or serial.

They assert that future research should be undertaken that enables academics to assist practitioners discern the most appropriate relationship management strategies available given the power and leverage circumstances they are in.

Chapter 18: Signals from site – embodied logic and management accounting on construction projects

Andrew Ross and David Hugill quote Walsham (1993) who suggested that accounting information systems are only one way of looking at the world and that they institutionalise certain boundaries and ascribe a privileged position to numeric data which can exert pressures that may be to the detriment of communities of practice. They propose a research agenda that seeks to define these boundaries and describe how site communities protect their autonomy by signalling using codified systems, and how management use and respond to these signals. Specifically, they suggest that research should be undertaken to:

- Identify the factors that influence the design of systems, the compliance with organisational processes and the uncovering of opportunities to use the data generated to develop data sets that may be used for future predictive purposes.
- Develop processes for data capture and reporting that uncover resource costs which are based on greater transparency, will increase the portability of information from sites and organisations, and allow design and production economics to benefit from the massive improvements in information analysis.
- Develop cost models based on these new data collection processes.

Core challenges for commercial management

The foregoing summary of research topics reveals several areas of synergy and recurring themes. These include:

- *Relationship management*: For example, the management of internal interfaces between different units within the organisation and external interfaces with purchasers, suppliers and other collaborators and/or partners.
- *Value and value management*: For example, the implications and opportunities derived from the introduction of Best Value legislation on the development of value management, procurement strategies based on value optimisation rather than price minimisation, and bidding models that allow the assessment of alternative value and price combinations.
- *Risk and uncertainty management*: This was a recurring subject in many of the chapters, for example, in terms of managing risks associated with relationship breakdown, financing life-cycle solutions and project delivery, and in terms of risk aversion and performance measurement.

- *Performance measurement*: This was linked to performance-based contracts, performance and/or continuous improvement, competitive advantage, customer focus, cultural differences and the effectiveness of marketing.
- *Procurement and supply chain management*: The introduction of integrated logistics, the retention and development of an effective procurement function following outsourcing, marketing and organisational structures aimed at specific procurement routes, as a driver of trust, the performance measurement of members of integrated supply chains, and the need for new financial structures for companies in the supply chain. Other allied and recurring themes include partnering and the influence of corporate social responsibility.
- *Contract management*: The development of new contractual mechanisms to give effect to emerging business models, the implications of the transactional cost perspective on the development of contracts, partnering and performance based contracts, the development of service level agreements and other contractual documents associated with outsourcing, and the implications of culture on the application of specific contract terms.
- *Dispute resolution and conflict management*: The development of predictive tools based on an aetiological approach, relationship management, in particular, the implications of relative power positions within supply chains, government intervention to introduce mandatory arbitration, and the implications of culture on the willingness of parties to pursue claims under a contract.
- *Cost management*: In terms of developing appropriate cost–value reconciliation mechanisms that can also be integrated into performance measurement systems, and support the development of a transactional cost perspective, more appropriate cost estimating and project selection.
- *Organisational learning and knowledge management*: In particular, the instigation of post-project evaluation, is associated with the development of competitive advantage, procurement processes, and performance indicators. Further applications are in the development of risk, bidding and dispute prediction models.

Further areas of commonality include trust, competition, marketing, lean thinking, bidding, and the development of new professional roles and responsibilities.

There is agreement that there should be more interaction between academics and practitioners in order to develop innovative and appropriate models, systems and practices to support the emerging practice of commercial management. Moreover, interaction is needed to ensure that the business models presented actually reflect what goes on in practice and that the context (commercial environment) in which organisations operate is accounted for. For example, commercial managers generally exist to safeguard the interests of organisations, as the systems currently in place are not perfect. This proposal for co-operation is supported by Eraut (1994) who suggests that 'the more time researchers spend listening to practitioners, the more their research will attend to practitioners' perspectives and concerns... [while there is a]... recognition that "leading-edge" professionals develop new knowledge in practice rather than through formally designed research'. This interaction between research and practice is represented in Fig. 19.1. In the case of commercial management, the practitioner or

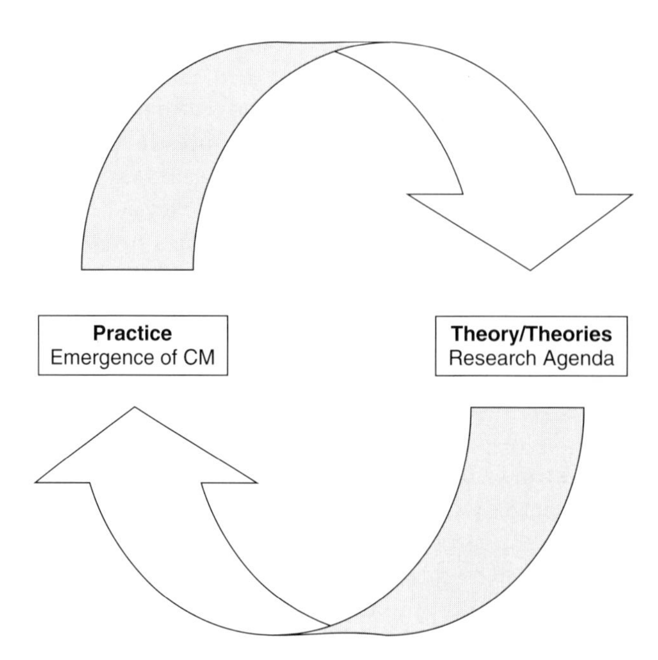

Figure 19.1 Commercial management: developing a research agenda.

practice exists; however, there is a need to develop an underlying theory or theories, again supporting the need for academic–practitioner interaction.

Finally, in terms of an underlying theory of commercial management, Winch (Chapter 14) advocates that if the transactional cost approach can be empirically realised, commercial management will have been placed on a sound empirical and theoretical foundation. However, due to the eclectic nature of commercial management, there are other potential sources for a theoretical foundation to the discipline; for example, arising from the legal aspects of the function: contract formulation, negotiation and deal creation. It is more likely, therefore, that there are several interlinking theories underpinning the discipline.

Relevance of the research themes to practice

Lowe and Leiringer's (2005) practitioner survey (reported in Chapter 1) identifies the following as potential areas for targeted research: the role of commercial management (establishing a BOK), risk management, contracts and contract management, value management, bidding, procurement – particularly partnering arrangements, cost management and dispute resolution, although dispute resolution appears to be of special importance to the construction sector. There is much similarity between these results and the foregoing summary, therefore providing verification of the selected research topics and perhaps the emergence of a cohesive integrated research base to underpin the professionalisation of the commercial management function.

A research network in commercial management

To support this formative research agenda and to underpin the development of the emerging discipline, it is proposed that a new research network in commercial management be established to:

- Create a new multi-disciplinary network of academics, researchers and practitioners interested in developing the subject of commercial management and improving commercial practice.
- Identify and define a new interdisciplinary research agenda as part of a wider community of practice for commercial managers.

In addition to the specific topics identified above, this research agenda should aim to:

- Investigate the developing practice of commercial management across different industries and sectors within the context of the increasing significance of 'projects' to organisations and markets, and the application of programme and portfolio management; the increasing complexity of engineering projects; emerging project concepts and practices; and the growing recognition of the need to raise the status of commercial managers.
- Advance a theory or theories of commercial management, bearing in mind the developing theories of the 'project management', the 'management of project portfolios', 'programme management' and, in particular, Morris's concept of the 'management of projects'.
- Establish a commercial management BOK, bearing in mind the criticisms that have been levelled at the content of the various project management BOKs and the mechanisms and techniques used to establish them, in particular their inability to explain the behaviour of complex projects.
- Apply a wide range of research perspectives that encompass both 'hard' and 'soft' systems thinking and systems dynamics.
- Generate and communicate new research programmes and topics pertinent to the emerging discipline of commercial management.

As an initial step, a development workshop was held as part of the International Commercial Management Symposium hosted by the University of Manchester, UK, on 7 April 2005. The workshop facilitated a discussion between key academics in the field and senior practitioners from major project-based industries. One outcome of the meeting was a consensus on the need to establish an underlying body of knowledge for commercial management. A second was the acknowledgement that, in the case of commercial management, the majority of the underlying knowledge resides with practitioners, not academics.

Conclusion

Undoubtedly, 'commercial management' as a function clearly exists across a spectrum of industries, especially those that are primarily project based. Further, it has been established that, despite some differences in approach and application, both between and within industry sectors, there is sufficient similarity and synergy in practice to identify an emerging discrete discipline. As commercial managers and their employers endeavour to raise the status of commercial management, parallels have been drawn with the emergence of project management in the late 1950s and 1960s, and its development during the latter half of the twentieth century to become a widely recognised core competency with aspirations towards professional status.

This chapter has sought to validate the establishment of a research network and an associated research agenda to support both the practice and the professionalisation of commercial management. It recommends the development of a body of knowledge for commercial management and proposes key areas for targeted research to underpin the discipline. These areas include: relationship management, value and value management, risk and uncertainty management, performance measurement, procurement and supply chain management, contract management, dispute resolution and conflict management, cost management, organisational learning and knowledge management, and bidding.

Finally, this book reviews the current state of knowledge concerning the commercial management of projects, based upon existing research encompassing a wide range of commercial issues. Collectively the chapters constitute a step in the direction of systematic knowledge building: a step towards the creation of a body of knowledge and a research agenda for the development of an underlining theory (or theories) relevant to commercial management practice.

References

Boyatzis, R.E. (1982) *The Competent Manager: A Model for Effective Performance*. John Wiley and Sons, New York.

Caupin, G., Knofel, H., Morris, P.W.G., Motzel, E. & Pannenbacker, O. (1998) *ICB IPMA Competence Baseline*. International Project Management Association, Zurich.

Crawford, L. (2004a) Global body of project management knowledge and standards. In: *The Wiley Guide to Managing Projects* (eds P.W.G. Morris & J.K. Pinto). John Wiley and Sons, Hoboken, NJ; pp. 1150–1195.

Crawford, L. (2004b) Professional associations and global initiatives. In: *The Wiley Guide to Managing Projects* (eds P.W.G. Morris & J.K. Pinto), John Wiley and Sons, Hoboken, NJ; pp. 1389–1402.

Dixon, M. (2000) *Project Management Body of Knowledge* (4th edn). Association of Project Management, High Wycombe.

Eraut, M. (1994) *Developing Professional Knowledge and Competence*. Falmer Press, London.

Gale, A. (2004) Competencies: organizational and personal. In: *The Wiley Guide to Managing Projects* (eds P.W.G. Morris & J.K. Pinto). John Wiley and Sons, Hoboken, NJ; pp. 1087–1111.

Lave, J. & Wenger, E. (1991) *Situated Learning: Legitimate Peripheral Participation*. Oxford University Press, Oxford.

Lowe, D.J. & Leiringer, R. (2005) Commercial management in project-based organisations. *Journal of Financial Management of Property and Construction*, **10**(1), 4–18.

Morgan, G. (1986) *Images of Organisations*. Sage, London.

Morris, P.W.G., Patel, M.B. & Wearne, S.H. (2000) Research into revising the APM Project Management Body of Knowledge. *International Journal of Project Management*, **18**(3), 155–164.

Morris, P.W.G. (1994) *The Management of Projects*. Thomas Telford, London.

Morris, P.W.G. (2001) Updating the project management bodies of knowledge. *Project Management Journal*, **32**, 21–30.

Morris, P.W.G. (2004) The validity of knowledge in project management and the challenge of learning and competency development. In: *The Wiley Guide to Managing Projects* (eds P.W.G. Morris and J.K. Pinto). John Wiley and Sons, Hoboken, NJ; pp. 1137–1149.

PMI (2000) *A Guide to the Project Management Body of Knowledge*. Project Management Institute, Newtown Square, PA.

RICS (1992) *The Core Skills and Knowledge Base of the Quantity Surveyor*. Royal Institution of Chartered Surveyors, London.

RICS (2002) *APC/ATC Requirements and Competencies*. Royal Institution of Chartered Surveyors, London.

Runeson, G. & Skitmore, M. (1999) Tendering theory revisited. *Construction Management and Economics*, **17**, 285–296.

Walsham, G. (1993) *Organizational Metaphors: Interpreting Information Systems in Organizations*. John Wiley and Sons, Chichester.

Index